New Perspectives in Magnetism of Metals

New Perspectives in Magnetism of Metals

Duk Joo Kim

Late of Aoyama Gakuin University
Tokyo, Japan

Kluwer Academic / Plenum Publishers
New York, Boston, Dordrecht, London, Moscow

PHYS

ISBN: 0-306-46209-5

233 Spring Street, New York, N.Y. 10013

10 9 8 7 6 5 4 3 2 1

A C.I.P. record for this book is available from the Library of Congress

Printed in the United States of America

Preface

The fundamental physics of metallic magnetism is not yet satisfactorily understood and continues to be interesting. For instance, although the detail is yet to be clarified, magnetism is anticipated to be playing a principal role in producing the high T_C superconductivity of the oxides. This book has two major objectives. First, it intends to provide an introduction to magnetism of metals in a broad sense. Besides pursuing the mechanism of metallic magnetism itself, it attempts to find and actively analyze magnetic causes hidden hitherto unnoticed behind various physical phenomena. My foremost goal is to expose the fundamental role played by phonons in the mechanism of metallic magnetism. I demonstrate how such a view also helps to elucidate a broad spectrum of other observations. The second objective is to concisely introduce the standard many-body points of view and techniques necessary in studying solid physics in general.

The book is intended to be self-contained and starts with Chapter 1 containing a brief summary on the rudiments of quantum mechanics and statistical mechanics including the method of second quantization. In the same spirit, the foundation of magnetism in general is summarized in Chapter 2 and that for metals in particular, the Stoner theory, in Chapter 3.

In Chapter 4, various linear responses of metallic electrons are systematically discussed with emphasis on the role of magnetism in them. For instance, it is demonstrated how the screening constants of electron-ion and ion-ion interactions become different, in an important manner, when the exchange effects are taken into account. Discussions leading to very fundamental but not widely known results are given, with particular emphasis on the linear responses in ferromagnetic state.

It has long been noted that the phonon or elastic properties of a metal are closely related to its magnetic properties. A most natural approach to explore such a relation should be to start from the electron-phonon interaction. By taking such a standard approach and based on the result of Chapter 4, particularly concerning the screening of the ion-ion interaction, in Chapter 5 we find how the phonon frequency in the paramagnetic state of a metal is directly related to its magnetic susceptibility. As for phonon frequency in the ferromagnetic state of a metal, we show how it depends on its magnetization, or, the spin splitting of electron energy bands.

Although the findings of Chapter 5 open up a new perspective in this long standing problem on the relation between elasticity behavior and magnetism, they turn out to be not quite satisfactory. Most notably, with the result of this chapter we can not explain the observed temperature dependences of elastic constants in the paramagnetic state of metals. Our final answer to the subject of this chapter will be given in Chapter 7.

The subject of Chapter 6 is the opposite of Chapter 5: We demonstrate how important the effect of the electron-phonon interaction (EPI) can be in magnetism of metals. Such a possibility was repeatedly raised in the past, but always failed to withstand refutation. As the result of such a history of controversy, now almost no one even seem to pay attention to this problem. In order to overcome such widespread skepticism we need a very well-founded, clear-cut argument.

According to the Landau theory of the 2nd order phase transition, the magnetic susceptibility of a metal is determined from the magnetization dependence of its free energy. We adopt this Landau procedure. Then in obtaining the free energy, we treat a metal as an interacting electron-phonon system, not as an electron only system. The magnetization dependence of the phonon part of the free energy comes from that of phonon frequency, as discussed in detail in Chapter 5. Herring and Hopfield previously took such an approach in estimating the size of effect of the EPI on the spin susceptibility of a metal, and correctly concluded the relative size of the effect on the Pauli spin susceptibility or the effective exchange interaction to be of the order of $\hbar\omega_D/\varepsilon_F$, ω_D and ε_F being, respectively, the phonon Debye frequency and the electron Fermi energy. However, Herring and Hopfield could not go beyond an order of magnitude estimate. With such a small size of the effect, it can not be important except in some very special situations. However, even such a small effect was doubted since Knapp et al. failed to observe an isotope effect in T_C of $ZrZn_2$ as large as Hopfield's prediction.

In Chapter 6, however, we show that the effect of the EPI can be much more important than earlier theories had envisaged. One of the most important new development may be the discovery that the effect of EPI is temperature dependent. With such a result we have successfully elucidated a number of observations concerning the magnetic susceptibility of a metal which were difficult to understand otherwise. Among others, we find that the EPI renders the temperature dependence of the magnetic susceptibility of a ferromagnet Curie-Weiss form.

We also find the EPI to play an important role in determining spontaneous magnetization below T_C of a ferromagnetic metal, through its effect on the spin splitting of the electron energy bands. Further, most surprisingly, we demonstrate that phonon carries magnetization in the ferromagnetic state of a metal.

In Chapter 7 we explore how magnetism is involved in the volume and elasticity behavior in a ferromagnetic metal by using the Landau procedure with the same free energy as in Chapter 6. While now what matters is the volume dependence of the free energy, we first note that the result of Chapter 5 on the elastic constants is reproduced by considering only the electron part in that free energy. Then, an improvement over the result of Chapter 5 can be immediately achieved by including the contribution from thermal phonons. Such a result will be found to resolve earlier difficulties with the result of Chapter 5 concerning the temperature dependence of elastic constants.

The volume behavior of ferromagnetic metals is often very anomalous, and its origin is not yet fully understood. Working with the same free energy as above we find, quite surprisingly, that the phonon part of the free energy can give rise to enough variety and size of effect on volume to account for actual observations. No one has previously anticipated such a possibility.

Up to Chapter 7 we employed only the elementary methods described in Chapter 1. In order to proceed further in our study, in Chapter 8 we introduce the more sophisticated method of Green's functions and Feynman diagrams. We tried to make this short introduction self-contained and as accessible as possible.

In the remaining Chapters 9 and 10, this method of Green's functions and Feynman diagrams is fully used to discuss a number of fundamental problems related to magnetism of metals. In Chapter 9, we discuss the effects of EPI and spin fluctuation on electrons, and the constructive effect of magnetism on superconductivity, among other phenomena. Also we critically summarize various earlier views on the role of EPI in itinerant electron magnetism.

In Chapter 10, we explore how the free energy of a metal can be derived microscopically in terms of electron charge and spin fluctuations and phonons. We obtain the result exactly in the same form as that used in Chapters 6 and 7, reconfirming the validity of our conclusions there. Recently spin fluctuation has been advocated as the principal determinants of the magnetic behavior of a metal. In this connection I raise some fundamental questions concerning the current treatment of spin fluctuation in the physics of metals.

Throughout this book I based my discussion on the jellium model with some extension. It was necessary to use such a model to make the discussion as simple and transparent as possible ; the jellium model is a physical model of a metal with perhaps the least number of arbitrary parameters. Before embarking on a much more laborious calculation on a real metal, such a model calculation may be a necessary prerequisite. What I did, then, is to treat such a model with the mean field, or, random phase approximation. The feature of the work in this volume that distinguishes it from that of many others is that the electrons and phonons are consistently treated on the same footing. Ambiguities are avoided in such a straightforward theory. Note that exactly the same approach was essential in understanding the origin of the superconductivity. The results of this book seems to suggest that such an attitude may be required also in understanding magnetism of metals.

This book is written as a text and reference book for a specialty course of the given title at the graduate level. However, an undergraduate with an elementary background in quantum mechanics, statistical mechanics and solid state physics can easily understand its contents. This conclusion is drawn from experience; I have used the material of Chapters 1–4 for undergraduates on several occasions.

The theory of magnetism of metals is an ever expanding field with many unresolved fundamental difficulties, and beyond the ideas of Chapters 3 there is no one generally accepted theory. It is beyond my competence to describe and fairly evaluate all the current attempts, and such an endeavor is not attempted. Instead, beyond the rather standard material of Chapters 1–3 and 8, I concentrated on a few selected aspects in which I am most interested. Many things are my own. However, a consistent effort is devoted to maintain the objectivity needed in a textbook. The reader will find that the subjects I have chosen are mostly fundamental and general ones. I have included some exercise problems at the end of each chapter. Note, however, that more important, real problems are scattered throughout the book.

This book was written originally in Korean and published by Minumsa as a volume of the Daewoo Series. I wrote this book for my wife, Kap Soon, whose entire family left for North Korea 30 years ago, in the hope that at least this book written by her husband would go to the country she cannot visit. I would like to thank Prof. Byung Ho Lee and Prof. Hwe Ik Zhang for their obliging me to toil with this work. I would like to thank Prof. Jae Il Lee and Dr. An Sung Choe for their collaboration on the Korean version.

This English version was undertaken with the strong encouragement of Prof. Arthur J. Freeman and Prof. Brian B. Schwartz. It is my pleasure to thank them for their warm friendship on this occasion. The preparation of this English version was started in the summer of 1985 with the translation of Chapter 1 by Prof. Keum Hwi Lee. Prof. Lee intended to carry out the translation of the entire volume, but he soon became to busy to continue with this project. However, without his initial enthusiasm and collaboration this English version would not exist. Prof. Marvin M. Antonoff read the first seven chapters of an earlier version very thoroughly, and contributed numerous criticisms and suggestions. Prof. Leon Gunther and Prof. Kazumi Maki read, respectively, Chapter 8 and Chapter 9. Prof. Sung G. Chung compiled an extensive errata for the Korean version, which was useful in preparing the present version. I owe much to these friends in preparing this book. However, the final responsibility for the content of the book lies solely with me.

I began my career in physics as a graduate student of Prof. Ryogo Kubo. I would like to thank him for his help and encouragement over the years. I would like also to thank many colleagues and friends on this occasion for their cooperation, encouragement, and friendship.

Most of the figures in this volume were prepared by Mr. Shinji Fukumoto; without his cooperation it would have been impossible to produce the English version.

This manuscript, and all its many revisions, were typed solely by Kap Soon Kim, as were most of my papers cited in the book. This was no easy task; she collaborated with me in revising my papers in response to the comments and suggestions of referees and editors. She was my comrade in the truest sense of the word.

Duk Joo Kim[†]

[†]Deceased.

Contents

New Perspectives in Magnetism of Metals

Chapter 1

QUANTUM MECHANICS AND STATISTICAL MECHANICS OF AN ELECTRON GAS

This book treats the physics of metals in which a major role is played by electrons. The most basic model of metallic electrons, an electron gas in a box is treated in this chapter. Also included is a brief review of the basics of the quantum mechanics and statistical mechanics of electrons which are needed later in the book.

1.1 ONE ELECTRON IN A BOX

Consider one electron in a cubic box of side L. In this case, the electron energy eigenvalue ε and the wave function $\varphi(\boldsymbol{r})$ are obtained by solving the time-independent Schrödinger equation

$$H_0 \varphi(\boldsymbol{r}) = \varepsilon \varphi(\boldsymbol{r}), \tag{1.1.1}$$

where $\boldsymbol{r} = (x, y, z)$ is the position coordinate of electron and

$$H_0 = -\frac{\hbar^2}{2m}\left(\frac{\partial^2}{\partial x^2} + \frac{\partial^2}{\partial y^2} + \frac{\partial^2}{\partial z^2}\right) \tag{1.1.2}$$

is the Hamiltonian operator representing the kinetic energy of an electron, m being the electron mass.

In solving (1.1.1), we usually impose the *periodic boundary condition*

$$\varphi(x + L, y, z) = \varphi(x, y + L, z) = \varphi(x, y, z + L) = \varphi(x, y, z) \tag{1.1.3}$$

on the wave function. Under this condition, the solution of (1.1.1) is obtained as

$$\varepsilon_{\boldsymbol{k}} = \frac{\hbar^2}{2m}\boldsymbol{k}^2, \tag{1.1.4}$$

$$\phi_{\boldsymbol{k}}(\boldsymbol{r}) = \frac{1}{\sqrt{V}}\, e^{i\boldsymbol{kr}} \tag{1.1.5}$$

with $\boldsymbol{kr} = k_x x + k_y y + k_z z$, where $V = L^3$ is volume of the box and each component of the wave vector

$$\boldsymbol{k} = (k_x, k_y, k_z), \tag{1.1.6}$$

which satisfies the condition (1.1.3), is given by

$$k_\nu = \left(\frac{2\pi}{L}\right) n_\nu, \quad n_\nu = \text{integer}, \quad \nu = x, y, \text{ or } z. \tag{1.1.7}$$

Because of (1.1.7), the kinetic energy of an electron becomes discrete, i.e. quantized.

The plane wave (1.1.5) satisfies the normalization condition

$$\int_V d\boldsymbol{r}\; \phi_{\boldsymbol{k}}^*(\boldsymbol{r})\phi_{\boldsymbol{k}}(\boldsymbol{r}) \equiv (\phi_{\boldsymbol{k}}, \phi_{\boldsymbol{k}}) = 1, \tag{1.1.8}$$

where $d\boldsymbol{r} = dxdydz$ is the volume element, and * implies the complex conjugate. We can also prove the orthogonality $(\phi_{\boldsymbol{k}}, \phi_{\boldsymbol{k}'}) = 0$ for $\boldsymbol{k} \neq \boldsymbol{k}'$, which together with (1.1.8) leads to the orthonormality condition,

$$(\phi_{\boldsymbol{k}}, \phi_{\boldsymbol{k}'}) = \delta_{\boldsymbol{k},\boldsymbol{k}'}, \tag{1.1.9}$$

with the Kronecker delta, $\delta_{\boldsymbol{k},\boldsymbol{k}'}$.

In addition to the space motion discussed above, an electron has the *spin* degree of freedom. We often try to visualize it as an electron spinning around its own axis. This spin degree of freedom is represented by the dimensionless spin operator, $\boldsymbol{s} = (s_x, s_y, s_z)$, such that $\hbar\boldsymbol{s}$ possesses the properties of an angular momentum satisfying the commutation relation,

$$\boldsymbol{s} \times \boldsymbol{s} = i\boldsymbol{s}, \tag{1.1.10}$$

or, in component form,

$$s_x s_y - s_y s_x \equiv [\, s_x, s_y\,] = i s_z, \text{ etc. }, \tag{1.1.10'}$$

where $\times$ is the vector product symbol and [,] stands for a commutator. In units of $\hbar$, the electron spin angular momentum has magnitude of 1/2, i.e.,

$$\boldsymbol{s}^2 = s_x^2 + s_y^2 + s_z^2 = [\, s(s+1)\,]_{s=1/2} = \frac{1}{2}\left(\frac{1}{2}+1\right). \tag{1.1.11}$$

Since s_x, s_y, and s_z do not commute with one another, we can choose a representation (i.e., a coordinate system) in which only one of them is diagonal. We usually choose a coordinate system in which s_z is diagonal with its eigenvalues $\pm 1/2$. With such a spin degree of freedom included, the electron wave function can be rewritten as $\phi_{\boldsymbol{k}}(\boldsymbol{r}, \pm 1/2)$ satisfying the eigenvalue equation,

$$s_z\, \phi_{\boldsymbol{k}}(\boldsymbol{r}, \pm 1/2) = \pm\, 1/2\; \phi_{\boldsymbol{k}}(\boldsymbol{r}, \pm 1/2), \tag{1.1.12}$$

for the spin operator s_z.

As the wave function satisfying the requirement (1.1.12) in addition to (1.1.1), we can take the product of an orbital wave function and a spin eigenfunction,

$$\phi_{\boldsymbol{k}}(\boldsymbol{r}, s_z) = \phi_{\boldsymbol{k}}(\boldsymbol{r})\, \chi_\sigma = \phi_{\boldsymbol{k}\sigma}(\boldsymbol{r}) \equiv \phi_k(\boldsymbol{r}), \tag{1.1.13}$$

with the spin eigenfunction χ_σ, $\sigma = +$ or $-$, satisfying the following relation,

$$s_z\, \chi_+ = \frac{1}{2}\chi_+, \quad s_z\, \chi_- = -\frac{1}{2}\chi_- . \tag{1.1.14}$$

We often use the abbreviation $(\boldsymbol{k}, \sigma) = (k)$.

The electron spin $\boldsymbol{s}$ may be represented by the *Pauli spin matrices* $\boldsymbol{\sigma} = (\sigma_x, \sigma_y, \sigma_z)$:

$$\boldsymbol{s} = \frac{1}{2}\boldsymbol{\sigma}, \tag{1.1.15}$$

$$\sigma_x = \begin{bmatrix} 0 & 1 \\ 1 & 0 \end{bmatrix}, \quad \sigma_y = \begin{bmatrix} 0 & -i \\ i & 0 \end{bmatrix}, \quad \sigma_z = \begin{bmatrix} 1 & 0 \\ 0 & -1 \end{bmatrix}. \tag{1.1.16}$$

We can easily confirm that the Pauli spin matrices satisfy the commutation relation of (1.1.10), or,

$$\boldsymbol{\sigma} \times \boldsymbol{\sigma} = 2i\boldsymbol{\sigma}. \tag{1.1.17}$$

Correspondingly, the spin eigenfunctions are given in the following form

$$\chi_+ = \begin{bmatrix} 1 \\ 0 \end{bmatrix}, \quad \chi_- = \begin{bmatrix} 0 \\ 1 \end{bmatrix}. \tag{1.1.18}$$

They satisfy

$$\sigma_z \chi_\pm = \pm \chi_\pm , \tag{1.1.19}$$

corresponding to (1.1.14), and the following orthonormality relation,

$$\chi_\sigma^+ \chi_{\sigma'} = (\chi_\sigma , \chi_{\sigma'}) = \delta_{\sigma,\sigma'} , \tag{1.1.20}$$

where $^+$ implies the Hermite conjugate. The orthonormality of $\phi_{k\sigma}(\boldsymbol{r})$ is given as

$$(\phi_{k\sigma} , \phi_{k'\sigma'}) = \int \phi_k^*(\boldsymbol{r})\, \phi_{k'}(\boldsymbol{r})\, d\boldsymbol{r}\, (\chi_\sigma , \chi_{\sigma'})$$

$$\equiv \int \phi_{k\sigma}^+(\boldsymbol{r})\, \phi_{k'\sigma'}(\boldsymbol{r})\, d\boldsymbol{r} = \delta_{k,k'}\, \delta_{\sigma,\sigma'} . \tag{1.1.21}$$

The quantity $| \phi_{k\sigma}(\boldsymbol{r}) |^2$ gives the probability density of an electron with wave vector $\boldsymbol{k}$ and spin $s_z = \sigma/2$ ($\sigma = \pm$) being found at $\boldsymbol{r}$.

With $s_\pm$ defined by

$$s_\pm = \frac{1}{2}(s_x \pm i s_y), \tag{1.1.22}$$

or, equivalently,

$$\sigma_\pm = \frac{1}{2}(\sigma_x \pm i \sigma_y), \tag{1.1.23}$$

it is easy to verify that

$$s_\pm \chi_\pm = \sigma_\pm \chi_\pm = 0 ; \quad s_\pm \chi_\mp = 1/2\, \chi_\pm , \quad \sigma_\pm \chi_\mp = \chi_\pm . \tag{1.1.24}$$

When there is no external magnetic field present, the electron energy ε_k does not depend on its spin state.

1.2 MANY ELECTRONS IN A BOX

Consider a system of n electrons ($n >> 1$) in the same cubic box of volume $V = L^3$ discussed in the preceding section. The free electron gas model neglects the inter-electron interactions and takes into account the electronic kinetic energies only. In this case, the Schrödinger equation reads

$$H_o \Psi(\boldsymbol{x}_1, \boldsymbol{x}_2, \cdots, \boldsymbol{x}_n) = E\Psi(\boldsymbol{x}_1, \boldsymbol{x}_2, \cdots, \boldsymbol{x}_n), \tag{1.2.1}$$

$$H_o = -\frac{\hbar^2}{2m}\sum_{i=1}^{n}\nabla_i^2 = -\frac{\hbar^2}{2m}\sum_{i=1}^{n}\left(\frac{\partial^2}{\partial x_i^2} + \frac{\partial^2}{\partial y_i^2} + \frac{\partial^2}{\partial z_i^2}\right), \tag{1.2.2}$$

where

$$\boldsymbol{x}_i = (\boldsymbol{r}_i, s_{iz}) \tag{1.2.3}$$

is the coordinate of the i-th electron including spin; $\boldsymbol{r}_i = (x_i, y_i, z_i)$.

When the periodic boundary condition on $\Psi(\boldsymbol{x}_1, \boldsymbol{x}_2, \cdots, \boldsymbol{x}_n)$ for every $\boldsymbol{r}_i$ of type (1.1.3) is imposed, a solution of (1.2.1) is obtained as

$$E_{k_1, k_2, \cdots, k_n}, = \frac{\hbar^2}{2m}\sum_{i=1}^{n} \boldsymbol{k}_i^2, \tag{1.2.4}$$

$$\begin{aligned}\Psi_{k_1 k_2 \cdots k_n}(\boldsymbol{x}_1, \boldsymbol{x}_2, \cdots, \boldsymbol{x}_n) &\equiv \Psi_{k_1 k_2 \cdots k_n}(\boldsymbol{r}_1, \boldsymbol{r}_2, \cdots, \boldsymbol{r}_n) \\ &= \phi_{k_1\sigma_1}(\boldsymbol{r}_1)\, \phi_{k_2\sigma_2}(\boldsymbol{r}_2) \cdots \phi_{k_n\sigma_n}(\boldsymbol{r}_n) \\ &= \phi_{k_1}(\boldsymbol{r}_1)\, \phi_{k_2}(\boldsymbol{r}_2) \cdots \phi_{k_n}(\boldsymbol{r}_n),\end{aligned} \tag{1.2.5}$$

with each $\boldsymbol{k}_i$ satisfying the condition (1.1.7) and $\phi_{k_i}(\boldsymbol{x}_i) = \phi_{k_i\sigma_i}(\boldsymbol{r}_i)$ given in (1.1.13). It is easy to prove the orthonormality,

$$\begin{aligned}(\Psi_{k_1 k_2 \cdots k_n}, \Psi_{k'_1 k'_2 \cdots k'_n}) &= (\phi_{k_1}, \phi_{k'_1})(\phi_{k_2}, \phi_{k'_2}) \cdots (\phi_{k_n}, \phi_{k'_n}) \\ &= \delta_{(k_1, k_2, \cdots, k_n), (k'_1, k'_2, \cdots, k'_n)},\end{aligned} \tag{1.2.6}$$

of these n-electron wave functions Ψ, corresponding to (1.1.21) for the one-electron case.

Note that

$$\Psi_{k_1 k_2 \cdots k_n}(\boldsymbol{r}_2, \boldsymbol{r}_1, \cdots, \boldsymbol{r}_n) = \phi_{k_1}(\boldsymbol{r}_2)\, \phi_{k_2}(\boldsymbol{r}_1)\, \phi_{k_3}(\boldsymbol{r}_3) \cdots \phi_{k_n}(\boldsymbol{r}_n) \tag{1.2.7}$$

also is a solution of (1.2.1) with the same energy eigenvalue (1.2.4). Both of these wave functions, (1.2.5) and (1.2.7), represent n electrons occupying the states k_1, k_2, ... , and k_n, one in each $k_i = (\boldsymbol{k}_i,\sigma_i)$. Since the electrons are identical particles, these two wave functions should have the same mathematical expression except for a constant factor, i.e.,

$$\Psi_{k_1\,k_2\,\cdots\,k_n}(\boldsymbol{r}_2, \boldsymbol{r}_1, \cdots, \boldsymbol{r}_n) = c\Psi_{k_1\,k_2\,\cdots\,k_n}(\boldsymbol{r}_1, \boldsymbol{r}_2, \cdots, \boldsymbol{r}_n). \qquad (1.2.8)$$

With one more exchange of $\boldsymbol{r}_1$ and $\boldsymbol{r}_2$ in the right hand side of (1.2.8), we get

$$\Psi_{k_1\,k_2\,\cdots\,k_n}(\boldsymbol{r}_2, \boldsymbol{r}_1, \cdots, \boldsymbol{r}_n) = c^2\Psi_{k_1\,k_2\,\cdots\,k_n}(\boldsymbol{r}_2, \boldsymbol{r}_1, \cdots, \boldsymbol{r}_n).$$

Thus, it is required that $c^2 = 1$, or

$$c = \pm 1. \qquad (1.2.9)$$

The above argument leading to (1.2.9) applies to any system of identical particles. The particles with $c = +1$ and -1 are called *bosons* (or *Bose particles*) and *fermions* (or *Fermi particles*), respectively. The electron is a fermion and the phonon, to be discussed later, is a boson.

The wave function given in (1.2.5) does not satisfy the fermion requirement, (1.2.8) with $c = -1$. An n-electron wave function which satisfies such a fermion requirement is given by the *Slater determinant*,

$$\Phi_{k_1k_2\,\cdots\,k_n}(\boldsymbol{r}_1, \boldsymbol{r}_2, \cdots, \boldsymbol{r}_n) = \frac{1}{\sqrt{n!}}\begin{vmatrix} \phi_{k_1}(r_1) & \phi_{k_1}(r_2) & \cdots & \phi_{k_1}(r_n) \\ \phi_{k_2}(r_1) & \phi_{k_2}(r_2) & \cdots & \phi_{k_2}(r_n) \\ \vdots & \vdots & & \vdots \\ \phi_{k_n}(r_1) & \phi_{k_n}(r_2) & \cdots & \phi_{k_n}(r_n) \end{vmatrix}. \qquad (1.2.10)$$

It is easy to verify that this Slater determinant is the solution of Schrödinger equation (1.2.1) with energy eigenvalue (1.2.4). It also satisfies the orthonormality condition (1.2.6).

From (1.2.10), it immediately follows that these electrons obey the *Pauli exclusion principle*. Substitution of $k_1 = k_2 = k$ obviously gives

$$\Phi_{k\,k\,k_3\,\cdots\,k_n}(\boldsymbol{r}_1, \boldsymbol{r}_2, \boldsymbol{r}_3, \cdots, \boldsymbol{r}_n) = 0. \qquad (1.2.11)$$

Thus, the Slater determinant ensures that the same state $k = (\boldsymbol{k},\sigma)$ cannot be occupied by more than one electron. It is easy to observe that the Pauli principle does not apply to bosons.

1.3 GROUND STATE OF A FREE ELECTRON GAS

What we studied in the preceding section 1.2 is the possible states for a system of n electrons in a box. A state represented by the wave function (1.2.10) with energy eigenvalue (1.2.4) is possible for any set of $(k_1, k_2, \cdots, k_n)$ not violating the Pauli principle. However, in the case of absolute zero temperature, the state which actually occurs is that with the minimum total energy, i.e., the ground state.

It is convenient to use the $\boldsymbol{k}$ space of Fig.1.1 to visualize this ground state at $T = 0$ K. In this $\boldsymbol{k}$ space, the states electrons can occupy are the lattice points given by (1.1.7). Starting from the origin $\boldsymbol{k} = 0$ with the lowest one-electron energy $\varepsilon_{\boldsymbol{k}} = 0$, n electrons are put into these lattice points one by one in order of increasing energy $\varepsilon_{\boldsymbol{k}}$. The finished constant energy surface of $\boldsymbol{k}$ space volume occupied by these n electrons is called the *Fermi surface*. For the free electrons, it is the *Fermi sphere* as shown in Fig.1.1 and its radius is called the *Fermi wave number* k_F. Since the unit cell volume of the simple cubic lattice of $\boldsymbol{k}$ space defined by (1.1.7) is $(2\pi/L)^3$, it is obvious that

$$n = 2\sum_{\boldsymbol{k}=0}^{|\boldsymbol{k}|\le k_F} 1 = 2\int_0^{k_F} dk\, 4\pi k^2 \Big/ \left(\frac{2\pi}{L}\right)^3 = \frac{V}{3\pi^2} k_F^3, \qquad (1.3.1)$$

with $V = L^3$, where the factor 2 comes from two possible spin states ($\sigma = \pm$) for each lattice point $\boldsymbol{k}$. Replacement of the $\boldsymbol{k}$ space sum by a corresponding integral is justified since $k_F \gg 2\pi/L$ for metals, in which typically $n/V \sim 10^{22}$ cm^{-3} and hence $k_F \sim 10^8$ cm^{-1}.

The total energy of electrons in this ground state is

$$E_o = 2\sum_{\boldsymbol{k}=0}^{|\boldsymbol{k}|\le k_F} \varepsilon_{\boldsymbol{k}} = 2\frac{V}{(2\pi)^3}\int_0^{k_F} dk\, 4\pi k^2 \frac{\hbar^2 k^2}{2m} = \frac{3}{5} n\varepsilon_F, \qquad (1.3.2)$$

with the *Fermi energy*

$$\varepsilon_F = \frac{\hbar^2}{2m} k_F^2, \qquad (1.3.3)$$

defined such that all the states with energy up to ε_F are occupied at $T = 0$ K.

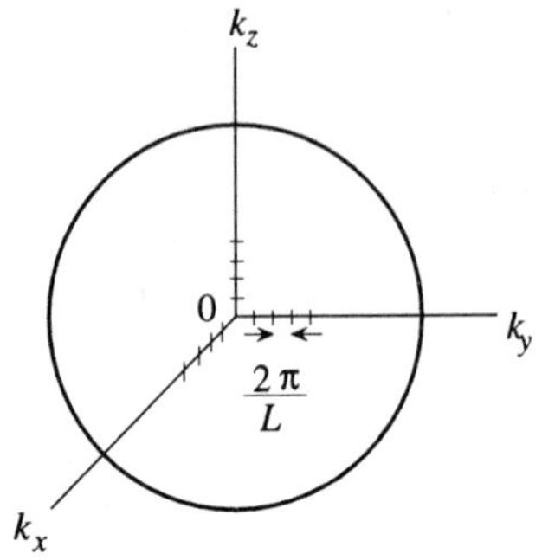

Fig.1.1 The $\boldsymbol{k}$ space for an electron gas in a cubic box of volume $V = L^3$.

For a calculation of the type in (1.3.2), it is convenient to introduce the *electronic density of states* $N(\varepsilon)$ defined by

$$E_0 = 2\int_0^{\varepsilon_F} d\varepsilon\, N(\varepsilon)\, \varepsilon\,, \tag{1.3.4}$$

where, for free electrons,

$$N(\varepsilon) = \frac{V}{(2\pi)^3}\, 4\pi k^2 \left(\frac{dk}{d\varepsilon}\right) = \frac{V}{4\pi^2}\left(\frac{2m}{\hbar^2}\right)^{3/2}\sqrt{\varepsilon}\,. \tag{1.3.5}$$

This density of state $N(\varepsilon)$ is expressed in units of "states per energy and spin" and represents the number of electron energy levels per unit energy interval for each spin. In general, $N(\varepsilon)$ can be used in converting a $\boldsymbol{k}$ space lattice sum of $\varepsilon_{\boldsymbol{k}}$ dependent quantity into a corresponding integral over energy ε :

$$\sum_{\boldsymbol{k}} g(\varepsilon_{\boldsymbol{k}}) = \int d\varepsilon\, N(\varepsilon)\, g(\varepsilon). \tag{1.3.6}$$

The density of states defined here is proportional to the volume. Often it is defined for unit volume or one atom. However, $N(\varepsilon)$ defined as (1.3.4) or (1.3.6) will be used throughout this book.

We should note that the ground state energy (1.3.2) can be directly calculated by

$$E_o = \int dr_1 dr_2 \cdots dr_n\, \Phi^+_{k_1\sigma_1 k_2\sigma_2 \cdots k_n\sigma_n}(r_1, r_2, \cdots, r_n)$$

$$\times\left(\sum_{i=1}^{n} \frac{-\hbar^2}{2m} \nabla_i^2\right) \Phi_{k_1\sigma_1 k_2\sigma_2 \cdots k_n\sigma_n}(r_1, r_2, \cdots, r_n), \qquad (1.3.7)$$

using the ground state Slater determinant of (1.2.10) for Φ.

1.4 FUNDAMENTALS OF STATISTICAL MECHANICS : DENSITY MATRIX

In the real world the interactions between electrons or between electron and vibrating ion cannot be ignored, and often a system is under a time dependent external perturbation such as magnetic field. In such a situation, instead of (1.2.1), we have to solve the time-dependent Schrödinger equation with such an interaction or external perturbation represented by H':

$$i\hbar \frac{\partial}{\partial t} \Psi(x_1, x_2, \cdots, x_n, t) = H\Psi(x_1, x_2, \cdots, x_n, t)$$

$$= (H_o + H')\Psi(x_1, x_2, \cdots, x_n, t). \qquad (1.4.1)$$

When $H' = 0$, it reduces to the eigenvalue problem for H_o, which has already been worked out in 1.2: the energy eigenvalue E_ν with $\nu = (k_1, k_2, \ldots, k_n) = (k_1\sigma_1, k_2\sigma_2, \ldots, k_n\sigma_n)$ given in (1.2.4) and the corresponding eigenfunction (i.e., the Slater determinant) Φ_ν of (1.2.10).

To solve (1.4.1), we expand the wave function Ψ in terms of Slater determinants Φ_ν as

$$\Psi(x_1, x_2, \cdots, x_n, t) = \sum_{\nu'} c_{\nu'}(t)\, \Phi_{\nu'}(x_1, x_2, \cdots, x_n). \qquad (1.4.2)$$

Substituting this into (1.4.1), we multiply Φ_ν^+ to the result from left, and then integrate over $(r_1, r_2, \cdots, r_n)$ taking into account the orthonormality (1.2.6) to get

$$i\hbar \frac{\partial}{\partial t} c_\nu(t) = \sum_{\nu'} H_{\nu\nu'} c_{\nu'}(t), \qquad (1.4.3)$$

where

$$H_{\nu\nu'} = \int dx_1 dx_2 \cdots dx_n \, \Phi_\nu^+(x_1, x_2, \cdots, x_n) \left(H_o + H'\right) \Phi_{\nu'}(x_1, x_2, \cdots, x_n)$$

$$\equiv (\Phi_\nu , H \, \Phi_{\nu'}) \tag{1.4.4}$$

is the $\nu\nu'$-th matrix element of total Hamiltonian $H = H_0 + H'$. Thus, the solution Ψ for (1.4.1) is obtained via the expansion (1.4.2) by solving (1.4.3) for c_ν first, under an appropriate initial condition for them. Then the quantum mechanical expectation value of a physical operator A in this state Ψ is given by

$$(\Psi, A\Psi) = \sum_{\nu,\nu'} c_\nu^*(t) c_{\nu'}(t) A_{\nu\nu'}, \tag{1.4.5}$$

where

$$A_{\nu\nu'} = (\Phi_\nu , A\Phi_{\nu'}) \tag{1.4.6}$$

is the $\nu\nu'$-th matrix element of A defined in the same way as (1.4.4).

However, a quantum mechanical calculation as outlined above is practically impossible to carry out for a metal, when there are extremely large number ($\sim 10^{22}$ cm^{-3}) of electrons. To solve (1.4.3) we need the initial values at a certain time of all those $c_\nu(t)$, it is practically impossible. Fortunately, what we actually observe in macroscopic world is the statistical expectation, namely,

$$\langle A \rangle = \sum_{\nu,\nu'} \left\langle c_\nu^*(t) \, c_{\nu'}(t) \, A_{\nu\nu'} \right\rangle = \sum_{\nu,\nu'} \left\langle c_\nu^*(t) \, c_{\nu'}(t) \right\rangle A_{\nu\nu'}, \tag{1.4.7}$$

where $\langle \cdots \rangle$ indicates to take the statistical average over the initial values of $c_\nu(t)$ (the ensemble average).

If we introduce the density matrix ρ defined by its matrix elements as

$$\rho_{\nu'\nu}(t) = \left\langle c_\nu^*(t) \, c_{\nu'}(t) \right\rangle, \tag{1.4.8}$$

(1.4.7) can be written in the form

$$\langle A \rangle = \mathrm{tr}(\rho A), \tag{1.4.9}$$

where tr stands for trace (or Spur) implying summation of the diagonal elements.

Thus, for calculation of $\langle A \rangle$, we need to find ρ. The equation of motion for ρ is derived from (1.4.3) and its complex conjugate

$$- i\hbar \frac{\partial}{\partial t} c_\nu^*(t) = \sum_{\nu'} H_{\nu\nu'}^* c_{\nu'}^*(t) = \sum_{\nu'} c_{\nu'}^*(t) H_{\nu'\nu} , \qquad (1.4.3')$$

where the last equality is obtained by using the Hermite property of the total Hamiltonian H, i.e., $H_{\nu\nu'}^* = H_{\nu'\nu}^+ = H_{\nu'\nu}$. From (1.4.3) and (1.4.3'), we obtain

$$i\hbar \frac{\partial}{\partial t} \rho_{\nu\nu'} = \sum_{\nu''} (H_{\nu\nu''}\rho_{\nu''\nu'} - \rho_{\nu\nu''} H_{\nu''\nu'}),$$

or equivalently,

$$i\hbar \frac{\partial}{\partial t} \rho = [H, \rho]. \qquad (1.4.10)$$

This is called the *Liouville*, or *von Neumann equation.* As will be seen in Chapter 4, the *von Neumann* equation (1.4.10) becomes useful for handling a time dependent ρ in a non-equilibrium state. In thermal equilibrium, $\langle A \rangle$ and hence ρ are independent of time. In such a situation, (1.4.10) implies only that ρ should commute with H. With such a condition alone we can not determine the density matrix of an equilibrium state.

We know that when the system is in equilibrium through energy exchange with a heat bath of absolute temperature T (canonical ensemble), the density matrix ρ is given by

$$\rho \propto \exp(-\beta H), \qquad (1.4.11)$$

where $\beta = 1/k_B T$ with the Boltzmann constant k_B and H is the Hamiltonian of the system. In particular, when there is no interaction (i.e., $H = H_o$), (1.4.11) gives

$$\rho_{\nu\nu'} \propto \delta_{\nu\nu'} \exp(-\beta E_\nu), \qquad (1.4.12)$$

with eigenvalue E_ν of H_o; this is the familiar Boltzmann distribution for the probability of finding the system in the state with E_ν. The expression (1.4.11) is a generalization of (1.4.12). When there is interaction between particles, generally it is impossible to obtain the eigenvalue for H and to reduce (1.4.11) to the form of (1.4.12).

In studying a many-body system such as the electron gas, it is convenient to assume that the system is in equilibrium with a large bath not only through energy exchange but through particle exchange as well (the grand canonical ensemble). In this case, the density matrix is given by

$$\rho = \exp[-\beta(H - \mu\tilde{n})]/\Xi, \qquad (1.4.13)$$

with the grand partition function,

$$\Xi = \mathrm{tr}\{\exp[-\beta(H - \mu\tilde{n})]\}, \qquad (1.4.14)$$

instead of (1.4.11), where $\tilde{n}$ is an operator representing the total number of particles and μ is the chemical potential. This density matrix is normalized as

$$\mathrm{tr}\,\rho = 1. \qquad (1.4.15)$$

We call this distribution the grand canonical distribution.

When the number of particles is fixed, (1.4.13) reduces to (1.4.11). In this case, the normalized density matrix is given as

$$\rho = \frac{1}{Z}\exp(-\beta H), \qquad (1.4.11')$$

$$Z = \mathrm{tr}[\exp(-\beta H)]. \qquad (1.4.16)$$

We call this ρ the canonical distribution and Z the canonical partition function.

Connection between statistical mechanics and thermodynamics is achieved through the relations,

$$F = -k_B T \ln Z, \qquad (1.4.17)$$

$$\Omega = -k_B T \ln \Xi, \qquad (1.4.18)$$

with the (Helmholtz) free energy F and the thermodynamic potential Ω. There is a relation between F and Ω,

$$F = \Omega + n\mu, \qquad (1.4.19)$$

for a system of n particles. This relation can be obtained as follows. If we identify the given n with the expectation value of $\tilde{n}$ in the grand canonical ensemble, the right hand side of (1.4.14) can be rewritten as

$$\mathrm{tr}\{\exp[-\beta(H - \mu\tilde{n})]\} = \exp(\beta\mu n)\,\mathrm{tr}[\exp(-\beta H)],$$

where the trace on the right hand side is to be taken only for the fixed particle number n, whereas the trace on the left hand side is for all the possible different particle numbers. This is because the probability distribution of the particle number is very sharply peaked at n. Then from (1.4.18) we obtain

$$\Omega = -n\mu - \frac{1}{\beta} \ln \mathrm{tr}[\exp(-\beta H)], \tag{1.4.20}$$

which is identical to (1.4.19). To calculate F of an electron gas, it is more convenient to obtain Ω first and then use (1.4.19). The difference between $F(T, V, n)$ and $\Omega(T, V, \mu)$ lies in their third independent variables.

1.5 SECOND QUANTIZATION

In working with the rewritten Schrödinger equation (1.4.3) or the statistical average given in (1.4.9) for a many-body system, it is necessary to calculate the matrix elements involved. For example, (1.4.9) can be rewritten as

$$\langle A \rangle = \sum_{\nu\nu'} \rho_{\nu\nu'} A_{\nu'\nu}, \tag{1.5.1}$$

where $\rho_{\nu\nu'}$ and $A_{\nu'\nu}$ are the matrix elements defined as in (1.4.6) with respect to any chosen complete orthonormal set $\{\Psi_\nu\}$. If the system is an electron gas (consisting of n electrons), we use as Ψ_ν the $n \times n$ Slater determinant given in (1.2.10), as in (1.4.6). Then, calculation of matrix element $A_{\nu\nu'}$ becomes practically impossible for such a large n (e.g., $n \sim 10^{22}$).

The method of *second quantization* drastically simplifies calculations involving such matrix elements. In this method we define a new state vector (wave function) $\tilde{\Psi}_\nu$ and a physical operator $\tilde{A}$ corresponding to the old Ψ_ν and A by a simple rule to satisfy

$$(\Psi_\nu, A\Psi_{\nu'}) = (\tilde{\Psi}_\nu, \tilde{A}\,\tilde{\Psi}_{\nu'}), \tag{1.5.2}$$

and to make the right-hand side much easier to handle than the left.

First let us discuss how to construct $\tilde{\Psi}_\nu$. We recall that the fundamental information carried by the Slater determinant Φ_ν in (1.2.10) is how the one-particle states are occupied, namely the knowledge of $\nu = (k_1, k_2, \cdots, k_n)$ only. Reflecting this fact, a new state vector $\tilde{\Phi}_\nu$ corresponding to the Slater determinant Φ_ν is introduced by

$$\tilde{\Phi}_{k_1 k_2 \cdots k_n} = a^+_{k_n} \cdots a^+_{k_2} a^+_{k_1} | \text{vac} > = | k_n, \cdots, k_2, k_1 >, \tag{1.5.3}$$

where $|\cdots\rangle$ is the Dirac ket symbol, $|\,\text{vac}\,\rangle$ is the vacuum state vector without any particle (i.e., no electron) in it, and a_k^+ is the *creation operator* putting one electron in the state $k = (\boldsymbol{k},\sigma)$. Thus, (1.5.3) represents the state obtained by putting k_1, k_2, $\cdots$, and k_n electrons one by one starting from the vacuum. For definiteness, we adopt the order of putting electrons to be that of increasing energy such that

$$\varepsilon_{k_n} \geq \cdots \geq \varepsilon_{k_2} \geq \varepsilon_{k_1} . \tag{1.5.4}$$

What will be the consequences of changing order of the two creation operators $a_{k_1}^+$ and $a_{k_2}^+$ in (1.5.3)? Since $\Phi_{k_2 k_1 \cdots k_n} = -\,\Phi_{k_1 k_2 \cdots k_n}$ for two Slater determinants, it is required to have

$$\widetilde{\Phi}_{k_2\, k_1 \cdots\, k_n} = -\,\widetilde{\Phi}_{k_1\, k_2 \cdots\, k_n} ,$$

or equivalently $a_{k_1}^+ a_{k_2}^+ = -\,a_{k_2}^+ a_{k_1}^+$, i.e.,

$$a_{k_1}^+ a_{k_2}^+ + a_{k_2}^+ a_{k_1}^+ \equiv \{\, a_{k_1}^+ ,\, a_{k_2}^+ \,\} = 0. \tag{1.5.5a}$$

$\{\ ,\ \}$ stands for an anticommutator. A special case of (1.5.5a) for $k_1 = k_2 = k$ is

$$a_k^+ a_k^+ = -\,a_k^+ a_k^+ = 0 .$$

This represents the Pauli principle's disallowance of any one-electron level being occupied by more than one electron.

The Hermite conjugate $a_k = (a_k^+)^+$ of a_k^+ is the *annihilation* (or *destruction*) *operator*. To see such property of a_k, we first recall that the Hermite conjugate of a ket, $|\,\rangle$, is its corresponding bra, $\langle\,| = (\,|\,\rangle\,)^+$. Then, the normalization condition on $|\,k\,\rangle = a_k^+\,|\,\text{vac}\rangle$ leads to

$$1 = <k\,|\,k> = <k\,|\,|\,k> = (\,|\,k>)^+\,|\,k> = (\,a_k^+\,|\,\text{vac}>)^+\,|\,k>$$

$$= <\text{vac}\,|\,a_k\,|\,k> .$$

Noting that $\langle\text{vac}\,|\,\text{vac}\rangle = 1$, we have

$$a_k\,|\,k> = |\,\text{vac}> . \tag{1.5.6}$$

This justifies the name of annihilation operator for a_k. Since a_k is defined as the Hermite conjugate of a_k^+, (1.5.5a) implies

$$a_{k_1} a_{k_2} + a_{k_2} a_{k_1} = \{ a_{k_1}, a_{k_2} \} = 0. \tag{1.5.5b}$$

Its special case is $a_k a_k = 0$.

Finally, we leave it as an exercise for the reader to prove (*Problem* 1.1)

$$a_{k_1} a_{k_2}^+ + a_{k_2}^+ a_{k_1} = \{ a_{k_1}, a_{k_2}^+ \} = \delta_{k1,\, k2}\,. \tag{1.5.5c}$$

These (1.5.5a)–(1.5.5c) are the fundamental commutation relations for electron (or fermion in general) creation and annihilation operators. We have now completed discussion on the nature of the new state vectors $\tilde{\boldsymbol{\Phi}}_\nu$ in the second quantized form.

Next, let us proceed to study the operator $\tilde{A}$ representing a physical quantity in the second quantized form. Obviously it must be expressed in terms of a_k's and a_k^+'s. Let us start with the particle number operator

$$\tilde{n}_k = a_k^+ a_k \tag{1.5.7}$$

for the state k. $\tilde{n}_k$ has an eigenvalue of 1 or 0, as the Pauli principle dictates. We can prove it by operating this $\tilde{n}_k$ directly on a ket with occupied or empty one-electron state k, or, alternatively, by noting the relation

$$\tilde{n}_k (1 - \tilde{n}_k) = 0, \tag{1.5.8}$$

which can be derived from the commutation relations (1.5.5).

In presenting the general prescription of obtaining $\tilde{A}$ from A, we introduce the *quantized wave function*, or, *field operator*,

$$\psi_\sigma(\boldsymbol{r}) = \sum_{\boldsymbol{k}} a_{\boldsymbol{k}\sigma}\, \varphi_{\boldsymbol{k}\sigma}(\boldsymbol{r}) = \sum_{\boldsymbol{k}} a_{\boldsymbol{k}\sigma}\, \varphi_{\boldsymbol{k}}(\boldsymbol{r})\, \chi_\sigma\,, \tag{1.5.9}$$

and its Hermite conjugate

$$\psi_\sigma^+(\boldsymbol{r}) = \sum_{\boldsymbol{k}} a_{\boldsymbol{k}\sigma}^+\, \varphi_{\boldsymbol{k}}^*(\boldsymbol{r})\, \chi_\sigma^+\,, \tag{1.5.9'}$$

where $\varphi_{\boldsymbol{k}\sigma}(\boldsymbol{r})$ is the one-particle (i.e., one-electron) wave function and $a_{\boldsymbol{k}\sigma}^+$ and $a_{\boldsymbol{k}\sigma}$ are the creation and annihilation operators. More generally, we can use any complete orthonormal set for $\{\varphi_{\boldsymbol{k}\sigma}\}$. In almost all cases of physical interest, a general physical quantity A (e.g., Hamiltonian) consists of one-body quantities (e.g., kinetic energy) and two-body quantities (e.g., Coulomb interaction between electrons):

$$A(\boldsymbol{r}_1\sigma_1, \boldsymbol{r}_2\sigma_2, \cdots) = \sum_i A^{(1)}(\boldsymbol{r}_i\sigma_i) + \sum_{i \neq j} A^{(2)}(\boldsymbol{r}_i\sigma_i, \boldsymbol{r}_j\sigma_j). \quad (1.5.10)$$

Then, with

$$\psi(\boldsymbol{r}) = \sum_\sigma \psi_\sigma(\boldsymbol{r}), \quad (1.5.11)$$

the corresponding $\tilde{A}^{(1)}$ and $\tilde{A}^{(2)}$ in $\tilde{A} = \tilde{A}^{(1)} + \tilde{A}^{(2)}$ are given as

$$\tilde{A}^{(1)} = \int d\boldsymbol{r}\psi^+(\boldsymbol{r})\, A^{(1)}(\boldsymbol{r}\sigma)\psi(\boldsymbol{r}), \quad (1.5.12)$$

$$\tilde{A}^{(2)} = \int d\boldsymbol{r}\, d\boldsymbol{r}'\psi^+(\boldsymbol{r})\, \psi^+(\boldsymbol{r}')\, A^{(2)}(\boldsymbol{r}\sigma, \boldsymbol{r}'\sigma')\, \psi(\boldsymbol{r}')\, \psi(\boldsymbol{r}). \quad (1.5.13)$$

As an example, consider the kinetic energy H_0 of the system of electrons given in (1.2.2). It is obviously a sum of one-body quantities. If we use the plane wave (1.1.5) for $\varphi_{\boldsymbol{k}}(\boldsymbol{r})$ in (1.5.9), from the prescription (1.5.12) we obtain

$$\begin{aligned}\tilde{H}_0 &= \sum_{\boldsymbol{k}\sigma\boldsymbol{k}'\sigma'} a^+_{\boldsymbol{k}\sigma}\, a_{\boldsymbol{k}'\sigma'} \int d\boldsymbol{r}\, \phi^*_{\boldsymbol{k}}(\boldsymbol{r}) \left(\frac{-\hbar^2}{2m}\nabla^2\right) \phi_{\boldsymbol{k}'}(\boldsymbol{r})\, \chi^+_\sigma\, \chi_{\sigma'} \\ &= \sum_{\boldsymbol{k}\sigma} \varepsilon_{\boldsymbol{k}}\, a^+_{\boldsymbol{k}\sigma}\, a_{\boldsymbol{k}\sigma} = \sum_{\boldsymbol{k}} \varepsilon_{\boldsymbol{k}}\, \tilde{n}_{\boldsymbol{k}}, \end{aligned} \quad (1.5.14)$$

with $\varepsilon_{\boldsymbol{k}}$ given in (1.1.4). This expression is very intuitive in view of the fact that the total energy E_0 of the free electron gas should be

$$E_0 = \sum_{\boldsymbol{k}} \varepsilon_{\boldsymbol{k}}\, n_{\boldsymbol{k}}, \quad (1.5.14')$$

when the level with energy $\varepsilon_{\boldsymbol{k}}$ has $n_{\boldsymbol{k}}$ electrons ($n_{\boldsymbol{k}} = 1$ or 0). Hence, it is obvious that

$$(\tilde{\Phi}_{k_1k_2\cdots k_n}, \tilde{H}_0\tilde{\Phi}_{k_1k_2\cdots k_n}) = (\Phi_{k_1k_2\cdots k_n}, H_0\Phi_{k_1k_2\cdots k_n}). \quad (1.5.15)$$

In summary, if we construct a new state vector $\tilde{\Psi}_\nu$ and a new physical quantity $\tilde{A}$ following the rules (1.5.3) and (1.5.10)–(1.5.13), respectively, we observe that the equality (1.5.2) holds. We have shown this in the above for the kinetic energy of an electron gas. In the same way, we can generally prove the validity of such a prescription including the case of two-body quantities. See, for instance, Refs. [1.1–1.6]. In this book, however, we concentrate on illustrating how to use this convenient method of second quantization without giving such a general proof. As for the two-body quantities, we will give an example later in this section.

Using (1.5.5a)–(1.5.5c), we can verify the following commutation relations for $\psi_\sigma^+(\boldsymbol{r})$ and $\psi_\sigma(\boldsymbol{r})$:

$$\{\psi_\sigma^+(\boldsymbol{r}), \psi_{\sigma'}^+(\boldsymbol{r}')\} = \psi_\sigma^+(\boldsymbol{r})\,\psi_{\sigma'}^+(\boldsymbol{r}') + \psi_{\sigma'}^+(\boldsymbol{r}')\,\psi_\sigma^+(\boldsymbol{r}) = 0, \tag{1.5.16a}$$

$$\{\psi_\sigma(\boldsymbol{r}), \psi_{\sigma'}(\boldsymbol{r}')\} = 0, \tag{1.5.16b}$$

$$\{\psi_\sigma^+(\boldsymbol{r}), \psi_{\sigma'}(\boldsymbol{r}')\} = \sum_{\boldsymbol{k}\,\boldsymbol{k}'} \{a_{\boldsymbol{k}\sigma}^+, a_{\boldsymbol{k}'\sigma'}\}\, \varphi_{\boldsymbol{k}\sigma}^*(\boldsymbol{r})\, \varphi_{\boldsymbol{k}'\sigma'}(\boldsymbol{r}'),$$

$$= \sum_{\boldsymbol{k}} \varphi_{\boldsymbol{k}\sigma}^*(\boldsymbol{r})\, \varphi_{\boldsymbol{k}\sigma'}(\boldsymbol{r}')\, \delta_{\sigma,\sigma'} = \delta(\boldsymbol{r}-\boldsymbol{r}')\delta_{\sigma,\sigma'}, \tag{1.5.16c}$$

where $\delta(\boldsymbol{r} - \boldsymbol{r}') = \delta(x - x')\delta(y - y')\delta(z - z')$ is the three dimensional Dirac delta function and the last step follows from completeness assumed for the function set $\{\varphi_{\boldsymbol{k}\sigma}(\boldsymbol{r})\}$. In particular, when $\varphi_{\boldsymbol{k}}(\boldsymbol{r})$'s are plane waves, we can directly verify that

$$\sum_{\boldsymbol{k}} \phi_{\boldsymbol{k}}^*(\boldsymbol{r})\, \phi_{\boldsymbol{k}}(\boldsymbol{r}') = \frac{1}{V}\sum_{\boldsymbol{k}} e^{i\boldsymbol{k}(\boldsymbol{r}'-\boldsymbol{r})} = \frac{1}{(2\pi)^3}\int d\boldsymbol{k}\, e^{i\boldsymbol{k}(\boldsymbol{r}'-\boldsymbol{r})}$$

$$= \delta(\boldsymbol{r}'-\boldsymbol{r}). \tag{1.5.17}$$

In ordinary quantum mechanics, the quantization is achieved by imposing commutation relations

$$[x_\mu, x_\nu] = [p_\mu, p_\nu] = 0 \quad \text{and} \quad [x_\mu, p_\nu] = i\hbar\,\delta_{\mu\nu} \tag{1.5.18}$$

on the operators representing physical quantities such as position $\boldsymbol{r} = (x, y, z)$ and momentum $\boldsymbol{p} = (p_x, p_y, p_z)$. The wave functions are ordinary numbers, not requiring any quantum mechanical commutation relation. However, the commutation relations (1.5.16a)–(1.5.16c) in a sense represent quantization of the wave functions. Thus, the above formulation of quantum mechanics involving the creation and annihilation operators is referred as the *second quantization*.

So far, the second quantization is introduced primarily as a convenient method to make it easier to handle the matrix elements required in quantum mechanics, (1.4.3), and statistical mechanics, (1.4.8) or (1.5.1). However, we saw that the second quantized representation of the kinetic energy of (1.5.14) can be intuitively understood. What, then, is the meaning of the quantized wave functions? According to quantum mechanics, an electron has the dual property of wave and particle. The second quantized wave function $\psi_\sigma(\boldsymbol{r})$ of (1.5.9) embodies such a property. It is a superposition of the *waves*, $\varphi_{k\sigma}$, and its squared amplitude $|a_{k\sigma}|^2 = a^+_{k\sigma} a_{k\sigma} = n_{k\sigma}$ represents the number of corresponding *particles*. Thus,

$$\int d\boldsymbol{r}\ \psi^+(\boldsymbol{r})\psi(\boldsymbol{r}) = \sum_{k\sigma} a^+_{k\sigma} a_{k\sigma} = \sum_k \tilde{n}_k \tag{1.5.19}$$

gives the total number of electrons in this system. Obviously

$$\tilde{n}(\boldsymbol{r}) = \psi^+(\boldsymbol{r})\ \psi(\boldsymbol{r}) = \sum_{k\,k'\sigma} a^+_{k\sigma}\ a_{k'\sigma}\ \varphi^*_k(\boldsymbol{r})\ \varphi_{k'}(\boldsymbol{r}) \tag{1.5.20}$$

represents the spatial number density of electrons in the system.

As an example for the second-quantized representation of two-body quantities, consider the Coulomb interaction between electrons,

$$H_{\mathrm{C}} = \frac{1}{2} \sum_{i \neq j} \frac{e^2}{|\boldsymbol{r}_i - \boldsymbol{r}_j|}. \tag{1.5.21}$$

The sum of this and the kinetic energy H_0 of (1.2.2) constitutes a proper Hamiltonian for an electron gas. Abbreviating as $(\boldsymbol{k}, \sigma) \equiv k$, we have, according to (1.5.13)

$$\tilde{H}_{\mathrm{C}} = \frac{1}{2} \sum_{\sigma,\sigma'} \int d\boldsymbol{r}\, d\boldsymbol{r}'\, \psi^+_\sigma(\boldsymbol{r})\ \psi^+_{\sigma'}(\boldsymbol{r}')\ \frac{e^2}{|\boldsymbol{r}-\boldsymbol{r}'|}\ \psi_{\sigma'}(\boldsymbol{r}')\ \psi_\sigma(\boldsymbol{r})$$

$$= \frac{1}{2} \sum_{k,k',l,l'} a^+_k a^+_l a_{l'} a_{k'} \int d\boldsymbol{r}\, d\boldsymbol{r}' \frac{e^2}{|\boldsymbol{r}-\boldsymbol{r}'|}\ \phi^+_k(\boldsymbol{r})\ \phi^+_l(\boldsymbol{r}')\ \phi_{l'}(\boldsymbol{r}')\ \phi_{k'}(\boldsymbol{r}), \tag{1.5.22}$$

where $\phi_k(\boldsymbol{r})$ is the plane wave given by (1.1.13) and (1.1.5). If we introduce the Fourier transform of the Coulomb potential as

$$\frac{1}{V}\int dr\,\frac{e^2}{r}\,e^{-i\kappa r} \equiv \frac{1}{V}\int dr\;\mathrm{v}(\boldsymbol{r})\,e^{-i\kappa r} = \mathrm{v}(\kappa)$$

$$= \frac{1}{V}\frac{4\pi e^2}{\kappa^2}, \tag{1.5.23a}$$

(1.5.22) reduces to

$$\widetilde{H}_{\mathrm{C}} = \frac{1}{2}\sum_{k,\sigma,\sigma',\kappa}{}' \mathrm{v}(\kappa)\, a_{k\sigma}^{+}\, a_{l\sigma'}^{+} a_{l+\kappa,\sigma'} a_{k-\kappa,\sigma}, \tag{1.5.24}$$

where we used the relation

$$\int_V dr\, e^{-ikr} = V\delta_{k,0} \tag{1.5.25a}$$

for $\boldsymbol{k}$ satisfying (1.1.7): also note

$$\sum_k e^{ikr} = V\delta(\boldsymbol{r}). \tag{1.5.25b}$$

Confirm the Fourier inverse transform of $\mathrm{v}(\kappa)$

$$\mathrm{v}(\boldsymbol{r}) = \sum_{\kappa} \mathrm{v}(\kappa)\, e^{i\kappa r}. \tag{1.5.23b}$$

The result of (1.5.24) can be represented as in Fig 1.2. It offers a very intuitive picture: Two electrons at the states $(\boldsymbol{k}-\kappa, \sigma)$ and $(\boldsymbol{l}+\kappa, \sigma')$ interact with an exchange of momentum $\hbar\kappa$ to make a transition to the states the $(\boldsymbol{k}, \sigma)$ and $(\boldsymbol{l}, \sigma')$. Since the Coulomb interaction is spin independent, the spin states of the particles remain unchanged. As for this Coulomb interaction between electrons we will discuss it extensively later in this book.

As another important example of the second quantization method, let us derive the Hamiltonian of electrons interacting with an impurity in a metal. Assume that the impurity is at the origin of the coordinates and that the impurity potential is of the form of

$$H_{\mathrm{imp}} = \sum_j U(\boldsymbol{r}_j), \tag{1.5.26}$$

where $\boldsymbol{r}_j$ is the coordinate of the j-th electron. Then, according to the

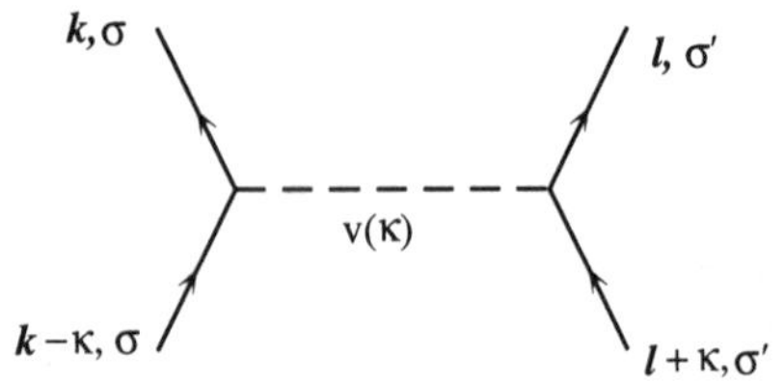

Fig.1.2 The Coulomb interaction (broken line) between electrons. The solid lines represent the propagation of electrons.

prescription (1.5.12), its second quantized representation is obtained as

$$\tilde{H}_{\text{imp}} = \sum_{\sigma,\sigma'} \int d\boldsymbol{r}\ \psi_\sigma^+(\boldsymbol{r})\ U(\boldsymbol{r})\ \psi_{\sigma'}(\boldsymbol{r})$$

$$= \sum_{k,\sigma,k',\sigma'} a_{k\sigma}^+ a_{k'\sigma'} \int d\boldsymbol{r}\ \phi_k^*(\boldsymbol{r}) U(\boldsymbol{r})\ \phi_{k'}(\boldsymbol{r})\ \chi_\sigma^+ \chi_{\sigma'}$$

$$= \sum_{k,\sigma,\kappa} U(\kappa)\ a_{k\sigma}^+ a_{k-\kappa,\sigma}\,, \tag{1.5.27}$$

with

$$U(\kappa) = \frac{1}{V} \int d\boldsymbol{r}\, U(\boldsymbol{r})\, e^{-i\kappa \boldsymbol{r}}. \tag{1.5.28}$$

This result can also be intuitively represented as in Fig.1.3.

If we use the electron density operator (1.5.20), the above result can be rewritten as

$$\tilde{H}_{\text{imp}} = \int d\boldsymbol{r}\ U(\boldsymbol{r})\, \tilde{n}(\boldsymbol{r}), \tag{1.5.27'}$$

which is also very intuitive.

Now let us define the Fourier transform of the electron density as

$$\tilde{n}(\kappa) = \int_V d\boldsymbol{r}\ \tilde{n}(\boldsymbol{r})\ e^{-i\kappa \boldsymbol{r}}\,, \tag{1.5.29a}$$

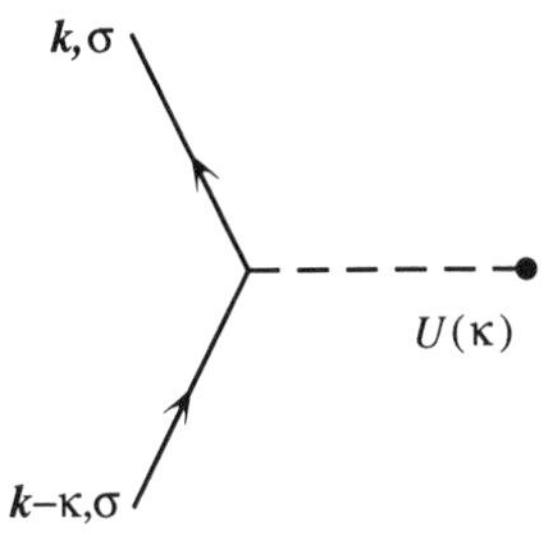

Fig.1.3 The scattering of an electron by an impurity potential $U(\kappa)$.

with its inverse

$$\tilde{n}(\boldsymbol{r}) = \frac{1}{V}\sum_{\kappa} \tilde{n}(\kappa)\, e^{i\kappa r} . \tag{1.5.29b}$$

Note that the Fourier transform (1.5.29a) of $\tilde{n}(\boldsymbol{r})$ is defined without the $1/V$ factor, in contrast with those for $\mathrm{v}(\boldsymbol{r})$ and $U(\boldsymbol{r})$ in (1.5.23a) and (1.5.28). This convention is adopted to avoid repeated appearance of the volume V in the subsequent expressions. The Fourier transforms of quantities independent of V (such as $\mathrm{v}(\boldsymbol{r})$ and $U(\boldsymbol{r})$) are defined with the $1/V$ factor, whereas those depending on V (such as the electron number, or, magnetization densities) are defined without it.

With the Fourier transform as described above, (1.5.27') can be rewritten as

$$\tilde{H}_{\mathrm{imp}} = \sum_{\kappa} U(\kappa)\, \tilde{n}(-\kappa) , \tag{1.5.27''}$$

with the Fourier transform of the electron density

$$\tilde{n}(\kappa) = \sum_{\kappa,\sigma} a^{+}_{k\sigma}\, a_{k+\kappa,\sigma} \tag{1.5.30}$$

which is obtained from (1.5.29a) by using the plane wave for $\varphi_k(\boldsymbol{r})$ in (1.5.20).

The usefulness of the second quantization method will be demonstrated throughout this book. In discussing spin waves and phonons, we need to know how to treat the Bose particles in the second quantized formalism. It will be discussed later when such a need actually arises.

1.6 A FREE ELECTRON GAS AT FINITE TEMPERATURE

The most fundamental quantity in studying an electron gas is the statistical expectation of the electron number operator given in (1.5.7). In calculating it, we neglect at first the Coulomb interaction between electrons and take into account only the kinetic energy H_o given in (1.5.14). Then, from (1.4.9) and (1.4.13), it follows that

$$\langle \tilde{n}_k \rangle = \frac{1}{\Xi_o} \operatorname{tr}\{ \exp[-\beta \sum_p (\varepsilon_p - \mu)\, \tilde{n}_p]\, \tilde{n}_k \}, \qquad (1.6.1)$$

with the grand partition function

$$\Xi_o = \operatorname{tr}\{\exp[-\beta \sum_p (\varepsilon_p - \mu)\tilde{n}_p] \}. \qquad (1.6.2)$$

In calculating the trace with respect to state vectors of the form of (1.5.3), we use the fact that such a state is an eigenstate of $\tilde{n}_k$ with the eigenvalues $n_k = 1$ or 0, as well as the commutativity $[\tilde{n}_p, \tilde{n}_k] = 0$ of two number operators $\tilde{n}_p$ and $\tilde{n}_k$. Thus, the grand partition function for the free electron gas is calculated as

$$\begin{aligned}\Xi_o &= \sum_{n_p = 0,1} \exp[-\beta \sum_p \{(\varepsilon_p - \mu) n_p \}] \\ &= \prod_p \{1 + \exp[-\beta(\varepsilon_p - \mu)]\}. \qquad (1.6.3)\end{aligned}$$

Since the numerator of (1.6.1) can be rewritten as

$$\begin{aligned}&\sum_{n_k = 0,1} \sum_{n_p = 0,1} \{ \exp[-\beta \sum_p (\varepsilon_p - \mu) n_p]\, n_k \} \\ &= \prod_{p(\neq k)} \{1 + \exp[-\beta(\varepsilon_p - \mu)]\} \exp[-\beta(\varepsilon_k - \mu)], \qquad (1.6.4)\end{aligned}$$

(1.6.1) reduces to

$$\langle \tilde{n}_k \rangle = \frac{1}{\exp[\beta(\varepsilon_k - \mu)] + 1} \equiv f(\varepsilon_k), \qquad (1.6.5)$$

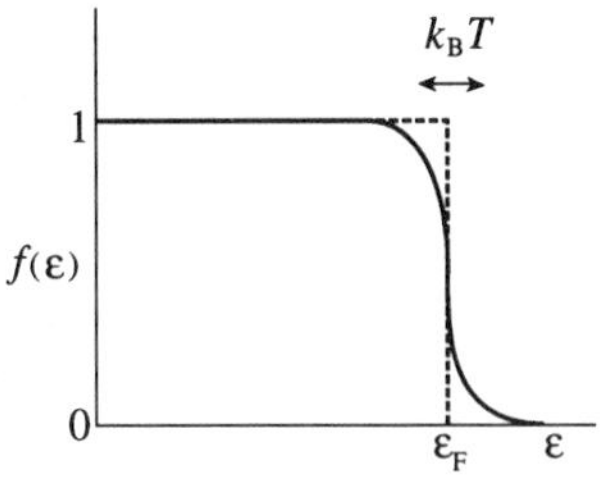

Fig.1.4 The Fermi distribution for $T = 0$ (dotted line) and for $T > 0$ (solid line).

which is the familiar *Fermi distribution function* shown in Fig.1.4. For $T = 0$, the Fermi distribution function behaves as given by the dotted line corresponding to

$$f(\varepsilon_k)_{\text{for } T=0} = \begin{cases} 1 & \text{for } \varepsilon_k \leq \varepsilon_F, \\ 0 & \text{for } \varepsilon_k > \varepsilon_F. \end{cases} \tag{1.6.6}$$

This represents the ground state discussed in Section 1.3. For $T > 0$, it deviates from (1.6.6) in the neighborhood of order k_BT around the Fermi energy ε_F as shown by the solid line. However, the dotted line expressed by (1.6.6) is a good approximation in the temperature region

$$k_BT \ll \varepsilon_F. \tag{1.6.7}$$

The chemical potential μ in Fermi distribution function is determined from the constancy condition on the total number of electrons, namely,

$$n = \sum_k f(\varepsilon_k) = \sum_{k,\sigma} f(\varepsilon_{k\sigma}) \,. \tag{1.6.8}$$

Hence μ is a function of temperature T. Obviously, at $T = 0$, $\mu = \varepsilon_F$.

The magnitude of $\partial f(\varepsilon)/\partial\varepsilon$ is large only around μ, as can be seen from Fig.1.4. This property is useful for the evaluation of integrals involving the Fermi distribution function as demonstrated below. If we put

$$G(\varepsilon) = \int_{-\infty}^{\varepsilon} d\varepsilon' g(\varepsilon'), \tag{1.6.9}$$

we obtain

$$\int_{-\infty}^{\infty} d\varepsilon\, g(\varepsilon) f(\varepsilon) = \Big[G(\varepsilon) f(\varepsilon)\Big]_{-\infty}^{\infty} - \int_{-\infty}^{\infty} d\varepsilon\, G(\varepsilon) \frac{\partial f(\varepsilon)}{\partial \varepsilon}, \qquad (1.6.10)$$

integrating by parts. Then, for the second term on the right hand side of (1.6.10), it can be shown (see, for instance, Ref. [1.3]) that

$$\int_{-\infty}^{\infty} d\varepsilon\, G(\varepsilon) \left(- \frac{\partial f(\varepsilon)}{\partial \varepsilon} \right) = G(\mu) + \frac{\pi^2}{6} (k_B T)^2 G''(\mu) + \cdots . \qquad (1.6.11)$$

This expansion in $(k_B T/\varepsilon_F)^2$ is very useful at low temperatures. It also implies that $-\partial f(\varepsilon)/\partial\varepsilon \cong \delta(\varepsilon - \varepsilon_F)$ at low temperatures. For ordinary metals, we have $\varepsilon_F = (1 \sim 10)$ eV and hence $\varepsilon_F \sim (10^4 \sim 10^5)\, k_B$. Thus the low temperature condition (1.6.7) becomes $T \ll (10^4 \sim 10^5)$ K.

By using (1.6.11), we can determine the temperature dependence of the chemical potential $\mu(T)$. By rewriting (1.6.8) with the electronic density of states of (1.3.5) as in (1.3.6), and then applying (1.6.9)–(1.6.11), we obtain

$$\mu(T) = \varepsilon_F - \frac{\pi^2}{6} (k_B T)^2 \frac{N'(\varepsilon_F)}{N(\varepsilon_F)} + \cdots . \qquad (1.6.12)$$

Substitution of (1.6.12) into (1.6.11) gives

$$\int_{-\infty}^{\infty} d\varepsilon\, G(\varepsilon) \left(- \frac{\partial f(\varepsilon)}{\partial \varepsilon} \right)$$

$$= G(\varepsilon_F) + \frac{\pi^2}{6} (k_B T)^2 \left(G''(\varepsilon_F) - G'(\varepsilon_F) \frac{N'(\varepsilon_F)}{N(\varepsilon_F)} \right) + \cdots . \qquad (1.6.13)$$

In particular, when $G(\varepsilon)$ is the electronic density of states, $N(\varepsilon)$, it becomes

$$\int_{-\infty}^{\infty} d\varepsilon\, N(\varepsilon) \left(- \frac{\partial f(\varepsilon)}{\partial \varepsilon} \right) = \sum_{k} - \frac{\partial f(\varepsilon_k)}{\partial \varepsilon_k}$$

$$= N(\varepsilon_F) \left[1 - \frac{\pi^2}{6} (k_B T)^2 \left\{ \left(\frac{N'(\varepsilon_F)}{N(\varepsilon_F)} \right)^2 - \frac{N''(\varepsilon_F)}{N(\varepsilon_F)} \right\} \right] + \cdots$$

$$\equiv N(\varepsilon_F)\left[1-\alpha\left(\frac{T}{T_F}\right)^2\right]+\cdots, \tag{1.6.14}$$

where $|\alpha| = O(1)$ (see (3.3.17)) and we put $k_B T_F = \varepsilon_F$; T_F is called the *Fermi temperature*. Later on in this book, we put $N(\varepsilon_F) \equiv N(0)$, $N'(\varepsilon_F) \equiv N'(0)$, etc.; we make the chemical potential or the Fermi energy the origin of measuring energy.

For $T > 0$, the energy of the free electron gas is given by

$$E = \langle H_o \rangle = 2\sum_k \varepsilon_k f(\varepsilon_k) = 2\int_{-\infty}^{\infty} d\varepsilon\, N(\varepsilon) f(\varepsilon)\, \varepsilon\,. \tag{1.6.15}$$

This result reduces to (1.3.2) at $T = 0$.

Let us calculate the specific heat of the free electron gas. If we put $g(\varepsilon) = 2\varepsilon N(\varepsilon)$ into (1.6.9)–(1.6.13), then (1.6.15) is rewritten as

$$E = 2\int_{-\infty}^{\varepsilon_F} d\varepsilon\, N(\varepsilon)\, \varepsilon + 2\frac{\pi^2}{6}(k_B T)^2\left[(\varepsilon N(\varepsilon))' - \varepsilon N(\varepsilon)\frac{N'(\varepsilon)}{N(\varepsilon)}\right]_{\varepsilon=\varepsilon_F} + \cdots$$

$$= E_o + \frac{\pi^2}{3} N(0)\,(k_B T)^2 + \cdots. \tag{1.6.16}$$

Hence the corresponding specific heat is given by

$$C_{el}^o = \frac{dE}{dT} = \frac{2}{3}\pi^2 k_B^2 N(0)\, T + \cdots \equiv \gamma T + \cdots. \tag{1.6.17}$$

It is an important fact that the low-temperature specific heat of metallic electrons takes the form of γT and the constant γ is proportional to the electronic density of states at the Fermi surface.

PROBLEMS FOR CHAPTER 1

1.1 Prove the Fermion anticommutation relation of (1.5.5c). Note that it is sufficient to prove that the relation holds when applied to each of the states $|\,0, 0>$, $|\,0, k_1>$, $|\,k_2, 0>$, and $|\,k_2, k_1>$ individually for the case $k_1 \neq k_2$, and for $|\,0>$ and $|\,k>$ for the case $k_1 = k_2 = k$.

1.2 Prove (1.5.25a) and (1.5.25b). Note that $\boldsymbol{k}$ satisfies the condition (1.1.7).

1.3 Generalize the result of (1.5.27) for the case of N (>1) impurity atoms;denote the coordinates of impurities as $\boldsymbol{R}_1, \boldsymbol{R}_2, \ldots \boldsymbol{R}_N$.

1.4 Represent the z-component of the spin density of an electron gas in the second quantized form. (See 3.1).

1.5 Find out how to derive the result of (1.6.11).

Chapter 2

BASICS OF MAGNETISM

In this chapter we briefly present an introduction to magnetism in general. Although in this book our interest is confined to magnetism of metals, the discussion of this chapter is focused on magnetism of insulators. An introduction to magnetism of metals will be given separately in the next chapter. For a more extensive introduction to magnetism, see, for instance, Refs. [1.1,1.6, 2.1–2.13].

2.1 MAGNETIZATION AND MAGNETIC FIELD

The motion of electrons in an atom, just outside of its closed shell, or a molecule, or a solid can be described by the following Hamiltonian,

$$H = \sum_i \left[\frac{1}{2m} \left\{ \boldsymbol{p}_i + \frac{e}{c} \boldsymbol{A}(\boldsymbol{r}_i) \right\}^2 - e V(\boldsymbol{r}_i) \right], \tag{2.1.1}$$

where $-e$ is the electronic charge, c is the light speed, m, $\boldsymbol{r}_i$ and $\boldsymbol{p}_i$ are, respectively, the mass, position and momentum of the i-th electron localized around an atom; in this chapter we consider the case of an insulator. We use the cgs unit system in this book. For a spatially uniform magnetic field $\boldsymbol{H}$ we can choose the vector potential in the following form

$$\boldsymbol{A}(\boldsymbol{r}) = \frac{1}{2} (\boldsymbol{H} \times \boldsymbol{r}). \tag{2.1.2}$$

$V(\boldsymbol{r})$ is an electrostatic potential for electrons; in the case of a solid it is the periodic lattice potential, and in an atom it is the atomic potential. Here we do not consider the interaction between electrons.

If we put (2.1.2) into (2.1.1) we obtain terms which are first and second orders in the magnetic field. First, for the first order term we have

$$H^{(1)} = \frac{e}{mc}\sum_i \boldsymbol{p}_i \boldsymbol{A}(\boldsymbol{r}_i) = \frac{e}{2mc}\sum_i \boldsymbol{p}_i\,(\boldsymbol{H}\times\boldsymbol{r}_i)$$

$$= \frac{e}{2mc}\sum_i (\boldsymbol{r}_i\times\boldsymbol{p}_i)\,\boldsymbol{H} = \sum_i \mu_{\mathrm{B}}\,\boldsymbol{l}_i\,\boldsymbol{H}, \qquad (2.1.3)$$

where

$$\hbar\,\boldsymbol{l}_i = \boldsymbol{r}_i\times\boldsymbol{p}_i \qquad (2.1.4)$$

is the orbital angular momentum of the i-th electron, and

$$\mu_{\mathrm{B}} = \frac{e\hbar}{2mc} = 0.927\times 10^{-20}\ \mathrm{erg}\,/\,\mathrm{Gauss} \qquad (2.1.5)$$

is the *Bohr magneton*. As shown in (2.1.4), the orbital angular momentum is quantized in units of the Planck constant $\hbar$. On this angular momentum operator $\boldsymbol{l}_i$ we impose the same commutation relation as that of the spin angular momentum, (1.1.10). Differing from the case of spin, however, the magnitude of $\boldsymbol{l}_i$ takes only integral values (including zero).

If we introduce the *magnetization*, $\boldsymbol{m}_i$, associated with the angular momentum $\boldsymbol{l}_i$ of the i-th electron as

$$\boldsymbol{m}_i = -\mu_{\mathrm{B}}\boldsymbol{l}_i\,, \qquad (2.1.6)$$

(2.1.3) is rewritten as

$$H^{(1)} = -\sum_i \boldsymbol{m}_i\,\boldsymbol{H} = -\boldsymbol{M}\boldsymbol{H}, \qquad (2.1.7)$$

where $\boldsymbol{M}$ is the magnetization of the entire system of electrons.

Equation (2.1.7) is the quantitative definition of magnetization. If the energy of an electron, or, a system of electrons changes with a magnetic field linearly as in (2.1.7), then the magnetization is defined as the coefficient of the proportionality. In the cgs unit system, magnetization is measured in units of emu (electromagnetic unit). An appreciation of the size of the unit may be gained by noting that (2.1.5) can be rewritten as

$$\mu_{\mathrm{B}} = 0.927\times 10^{-20}\ \mathrm{emu}\,. \qquad (2.1.8)$$

As can be seen from (2.1.6), μ_B has the same physical dimension as that of magnetization.

An important point to note in (2.1.6) is that the direction of magnetization is opposite to that of corresponding (orbital) angular momentum.

Electricity is generally more familiar to us than magnetism. The quantity which corresponds to the magnetic field $\boldsymbol{H}$ is the electric field $\boldsymbol{E}$, while that which corresponds to the magnetization $\boldsymbol{m}_i$ or $\boldsymbol{M}$ is not the electric charge but the electric dipole moment. An electron in a circular motion with radius r_i and velocity $v_i = p_i/m$ around an atomic nucleus, gives rise to the magnetic dipole moment of (2.1.6). We often, thus, call $\boldsymbol{m}_i$ the *magnetic moment,* or, simply, *moment.*

The term which is second order in $\boldsymbol{A}$ or $\boldsymbol{H}$ of (2.1.1) is rewritten as

$$H^{(2)} = \frac{e^2}{8mc^2}\sum_i (\boldsymbol{r}_i \times \boldsymbol{H})^2 = \frac{e^2}{8mc^2}\sum_i H^2 r_{i\perp}^2 , \tag{2.1.9}$$

where $r_{i\perp}$ is the radius of the electron orbit projected to the plane perpendicular to the magnetic field. This second order term is a consequence of the Lenz's law, or, the electromagnetic induction. Suppose an electron is making a circular motion around a nucleus with a radius $r_{i\perp}$. If we apply a magnetic field $\boldsymbol{H}$ in the direction perpendicular to the plane of the electron orbit, an electric field $\boldsymbol{E}$ is induced in the tangential direction of the circular orbit which is given by

$$E \cdot 2\pi r_{i\perp} = -\frac{1}{c}\pi r_{i\perp}^2 \frac{d}{dt}H, \quad \text{or} \quad E = -\frac{r_{i\perp}}{2c}\frac{d}{dt}H.$$

The torque due to this induced electric field leads to the following equation of motion for the orbital motion of the (i-th) electron

$$\frac{d(\hbar\, l_{iz})}{dt} = \frac{e}{2c} r_{i\perp}^2 \frac{d}{dt} H, \tag{2.1.10}$$

where we assumed that the electron orbit lies in the x-y plane, and, therefore, $\boldsymbol{H}$ is in the direction of the z-axis. From (2.1.10) and (2.1.6), for the change in the magnetization, Δm_{iz}, due to an increase in the magnetic field, ΔH, we have

$$\Delta m_{iz} = -\mu_B \Delta l_{iz} = -\frac{e\mu_B}{2c\hbar} r_{i\perp}^2 \Delta H. \tag{2.1.11}$$

With this, (2.1.9) can be rewritten as

$$H^{(2)} = -\sum_i \int_0^H m_{iz}(H')\,dH'. \tag{2.1.12}$$

If m_{iz} does not depend on the magnetic field, (2.1.12) reduces to the form of (2.1.7). As can be seen from (2.1.11), however, m_{iz} is proportional to the magnetic field with a negative coefficient. Thus, the magnetization induced by the external magnetic field is in the direction opposite to the magnetic field. This phenomenon is called *diamagnetism.*

In the case of (2.1.6), we have the magnetic moment of an electron independent of an applied magnetic fields, and the effect of the magnetic field is to align this magnetic moment in the direction of the field. Such property is called *paramagnetism.*

Note that the size of the atomic diamagnetism is much smaller than that of the atomic paramagnetism. If we introduce the Bohr radius, $a_B = \hbar^2/me^2$, (2.1.11) can be rewritten as

$$m_{iz} = -\left(\frac{\hbar^2}{2ma_B^2}\right)\frac{a_B^2}{e^4}\left(\frac{r_{i\perp}}{a_B}\right)^2 \mu_B{}^2 H \cong -\mu_B\left(\frac{\mu_B H}{\varepsilon_R}\right), \tag{2.1.13}$$

where we put

$$\frac{\hbar^2}{2ma_B^2} = \frac{e^2}{2a_B} \equiv \varepsilon_R,$$

and noted $r_{i\perp} \cong a_B$. Since $\varepsilon_R = 1$ Ryd $\cong 13.6$ eV, we require $\sim 10^9$ Gauss to make $\mu_B H/\varepsilon_R \cong 1$ in (2.1.13). Thus, the atomic diamagnetism becomes important only when the paramagnetism due to the orbital and spin (which we discuss below) angular momenta of electrons is entirely absent.

A term that is most important is missing in (2.1.1). It is the magnetic energy associated with the electron spin,

$$H_Z = g\mu_B \sum_i \boldsymbol{s}_i \boldsymbol{H}, \tag{2.1.14}$$

where $\boldsymbol{s}_i$ is the spin of the i-th electron which was introduced in Chapter 1. Such an energy associated with an orbital angular momentum, or a spin in a magnetic field is called the *Zeeman energy.* The quantity g, which is called the g-factor, takes the value of $2.0023 \cong 2$ for a free electron in vacuum. Both of $\boldsymbol{l}_i$ and $\boldsymbol{s}_i$ are angular momenta, but in producing magnetization the latter is nearly twice as effective as the former. Putting these two contributions together we have

$$\boldsymbol{m}_i = -\mu_B(\boldsymbol{l}_i + 2\boldsymbol{s}_i). \tag{2.1.15}$$

The operator relation of (2.1.15) is valid for any individual electrons in either an atom or a solid. However, the states of electrons are different for different situations, and what we observe is the quantum mechanical expectation in each case. For an isolated atom, the spin and the orbital angular momenta of the electrons outside of the closed electronic shell couple to form the resultant total spin $\boldsymbol{S}$, orbital angular momenta $\boldsymbol{L}$, and the total angular momenta $\boldsymbol{J} = \boldsymbol{S} + \boldsymbol{L}$ of the atom following Hund's rules (see 2.3 below).

In the case of a solid, we will have atomic spin $\boldsymbol{S}_j$ and orbital angular momenta $\boldsymbol{L}_j$ on each atom j, if there is no interaction between different atoms. In actuality, however, even in an insulator an electron on an atom experiences the electrostatic potential of neighboring atoms, called the crystal field. Thus, in a solid the potential of an electron is not spherically symmetric. Within classical mechanics, in such a situation the orbital angular momentum of the electron is not conserved. Correspondingly, in quantum mechanics, the expectation value of the orbital angular momentum vanishes [2.1–2.9]. Thus, in a solid we may put $\boldsymbol{L}_j \cong 0$. One well-known exception, however, is the case of rare earth metals. In a rare earth metal since the wave function of the incomplete $4f$ electrons is strongly localized around each nucleus, the effect of neighboring atoms is much smaller than in a transition metal. Thus, even in a solid a rare earth atom preserves its atomic orbital angular momentum.

In the following, as the source of atomic magnetization we consider only the atomic spin. Corresponding to the atomic spin $\boldsymbol{S}_j$, the atomic magnetization is given as

$$\boldsymbol{M}_j = -g\mu_B \boldsymbol{S}_j . \tag{2.1.16}$$

2.2 PARAMAGNETISM OF NON-INTERACTING LOCALIZED SPINS: CURIE SUSCEPTIBILITY

Consider a crystal in which each atom has an atomic spin $\boldsymbol{S}_j$. Let us assume that there are no interactions among these spins. Then, without an external magnetic field the directions of these spins would be completely disordered. Let us investigate how these spins will respond to a magnetic field.

If we assume a magnetic field $\boldsymbol{H}$ is applied in the direction of the z-axis, the Hamiltonian of this spin system is given as

$$H_Z = \sum_{i=1}^{N} g\mu_B \boldsymbol{H}\boldsymbol{S}_i = g\mu_B H \sum_{i=1}^{N} S_{iz}, \tag{2.2.1}$$

where N is the total number of spins (atoms). We now proceed to calculate the thermal expectation value of the magnetization,

$$\langle M_z \rangle = -g\mu_B \sum_i \langle S_{iz} \rangle = -Ng\mu_B \langle S_z \rangle , \tag{2.2.2}$$

where we assumed that $\langle S_{iz} \rangle$ is independent of the atomic site i, and set it equal to $\langle S_z \rangle$. The thermal expectation value is calculated as

$$\langle M_z \rangle = -Ng\mu_B \sum_{S_z=-S}^{S} S_z\, e^{-\beta g\mu_B H S_z}/Z, \tag{2.2.3}$$

$$Z = \sum_{S_z=-S}^{S} e^{-\beta g\mu_B H S_z}, \tag{2.2.4}$$

by the procedure of (1.4.9), (1.4.11') and (1.4.16). S_z is the eigenvalue of S_{iz} which can take the following values

$$S_z = S,\ S-1,\ \cdots,\ -S, \tag{2.2.5}$$

S being the magnitude of $\mathbf{S}_i$.

If we employ the free energy defined in (1.4.17), the result of (2.2.3) can be rewritten as

$$\langle M_z \rangle = -\frac{\partial}{\partial H} F. \tag{2.2.6}$$

It is straightforward to obtain

$$F = -k_B T N \ln \left[\sinh\left(\frac{2S+1}{2} y \right) / \sinh \frac{y}{2} \right], \tag{2.2.7}$$

with $y \equiv g\mu_B H/k_B T$, and, then

$$\langle M_z \rangle = Ng\mu_B S\, \mathrm{B}_S(g\mu_B H S/k_B T), \tag{2.2.8}$$

where

$$\mathrm{B}_S(x) = \frac{2S+1}{2S} \coth\left(\frac{2S+1}{2S} x \right) - \frac{1}{2S} \coth \frac{x}{2S} \tag{2.2.9}$$

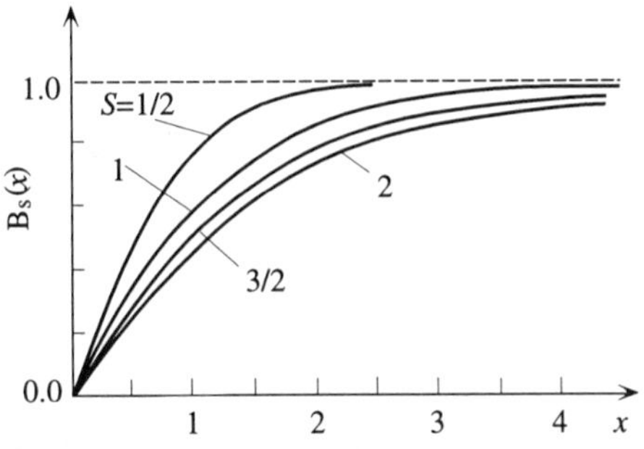

Fig.2.1 The Brillouin function for different S's .

is the *Brillouin function*. As shown in Fig.2.1, for $x = Sy = g\mu_B HS/k_B T \to \infty$, we have $B_S \to 1$, and, then, $\langle M_z \rangle = Ng\mu_B S$; this is the maximum possible magnetization and it is called the saturation magnetization.

When either the temperature is high or the magnetic field is weak, we have

$$\lim_{x \to 0} B_S(x) = (S+1)x/3S, \tag{2.2.10}$$

and, correspondingly,

$$\langle M_z \rangle = Ng^2 \mu_B^2 S(S+1)H/3k_B T. \tag{2.2.11}$$

The *magnetic susceptibility* is defined as

$$\chi = \lim_{H \to 0} \frac{\langle M_z \rangle}{H}. \tag{2.2.12}$$

Then, from (2.2.11) we obtain the *Curie susceptibility*

$$\chi_C = C/T, \tag{2.2.13}$$

where

$$C = N(g\mu_B)^2 S(S+1)/3k_B \tag{2.2.14}$$

is called the *Curie constant*.

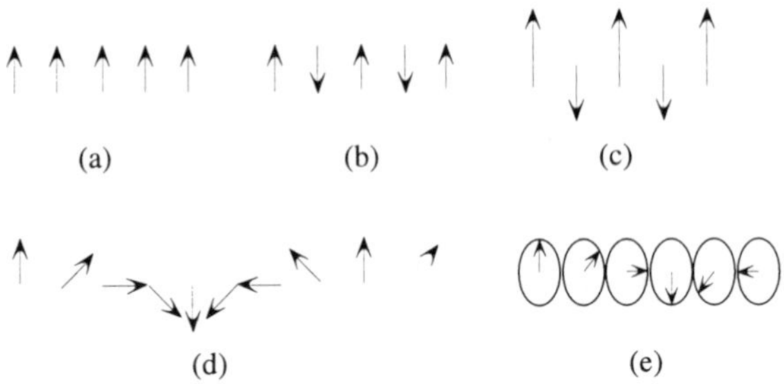

Fig.2.2 Various spin orderings: ferromagnetism, (a); antiferromagnetism, (b); ferrimagnetism, (c) ; sinusoidal, (d) ; helical, (e).

The magnetic susceptibility is among the most important quantities which characterize the magnetic properties of a material, particularly, in the paramagnetic state. It will be discussed extensively later in this book. One thing to note about the Curie susceptibility of (2.2.13)–(2.2.14) is that it applies to non-interacting atomic spins of an insulator, but not to a metal. The magnetic susceptibility of metallic electrons is fundamentally different from that of localized electrons of an insulator, as will be shown in 3.1.

According to (2.2.8) or (2.2.10), we have $\langle M_z \rangle = 0$ for $H = 0$. If there is interaction between spins, however, we can have $\langle M_z \rangle \neq 0$ even for $H = 0$. Spins can align spontaneously as shown in Fig.2.2(a). Such spontaneous spin ordering, without the effect of an external field, is called *ferromagnetism.*

There are other kinds of spin orderings as shown in Fig.2.2; (b) is *antiferromagnetism.* The *ferrimagnetism* of Fig.2.2(c) is understood as the antiferromagnetism of two kinds of spins, one is larger than the other. There are more complicated forms of spin ordering such as shown in (d) and (e). In the *sinusoidal spin ordering* of (d), the direction of spin rotates on the common plane, but in the *helical spin ordering* of (e), the periodical rotation of spins takes place on planes perpendicular to the chain of spins.

Let us now proceed to discuss the origin of the interaction between spins in the next section.

2.3 INTERACTION BETWEEN SPINS: EXCHANGE INTERACTION

Suppose we have two localized electron spins, separated by a distance *a*. The two electrons may be either in the same atom, or in the neighboring two different atoms. To be definite let us first assume the latter situation. Within classical electromagnetism we have the magnetic dipole-dipole interaction between them, whose size is given as

$$O\left(\frac{\mu_B^2}{a^3}\right) \cong \frac{\left(10^{-20}\right)^2}{\left(10^{-8}\right)^3} \text{ erg} \cong 10^{-16} \text{ erg}$$

$$\cong 10^{-4} \text{ eV} \cong 1\, k_B T, \tag{2.3.1}$$

where we assumed $a \cong 10^{-8}$ cm. As we will see shortly, the size of the Curie temperature T_C at which a magnetic order begins is given by the size of the interaction between neighboring two spins. Then the result of (2.3.1) says that the magnetic dipole-dipole interaction can not account for the observed T_C of 10^2 ~10^3 K. There must be, and there is a much stronger interaction between spins. It is the *exchange interaction* found by Heisenberg [2.14]. Let us see how the exchange interaction arises in quantum mechanics.

Suppose that there are two different electronic orbitals a, b, with the wave functions $\varphi_a(\boldsymbol{r})$ and $\varphi_b(\boldsymbol{r})$, respectively. These orbitals may belong either to the same atom, or to the neighboring two different atoms. Then the wave function for the two electrons is given by the Slater determinant

$$\Phi_{\sigma_a\sigma_b}(\boldsymbol{r}_1, \boldsymbol{r}_2) = \frac{1}{\sqrt{2}} \begin{vmatrix} \varphi_a(\boldsymbol{r}_1)\chi_{\sigma_a}(1) & \varphi_b(\boldsymbol{r}_1)\chi_{\sigma_b}(1) \\ \varphi_a(\boldsymbol{r}_2)\chi_{\sigma_a}(2) & \varphi_b(\boldsymbol{r}_2)\chi_{\sigma_b}(2) \end{vmatrix}, \tag{2.3.2}$$

where σ_a and σ_b, which are either +1 or −1, are the values of σ_z's of the electrons in φ_a and φ_b. $\chi_\sigma(i)$ is the spin eigenfunction of the i-th electron, i = 1, 2. Depending upon the spins of two electrons, there are 4 different states, Φ_{++}, Φ_{--}, Φ_{+-} and Φ_{-+}.

If there is no interaction between these two electrons, the energy of this system, E_o, is the sum of two electronic energies,

$$E_o = \varepsilon_a + \varepsilon_b. \tag{2.3.3}$$

Since the energy is independent of the direction of the spins, σ_a and σ_b, the four states are degenerate.

There is, however, a Coulomb interaction between these two electrons,

$$\mathrm{v}(\boldsymbol{r}_1 - \boldsymbol{r}_2) = \frac{e^2}{\left|\boldsymbol{r}_1 - \boldsymbol{r}_2\right|}. \tag{2.3.4}$$

It is an interplay of this Coulomb interaction and the Pauli principle that makes the energy of these two electrons dependent on whether their spins are parallel or antiparallel. Let us see how such a result comes out.

We follow the ordinary 1st order perturbation procedure for degenerate states. If we denote the matrix elements of the Coulomb interaction between the states $\Phi_{\sigma_a\sigma_b}$'s as

$$(\sigma_a\,\sigma_b\,|v|\,\sigma_a'\sigma_b') = \int d\boldsymbol{r}_1\, d\boldsymbol{r}_2\, \Phi^{+}_{\sigma_a\,\sigma_b}(\boldsymbol{r}_1\,,\boldsymbol{r}_2)\frac{e^2}{|\boldsymbol{r}_1-\boldsymbol{r}_2|}\,\Phi_{\sigma_a'\sigma_b'}(\boldsymbol{r}_1\,,\boldsymbol{r}_2), \quad (2.3.5)$$

we have

$$(+-|v|+-) = (-+|v|-+)$$

$$= \int d\boldsymbol{r}_1\, d\boldsymbol{r}_2\, \frac{e^2}{|\boldsymbol{r}_1-\boldsymbol{r}_2|}\,\frac{1}{2}\Big[\,|\,\varphi_a(\boldsymbol{r}_1)\,|^2\,|\,\varphi_b(\boldsymbol{r}_2)\,|^2 + |\,\varphi_b(\boldsymbol{r}_1)\,|^2\,|\,\varphi_a(\boldsymbol{r}_2)\,|^2\Big]$$

$$= \int d\boldsymbol{r}_1\, d\boldsymbol{r}_2\, \frac{e^2}{|\boldsymbol{r}_1-\boldsymbol{r}_2|}\,|\,\varphi_a(\boldsymbol{r}_1)\,|^2\,|\,\varphi_b(\boldsymbol{r}_2)\,|^2 \equiv K, \quad (2.3.6)$$

$$(+-|v|-+) = (-+|v|+-)$$

$$= -\int d\boldsymbol{r}_1\, d\boldsymbol{r}_2\,\frac{e^2}{|\boldsymbol{r}_1-\boldsymbol{r}_2|}$$

$$\times\,\frac{1}{2}\left\{\varphi_a^*(\boldsymbol{r}_1)\varphi_b^*(\boldsymbol{r}_2)\varphi_b(\boldsymbol{r}_1)\varphi_a(\boldsymbol{r}_2)+\varphi_b^*(\boldsymbol{r}_1)\varphi_a^*(\boldsymbol{r}_2)\varphi_a(\boldsymbol{r}_1)\varphi_b(\boldsymbol{r}_2)\right\}$$

$$= -\int d\boldsymbol{r}_1\, d\boldsymbol{r}_2\,\frac{e^2}{|\boldsymbol{r}_1-\boldsymbol{r}_2|}\,\mathrm{Re}\left\{\varphi_a^*(\boldsymbol{r}_1)\,\varphi_b^*(\boldsymbol{r}_2)\,\varphi_b(\boldsymbol{r}_1)\,\varphi_a(\boldsymbol{r}_2)\right\} \equiv -J, \quad (2.3.7)$$

and

$$(++|v|++) = (--|v|--) = K - J. \quad (2.3.8)$$

All the other matrix elements vanish. The quantity K, which is called the *direct Coulomb interaction energy*, is the electrostatic interaction energy between the electronic charge distributions. As to the non-diagonal matrix element, J, which is called the *exchange energy*, we will discuss its meaning shortly.

Now, we find the new basis which diagonalizes the above matrix (2.3.5), $(\sigma_a\sigma_b\,|v|\,\sigma_a'\sigma_b')$,

$$\Psi_s(\boldsymbol{r}_1, \boldsymbol{r}_2) = \frac{1}{\sqrt{2}}\left[\Phi_{+-}(\boldsymbol{r}_1, \boldsymbol{r}_2) - \Phi_{-+}(\boldsymbol{r}_1, \boldsymbol{r}_2)\right]$$

$$= \frac{1}{\sqrt{2}}\left[\varphi_a(\boldsymbol{r}_1)\,\varphi_b(\boldsymbol{r}_2) + \varphi_b(\boldsymbol{r}_1)\,\varphi_a(\boldsymbol{r}_2)\right]$$

$$\times \frac{1}{\sqrt{2}}\left[\chi_+(1)\,\chi_-(2) - \chi_-(1)\,\chi_+(2)\right], \tag{2.3.9}$$

with

$$\int d\boldsymbol{r}_1 d\boldsymbol{r}_2\; \Psi_s^+(\boldsymbol{r}_1, \boldsymbol{r}_2)\, \frac{e^2}{|\boldsymbol{r}_1 - \boldsymbol{r}_2|}\, \Psi_s(\boldsymbol{r}_1, \boldsymbol{r}_2) = K + J, \tag{2.3.10}$$

and

$$\Psi_t(\boldsymbol{r}_1, \boldsymbol{r}_2) = \begin{cases} \Phi_{++}(\boldsymbol{r}_1, \boldsymbol{r}_2) \\ \frac{1}{\sqrt{2}}\left[\Phi_{+-}(\boldsymbol{r}_1, \boldsymbol{r}_2) + \Phi_{-+}(\boldsymbol{r}_1, \boldsymbol{r}_2)\right] \\ \Phi_{--}(\boldsymbol{r}_1, \boldsymbol{r}_2) \end{cases}$$

$$= \frac{1}{\sqrt{2}}\left[\varphi_a(\boldsymbol{r}_1)\,\varphi_b(\boldsymbol{r}_2) - \varphi_b(\boldsymbol{r}_1)\,\varphi_a(\boldsymbol{r}_2)\right]$$

$$\times \begin{cases} \chi_+(1)\,\chi_+(2) \\ \frac{1}{\sqrt{2}}\left[\chi_+(1)\,\chi_-(2) + \chi_-(1)\,\chi_+(2)\right] \\ \chi_-(1)\,\chi_-(2)\,, \end{cases} \tag{2.3.11}$$

with

$$\int d\boldsymbol{r}_1 d\boldsymbol{r}_2\; \Psi_t^+(\boldsymbol{r}_1, \boldsymbol{r}_2)\, \frac{e^2}{|\boldsymbol{r}_1 - \boldsymbol{r}_2|}\, \Psi_t(\boldsymbol{r}_1, \boldsymbol{r}_2) = K - J. \tag{2.3.12}$$

With these wave functions, it is straightforward to check the following properties

$$(\Psi_s, (\boldsymbol{S}_1 + \boldsymbol{S}_2)^2 \Psi_s) = 0, \tag{2.3.13}$$

$$(\Psi_t, (S_1 + S_2)^2 \Psi_t) = 2, \tag{2.3.14}$$

and

$$(S_{1z} + S_{2z}) \frac{1}{\sqrt{2}} [\chi_+(1)\chi_-(2) - \chi_-(1)\chi_+(2)] = 0, \tag{2.3.15}$$

$$(S_{1z} + S_{2z}) \begin{cases} \chi_+(1)\chi_+(2) \\ \frac{1}{\sqrt{2}} [\chi_+(1)\chi_-(2) + \chi_-(1)\chi_+(2)] \\ \chi_-(1)\chi_-(2) \end{cases} = \begin{cases} \chi_+(1)\chi_+(2) \\ 0 \\ -\chi_-(1)\chi_-(2) . \end{cases} \tag{2.3.16}$$

Thus, we identify the state Ψ_s as that in which the spins of two electrons are coupled antiparallelly to result in $S_1 + S_2 = 0$. We call such a state the *singlet* state. On the other hand, in Ψ_t the spins of two electrons are parallel to form a coupled spin of $S = 1$; since such a spin state can have three different values for S_z, 1, 0, and –1, as shown in (2.3.16), it is called the *triplet* state.

From (2.3.10) and (2.3.12) we find

$$(\Psi_s | v | \Psi_s) - (\Psi_t | v | \Psi_t) = 2J. \tag{2.3.17}$$

Thus, energy of two electrons depends upon whether the spins of the two electrons are parallel or antiparallel.

What will be the sign and magnitude of J? For $\boldsymbol{r}_1 = \boldsymbol{r}_2$, the integrands of (2.3.7) become nearly the same as that of (2.3.6). Since the dominant contribution to the integral comes from the region $\boldsymbol{r}_1 \cong \boldsymbol{r}_2$ both in (2.3.6) and (2.3.7), this implies $J \cong K$; the sign of J is positive and its size is of the same order as the Coulomb repulsion,

$$J \cong K \cong O\left(\frac{e^2}{a}\right) \cong \frac{(4.8 \times 10^{-10})^2}{10^{-8}} \text{ erg} \cong 10^{-11} \text{ erg} \cong 10 \text{ eV}, \tag{2.3.18}$$

where $a \cong 10^{-8}$ cm is the atomic radius or the inter-atomic distance. This exchange energy is much larger than that of the magnetic dipole-dipole interaction of (2.3.1).

What, then, is the physical origin of the exchange energy? Note that

$$\Psi_t(\boldsymbol{r}, \boldsymbol{r}) = 0. \tag{2.3.19}$$

The probability of finding two electrons at the same position, $\boldsymbol{r}_1 = \boldsymbol{r}_2 = \boldsymbol{r}$, is zero when their spins are parallel. On the other hand, if two electrons have antiparallel spins, they can be found at the same place,

$$\Psi_s(\boldsymbol{r}, \boldsymbol{r}) \neq 0. \tag{2.3.20}$$

The spatial behavior of two-electron wave function is drastically different depending upon whether the spins of these electrons are parallel or antiparallel. This is a consequence of the Pauli principle. Thus, the quantum mechanical expectation value of (2.3.4) is lower for the electrons with parallel spins than for those with antiparallel spins. The exchange energy J comes out in this way as the effect of the Pauli principle on the Coulomb interaction between electrons.

We can express the above result in the form of a Hamiltonian involving the spin operator. From (2.3.13) and (2.3.14), we find that in (2.3.17) we can put $-J(\boldsymbol{S}_1 + \boldsymbol{S}_2)^2 = -J(\boldsymbol{S}_1{}^2 + \boldsymbol{S}_2{}^2 + 2\boldsymbol{S}_1\boldsymbol{S}_2)$ in place of v. Since $\boldsymbol{S}_1{}^2 + \boldsymbol{S}_2{}^2$ is a constant, we may further replace it by

$$H = -2J\boldsymbol{S}_1\boldsymbol{S}_2 .$$

This result can be extended to the more general situation in which there are N atomic spins $\boldsymbol{S}_i$ ($i = 1, 2, \cdots, N$) with magnitude larger than 1/2 as,

$$H = -2\sum_{i<j} J_{ij}\boldsymbol{S}_i\boldsymbol{S}_j , \tag{2.3.21}$$

where J_{ij} is the exchange interaction energy between the spins at the i-th and the j-th atomic sites. A spin system described by the Hamiltonian of (2.3.21) is called the *Heisenberg model.*

In the above discussion we neglected the possibility of an electron hopping to the neighboring atoms. If such an electron hopping is considered, the assumption that E_0 of (2.3.3) is independent of the spin state breaks down. As we will see in the next chapter, E_0 is lower for the spin antiparallel, that is, singlet spin state. If this effect, which is called the *dynamic exchange*, dominates over the Heisenberg direct exchange effect, the resultant effective interaction between spins becomes antiferromagnetic. One classical example of such a case is hydrogen molecule; the ground state of hydrogen molecule is a singlet state.

The place where the Heisenberg exchange interaction shows up most directly may be within an atom; it is the origin of the Hund's rule. In this case the two states a and b represent two different orbital states of electrons. Because of the Pauli principle, two electrons with the same spins can not occupy the same orbital state. Thus, to be occupied by two electrons with parallel spins, the state a and b must be different orbital states, with the orbital energies ε_a and ε_b that can be different. Note that two electrons with antiparallel spins can occupy the lower orbital energy state simultaneously. It is the competition between the orbital energy and the Heisenberg exchange energy that decides the electronic configuration of an atom. This leads to the Hund rule spin coupling in an atom.

The exchange interaction we discussed in the above is for electrons localized on each atom. How will the exchange interaction be different for itinerant electrons of a metal? As we will show below, the exchange interaction exists also between itinerant electrons and it plays a role as important as in the case of localized electrons.

Consider two electrons in a box of volume V. If we assume they are in the plane wave states $(\boldsymbol{k}_1,\sigma_1)$ and $(\boldsymbol{k}_2,\sigma_2)$, their wave function is obtained in the form of the Slater determinant, (2.3.2) or (1.2.10). Then we obtain the electron density taking trace over spin degrees of freedom as

$$\left|\Phi_{\boldsymbol{k}_1+,\,\boldsymbol{k}_2-}(\boldsymbol{r}_1,\boldsymbol{r}_2)\right|^2 = \frac{1}{V^2}, \tag{2.3.22}$$

$$\left|\Phi_{\boldsymbol{k}_1+,\,\boldsymbol{k}_2+}(\boldsymbol{r}_1,\boldsymbol{r}_2)\right|^2 = \frac{1}{V^2}\left[1-\cos\left\{(\boldsymbol{k}_1-\boldsymbol{k}_2)(\boldsymbol{r}_1-\boldsymbol{r}_2)\right\}\right]. \tag{2.3.23}$$

When two electrons have different spins there is no correlation between the spatial distribution of these two electrons as can be seen from (2.3.22). When two electrons have the same spins, however, their spatial distribution becomes strongly correlated. As can be seen from (2.3.23), for $\boldsymbol{r}_1=\boldsymbol{r}_2=\boldsymbol{r}$, we have $|\Psi_{\boldsymbol{k}_1+,\,\boldsymbol{k}_2+}(\boldsymbol{r},\boldsymbol{r})|^2=0$; because of the Pauli principle two electrons with the same spins can not come close within a distance of $\sim 1/k_{\mathrm{F}}$. Such region of electron deficiency around each electron is called the *Fermi hole*. Thus, if we calculate the energy due to the Coulomb repulsion between two electrons it depends on the spins of the electrons, and we have

$$E_{\boldsymbol{k}_1+,\,\boldsymbol{k}_2-} - E_{\boldsymbol{k}_1+,\,\boldsymbol{k}_2+} = \frac{1}{V^2}\int d\boldsymbol{r}_1 d\boldsymbol{r}_2 \frac{e^2}{|\boldsymbol{r}_1-\boldsymbol{r}_2|}\cos\left\{(\boldsymbol{k}_1-\boldsymbol{k}_2)(\boldsymbol{r}_1-\boldsymbol{r}_2)\right\}$$

$$= \frac{1}{V}\int d\boldsymbol{r}\,\frac{e^2}{r}\cos\left\{(\boldsymbol{k}_1-\boldsymbol{k}_2)\boldsymbol{r}\right\} = \frac{4\pi e^2}{V(\boldsymbol{k}_1-\boldsymbol{k}_2)^2} = \mathrm{v}(\boldsymbol{k}_1-\boldsymbol{k}_2). \tag{2.3.24}$$

The interaction energy is lower for itinerant electrons with parallel spins than for those with antiparallel spins, the same as in the case of localized electrons. Note, however, that in an electron gas with the total number of electrons n, $\mathrm{v}(\boldsymbol{k}_1-\boldsymbol{k}_2)=O(J/n)$ for $|\boldsymbol{k}_1-\boldsymbol{k}_2|\cong O(k_{\mathrm{F}})$, where J is the exchange interaction between a pair of neighboring localized electrons. In a metallic system too, however, the exchange energy of an electron attains the magnitude of ~ 1 eV by interacting with all the other electrons in the system. Thus the exchange energy per electron is as large as the kinetic energy. The importance of taking into account the exchange energy of metallic electrons is not confined to problems of magnetism in a narrow sense, as we will see later in this book.

2.4 FERROMAGNETISM OF LOCALIZED SPINS

We saw that the interaction between atomic localized spins can be written in the following form (see (2.3.21)),

$$H = -2 \sum_{i<j} J_{ij}\, \boldsymbol{S}_i \boldsymbol{S}_j = -\sum_{i \neq j} J_{ij}\, \boldsymbol{S}_i \boldsymbol{S}_j \,. \tag{2.4.1}$$

It is important not to count the interaction between a pair of spins more than once.

The above Heisenberg Hamiltonian can be simplified by assuming that $S_i = 1/2$ and that the spin has only a z component:

$$H = -\frac{1}{2} \sum_{i<j} J_{ij}\, \sigma_i \sigma_j \,, \tag{2.4.2}$$

where σ_i can take the value of either +1 or −1. We call this model described by (2.4.2) the *Ising model.*

In this section, we discuss the ferromagnetism of the Heisenberg model with the *mean field approximation.* The spirit of the mean field approximation is that when we are interested in the behavior of the i-th spin, $\boldsymbol{S}_i$, we replace each $\boldsymbol{S}_j$ for $j \neq i$, by their thermal expectations $\langle \boldsymbol{S}_j \rangle$. Then, (2.4.1) reduces to

$$H \cong -\sum_{i \neq j} J_{ij}\, \boldsymbol{S}_i \langle \boldsymbol{S}_j \rangle \,. \tag{2.4.3}$$

If we introduce

$$\boldsymbol{H}_{\mathrm{mf}} = -\frac{2}{g\mu_{\mathrm{B}}} \sum_{j(\neq i)} J_{ij} \langle \boldsymbol{S}_j \rangle \,, \tag{2.4.4}$$

(2.4.3) is rewritten as

$$H \cong g\mu_{\mathrm{B}} \sum_i \boldsymbol{H}_{\mathrm{mf}} \boldsymbol{S}_i \,. \tag{2.4.5}$$

The meaning of $\boldsymbol{H}_{\mathrm{mf}}$ may become clear if we compare (2.4.5) with the Zeeman energy of a spin given in (2.2.1). Thus it is called the *molecular field* or the *mean field.*

An important remark on (2.4.5) is that when we use it to calculate the contribution of the interaction to the total energy we must put a factor 1/2 on the right hand side, not to count doubly the interaction energy of a pair of spins.

After these preparations, now let us explore how the magnetic susceptibility of a localized spin system will be affected by the interaction of (2.4.1). If an external magnetic field $\boldsymbol{H}$ is applied in the z-direction the effective magnetic field a spin $\boldsymbol{S}_i$ will feel, including the mean field due to surrounding other spins, will be in the same direction with the magnitude

$$H_{\text{eff}} = H + H_{\text{mf}} = H - \frac{2}{g\mu_{\text{B}}} zJ\langle S_z\rangle . \tag{2.4.6}$$

Here we assumed the interaction to be non-zero only between nearest neighbor pairs of spins, with $J_{ij} = J$, that each spin has z nearest neighbor spins and that $\langle S_{iz}\rangle = \langle S_z\rangle$, independent of the spin site i.

Within above mean field approximation, we can view each spin as independently immersed in the effective field of (2.4.6). Then, by employing the Curie susceptibility of (2.2.13) we have

$$\begin{aligned} \langle M_z\rangle &= \chi_{\text{C}} H_{\text{eff}} = \chi_{\text{C}}\left[H + H_{\text{mf}}\right] \\ &= \chi_{\text{C}}\left[H + \frac{2zJ}{(g\mu_{\text{B}})^2}\frac{\langle M_z\rangle}{N}\right] \\ &= \chi H, \end{aligned} \tag{2.4.7}$$

where χ is the magnetic susceptibility defined with respect to the external field as in (2.2.12). Thus we obtain

$$\chi = \frac{\chi_{\text{C}}}{1 - 2\frac{z}{N}\frac{1}{(g\mu_{\text{B}})^2}J\chi_{\text{C}}} . \tag{2.4.8}$$

This result is rewritten as

$$\chi = \frac{C}{T - T_{\text{C}}} \equiv \chi_{\text{CW}}, \tag{2.4.9}$$

$$T_{\text{C}} = \frac{2z}{3k_{\text{B}}} S(S+1)J, \tag{2.4.10}$$

where C is Curie constant of (2.2.14). We call T_{C} the *Curie temperature* or *Curie point*. A magnetic susceptibility with the temperature dependence of the form of (2.4.9) is called the *Curie-Weiss susceptibility*.

The result of (2.4.8) shows how the Curie susceptibility is enhanced by the effect of the interaction between spins. As the temperature is lowered toward T_C the enhancement increases and finally at T_C the susceptibility diverges to infinity. An infinite magnetic susceptibility implies that $\langle M_z \rangle \neq 0$ for $H = 0$; the ordering of the spins takes place in the absence of an external magnetic field. The ferromagnetic phase transition starts at the Curie temperature T_C. The magnetization increases with decreasing temperature as shown in Fig.2.3, and at $T = 0$ it attains the maximum possible value $M_0 = Ng\mu_B S$, the saturation magnetization.

Let us see how the magnetization near $T = 0$ decreases with increasing temperature within the present mean field approximation. The temperature dependence of the magnetization can be calculated from the relation (2.2.8) with H replaced by H_{eff} of (2.4.6). Note that at low temperatures the Brillouin function of (2.2.9) behaves as $B_S(x) = 1 - \exp(-x/S)/S$ for $x \to \infty$, where in the present problem $x = g\mu_B S H_{\text{eff}}/k_B T = 2zSJ\langle S_z \rangle/k_B T \cong 2zJS^2/k_B T$. Thus, for $\langle M_z \rangle = M(T)$ near $T = 0$ we have

$$\frac{M(T)}{M_0} \cong 1 - \frac{1}{S}\exp\left(-\frac{3}{S+1}\frac{T_C}{T}\right). \tag{2.4.11}$$

Experimentally [2.15], however, we observe a quite different behavior,

$$1 - M(T)/M_0 \propto T^{3/2}. \tag{2.4.12}$$

What does such a drastic discrepancy between theory and experiment imply?

Assume that all the spins are aligned parallel to the z-axis. Then, flipping of any one spin S_j, from $S_{jz} = S$ to $S_{jz} = S-1$, causes a decrease in magnetization. Such a transition costs an excitation energy of

$$\Delta = g\mu_B H_{\text{mf}} = 2zJS = 3k_B T_C/(S+1). \tag{2.4.13}$$

The probability of thermally overcoming a gap Δ is proportional to

$$\exp(-\Delta/k_B T) = \exp\left(-3T_C/\{(S+1)T\}\right). \tag{2.4.14}$$

Thus, the simple mean field approximation result of (2.4.11) reflects the presence of the spin excitation energy gap of (2.4.13).

The experimental observation of (2.4.12), however, denies the presence of such a gap. It implies that the actual excitation process causing a decrease in magnetization at low temperatures is not the independent flipping of individual spins. What kind of spin excitation process is, then, responsible for the actual decrease in magnetization as described by (2.4.12)? It is believed to be the *spin wave*, a collective excitation of spins, which we will discuss in the next section.

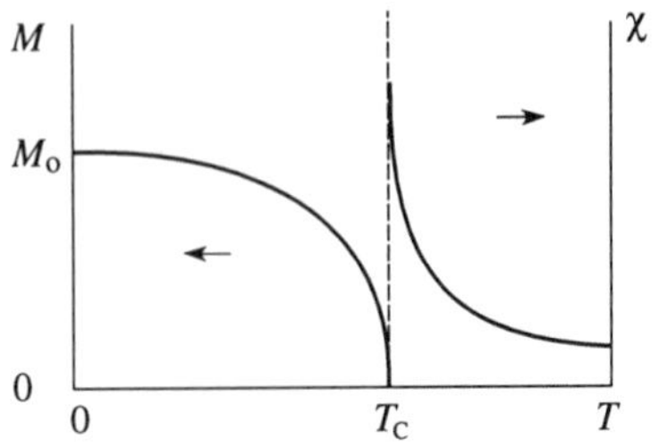

Fig.2.3 The temperature dependence of the spontaneous magnetization M below the Curie point and the magnetic susceptibility χ above the Curie point of a ferromagnet.

2.5 SPIN WAVE OF LOCALIZED SPINS

2.5.1 A Method to Obtain Elementary Excitations

Let us first present a general procedure to find the excitation spectrum of a many particle system. Assume that the many particle system we are going to study is in the state Φ_0; Φ_0 is not necessarily the ground state. This state (wave function) is assumed to satisfy the following Schrödinger equation.

$$i\hbar \frac{\partial}{\partial t} \Phi_0 = H\Phi_0 = E_0 \Phi_0 , \tag{2.5.1}$$

where H is the Hamiltonian of the system and E_0 is the energy of the state Φ_0. Let A be an operator which acting upon Φ_0 produces an eigenstate of the system with an excitation energy ε_A. Then, corresponding to (2.5.1), we require

$$i\hbar \frac{\partial}{\partial t}(A\Phi_0) = H(A\Phi_0) = (E_0 + \varepsilon_A)(A\Phi_0), \tag{2.5.2}$$

or,

$$(HA - AH)\Phi_0 = \varepsilon_A A\Phi_0 . \tag{2.5.3}$$

This final equation is often rewritten as

$$[H, A] = \varepsilon_A A . \tag{2.5.3'}$$

Note, however, that (2.5.3') is not exactly equivalent to (2.5.3); ε_A depends on Φ_0.

Our problem is to find A and ε_A corresponding to the spin wave excitation. As a preparation, in the next subsection we summarize some necessary properties of the spin operator.

2.5.2 Fourier Transform of Spin Operators

We define the Fourier transform of the spin operators as

$$S(\boldsymbol{q}) = \sum_{i=1}^{N} S_i e^{-i\boldsymbol{q}\boldsymbol{R}_i}, \tag{2.5.4}$$

where $\boldsymbol{R}_i$ is the coordinate of the i-th atom (spin). As for the wave vector $\boldsymbol{q}$, if we assume a simple cubic lattice with inter-atomic distance a, in order to satisfy the periodic boundary condition, $\boldsymbol{q}$ is allowed to take the values

$$\boldsymbol{q} = \frac{2\pi}{la}(l_x, l_y, l_z), \tag{2.5.5}$$

where the l_ν's, ν being x, y, or z, are integers and $l^3 = N$ is the total number of atoms within the volume, $V = (la)^3$. The region in the wave vector space

$$-\frac{\pi}{a} < q_\nu \leq \frac{\pi}{a}, \tag{2.5.6}$$

or, $-l/2 < l_\nu \leq l/2$ in (2.5.5), is called the *1st Brillouin zone* (1BZ). Within the 1st Brillouin zone there are N different values of $\boldsymbol{q}$ which satisfies the condition (2.5.5). Note that if we define the *reciprocal lattice vector* $\boldsymbol{K}$ as

$$\boldsymbol{K} = \frac{2\pi}{a}(m_x, m_y, m_z) \tag{2.5.7}$$

with integers m_ν's, we have

$$e^{i(\boldsymbol{K}+\boldsymbol{q})\boldsymbol{R}_i} = e^{i\boldsymbol{q}\boldsymbol{R}_i}. \tag{2.5.8}$$

With this reciprocal lattice vector, (2.5.6) is rewritten as

$$-\frac{K_\nu^{\mathrm{o}}}{2} < q_\nu \leq \frac{K_\nu^{\mathrm{o}}}{2}, \tag{2.5.6'}$$

where $K_\nu{}^{\mathrm{o}} = 2\pi/a$ is the smallest non-zero reciprocal lattice vector in the ν-direction.

We can easily confirm that the inverse transformation of (2.5.4) is given as

$$S_i = \frac{1}{N} \sum_{q}^{1BZ} S(q) e^{iqR_i} , \tag{2.5.9}$$

by noting the relation

$$\sum_{q}^{1BZ} e^{iqR_i} = N \delta_{R_i,0} . \tag{2.5.10}$$

Note also the following relation

$$\sum_{i} e^{iqR_i} = N \delta_{q,K} , \tag{2.5.11}$$

K being a reciprocal lattice vector. Most of the above relations will be useful also in discussing lattice vibrations in Chapter 5.

The commutation relation of (1.1.10) can be generalized to

$$[S_{ix}, S_{jy}] = i S_{iz} \delta_{ij} , \text{ etc}, \tag{2.5.12}$$

or, in the Fourier transformed form

$$[S_+(q), S_-(q')] = \frac{1}{2} S_z (q + q'), \tag{2.5.13}$$

$$[S_z(q), S_\pm(q')] = \pm S_\pm (q + q'), \tag{2.5.14}$$

where $S_\pm(q)$ is the Fourier transform of

$$S_{i\pm} = \frac{1}{2} (S_{ix} \pm i S_{iy}) . \tag{2.5.15}$$

2.5.3 Spin Wave

Let Φ_0 be the state in which all of N localized spins are aligned in the direction of the z-axis, as in Fig.2.2(a):

$$S_z(0) \Phi_0 = \sum_{i=1}^{N} S_{iz} \Phi_0 = NS \Phi_0 . \tag{2.5.16}$$

Let us see, then, what kind of state $S_-(q)\Phi_0$ is. We find, by using (2.5.14), that

$$S_z(0)\left(S_-(\boldsymbol{q})\,\Phi_0\right) = [S_-(\boldsymbol{q})\,S_z(0) - S_-(\boldsymbol{q})]\,\Phi_0$$

$$= \left(NS - 1\right)\left(S_-(\boldsymbol{q})\Phi_0\right). \qquad (2.5.17)$$

While for Φ_0 the eigenvalue of $S_z(0) = \Sigma_i S_{iz}$ is NS, for $S_-(\boldsymbol{q})\Phi_0$ it is $(NS-1)$; $S_-(\boldsymbol{q})$, which involves N spin operators, reduces the magnitude of the z component of the total spin of the entire system by one unit. Note that $S_-(\boldsymbol{q})$ plays such a role also for states without complete alignment of spins.

Now we examine the excitation energy $\varepsilon_A = \hbar\omega_{\rm sw}(\boldsymbol{q})$ corresponding to the operator $A = S_-(\boldsymbol{q})$ of (2.5.3), namely,

$$[H, S_-(\boldsymbol{q})]\,\Phi_0 = \hbar\omega_{\rm sw}(\boldsymbol{q})\,S_-(\boldsymbol{q})\Phi_0\,. \qquad (2.5.18)$$

First note that the Heisenberg Hamiltonian of (2.3.21) can be rewritten as

$$H = -\sum_{\boldsymbol{q}} J(\boldsymbol{q})\,\boldsymbol{S}(\boldsymbol{q})\,\boldsymbol{S}(-\boldsymbol{q})$$

$$= -\sum_{\boldsymbol{q}} J(\boldsymbol{q})\Big[S_z(\boldsymbol{q})\,S_z(-\boldsymbol{q}) + 2\left\{S_+(\boldsymbol{q})\,S_-(-\boldsymbol{q}) + S_-(\boldsymbol{q})\,S_+(-\boldsymbol{q})\right\}\Big], \qquad (2.5.19)$$

where the summation over $\boldsymbol{q}$ is restricted to the 1st Brillouin zone and we put

$$J(\boldsymbol{q}) = \frac{1}{N}\sum_{j(\neq i)} J_{ij}\,e^{-i\boldsymbol{q}\,(\boldsymbol{R}_j - \boldsymbol{R}_i)}\,. \qquad (2.5.20)$$

(Note that the Fourier transform is defined differently in (2.5.4) and (2.5.20), in accord with our convention in Chapter 1. Such a convention avoids the appearance of N in (2.5.19).) Then, by using the commutation relation of (2.5.13) and (2.5.14), we find

$$[H, S_-(\boldsymbol{q})] = \sum_{\boldsymbol{q}'} J(\boldsymbol{q}')\,[S_-(\boldsymbol{q}+\boldsymbol{q}')\,S_z(-\boldsymbol{q}') + S_-(\boldsymbol{q}-\boldsymbol{q}')\,S_z(\boldsymbol{q}')$$

$$- S_-(-\boldsymbol{q}')\,S_z(\boldsymbol{q}+\boldsymbol{q}') - S_-(\boldsymbol{q}')\,S_z(\boldsymbol{q}-\boldsymbol{q}')]. \qquad (2.5.21)$$

If the state Φ_0 is that given in (2.5.16), we have

$$S_z(\boldsymbol{q})\Phi_0 = NS\,\Phi_0\,\delta_{\boldsymbol{q},0}\,. \qquad (2.5.22)$$

If Φ_0 is not such a ground state, with the eigenvalue, $\langle S_z(0)\rangle$, of $S_z(0)$ smaller than NS, we may put, in place of (2.5.22),

$$S_z(\boldsymbol{q})\Phi_0 = \langle S_z(0)\rangle \Phi_0\, \delta_{\boldsymbol{q},0}\,. \tag{2.5.22'}$$

Thus, for $\hbar\omega_{\mathrm{sw}}(\boldsymbol{q})$ of (2.5.18) we obtain

$$\hbar\omega_{\mathrm{sw}}(\boldsymbol{q}) = 2\langle S_z(0)\rangle [\, J(0) - J(\boldsymbol{q})\,]\,, \tag{2.5.23}$$

where we used the relation $J(-\boldsymbol{q}) = J(\boldsymbol{q})$.

We call this operator $S_-(\boldsymbol{q})$, with the excitation energy of (2.5.23), the spin wave creation operator. According to the definition of $S_-(\boldsymbol{q})$, namely,

$$S_-(\boldsymbol{q}) = \sum_i S_{i-}\, e^{-i\boldsymbol{q}\boldsymbol{R}_i}\,, \tag{2.5.24}$$

a spin wave may be visualized as the wave of spin reversals or lowerings.

An important property of the spin wave is that, contrary to the case of individual spin flipping, its excitation energy does not have a gap:

$$\lim_{q\to 0} \omega_{\mathrm{sw}}(\boldsymbol{q}) = 0. \tag{2.5.25}$$

For $q \ll \pi/a$, we have

$$\hbar\omega_{\mathrm{sw}}(\boldsymbol{q}) = Dq^2, \tag{2.5.26}$$

where D is called the *exchange stiffness*. If

$$J(0) > J(\boldsymbol{q}), \tag{2.5.27}$$

we have $D > 0$. If (2.5.27) does not hold, we have $D < 0$, and it implies that the assumed ferromagnetic state is not a stable state; as will be shown in 2.6, in this case we have a spin ordering such as given in Fig.2.2 (d)–(e).

Let us derive an explicit expression for D. If we assume that the localized spins form a simple cubic lattice, with the interaction $J_{ij} = J\,(>0)$ non-zero only between nearest neighbor pair of spins, we have

$$J(\boldsymbol{q}) = 2(J/N)\,[\cos q_x a + \cos q_y a + \cos q_z a]\,, \tag{2.5.28}$$

where a is the lattice constant. Then for $|q_x a|, |q_y a|, |q_z a| \ll 1$ we have

$$D = 2\,(J/N)\langle S_z(0)\rangle a^2, \tag{2.5.29}$$

or, by putting $\langle S_z(0)\rangle = NS$,

$$\hbar\omega_{\rm sw}(\boldsymbol{q}) = 2JSa^2q^2. \tag{2.5.29'}$$

If the decrease in magnetization with increasing temperature is produced by the excitation of such spin waves, the result of (2.4.11) is not valid. Instead, the spin wave theory leads to the observed behavior of (2.4.12), as we will see in the next subsection.

2.5.4 Spin Wave Excitation as a boson: Temperature Dependence of Magnetization

As can be seen from (2.5.24), the spin wave excitation is very similar to the lattice vibration, or, phonon to be discussed in Chapter 5. The spin wave is also called a *magnon,* and it is a boson as is the phonon, as we will show shortly.

Let us briefly summarize how bosons are to be treated in the second quantization formalism. The most fundamental property of a boson is the following commutation relation for its creation and annihilation operators for the states $\boldsymbol{l}$ and $\boldsymbol{l}'$,

$$\begin{aligned} [\,b_l,\, b_{l'}^{+}] &= b_l\, b_{l'}^{+} - b_{l'}^{+}\, b_l = \delta_{l,l'}\,; \\ [\,b_l,\, b_{l'}] &= [\,b_l^{+},\, b_{l'}^{+}] = 0. \end{aligned} \tag{2.5.30}$$

While the boson number operator $\tilde{n}_l = b_l^{+} b_l$ is of the same form as that of fermion, the eigenvalue n_l of the boson number operator can be larger than 1, quite unlike with the fermion case. That is, for bosons

$$n_l = 0,\, 1,\, 2, \cdots . \tag{2.5.31}$$

Let us see how such a property of the boson number operator comes about. Assume that $|\, n_l\rangle$ is a state in which n_l of boson particles with momentum $\boldsymbol{l}$ are present,

$$\tilde{n}_l |\, n_l\rangle = n_l |\, n_l\rangle . \tag{2.5.32}$$

Then, by using the commutation relation of (2.5.30), we can prove

$$\tilde{n}_l\, b_l^{+} |\, n_l\rangle = (n_l + 1)\, b_l^{+} |\, n_l\rangle , \tag{2.5.33}$$

or, by considering normalization,

$$| \, n_l + 1 \, \rangle = \frac{1}{\sqrt{n_l + 1}} \, b_l^\dagger \, | \, n_l \, \rangle . \qquad (2.5.33')$$

If the spin wave is to be a boson, its creation and annihilation operator must satisfy the commutation relation of (2.5.30). We already know the spin wave excitation operator is given by $S_-(\boldsymbol{q})$. Then, the annihilation operator should be given by its Hermite conjugate, $S_+(-\boldsymbol{q})$. Thus, we set

$$b_q^+ = \sqrt{\frac{2}{\langle S_z(0) \rangle}} \, S_-(\boldsymbol{q}); \quad b_q = \sqrt{\frac{2}{\langle S_z(0) \rangle}} \, S_+(-\boldsymbol{q}). \qquad (2.5.34)$$

These operators satisfy the commutation relation of (2.5.30) if we use the approximation of putting $S_z(\boldsymbol{q}) = \langle S_z(0) \rangle \delta_{\boldsymbol{q},0}$.

If we recall that the excitation of one spin wave reduces the z component of the total spin of a system by one unit, for the temperature dependence of the magnetization at low temperatures we have

$$M(T) = M_0 - g \mu_B \sum_q \langle \tilde{n}_q \rangle , \qquad (2.5.35)$$

where $\langle \tilde{n}_q \rangle$ is thermal expectation of the number of the excited magnons.

The procedure for obtaining $\langle \tilde{n}_q \rangle$ for bosons is very similar to that for fermions, starting from (1.6.1); the only difference in the evaluation is to consider (2.5.31) in obtaining the grand partition function of (1.6.2). Thus, for bosons we have

$$\Xi_0 = \prod_l [\, 1 - e^{-\beta(\varepsilon_l - \mu)} \,]^{-1} \qquad (2.5.36)$$

in place of (1.6.3), where ε_l is the energy of a boson with momentum $\hbar \boldsymbol{l}$. Then, corresponding to (1.6.5), for bosons we have

$$\langle \tilde{n}_q \rangle = \frac{1}{e^{\beta(\varepsilon_q - \mu)} - 1} \equiv n(\varepsilon_q). \qquad (2.5.37)$$

We call this the *Bose distribution function.*

When the total number of particles is indefinite, as in the cases of magnons and phonons, we have $\mu = 0$ in the Bose distribution function. This is because we require that in the equilibrium state the free energy of a system should be minimized with respect to the particle number n,

$$\left.\frac{\partial F(T,\, V,\, n)}{\partial n}\right|_{V,\,T} = \mu = 0. \tag{2.5.38}$$

See (1.4.18) and below.

By putting the Bose distribution function into (2.5.35) we have

$$M(T) = M_o - g\,\mu_B \frac{V}{(2\pi)^3} \int_0^{\pi/a} \frac{4\pi q^2 dq}{e^{\beta D q^2} - 1}, \tag{2.5.39}$$

where we approximated the $\boldsymbol{q}$ sum over the 1st Brillouin zone by that over a sphere with a radius π/a, and assumed the relation (2.5.26) for the entire $\boldsymbol{q}$-space by noting that the dominant contribution comes from small q region. If we put $t^2 = \beta\, Dq^2$ in the integral of (2.5.39), we have

$$M(T) = M_o - g\,\mu_B \frac{V}{2\pi^2} \left(\frac{k_B T}{D} \right)^{3/2} \int_0^{\infty} \frac{t^2 dt}{e^{t^2} - 1}, \tag{2.5.40}$$

where the upper limit of the integral is set equal to infinity since $\beta\, D(\pi/a)^2 = 2\pi^2 SJ/k_B T = O(\,T_C/T\,) \to \infty$ for $T \to 0$. Clearly, (2.5.40) is of the desired form of (2.4.12). If in (2.5.40) we put the value of the integral $(\pi^{1/2}/4)\zeta(3/2) = \pi^{1/2}/4 \times 2.613$, ζ being the Riemann zeta function, and note $Na^3 = V$, we have

$$M(T) = M_o \left[1 - \frac{1}{S} \zeta\!\left(\frac{3}{2}\right) \left(\frac{k_B T}{8\pi J S} \right)^{3/2} \right]. \tag{2.5.40'}$$

So far in this and previous sections we considered the case of a three dimensional spin system. If we calculate the magnetic susceptibiliy for a 1- or, 2-dimensional systems in the mean field approximation we obtain the same result as given in (2.4.8); the difference in the dimensionality shows up only in the number of the nearest neighbor spins, z. Thus, within the simple mean field approximation both 1- and 2-dimensional Heisenberg systems can have a ferromagnetic phase transition.

Spin waves in these lower dimensional systems can also be discussed in the same way as in the 3-dimensional system. As for the spin wave excitation energy, the difference shows up again only through the difference in the number of nearest neighbor spins. Let us see, then, how magnetization decreases in the 1- and 2-dimensional system with excitation of spin waves. We can use the same (2.5.35) for that purpose. In transforming the sum over $\boldsymbol{q}$ to an integral, however, an important difference comes up. In the case of the 1- and 2-

dimensional systems, in the integral of (2.5.39) we have to replace $V/(2\pi)^3 \cdot 4\pi q^2$ by, respectively, $(L/2\pi)$ and $(L/2\pi)^2\, 2\pi q$, where $L^3 = V$. Then, we note that the integral diverges for both of these lower dimensional cases; magnetization is immediately destroyed by thermal excitation of spin waves. Thus, the conclusion of the mean field theory is reversed; if we consider the spin wave excitations, ferromagnetism becomes impossible in both 1- and 2-dimensional Heisenberg systems. Note, however, that the spin wave theory we used above is again an approximation. Very active efforts are currently being made from various directions to go beyond such an approximation. For a review and references, see, for instance, Ref. [1.6].

Finally, note that the fundamental equation concerning spin wave effect on the temperature dependence of magnetization, (2.5.35), which is universally used, is not that which is actually derived; it is, rather, intuitively postulated. Very recently, from a first-principle treatment of the problem, we found that (2.5.35) is not an exact relation; see 6.6.

2.5.5 Semi-classical Model of Spin Wave

Let us here briefly introduce the semi-classical picture of the spin wave. The Heisenberg equation of motion for the spin wave excitation operator $S_-(\boldsymbol{q})$ is given as,

$$i\hbar \frac{\partial}{\partial t} S_-(\boldsymbol{q}, t) = -\left[H,\, S_-(\boldsymbol{q}, t)\right], \tag{2.5.41}$$

where H is given in (2.5.19). If we put the results of (2.5.18) and (2.5.23) on the right hand side of (2.5.41), we have

$$i \frac{\partial}{\partial t} S_-(\boldsymbol{q}, t) = -\,\omega_{\mathrm{SW}}(\boldsymbol{q})\, S_-(\boldsymbol{q}, t). \tag{2.5.42}$$

This equation is solved as

$$S_-(\boldsymbol{q}, t) = S_-(\boldsymbol{q}, 0)\, e^{i\omega_{\mathrm{sw}}(\boldsymbol{q})\, t}\ . \tag{2.5.43}$$

By putting this into (2.5.9) we have

$$S_{ix}(t) - i S_{iy}(t) = \frac{1}{N}\left[S_x(\boldsymbol{q}, 0) - i S_y(\boldsymbol{q}, 0)\right] e^{i(\boldsymbol{q}\boldsymbol{R}_i + \omega_{\mathrm{sw}}(\boldsymbol{q})\, t)}\, . \tag{2.5.44}$$

We do not sum over $\boldsymbol{q}$ in the right hand side of the above expression since we are considering a situation where only spin waves with a particular wave number $\boldsymbol{q}$ are excited.

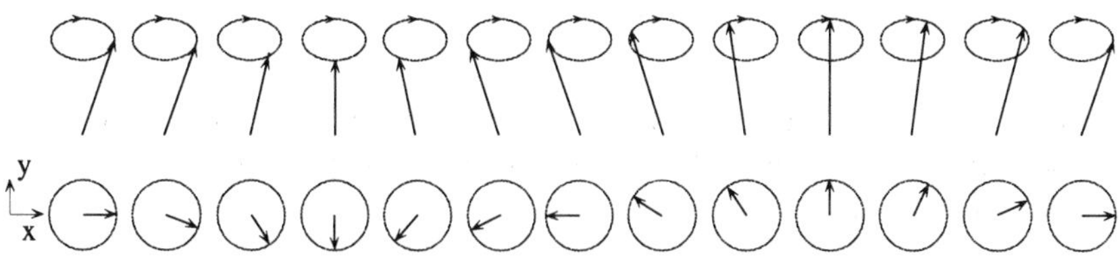

Fig.2.4 A semi-classical picture for a spin wave excitation. The spin wave is assumed to propagate in the direction of the x-axis. The top and bottom figures represent the motion of spins as viewed from the side and above, respectively.

As for the initial condition in (2.5.44), we put $S_x(\boldsymbol{q}, 0) \neq 0$, $S_y(\boldsymbol{q}, 0) = 0$, correspondingly to the situation of the left end figure of Fig.2.4. Then we have

$$
\begin{aligned}
S_{ix}(t) &= \frac{S_x(\boldsymbol{q}, 0)}{N} \cos(-\boldsymbol{q}\boldsymbol{R}_i - \omega_{\mathrm{sw}}(\boldsymbol{q})t), \\
S_{iy}(t) &= \frac{S_x(\boldsymbol{q}, 0)}{N} \sin(-\boldsymbol{q}\boldsymbol{R}_i - \omega_{\mathrm{sw}}(\boldsymbol{q})t).
\end{aligned}
\tag{2.5.45}
$$

This result describes the propagation of the precession of the spins as illustrated in Fig.2.4; each spin precesses clockwise in the x-y plane. The top figure shows the view from the above of the chain of spins.

Note, however, that the picture given in Fig.2.4 is not a fully justifiable one. As to a spin $\boldsymbol{S}_i$, if we take a representation which diagonalizes S_{iz}, we can no longer specify the values of S_{ix} or S_{iy}.

2.6 SPIN ORDER OTHER THAN FERROMAGNETISM

As illustrated in Fig.2.2, there are spin orderings other than ferromagnetism. A spin ordered state is characterized by a non-vanishing $\langle \boldsymbol{S}(\boldsymbol{Q}) \rangle$. For the ferromagnetic state, it is $\langle S_z(0) \rangle$, as we have already seen. Let us discuss the occurrence of a spin order with a wave vector $\boldsymbol{Q} \neq 0$.

Suppose that we apply a space dependent external magnetic field, $\boldsymbol{H}(\boldsymbol{q}) \exp(i\boldsymbol{q}\boldsymbol{r})$, to a localized spin system. The Zeeman energy of the spins in this field is given as

$$
H_{\mathrm{Z}} = g\mu_{\mathrm{B}} \boldsymbol{H}(\boldsymbol{q}) \sum_{i=1}^{N} e^{i\boldsymbol{q}\boldsymbol{R}_i} \boldsymbol{S}_i \tag{2.6.1}
$$

$$= g\mu_B \boldsymbol{H}(\boldsymbol{q})\, \boldsymbol{S}(-\boldsymbol{q})\,. \tag{2.6.1'}$$

For $\boldsymbol{q} = 0$, (2.6.1) reduces to (2.2.1). If we assume this spatially-varying magnetic field is polarized in the z-direction, then the expectation value of the z-component of the spin on site i is

$$-g\mu_B\langle S_{iz}\rangle = \frac{1}{N}\frac{C}{T} H(\boldsymbol{q})\, e^{i\boldsymbol{q}\boldsymbol{R}_i}, \tag{2.6.2}$$

where C is the Curie constant of (2.2.14). By multiplying both sides of this equation by $e^{-i\boldsymbol{q}\boldsymbol{R}_i}$ and summing over i, we obtain

$$\langle M_z(\boldsymbol{q})\rangle = \frac{C}{T} H(\boldsymbol{q}), \tag{2.6.3}$$

and then,

$$\chi(\boldsymbol{q}) = \frac{\langle M_z(\boldsymbol{q})\rangle}{H(\boldsymbol{q})} = \chi_C, \tag{2.6.4}$$

where χ_C is the Curie susceptibility of (2.2.13).

Now, let us take into account the exchange interaction between spins of (2.4.1), or, more conveniently, (2.5.19). Within the mean field approximation the effect of the exchange interaction is equivalent to modifying the external field to the following effective field,

$$H_{\text{eff}}(\boldsymbol{q}) = H(\boldsymbol{q}) - 2\frac{J(\boldsymbol{q})}{g\mu_B}\langle S_z(\boldsymbol{q})\rangle = H(\boldsymbol{q}) + \frac{2J(\boldsymbol{q})}{(g\mu_B)^2}\langle M_z(\boldsymbol{q})\rangle\,, \tag{2.6.5}$$

in (2.6.2) or (2.6.3), as we already discussed for the case of $\boldsymbol{q} = 0$ in 2.4. Note that since the magnetic field is applied in the direction of the z-axis, we have $\langle S_x(\boldsymbol{q}')\rangle = \langle S_y(\boldsymbol{q}')\rangle = 0$, $\langle S_z(\boldsymbol{q}')\rangle = \langle S_z(\boldsymbol{q})\rangle\delta_{\boldsymbol{q},\boldsymbol{q}'}$. By replacing $H(\boldsymbol{q})$ with the above $H_{\text{eff}}(\boldsymbol{q})$ in (2.6.3), we obtain

$$\langle M_z(\boldsymbol{q})\rangle = \chi_C\left[H(\boldsymbol{q}) + \frac{2J(\boldsymbol{q})}{(g\mu_B)^2}\langle M_z(\boldsymbol{q})\rangle\right], \tag{2.6.6}$$

or

$$\chi(\boldsymbol{q}) = \frac{\langle M_z(\boldsymbol{q})\rangle}{H(\boldsymbol{q})} = \frac{\chi_C}{1 - \dfrac{2J(\boldsymbol{q})}{(g\mu_B)^2}\chi_C} = \frac{C}{T - T_{\boldsymbol{q}}}, \tag{2.6.7}$$

where C is the Curie constant of (2.2.14) and we put,

$$T_{\boldsymbol{q}} = \frac{2}{3k_{\rm B}} S(S+1) J(\boldsymbol{q}). \tag{2.6.8}$$

For $q = 0$, (2.6.7) and (2.6.8) reduce, respectively, to (2.4.9) and (2.4.10).

Suppose that $T_{\boldsymbol{q}}$ is highest for $\boldsymbol{q} = \boldsymbol{Q}$ in a spin system. Then if we lower the temperature from above, $\chi(\boldsymbol{Q})$ diverges at $T = T_{\boldsymbol{Q}}$ and a spin ordering starts to make $\langle S_z(\boldsymbol{Q})\rangle \neq 0$. In this way we can have the spin ordering of the type shown in Fig.2.2.(d). It is the reader's exercise to find what kind of magnetic susceptibility is relevant to the spin ordering of the type shown in Fig.2.2(e).

Note that while $T_{\boldsymbol{q}} \propto J(\boldsymbol{q})$ according to (2.6.8), if the interaction between spins is restricted only to the nearest neighbor pairs, we always have $J(0) > J(\boldsymbol{q})$ as we saw in (2.5.28). Thus, when a localized spin system has a magnetic ordering with $q \neq 0$ an important role must be played by the interaction between spins beyond the nearest neighbors; i.e. the exchange interaction must be long ranged. Such a magnetic ordering with $q \neq 0$ is actually observed in rare earth metals, for instance. In rare earth metals the interaction between localized $4f$ spins is mediated by the conduction electrons of the $5d$ and $6s$ bands. As we will discuss later in 4.2.4, such an interaction, called the RKKY interaction, has long range.

Based on the discussion given in 2.3 on the (direct) exchange interaction, so far we assumed that $J > 0$ for the nearest neighbor pairs of localized spins. Note, however, that it is also possible to have $J < 0$; in this case the spins favor to align antiferromagnetically. There are a number of indirect exchange interaction mechanisms similar to the above RKKY interaction, called the superexchange interaction, for such an antiferromagnetic exchange interaction (see [2.2–2.4]).

2.1 Derive the free energy of non-interacting localized spins under a magnetic field in the form of (2.2.7). Then obtain the magnetization as given in (2.2.8) by the procedure of (2.2.6).

2.2 Confirm the relations of (2.3.13)–(2.3.16).

2.1 Estimate the energy difference, the exchange energy, of (2.3.24) by assuming both electrons are near the Fermi surface of a metal.

2.4 Derive the results of (2.5.10) and (2.5.11). Note that $\boldsymbol{q}$ satisfies the condition, (2.5.5).

2.5 Confirm that the operators defined in (2.5.34) actually satisfy the boson commutation relation of (2.5.30).

2.6 Show that the normalized state vector whose $\boldsymbol{l}$ state is occupied by n_l bosons is given by

$$| n_l \rangle = \frac{1}{\sqrt{n_l\,!}} (b_l^\dagger)^{n_l} | \mathrm{vac} \rangle .$$

2.7 Derive the Bose distribution function of (2.5.37) by the procedure suggested in the text.

Chapter 3

MEAN FIELD THEORY OF MAGNETISM OF METALS: THE STONER MODEL

In Chapter 2 we presented an introduction to magnetism, but the discussion was mostly concerned with insulators. The behavior of electrons in a metal is quite different from that in an insulator; in the metal an electron outside of the closed shell of an atom can move around the entire volume, but in the insulator such a free motion of an electron is not possible. Correspondingly, the magnetic properties of a metal are quite different from that of an insulator. In this chapter we present an introduction to magnetism of metals. For a more extensive introduction, see, for instance, Refs. [3.1–3.4]. See also, Refs. [1.4, 1.6, 2.4–2.9, 2.13].

3.1 PAULI SPIN SUSCEPTIBILITY OF METALLIC ELECTRONS

As we noted in Chapter 1, the simplest model to describe metallic electrons is the electron gas model. Let us calculate the magnetic susceptibility of an electron gas.

Although the orbital motion of an electron is affected by an external magnetic field, in this chapter we consider the effect of the magnetic field only on the spin of the electron. It is the spin of the electron that is principally responsible for magnetism in a metal, as it was in an insulator. As for the contribution coming from the orbital motion of electrons we briefly discuss it in 4.2.5.

If we apply an external magnetic field $\boldsymbol{H}$ in the z-direction, the electrons, whose total number is n, acquire the Zeeman energy,

$$H_Z = -HM_z\,, \tag{3.1.1}$$

with the z-component of the electron spin magnetization M_z, which is given in the second quantized form, as

$$\begin{aligned} M_z &= -\mu_B \int \psi^+(\boldsymbol{r})\sigma_z\psi(\boldsymbol{r})d\boldsymbol{r} \\ &= -\mu_B \sum_{k,k',\sigma,\sigma'} a^+_{k\sigma}\, a_{k'\sigma'} \int d\boldsymbol{r}\phi^*_k(\boldsymbol{r})\; \phi_{k'}(\boldsymbol{r})\left(\chi_\sigma\, \sigma_z\, \chi_{\sigma'}\right) \\ &= -\mu_B \sum_k \left[a^+_{k+}\, a_{k+} - a^+_{k-}\, a_{k-}\right]. \end{aligned} \tag{3.1.2}$$

The notation is the same as in Chapter 1. Combining this Zeeman energy with the kinetic energy of electrons given in (1.5.14), we have

$$H_o + H_Z = \sum_{k,\sigma} \varepsilon_{k\sigma}\, a^+_{k\sigma}\, a_{k\sigma}\,, \tag{3.1.3}$$

$$\varepsilon_{k\sigma} = \varepsilon_k + \sigma\mu_B H. \tag{3.1.4}$$

The thermal expectation of the magnetization under the magnetic field is calculated following the prescription of 1.6:

$$\langle M_z \rangle = -\mu_B \sum_k \left[\langle a^+_{k+}\, a_{k+}\rangle - \langle a^+_{k-}\, a_{k-}\rangle\right] = -\mu_B \sum_{k,\sigma} \sigma f(\varepsilon_{k\sigma}), \tag{3.1.5}$$

$$f(\varepsilon_{k\sigma}) = \frac{1}{e^{\beta(\varepsilon_{k\sigma} - \mu)} + 1}\,. \tag{3.1.6}$$

For $|\mu_B H|/\varepsilon_F \ll 1$, the Fermi distribution function under the magnetic field can be expanded as

$$f(\varepsilon_{k\sigma}) = f(\varepsilon_k) + \frac{\partial f(\varepsilon_k)}{\partial \varepsilon_k}\sigma\mu_B H + \cdots, \tag{3.1.7}$$

where $f(\varepsilon_k)$ is the Fermi distribution without the effect of the magnetic field. By putting (3.1.7) into (3.1.5) and recalling (1.6.14) we obtain

$$\chi = \lim_{H \to 0} \frac{\langle M_z \rangle}{H} = -2\mu_B^2 \sum_k \frac{\partial f(\varepsilon_k)}{\partial \varepsilon_k}$$

$$= 2\mu_B^2 N(0) \left[1 - \alpha \left(\frac{T}{T_F} \right)^2 + \cdots \right] \equiv \chi_P \cong 2\mu_B^2 N(0) , \qquad (3.1.8)$$

where we have written $N(0)$ for $N(\varepsilon_F)$. This is the *Pauli spin susceptibility.*

Let us rederive the Pauli susceptibility more illustratively. Recall that the direction of spin is opposite to that of magnetization (see (2.1.16)). Thus, if we apply a magnetic field in the positive z-direction we have $n_- = \Sigma_k f(\varepsilon_{k-}) > n_+$ and therefore $\langle M_z \rangle > 0$. Often we refer to the spin up electrons as the majority electrons in the magnetized state, as illustrated in Fig.3.1 (and Fig.3.5(a)):

$$n_+ - n_- \equiv \overline{M} > 0 . \qquad (3.1.9)$$

Correspondingly, often $\overline{M}$ is used as the magnetization in place of M_z $(= -\mu_B \overline{M})$. Here let us assume this situation. To produce such a state we have to apply a magnetic field in the negative direction of the z-axis; we have to put

$$H = -|H| \qquad (3.1.10)$$

in (3.1.4)–(3.1.5).

Under such a magnetic field, we have

$$n_\sigma = \sum_k \langle a_{k\sigma}^+ a_{k\sigma} \rangle = \int d\varepsilon \, N(\varepsilon) f(\varepsilon - \sigma\mu_B |H|)$$

$$= \int d\varepsilon \, N(\varepsilon + \sigma\mu_B |H|) f(\varepsilon) = \int d\varepsilon \, N_\sigma(\varepsilon) f(\varepsilon) , \qquad (3.1.11)$$

where

$$N_\sigma(\varepsilon) = N(\varepsilon + \sigma\mu_B |H|) \qquad (3.1.12)$$

is the density of states of electrons with spin σ (+ or −). The effect of the magnetic field is to spin split the electronic density of states as illustrated in Fig.3.1. The electrons which originally occupied the shaded region of $N_-(\varepsilon)$ are transferred to the shaded region of $N_+(\varepsilon)$. The number of the transferred electrons with spin σ, Δn_σ, is given by the area of the shaded region as

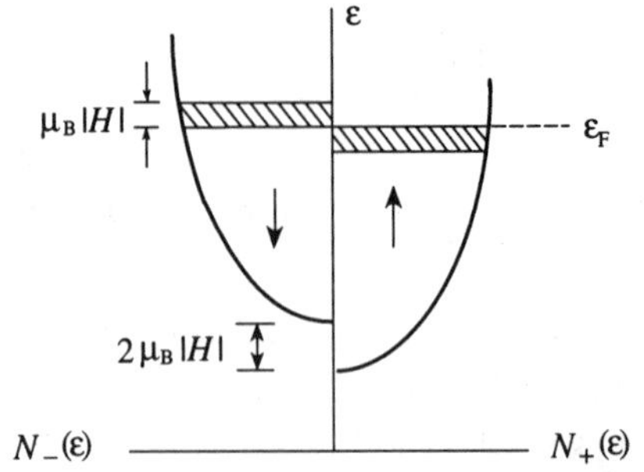

Fig.3.1 The spin splitting of the electronic density of states, $N_\sigma(\varepsilon)$, by an external magnetic field H. The magnetic field is applied so as to make $n_+ > n_-$, that is, in the negative direction of the z-axis as given in (3.1.10). While the shaded region in $N_+(\varepsilon)$ represents the increased number of +spin electrons, that in $N_-(\varepsilon)$ represents the decreased number of −spin electrons.

$$\Delta n_+ = -\Delta n_- \cong N(0)\mu_B|H| , \tag{3.1.13}$$

and then we obtain the magnetic susceptibility as

$$\chi = \frac{-\mu_B(n_+ - n_-)}{H} = \frac{\mu_B(\Delta n_+ - \Delta n_-)}{|H|} = 2\mu_B{}^2N(0) . \tag{3.1.14}$$

This result is the same as that of (3.1.8).

According to (3.1.8) the effect of temperature on the Pauli susceptibility is of the order of $O(k_BT/\varepsilon_F)^2$. Thus, since generally $\varepsilon_F \cong 1$ eV $\cong 10^4 k_B$, such a temperature dependence can be neglected. As for the size of the Pauli susceptibility, since

$$N(0) = O(n/\varepsilon_F) , \tag{3.1.15}$$

we have

$$\chi_P = O\left(\mu_B{}^2 n/\varepsilon_F\right) . \tag{3.1.16}$$

Compared to this, the size of the Curie susceptibility is estimated as (see (2.2.13) and (2.2.14))

$$\chi_C = O\left(\mu_B{}^2 n/k_BT\right), \text{ or, } \chi_P / \chi_C = O(k_BT/\varepsilon_F) . \tag{3.1.17}$$

At ordinary temperatures where $k_BT/\varepsilon_F \ll 1$, the Pauli susceptibility is much smaller than the Curie susceptibility. This result is understandable if we note that in a metal only those electrons which lie close to the Fermi surface can respond to an external perturbation. If they are to be within the distance ~ k_BT from the Fermi surface, the number of such electrons is estimated as $n_{\text{eff}} \cong O(N(0)k_BT)$. If these electrons behave like independent localized electrons, their susceptibility is estimated as $\chi \cong n_{\text{eff}}\,\mu_B{}^2/\,k_BT \cong O(\mu_B{}^2N(0)) \cong O(\chi_P)$.

Ferromagnetism is observed in many metals and alloys. However, it is not possible to explain the ferromagnetism of metals with the Pauli spin susceptibility. In order to have a ferromagnetic transition, it is necessary to invoke the exchange interaction which we introduced at the end of 2.3. In the next section we reformulate the effect of the exchange interaction between itinerant electrons.

3.2 THE MEAN FIELD APPROXIMATION ON THE COULOMB INTERACTION BETWEEN ELECTRONS

The interaction between electrons in a metal is the Coulomb repulsion. If we describe an electron by a plane wave the Coulomb interaction is given by H_C of (1.5.24). Then, in order to obtain the magnetic susceptibility including the effect of the Coulomb interaction, we are required to calculate

$$\langle a_{k\sigma}^{+}\, a_{k\sigma}\rangle = \operatorname{tr}\left[\, e^{-\beta(H_0 + H_C + H_Z - \mu\tilde{n})}\, a_{k\sigma}^{+} a_{k\sigma}\right] / \, \Xi, \tag{3.2.1}$$

$$\Xi = \operatorname{tr} e^{-\beta(H_0 + H_C + H_Z - \mu\tilde{n})}\,. \tag{3.2.2}$$

Unlike the case without H_C of the preceding subsection, however, it is not feasible to carry out this calculation exactly, since H_C is of the form of the product of 4 creation and annihilation operators.

As to how to approximately treat the effect of the Coulomb interaction we discuss it systematically later in Chapters 8–10. In this section we treat it most simply by the mean field approximation which we already encountered in 2.4.

The spirit of the mean field approximation is to put as

$$H_C \cong \sum_{k,l,\kappa,\sigma,\sigma'} \mathrm{v}(\kappa) \left\langle a_{l\sigma'}^{+}\, a_{l+\kappa,\sigma'}\right\rangle a_{k\sigma}^{+}\, a_{k-\kappa,\sigma}\,. \tag{3.2.3}$$

This approximation for the Coulomb interaction between electrons is also called the *Hartree approximation*. The factor 1/2 disappears since by replacing $a_{k\sigma}^{+}$ $a_{k-\kappa,\sigma}$ by its thermal expectation in H_C we obtain another contribution of the same form. (Note, however, that we need the factor 1/2 when we calculate the total interaction energy of the system. Recall our discussion for the case of the Heisenberg model).

Note that if we put $U(\kappa) = \mathrm{v}(\kappa)\Sigma_{l\sigma'}\langle a_{l\sigma'}^{+}\, a_{l+\kappa,\sigma'}\rangle = \mathrm{v}(\kappa)\langle n(\kappa)\rangle$, the right hand side of (3.2.3) becomes of the same form as that of the impurity potential of (1.5.27). (We omit the tilde on second quantized operators when there is no possibility of confusion). Thus, the Hartree approximation treats all the other electrons interacting with an electron as a medium to provide a potential $U(\kappa)$.

If a magnetic field is spatially uniform, then even in the presence of the magnetic field we have $\langle n(\kappa)\rangle = \langle n(0)\rangle \delta_{\kappa,0}$. The only non-vanishing term in (3.2.3) is that with $\kappa = 0$ and it diverges because $\lim_{\kappa \to 0} \mathrm{v}(\kappa) = \infty$. Fortunately, however, in H_C we have to exclude the contribution from the terms with $\kappa = 0$. Although we are pursuing only the behavior of electrons we should not forget the presence of a uniform positive (ionic) charge background to neutralize the system. Then, we have three kinds of Coulomb interactions: among electrons, among ions, and between electrons and ions. The uniform part of these three Coulomb interactions, corresponding to the $\kappa = 0$ contribution in the case of (3.2.3), cancel each other out. Thus we often put a prime on the summation in an expression such as (3.2.3) to indicate that the contribution from $\kappa = 0$ is to be excluded.

For the reason given in the above, in an electron gas the effect on the uniform magnetic susceptibility of the Hartree contribution of (3.2.3) is null. Does the Coulomb interaction, then, not have any effect on the magnetic susceptibility of a metal? Yes. There is another contribution from the Coulomb interaction between electrons, the exchange interaction, which we already discussed in 2.3. This contribution is derived from approximating the Coulomb repulsion in the following way,

$$H_C \cong - \sum_{k,l,\kappa,\sigma,\sigma'}{}' \mathrm{v}(\kappa) \left\langle a_{k\sigma}^{+} a_{l+\kappa,\sigma'} \right\rangle a_{l\sigma'}^{+} a_{k-\kappa,\sigma} . \tag{3.2.4}$$

The negative sign is the result of commuting $a_{l\sigma'}^{+}$ and $a_{l+\kappa,\sigma'}$ in H_C before introducing the mean field approximation. This contribution is also called the *Fock* term.

In a spatially uniform situation, since

$$\left\langle a_{k\sigma}^{+} a_{l+\kappa,\sigma'} \right\rangle = \left\langle a_{l+\kappa,\sigma}^{+} a_{l+\kappa,\sigma} \right\rangle \delta_{\sigma,\sigma'} \delta_{l+\kappa,k} , \tag{3.2.5}$$

the exchange contribution of (3.2.4) reduces to

$$H_{C,\,\mathrm{ex}} = - \sum_{l,\kappa,\sigma}{}' \mathrm{v}(\kappa) \left\langle a_{l+\kappa,\sigma}^{+} a_{l+\kappa,\sigma} \right\rangle a_{l\sigma}^{+} a_{l\sigma} . \tag{3.2.6}$$

Combining these two kinds of mean field procedures, (3.2.3) and (3.2.6), for the Coulomb interaction constitutes the *Hartree-Fock approximation*, or, simply, the mean field approximation.

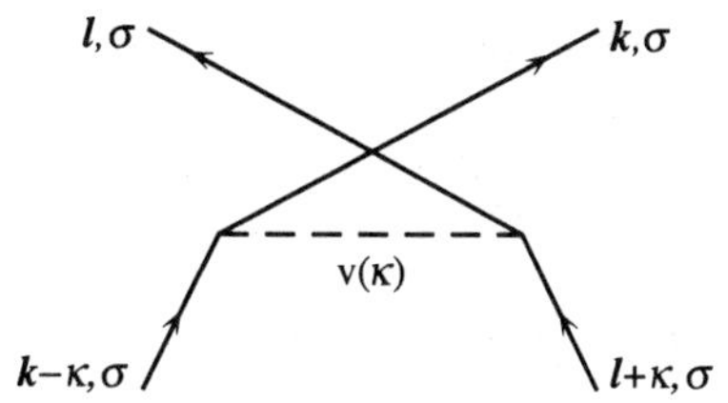

Fig.3.2 The exchange interaction between electrons.

The nature of the mean field approximation may be understood as follows. While Fig.1.2 represents the Coulomb interaction, the Hartree approximation replaces the right hand electron line by its average, the medium. On the other hand, in the exchange procedure, as illustrated in Fig.3.2, an electron is taken into the medium on interaction and then another electron emerges out of the medium instead and behaves as if it were the same electron which originally interacted with the medium. Note that to make such an exchange process possible two electrons must have the same spins.

As can be seen from (3.2.6), the exchange interaction is effectively an attractive interaction between two electrons of the same spins. This nature of the exchange interaction between metallic electrons is the same as that between localized electrons.

3.3 FERROMAGNETISM OF METALLIC ELECTRONS: THE STONER MODEL

Let us calculate the magnetic susceptibility of metallic electrons by the procedure of 3.1, now including the effect of the exchange interaction between electrons. First note that (3.2.6) is rewritten as

$$H_{\mathrm{C,\,ex}} = -\sum_{l,\kappa,\sigma}{}' \mathrm{v}(\kappa)\, f(\varepsilon_{l+\kappa,\sigma})\, a_{l\sigma}^{+}\, a_{l\sigma}\,, \qquad (3.3.1)$$

where $f(\varepsilon_{l+\kappa,\sigma})$ is the Fermi distribution to be obtained by replacing H_{C} in (3.2.1)–(3.2.2) by $H_{\mathrm{C,\,ex}}$. To obtain the Fermi distribution we require a self-consistent procedure since $H_{\mathrm{C,\,ex}}$ itself contains the Fermi distribution.

We simplify (3.3.1) as

$$H_{\mathrm{C,\,ex}} = -\widetilde{V}(0)\sum_{l,\kappa,\sigma} f(\varepsilon_{l+\kappa,\sigma})\, a_{l\sigma}^{+}\, a_{l\sigma} = \sum_{l,\sigma}\left(-\widetilde{V}(0)n_{\sigma}\right) a_{l\sigma}^{+}\, a_{l\sigma}\,, \qquad (3.3.2)$$

by introducing the *effective exchange interaction* $\widetilde{V}(0)$. Mathematically this is to invoke the mean value theorem in integration; both $\mathrm{v}(\kappa)$ and $\widetilde{V}(0)$ are positive.

$\widetilde{V}(0)$ represents the strength of the exchange potential produced by the other electrons in the medium with wave number zero. As to the effective exchange interaction, we shall leave further discussion of it for later.

With the simplification of (3.3.2), we have

$$H_0 + H_{C,\,ex} + H_Z = \sum_{k,\sigma} \left(\varepsilon_k - \widetilde{V}(0)n_\sigma + \sigma\mu_B H\right) a_{k\sigma}^{+} a_{k\sigma}$$

$$\equiv \sum_{k,\sigma} \varepsilon_{k\sigma} a_{k\sigma}^{+} a_{k\sigma}, \qquad (3.3.3)$$

for the total Hamiltonian of the system under the magnetic field. This implies that the energy of an electron $(\boldsymbol{k},\sigma)$ is given as

$$\varepsilon_{k\sigma} = \varepsilon_k + \sigma\mu_B H - \widetilde{V}(0)n_\sigma$$

$$= \varepsilon_k + \sigma\left[\mu_B H - \frac{\widetilde{V}(0)}{2}(n_+ - n_-)\right] - \frac{\widetilde{V}(0)}{2}(n_+ + n_-). \qquad (3.3.4)$$

If we neglect the last term, since the total number of electrons, $n_+ + n_- = n$, is a constant, it is further rewritten as

$$\varepsilon_{k\sigma} = \varepsilon_k + \sigma\mu_B H_{\text{eff}}, \qquad (3.3.5)$$

$$H_{\text{eff}} = H\left[1 + \frac{1}{2\mu_B^2}\widetilde{V}(0)\,\chi_S\right], \qquad (3.3.6)$$

where χ_S is the magnetic susceptibility including the effect of the exchange interaction which we are seeking. Then by putting (3.3.6) into

$$\langle M_z \rangle = \chi_P H_{\text{eff}} = \chi_S H, \qquad (3.3.7)$$

where χ_P is the Pauli susceptibility, we obtain

$$\chi_S = \frac{\chi_P}{1 - \dfrac{1}{2\mu_B^2}\widetilde{V}(0)\,\chi_P}. \qquad (3.3.8)$$

We call this χ_S the *Stoner susceptibility* and, more generally, the mean field theory of itinerant electron magnetism the *Stoner model* [3.5]. By the effect of the exchange interaction, the Pauli susceptibility of metallic electrons is enhanced to the Stoner susceptibility. Note that the result of (3.3.8) has the same structure as that of the Curie-Weiss susceptibility for a Heisenberg model, (2.4.8).

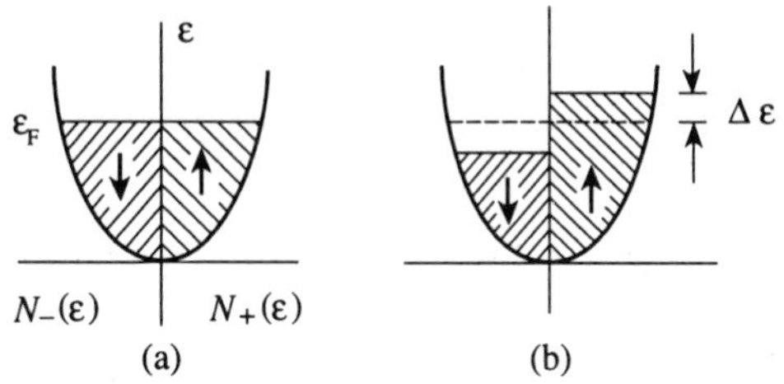

Fig.3.3 Electronic states without magnetization, (a), and with magnetization, (b). The shaded regions represent the occupied states.

For $T = 0$, (3.3.8) reduces to

$$\chi_S = \frac{2\mu_B^2 N(0)}{1 - \tilde{V}(0) N(0)} . \tag{3.3.9}$$

We consider now the meaning of

$$\tilde{V}(0) N(0) \geq 1 \tag{3.3.10}$$

in (3.3.9). For $\tilde{V}(0)\, N(0) = 1$, we have $\chi_S = \infty$, which implies a ferromagnetic transition at $T = 0$. If $\tilde{V}(0)\, N(0) > 1$, we have $\chi_S < 0$, which implies the instability of the paramagnetic state; in a stable state the magnetic susceptibility is positive (see the next subsection). Thus, (3.3.10) gives the condition for the occurrence of ferromagnetism in a metal. We call it the *Stoner condition.* Since the Stoner condition is derived within the mean field approximation it can not be taken as the condition for ferromagnetism in a rigorous sense. However, the quantity $\tilde{V}(0) N(0)$ serves as a useful measure of the magnetic tendency of a metal beyond such a limitation.

The Stoner condition can be derived more directly. Let us assume a metallic electron system is in the paramagnetic state such as illustrated in Fig.3.3(a). For this state to be stable, its energy must be lower than that of Fig.3.3(b). If the energy of the state of (b), E_b, is lower than that of (a), E_a, it implies that the stable state to be realized is the ferromagnetic state of (b).

Let us put

$$\Delta E = E_b - E_a = \Delta E_{kin} + \Delta E_{ex}, \tag{3.3.11}$$

where ΔE_{kin} and ΔE_{ex} are, respectively, the differences in the kinetic and the exchange energies between the states (b) and (a) in Fig.3.3. Then, it is evident that

$$\Delta E_{\rm kin} \cong \big(N(0)\,\Delta\varepsilon\big)\Delta\varepsilon > 0. \qquad (3.3.12)$$

$N(0)\Delta\varepsilon$ represents the number of the down spin electrons transferred to the up spin electron states.

As for the exchange energy, with the approximation of (3.3.2) we have

$$\Delta E_{\rm ex} = -\frac{\widetilde{V}(0)}{2}\left[\left(n_+^2 + n_-^2\right) - 2\left(\frac{n}{2}\right)^2\right] \cong -\widetilde{V}(0)\big(N(0)\Delta\varepsilon\big)^2 < 0. \qquad (3.3.13)$$

Note the appearance of the factor 1/2 in (3.3.13) which was absent in (3.3.2); it is included not to doubly count the interaction energy of a pair of electrons.

As can be seen from (3.3.12) and (3.3.13), the kinetic energy and the exchange energy play opposite roles; while the former opposes a ferromagnetic state, the latter favors it. Thus, the condition for the occurrence of a ferromagnetic state is given as

$$\Delta E = \left[1 - \widetilde{V}(0)N(0)\right]N(0)\Delta\varepsilon^2 \leq 0. \qquad (3.3.14)$$

This reproduces the Stoner condition of (3.3.10).

In the case of an interacting localized spin system, within the mean field approximation, ferromagnetism or some other spin ordering is always anticipated. In the case of an itinerant electron system, however, the Stoner condition is not always satisfied. This is a decisive difference between the magnetic properties of a metal and an insulator.

The temperature dependence of the Stoner susceptibility is obtained by putting the Pauli susceptibility, given in (3.1.8), into (3.3.8) to obtain

$$\chi_{\rm s}(T) \cong \frac{2\mu_{\rm B}^2 N(0)\left(1 - \alpha\,(T/T_{\rm F})^2\right)}{\left(1 - \overline{V}\right) + \overline{V}\alpha\,(T/T_{\rm F})^2}, \qquad (3.3.15)$$

where in the above and hereafter we set

$$\widetilde{V}(0)\,N(0) \equiv \widetilde{V}N(0) \equiv \overline{V}. \qquad (3.3.16)$$

As in the above we will often abbreviate $\widetilde{V}(0)$ by $\widetilde{V}$. α, defined in (1.6.14), is given explicitly as

$$\alpha = \frac{\pi^2}{6}\left[\left(\frac{N'(0)}{N(0)}\right)^2 - \frac{N''(0)}{N(0)}\right]\varepsilon_{\rm F}^2. \qquad (3.3.17)$$

According to the above result, if we know the electronic density of states near the Fermi surface of a metal, we can predict, or, reproduce the magnitude and temperature dependence of its magnetic susceptibility. Indeed, it has long been one of the principal motivations for carrying out detailed electronic structure calculations on transition metals. For a review and references, see Shimizu [3.6]. However, it is now widely recognized that the Stoner susceptibility can not be used for a quantitative analysis. There are other important physical mechanisms determining the magnetic susceptibility of a metal which are not considered in the Stoner susceptibility. Electronic structure calculations are more important than ever in understanding the magnetic properties of a metal. However, the relation between electronic structure and magnetic susceptibility is entirely different from that of the Stoner susceptibility, as we will see in Chapter 6.

When the Stoner condition for ferromagnetism is satisfied, the Curie point T_C is obtained from (3.3.15) by requiring $1/\chi_S(T_C) = 0$, so that

$$T_C = \left(\bar{V} - 1\right)^{1/2} T_F / \left(\bar{V}\alpha\right)^{1/2}, \tag{3.3.18}$$

where we assumed $\alpha > 0$. The Curie temperature may then be estimated to be

$$k_B T_C \cong O\left(\left(\bar{V} - 1\right)^{1/2} \varepsilon_F\right). \tag{3.3.19}$$

If we assume $\varepsilon_F \cong 1$ eV, and $\bar{V} = 1.1$, we have $T_C \cong 3 \times 10^3$ K. Such a value of T_C is generally too high and it is considered to be one of the serious faults of the Stoner model.

By using the above expression for T_C, (3.3.15) can be rewritten as

$$\chi_S = \frac{2\mu_B^2 N(0)}{(T + T_C)(T - T_C)} \frac{T_F^2}{\bar{V}\alpha} \left(1 - \alpha(T/T_F)^2\right). \tag{3.3.20}$$

For $(T/T_F)^2 \ll 1$, in a narrow temperature region above T_C this approximates the Curie-Weiss form. Experimentally, however, the Curie-Weiss like temperature dependence of the magnetic susceptibility is observed over much wider temperature range in many ferromagnetic metals; see, for instance, Ref. [3.7].

We have discussed the Stoner condition for the instability of paramagnetic state. In the next section we discuss the Stoner model in the ferromagnetic state.

3.4 STONER MODEL FROM THE LANDAU THEORY OF PHASE TRANSITIONS

3.4.1 Landau Theory

Let us assume that the free energy of a system, $F(M)$, at some temperature and volume is given as the function of the magnetization M. Then the equilibrium magnetization of the system is given by the value of M at which $F(M)$ becomes minimum. That is, the equilibrium magnetization is determined from the requirements

$$\frac{\partial F(M)}{\partial M} = 0, \qquad \frac{\partial^2 F(M)}{\partial M^2} > 0 . \tag{3.4.1}$$

As illustrated in Fig.3.4, for $T > T_C$ the minimum of $F(M)$ should be at $M = 0$, and for $T < T_C$ it should be at some $M \neq 0$. This is the very basis of the Landau theory of the second order phase transition as applied to magnetic phenomena.

In order to derive the relation between the magnetic susceptibility for $T > T_C$ and the free energy, let us expand $F(M)$ for a small M,

$$F(M) = F_o + \frac{1}{2!} a_2 M^2 + \frac{1}{4!} a_4 M^4 + \cdots - HM, \tag{3.4.2}$$

where F_o is independent of M, and H is an external magnetic field. Since, in the absence of the external field, $H = 0$, $F(M)$ has to be invariant to a change in the sign of magnetization, there do not appear any odd power terms of M, except the last term.

For a given H, the magnetization M to be realized is determined by putting (3.4.2) into (3.4.1):

$$0 = \frac{\partial F(M)}{\partial M} = a_2 M + \frac{1}{3!} a_4 M^3 + \cdots - H.$$

The magnetic susceptibility for $T > T_C$ is, then, obtained as

$$\chi = \lim_{H \to 0} \frac{M}{H} = \frac{1}{a_2}, \tag{3.4.3}$$

or, since $a_2 = \partial^2 F(M) / \partial M^2 |_{M=0}$,

$$\frac{1}{\chi} = \left. \frac{\partial^2 F(M)}{\partial M^2} \right|_{M=0} . \tag{3.4.4}$$

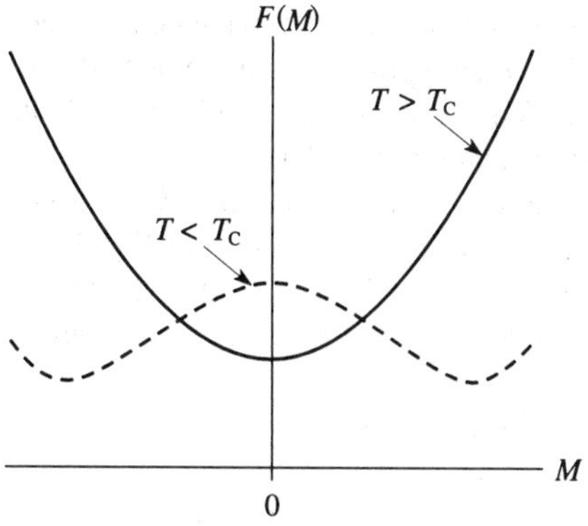

Fig.3.4 The schematic illustration of the free energy of a magnetic system for $T > T_C$, and $T < T_C$.

In deriving the high field susceptibility for $T < T_C$, where a system has spontaneous magnetization $M(T)$, we put $M = M(T) + \Delta M$, and expand $F(M)$ in terms of ΔM. The equilibrium value of ΔM under a magnetic field H is determined from the minimum of $F(M + \Delta M)$ with respect to ΔM. Then, from a procedure parallel to (3.4.1)–(3.4.3) we obtain

$$\frac{1}{\chi_{hf}} = \left.\frac{\partial^2 F(M(T) + \Delta M)}{\partial(\Delta M)^2}\right|_{\Delta M = 0}$$

$$= \left.\frac{\partial^2 F(M)}{\partial M^2}\right|_{M = M(T)} \tag{3.4.4'}$$

In the next subsection we apply these results to a metallic electron system.

3.4.2 Free Energy and Magnetism of Metallic Electrons

The thermodynamic potential of electrons without the Coulomb interaction, Ω_0, is given from (1.4.18) and (1.4.14) as

$$\Omega_0(M) = -\frac{1}{\beta}\ln \sum_{n_{k,\sigma}=0,1} e^{-\sum_{k,\sigma} \beta(\varepsilon_k - \mu_\sigma) n_{k,\sigma}}$$

$$= -\frac{1}{\beta}\ln \prod_{k,\sigma}\left[1 + e^{-\beta(\varepsilon_k - \mu_\sigma)}\right] = -\frac{1}{\beta}\sum_{k,\sigma}\ln\left[1 + e^{-\beta(\varepsilon_k - \mu_\sigma)}\right], \tag{3.4.5}$$

where μ_σ is the chemical potential of σ spin electrons measured from the bottom of the energy band of each spin electrons; when $M \neq 0$, that is when $n_+ \neq n_-$, we have $\mu_+ \neq \mu_-$. The free energy is obtained from the thermodynamic potential by using the relation of (1.4.19).

Note that the magnetization involved in Ω_0 is a variational one such as is required in the Landau procedure of (3.4.1) or Fig.3.4. Then, as we will show more in detail later in this subsection, the chemical potential μ_σ is to be determined so as to produce the given value of magnetization while keeping the total number of electrons constant.

As for the contribution of the Coulomb interaction to the thermodynamic potential we will give a detailed discussion in Chapters 8–10. Here we present, however, a simplified treatment corresponding to the discussion of the mean field approximation of 3.2.

The effect of the Coulomb interaction on the thermodynamic potential is obtained most simply by the following approximation,

$$\Omega = -\frac{1}{\beta} \ln \mathrm{tr}\left[e^{-\beta(H_0 + H_C - \mu\tilde{n})}\right] \cong \Omega_0 + \langle H_C \rangle_0 , \tag{3.4.6}$$

where $\langle \cdots \rangle_0$ indicates the thermal expectation value without the effect of the Coulomb interaction.

The above result is derived by using the following well-known expansion theorem,

$$e^{A+B} = e^A \left[1 + \int_0^1 d\lambda\, e^{-\lambda A} B e^{\lambda(A+B)}\right]$$

$$= e^A \left[1 + \int_0^1 d\lambda\, e^{-\lambda A} B e^{\lambda A} + \int_0^1 d\lambda \int_0^\lambda d\lambda'\, e^{-\lambda A} B e^{\lambda A} e^{-\lambda' A} B e^{\lambda' A} + \cdots \right], \tag{3.4.7}$$

where the operators A and B are not commutative, namely, $[A, B] \neq 0$. This theorem is proved as follows.

If we note $[A + B, \lambda(A + B)] = 0$, λ being a constant, we have

$$\frac{d}{d\lambda}\left(e^{-\lambda A} e^{\lambda(A+B)}\right) = e^{-\lambda A} B e^{\lambda(A+B)} .$$

By integrating this equation we obtain

$$e^{-\lambda A}\, e^{\lambda(A+B)} \;=\; \int_0^{\lambda} d\lambda'\, e^{-\lambda' A}\, B\, e^{\lambda'(A+B)} \;+\; 1\,.$$

If we put $\lambda = 1$ in the above equation and then multiply both sides with e^A, we arrive at the first equation of (3.4.7).

Now, if we use (3.4.7) we have

$$e^{-\beta(H_0 \,-\, \mu\tilde{n} \,+\, H_C)}$$

$$= e^{-\beta(H_0 \,-\, \mu\tilde{n})}\left[1 + \int_0^1 d\lambda\, e^{\lambda\beta(H_0 \,-\, \mu\tilde{n})}\left(-\,\beta H_C\right) e^{-\lambda\beta(H_0 \,-\, \mu\tilde{n})} \;+\cdots\right],$$

where we retained only up to the first order in H_C. In taking the trace of the both sides of the above equation we note the following cyclic property of the trace,

$$= e^{-\beta(H_0 \,-\, \mu\tilde{n})}\left[1 + \int_0^1 d\lambda\, e^{\lambda\beta(H_0 \,-\, \mu\tilde{n})}\left(-\,\beta H_C\right) e^{-\lambda\beta(H_0 \,-\, \mu\tilde{n})} \;+\cdots\right],$$

Thus, the right hand side of the first equality of (3.4.6) is approximated as

$$\Omega \cong -\frac{1}{\beta}\ln \mathrm{tr}\left[e^{-\beta(H_0 - \mu\tilde{n})}\left(1 - \beta H_C\right)\right]$$

$$= -\frac{1}{\beta}\ln\left[\,\Xi_0\left\{1 - \beta\frac{\mathrm{tr}\left[e^{-\beta(H_0 - \mu\tilde{n})}H_C\right]}{\Xi_0}\right\}\right] = -\frac{1}{\beta}\left[\,\ln \Xi_0 + \ln\left\{1 - \beta\left\langle H_C\right\rangle_0\right\}\right]$$

$$\cong -\frac{1}{\beta}\left[\,\ln \Xi_0 - \beta\left\langle H_C\right\rangle_0\right]$$

by retaining only up to first order, where Ξ_0 is the grand partition function corresponding to Ω_0, namely, $\Omega_0 = -\,(1/\beta)\ln \Xi_0$ (See (1.6.2)). This gives the final expression of (3.4.6), which we will use from now in this subsection. For a more thorough discussion, see, for instance, Ref. [1.4].

As we have seen, (3.4.6) is the result of the lowest order approximation with respect to H_C. Can we safely neglect the higher order contributions? No. Those neglected higher order contributions are by no means small. However, as we will see in Chapter 8, one of the principal effects of those higher order contributions is the screening of the bare Coulomb interaction in the last term of (3.4.6). Thus, the effect of such higher order contributions may be incorporated by appropriately choosing the value of $\widetilde{V}\,(=\widetilde{V}(0))$ in the mean field approximation for

$$\langle H_{\rm C} \rangle_{\rm o} = -\frac{1}{2}\widetilde{V}\sum_{\sigma} n_{\sigma}^{2} . \tag{3.4.9}$$

Note that we already used this result in (3.3.13).

Thus, finally we obtain the free energy to be used in the mean field theory of itinerant electron magnetism as

$$F(M) = \Omega_{\rm o} - \frac{1}{2}\widetilde{V}\sum_{\sigma} n_{\sigma}^{2} + \sum_{\sigma} n_{\sigma}\mu_{\sigma} . \tag{3.4.10}$$

For a given magnetization M, (μ_+ , μ_-) are determined by the condition

$$M = -\mu_{\rm B}\left(n_+ - n_-\right) = -\mu_{\rm B}\overline{M} , \tag{3.4.11}$$

under the constraint $n_+ + n_- = n$, with

$$n_{\sigma} = \int N(\varepsilon)\, f_{\sigma}(\varepsilon)\, d\varepsilon , \tag{3.4.12}$$

$$f_{\sigma}(\varepsilon) = \frac{1}{e^{\beta(\varepsilon - \mu_{\sigma})} + 1} . \tag{3.4.13}$$

Let us emphasize again that the magnetization appearing in (3.4.10)–(3.4.11) is not that which is actually observed. The magnetization to be realized is determined by imposing the condition (3.4.1) on (3.4.10).

If we note that it is n_{σ} and μ_{σ} that depends on M, we have

$$\begin{aligned} \frac{\partial}{\partial \overline{M}} F &= \sum_{\sigma} \mu_{\sigma} \frac{\partial}{\partial \overline{M}} n_{\sigma} - \widetilde{V}\sum_{\sigma} n_{\sigma} \frac{\partial}{\partial \overline{M}} n_{\sigma} \\ &= \frac{1}{2}\left(\mu_+ - \widetilde{V} n_+\right) - \frac{1}{2}\left(\mu_- - \widetilde{V} n_-\right), \end{aligned} \tag{3.4.14}$$

where we used the following relations,

$$\frac{\partial}{\partial \overline{M}} \Omega_{\rm o} = -\sum_{\sigma} n_{\sigma} \frac{\partial \mu_{\sigma}}{\partial \overline{M}} , \tag{3.4.15}$$

$$\frac{\partial}{\partial \overline{M}} n_\sigma = \frac{1}{2}\sigma, \tag{3.4.16}$$

which are derived from (3.4.5) and (3.4.11). Then from the first condition of (3.4.1) we obtain

$$\mu_+ - \mu_- = \widetilde{V}\overline{M} = \widetilde{V}(n_+ - n_-), \quad \text{or,} \quad \mu_- - \mu_+ = \widetilde{V}M/\mu_B. \tag{3.4.17}$$

In equilibrium, the chemical potential appearing in the Fermi distribution of (3.4.12) is required to satisfy the above condition of (3.4.17). Such a Fermi distribution is given as

$$f(\varepsilon_{k\sigma}) = \frac{1}{e^{\beta(\varepsilon_{k\sigma} - \mu)} + 1}; \quad \varepsilon_{k\sigma} = \varepsilon_k - \sigma \widetilde{V}\overline{M}/2 \tag{3.4.18}$$

with the equilibrium chemical potential, μ, which depends on the magnetization but not on spin.

The number of σ spin electrons in equilibrium is given as

$$n_\sigma = \sum_k f(\varepsilon_{k\sigma}) = \int N_\sigma(\varepsilon) f(\varepsilon)\, d\varepsilon, \tag{3.4.19}$$

where $N_\sigma(\varepsilon)$ is the electronic density of states shifted from $N(\varepsilon)$ by $\sigma\widetilde{V}M/2\mu_B$ as given (3.4.18) and, therefore, spin split as shown in Fig.3.5.

If it were to derive only the above result of Fig.3.5, or (3.4.18), the above procedure of this subsection would not be necessary; the result is already implied in (3.3.4). In the later part of this book, however, we explore how the electron-phonon interaction is involved in the magnetism of a metal. In order to discuss that subject we have to start from the most fundamental point. That is why we here present such a detailed discussion of the Stoner theory.

Now, the spontaneous magnetization is obtained by solving (3.4.18)–(3.4.19), or self-consistently for given $N(\varepsilon)$, n, $\widetilde{V}$, V and T. Generally it requires a numerical procedure, and we will present some examples in Chapter 6.

In the above procedure for determining magnetization, we used only the first condition of (3.4.1). According to (3.4.4), the second condition of (3.4.1) requires the positiveness of the magnetic susceptibility.

Let us here see how the magnetic susceptibility, particularly in the ferromagnetic state of a metal, is obtained by the procedure of (3.4.4).

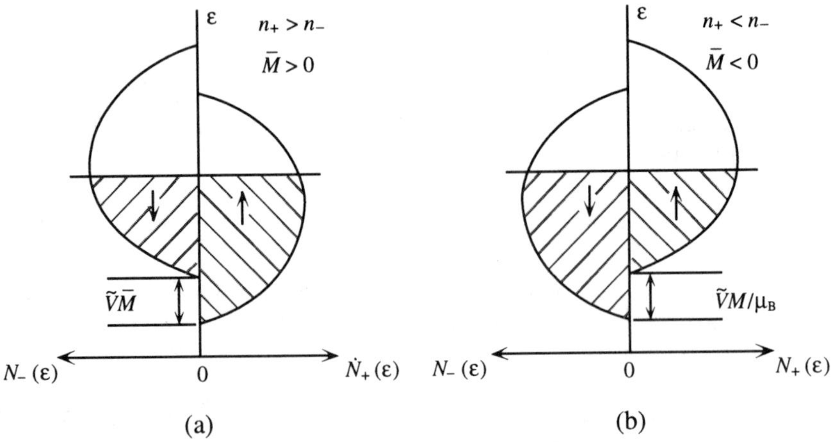

Fig.3.5 A ferromagnetic equilibrium state of electrons in a metal either with $\bar{M} = n_+ - n_- > 0$, (a) or with $M = -\mu_B \bar{M} > 0$, (b). Occupied states are shaded.

By differentiating both side of (3.4.14) once again, we obtain

$$\frac{\partial^2}{\partial \bar{M}^2} F(M) = \frac{1}{2} \sum_\sigma \sigma \left[\frac{\partial}{\partial \bar{M}} \mu_\sigma - \frac{1}{2} \sigma \tilde{V} \right], \tag{3.4.20}$$

where we used (3.4.16). The quantity $\partial \mu_\sigma / \partial \bar{M}$ required in (3.4.20) is obtained in the following way: If we put (3.4.12) together with (3.4.13) into the left hand side of (3.4.16), we have

$$\frac{\partial n_\sigma}{\partial \bar{M}} = \int d\varepsilon \, N(\varepsilon) \left(- \frac{\partial f_\sigma(\varepsilon)}{\partial \varepsilon} \right) \frac{\partial \mu_\sigma}{\partial \bar{M}} = \frac{1}{2} \sigma. \tag{3.4.21}$$

From this we obtain

$$\frac{\partial \mu_\sigma}{\partial \bar{M}} = \frac{\sigma}{2} \Big/ \int N(\varepsilon) \left(- \frac{\partial f_\sigma(\varepsilon)}{\partial \varepsilon} \right) d\varepsilon = \frac{\sigma}{2} \Big/ \int N_\sigma(\varepsilon) \left(- \frac{\partial f(\varepsilon)}{\partial \varepsilon} \right) d\varepsilon. \tag{3.4.21'}$$

Thus, the second condition of (3.4.1) is explicitly given as

$$\frac{\partial^2 F(M)}{\partial \overline{M}^2} = \frac{1}{4}\left[\sum_{\sigma} \frac{1}{\displaystyle\int N_\sigma(\varepsilon)\left(-\frac{\partial f(\varepsilon)}{\partial \varepsilon}\right)d\varepsilon} - 2\widetilde{V}\right] > 0. \tag{3.4.22}$$

For $M = 0$, this condition reduces to $\chi_s > 0$, or,

$$\widetilde{V}\int d\varepsilon\, N(\varepsilon)\left(-\frac{\partial f(\varepsilon)}{\partial \varepsilon}\right) < 1. \tag{3.4.23}$$

This reconfirms the Stoner condition for the instability of the paramagnetic state, (3.3.10).

From (3.4.22), the magnetic susceptibility in the ferromagnetic state with $M \neq 0$, $\chi_{\rm hf}$, is obtained as

$$\frac{1}{\chi_{\rm hf}} = \frac{1}{4\mu_{\rm B}^2}\left[\sum_{\sigma} \frac{1}{\displaystyle\int N_\sigma(\varepsilon)\left(-\frac{\partial f(\varepsilon)}{\partial \varepsilon}\right)d\varepsilon} - 2\widetilde{V}\right]$$

$$\cong \frac{1}{4\mu_{\rm B}^2}\left[\frac{1}{N_+(0)} + \frac{1}{N_-(0)} - 2\widetilde{V}\right]. \tag{3.4.24}$$

This susceptibility in the ferromagnetic state of a metal is called the *high field susceptibility*. It represents how a metal which is already in the ferromagnetic state is further magnetized by an external magnetic field. The condition (3.4.22) requires $\chi_{\rm hf} > 0$ for the stability of the ferromagnetic state.

3.5 THE HUBBARD MODEL

In discussing magnetism of a metal, we have thus far been using the electron gas model, in which a one electron state is described by a plane wave and the interaction between electrons is the long range Coulomb repulsion. Recently, however, the Hubbard model has been used more frequently than the electron gas model. In view of such a situation, in this section we introduce the Hubbard model and discuss the relation between these two models.

In a real metal, positively charged ions form a periodic lattice, and produce a periodic potential, $U(\boldsymbol{r})$, for electrons. In such a periodic potential the state of an electron is described by the *Bloch function* $\varphi_{\boldsymbol{k}\sigma}(\boldsymbol{r})$ which satisfies the following Schrödinger equation

$$\left[-\frac{\hbar^2}{2m}\nabla^2 + U(\boldsymbol{r})\right]\varphi_{k\sigma}(\boldsymbol{r}) = \varepsilon_k \varphi_{k\sigma}(\boldsymbol{r}). \tag{3.5.1}$$

It can be shown, then, that the Bloch function is represented as

$$\varphi_{k\sigma}(\boldsymbol{r}) = u_k(\boldsymbol{r}) e^{i\boldsymbol{kr}} \chi_\sigma , \tag{3.5.2}$$

where χ_σ is the spin eigenfunction (see Chapter 1) and $u_k(\boldsymbol{r})$ is a periodic function such that

$$u_k(\boldsymbol{r} + \boldsymbol{R}_i) = u_k(\boldsymbol{r}), \tag{3.5.3}$$

$\boldsymbol{R}_i$ being any lattice vector; the wave vector $\boldsymbol{k}$ can take only the discrete values within the 1st Brillouin zone such as given in (2.5.6) for the simple cubic lattice.

The free electron gas model of a metal which we have been using corresponds to taking $u_k(\boldsymbol{r}) = V^{-1/2}$ in (3.5.2), V being the volume of the system.

There is an alternative way of representing one electron states in a periodic potential. It is to use the *Wannier function* $w_{i\sigma}(\boldsymbol{r}) = w_i(\boldsymbol{r})\chi_\sigma$, i representing the lattice site $\boldsymbol{R}_i$, which is related to the Bloch function as

$$w_i(\boldsymbol{r}) = \frac{1}{\sqrt{N}} \sum_{\boldsymbol{k}}^{1\mathrm{BZ}} e^{-i\boldsymbol{kR}_i} \varphi_k(\boldsymbol{r}), \tag{3.5.4}$$

where N is the total number of ions. The sum over $\boldsymbol{k}$ is limited to the 1st Brillouin zone. The inverse transformation is given as

$$\varphi_k(\boldsymbol{r}) = \frac{1}{\sqrt{N}} \sum_j e^{i\boldsymbol{kR}_j} w_j(\boldsymbol{r}). \tag{3.5.5}$$

As we can see by putting $\varphi_k(\boldsymbol{r}) = e^{i\boldsymbol{kr}}/\sqrt{V}$ in (3.5.4), the Wannier function $w_i(\boldsymbol{r})$ is quite localized around the ionic site $\boldsymbol{R}_i$. The orthonormality relation

$$\int w_i^*(\boldsymbol{r})\, w_j(\boldsymbol{r})\, d\boldsymbol{r} = \delta_{ij}, \tag{3.5.6}$$

which can be easily confirmed by using that of Bloch functions, can be viewed as reflecting the localized nature of the Wannier functions. Thus we often write the Wannier function as $w(\boldsymbol{r} - \boldsymbol{R}_j)$ in place of $w_i(\boldsymbol{r})$. These Wannier functions form a complete set, as do the Bloch functions.

Let us proceed to obtain the second quantized form of the Hamiltonian of interacting electrons in a metal by using the Wannier functions as the basis. What is required is to put the quantized wave function in the following form

$$\psi(\boldsymbol{r}) = \sum_{i,\sigma} c_{i\sigma} w_i(\boldsymbol{r})\chi_\sigma \tag{3.5.7}$$

in the prescription of 1.5; $c_{i\sigma}$ is the annihilation operator of an electron in the state $w_i(\boldsymbol{r})\chi_\sigma$ which satisfies the following Fermion anticommutation relation,

$$\left\{c_{i\sigma}^{+}, c_{j\sigma'}\right\} = \delta_{ij}\delta_{\sigma\sigma'}, \ \left\{c_{i\sigma}, c_{j\sigma'}\right\} = \left\{c_{i\sigma}^{+}, c_{j\sigma'}^{+}\right\} = 0. \tag{3.5.8}$$

The relation between the creation and annihilation operators of the Wannier states and that of the Bloch states, obtained by noting the relations of (3.5.4) and (3.5.5) in the two different representations of the quantized wave functions, is

$$c_{i\sigma} = \frac{1}{\sqrt{N}}\sum_{k}^{\mathrm{1BZ}} e^{i\boldsymbol{k}\boldsymbol{R}_i} a_{\boldsymbol{k}\sigma}\,; \quad a_{\boldsymbol{k}\sigma} = \frac{1}{\sqrt{N}}\sum_{i=1}^{N} e^{-i\boldsymbol{k}\boldsymbol{R}_i} c_{i\sigma}\,. \tag{3.5.9}$$

The relations between the creation operators is obtained if we take the Hermite conjugates of the above relations. With (3.5.9) the anticommutation relation of (3.5.8) is derived from (1.5.5).

By carrying out with (3.5.7) the procedure of (1.5.12) for the one particle part of the electron energy, we obtain

$$H_0 = \sum_{i,\sigma} t_0\, c_{i\sigma}^{+} c_{i\sigma} + \sum_{i\neq j,\sigma} t_{ij}\, c_{i\sigma}^{+} c_{j\sigma}\,, \tag{3.5.10}$$

where we put

$$t_{ij} = \int w_i^{*}(\boldsymbol{r})\left[-\frac{\hbar^2}{2m}\nabla^2 + U(\boldsymbol{r})\right] w_j(\boldsymbol{r})\, d\boldsymbol{r}, \tag{3.5.11}$$

with $t_{ii} = t_0$. This t_{ij}, which is called the *transfer energy*, is related to the energy of the Bloch state $\varepsilon_{\boldsymbol{k}}$ as

$$\varepsilon_{\boldsymbol{k}} = \sum_{j} t_{ij}\, e^{-i\boldsymbol{k}(\boldsymbol{R}_i - \boldsymbol{R}_j)}, \tag{3.5.12}$$

$$t_{ij} = \frac{1}{N} \sum_{k}^{1BZ} \varepsilon_{k}\, e^{i\boldsymbol{k}(\boldsymbol{R}_i - \boldsymbol{R}_j)}, \tag{3.5.13}$$

as can be easily shown.

Similarly, for the Coulomb interaction between electrons we have

$$H_C = \frac{1}{2} \sum_{i,j,i',j',\sigma,\sigma'} \mathrm{v}_{ij,i'j'}\, c_{i\sigma}^{+} c_{j\sigma'}^{+}\, c_{j'\sigma'}\, c_{i'\sigma}, \tag{3.5.14}$$

where we put

$$\mathrm{v}_{ij,i'j'} = \int d\boldsymbol{r} d\boldsymbol{r}'\; w_i^*(\boldsymbol{r})\, w_j^*(\boldsymbol{r}')\, \mathrm{v}(\boldsymbol{r} - \boldsymbol{r}')\, w_{j'}(\boldsymbol{r}')\, w_{i'}(\boldsymbol{r}), \tag{3.5.15}$$

with $\mathrm{v}(\boldsymbol{r} - \boldsymbol{r}') = e^2/|\boldsymbol{r} - \boldsymbol{r}'|$.

Thus far, what we have done is simply to rewrite the Hamiltonian of an interacting electron system in the Wannier representation. Thus, if all the matrix elements of the Coulomb interaction are kept as they are, the new expression for the Hamiltonian, (3.5.10) and (3.5.14), will offer nothing new.

Here, a very drastic approximation is introduced. By taking into account the localized nature of the Wannier function we put

$$\mathrm{v}_{ij,i'j'} = I \delta_{ii'}\, \delta_{jj'}\, \delta_{ij}\,. \tag{3.5.16}$$

The assumption is that the interaction between electrons takes place only when two electrons are centered on the same Wannier site. With this approximation, (3.5.14) is greatly simplified, and, then, the total Hamiltonian takes the following form

$$H_H = \sum_{i,j,\sigma} t_{ij}\, c_{i\sigma}^{+} c_{j\sigma} + I \sum_{i} n_{i+} n_{i-}\,, \tag{3.5.17}$$

where $n_{i\sigma} = c_{i\sigma}^{+} c_{i\sigma}$ is the electron number operator at the state (i,σ). We call this simplified model the *Hubbard model* [3.8]. The Hubbard Hamiltonian is also often written as

$$H_H = \sum_{\boldsymbol{k},\sigma} \varepsilon_{\boldsymbol{k}}\, a_{\boldsymbol{k}\sigma}^{+} a_{\boldsymbol{k}\sigma} + I \sum_{i} n_{i+} n_{i-}\,. \tag{3.5.17'}$$

Thus, the Hubbard model may be viewed as the consequence of the simplest possible approximation of (3.5.16) in the Coulomb interaction of

(3.5.14); without the approximation of (3.5.16), (3.5.14) is the ordinary Coulomb interaction between electrons.

An often invoked justification for the approximation (3.5.16) is that while we are concerned with $3d$ electrons in a transition metal the Coulomb interaction between them is screened by $4s$ and other electrons. Note, however, that assuming $v(\boldsymbol{r}-\boldsymbol{r}') \propto \delta(\boldsymbol{r}-\boldsymbol{r}')$ in (3.5.14) does not straightforwardly lead to (3.5.16). Thus, in principle, the Hubbard model should be taken literally as a "model", as are the electron gas model and the jellium model.

Let us obtain the magnetic susceptibility for the Hubbard model with the mean field approximation. First, note that we can rewrite the interaction term of the Hubbard Hamiltonian as

$$I \sum_i n_{i+} n_{i-} = -\frac{1}{4} I \sum_i \left(n_{i+} - n_{i-}\right)^2 + \frac{1}{4} I \sum_i \left(n_{i+} + n_{i-}\right)^2. \quad (3.5.18)$$

While the magnetic part of the interaction is given by the first term on the right hand side, with the mean field approximation it is rewritten as

$$-\frac{1}{4} I \sum_{i=1}^{N} \left(n_{i+} - n_{i-}\right)^2 \cong \frac{1}{2\mu_{\mathrm{B}}} \frac{I}{N} \langle \overline{M} \rangle M_z , \quad (3.5.19)$$

where, corresponding to (3.1.9), we put

$$\overline{M} = \sum_i \left(n_{i+} - n_{i-}\right). \quad (3.5.20)$$

Then, by repeating the procedure of (3.3.3)–(3.3.8), we obtain

$$\chi = \frac{\chi_{\mathrm{P}}}{1 - \dfrac{1}{2\mu_{\mathrm{B}}^2} \dfrac{I}{N} \chi_{\mathrm{P}}} . \quad (3.5.21)$$

This result becomes identical to the result of (3.3.8), the Stoner susceptibility, if we put

$$I/N = \widetilde{V}. \quad (3.5.22)$$

As we will see later, such a correspondence holds in many cases between the electron gas model and the Hubbard model. However, there are also cases where these two models lead to vastly different conclusions, as we will see later in problem 4.7 and 4.10.

3.6 PROBLEMS WITH THE MEAN FIELD THEORY OF ITINERANT ELECTRON MAGNETISM

So far in this chapter we have shown how we can understand the magnetic properties of a metal from the mean field approximation treatment of metallic electrons. If we look into the problem more closely, however, this Stoner model fails to account for some of the observed basic experimental facts.

One of these difficulties is concerned with the temperature dependence of the spontaneous magnetization. Within the Stoner model we have

$$\overline{M}(T) = \sum_{k} \left[f(\varepsilon_{k+}) - f(\varepsilon_{k-}) \right], \tag{3.6.1}$$

where $f(\varepsilon_{k\sigma})$ is the Fermi distribution function of (3.4.18). Then, from the temperature dependence of the Fermi distribution such as given in (1.6.13) or (3.1.8), we expect, as we will explicitly show in 6.6.2,

$$M(T)/M(0) - 1 \propto T^2. \tag{3.6.2}$$

Experimentally, however, the change in magnetization is proportional to $T^{3/2}$, the same dependence observed in the case of a localized spin ferromagnet. For a review and references, see Shimizu [3.6]. This seems to imply the existence of spin wave excitations in a metallic ferromagnet. Spin waves, however, do not appear within the simple Stoner theory.

Thus, starting from the work of Herring [3.9], the possibility of having spin wave in an itinerant electron ferromagnet was actively discussed in various ways. A quite natural approach was to use the method of the equation of motion [3.10–3.12], quite analogously to the case of a Heisenberg ferromagnet as discussed in 2.5. In this approach first it was required to postulate an appropriate form of spin wave excitation operator, and then to handle with approximations the resulting equation of motion of the form of (2.5.3). Thus it was inevitable for the conclusion of such an approach to give an impression of not entirely compelling. An alternative demonstration was given by the present author in [3.14] based on the dynamical susceptibility of a metal; the spin wave spectrum was shown to be obtained from the pole of dynamical susceptibility, as we will see in Chapter 4. Unlike in the case of the above equation of motion approach, there is no room of arbitrariness in this new approach; the transverse dynamical magnetic susceptibility is uniquely defined and its pole is uniquely determined. Thus it appears much more convincing than the other one.

Neutron diffraction began to be used to study magnetism in the 1950's. Since the neutron has a magnetic moment it interacts with electrons magnetically. The presence of spin waves in a metallic ferromagnet is now firmly demonstrated by neutron diffraction. For a review and references, see Lowde and Windsor [3.15].

An early important discovery using neutron diffraction was the fact that neutrons can be magnetically scattered even for $T > T_C$ in a ferromagnetic metal [3.16]. For a localized spin ferromagnet it is not surprising to have neutrons magnetically scattered for $T > T_C$; each individual atom has a non-vanishing magnetic moment and, thus, can scatter neutrons magnetically. Let us see, however, why it is not obvious that itinerant electrons in the paramagnetic state can scatter a neutron magnetically.

In a diffraction experiment, we use neutrons with wave lengths of the order of the atomic radius or the inter-atomic distance, $a \cong 10^{-8}$ cm. Then, the velocity, v_N, of such a neutron is given in terms of the neutron mass M_N as

$$v_N \cong \hbar / (M_N a). \tag{3.6.3}$$

On the other hand, the electrons which can interact with these incident neutrons are those near the Fermi surface with wave number $k_F \sim 1/a$ and accordingly, velocity

$$v_F = \hbar k_F / m = \hbar / ma = v_N (M_N / m) \cong 2\times 10^3 v_N , \tag{3.6.4}$$

m being the electron mass.

Thus, during a neutron passage through one atomic region of volume of ~ a^3, the region is traversed by as many as ~ 10^3 electrons. Then, although each of the passing electrons has a spin in some direction, for $T > T_C$ the magnetic field sensed by the neutron will average to zero. It is, thus, difficult to explain the observed magnetic scattering of neutron for $T > T_C$ from an itinerant electron model. This difficulty posed a serious challenge to the validity of the itinerant electron model of metallic ferromagnetism and was taken as a strong evidence to support the localized electron model; for a review and references, see Marshall [3.17].

Is it not at all possible to explain the observed paramagnetic neutron scattering from an itinerant electron model? It is this naive question which first evoked the author's interest in the subject of metallic magnetism. Thus, I arrived at a positive answer to this question [3.13, 3.14]. I showed that it is the spin fluctuation, a collective excitation of electrons, that scatters neutrons in the paramagnetic state, but not the individual electron spins. The concept of the spin fluctuation in itinerant electron magnetism first appeared in this context.

However, all the basic problems in magnetism of metals were not completely resolved with the itinerant electron model. For instance, while the Curie-Weiss like temperature dependence is observed in the magnetic susceptibility of many ferromagnetic metals over a wide temperature range above T_C, the Stoner susceptibility of (3.3.15) does not predict this temperature dependence. Thus, during 1970's many efforts were made to derive the Curie-Weiss magnetic susceptibility within an itinerant electron model, most notably, by considerations involving the role of spin fluctuations [3.18, 3.19]. For a review and references concerning these recent efforts, see, for instance, Refs. [3.20–3.23].

Do we now understand all the fundamentals of metallic magnetism on the basis of the itinerant electron model? No, many important problems remain to be solved. We will face a number of these unsolved fundamental problems in this book. Even the problem of the mechanism of the temperature dependence of the magnetic susceptibility, including the Curie-Weiss law, is not yet fully understood. We will see in Chapter 6 how important a role the electron-phonon interaction plays in determining the temperature dependence of the magnetic susceptibility [3.23].

There is a fundamental problem also with the classic picture for the ferromagnetic state given in Fig.3.5. In Chapter 6 we will show that spin waves and phonons also contribute to the spin splitting of the electron energy bands; the size of their contributions is comparable to that of the Stoner exchange splitting.

In the next chapter we discuss various linear responses of a metallic electron system with particular emphasis on their relations to the magnetic properties of the system. The subject itself is a useful one but it also serves as a preparation for the discussion of the effects of the electron-phonon interaction in the succeeding chapters.

PROBLEMS FOR CHAPTER 3

3.1 Derive an explicit expression of α defined in (1.6.14) or (3.3.17) for an electron gas with the electronic density of states given by (1.3.5).

3.2 The magnetization near T_C of a ferromagnet can be determined if we know the expansion coefficients up to a_4 in (3.4.2). As for a_2, let us assume the magnetic susceptibility of the system is given by the Curie-Weiss law of (2.4.9). Then show that with $a_4 > 0$ we have

$$M(T) = \left[6(T_C - T)/Ca_4\right]^{1/2}.$$

3.3 Confirm the relations (3.5.9), (3.5.12), and (3.5.13) between the Bloch states and the Wannier states.

3.4 Derive the magnetic susceptibility in the mean field approximation with the Hubbard model given in (3.5.21).

Chapter 4

LINEAR RESPONSES OF METALLIC ELECTRONS

In this chapter we study the linear responses of a metal to external magnetic and electric fields. As for the response to a mechanical perturbation, namely, the elastic constant, we discuss it later in Chapters 5 and 7. We will be particularly interested in the responses in the ferromagnetic state of a metal. The results of this chapter may be useful by themselves but they are also essential in our later discussions on the electron-phonon interaction which is the central theme of this book.

First we introduce the Kubo theory [4.1], a general formalism for dynamical linear responses for $T > 0$. In calculating a linear response such as the magnetic susceptibility of a metal the Kubo formalism is not always the simplest approach. In calculating various linear responses in this chapter we will use a simpler alternative method.

4.1 KUBO THEORY OF LINEAR RESPONSE

4.1.1 Kubo Formula

Let the Hamiltonian of the natural motion of a many particle system be H, and that of an external perturbation such as a magnetic field be H'. The total Hamiltonian is then given as

$$H_{\text{tot}} = H + H'. \tag{4.1.1}$$

In discussing a linear response we let

$$H' = -Be^{-i\omega t}. \qquad (4.1.2)$$

Any time dependence can be realized by a superposition of time components of this form.

If the external perturbation is an oscillating magnetic field applied in the direction of the $\nu(= x, y, \text{or } z)$-axis, for example, we have

$$H' = -M_\nu H_\nu e^{-i\omega t}, \qquad (4.1.3)$$

where H_ν is the magnitude of the magnetic field and M_ν is the operator representing the ν-component of magnetization. For this case we have

$$B = M_\nu H_\nu \qquad (4.1.3')$$

in (4.1.2).

If we have the density matrix $\rho_{\text{tot}}(t)$ corresponding to the total Hamiltonian of (4.1.1), the thermal expectation of a physical quantity A is obtained by the procedure of (1.4.9). We may set

$$\rho_{\text{tot}}(t) = \rho + \rho'(t), \qquad (4.1.4)$$

corresponding to (4.1.1). The density matrix ρ is that of thermal equilibrium in the absence of the external perturbation given by (1.4.13), namely,

$$\rho = e^{-\beta(H-\mu\tilde{n})} / \Xi \equiv e^{-\beta K} / \Xi, \qquad (4.1.5)$$

where Ξ is the grand partition function of (1.4.14), and

$$K = H - \mu\tilde{n}. \qquad (4.1.6)$$

For simplicity we assume that without the external field the thermal expectation of A is zero,

$$\text{tr}(\rho A) = 0. \qquad (4.1.7)$$

We have, then,

$$\langle A \rangle = \text{tr}\big(\rho'(t)A\big) \equiv \langle A(t) \rangle. \qquad (4.1.8)$$

When the external perturbation is a magnetic field such as (4.1.3), the physical quantity we look for is the magnetization,

$$A = M_\mu. \tag{4.1.9}$$

The starting point of obtaining the density matrix required in (4.1.8) is the von Neumann equation. Here note that we can rewrite the von Neumann equation by using K of (4.1.6) as

$$i\hbar\frac{\partial}{\partial t}(\rho + \rho') = \left[H + H', \rho + \rho'\right]$$

$$= \left[K + H', \rho + \rho'\right]. \tag{4.1.10}$$

Going back to (1.4.1), from which the von Neumann equation was derived, we notice that this rewriting corresponds to measuring the one particle energy from the chemical potential. For later convenience in discussing Green's functions in Chapter 8, we use the von Neumann equation with K.

If in (4.1.10) we note that

$$i\hbar\frac{\partial}{\partial t}\rho = \left[K, \rho\right] = 0, \tag{4.1.11}$$

and neglect the second order term,

$$[H', \rho'] \cong 0,$$

by assuming that the external perturbation is weak, we have

$$i\hbar\frac{\partial}{\partial t}\rho' = \left[K, \rho'\right] + \left[H', \rho\right]. \tag{4.1.12}$$

This is the von Neumann equation in the linear approximation. If we can solve this equation for ρ', then we can calculate the thermal expectation of A, the linear response, by the procedure of (4.1.8).

In solving (4.1.12), we introduce the interaction representation of the density matrix as

$$\rho'_{\mathrm{I}}(t) = e^{iKt/\hbar}\rho'(t)e^{-iKt/\hbar}. \tag{4.1.13}$$

After differentiating both sides of the above equation with respect to t,

$$i\hbar\frac{\partial}{\partial t}\rho'_{\mathrm{I}}(t) = -\left[K, \rho'_{\mathrm{I}}(t)\right] + e^{iKt/\hbar}\left(i\hbar\frac{\partial}{\partial t}\rho'\right)e^{-iKt/\hbar},$$

we put (4.1.12) into the second term on the right hand side to obtain

$$i\hbar \frac{\partial}{\partial t} \rho'_{\mathrm{I}}(t) = e^{iKt/\hbar} \left[H', \rho\right] e^{-iKt/\hbar} \equiv k(t). \tag{4.1.14}$$

This differential equation is readily solved as

$$\rho'_{\mathrm{I}}(t) = \frac{1}{i\hbar} \int_{-\infty}^{t} k(t')\, dt' + c\,, \tag{4.1.15}$$

where c is a constant. We assume that at the infinite past the system was without the external perturbation and, therefore, in an equilibrium state:

$$H' = 0 \qquad \text{for} \quad t = -\infty, \tag{4.1.16}$$

and, accordingly,

$$\rho'(t) = \rho'_{\mathrm{I}}(t) = 0 \qquad \text{for} \quad t = -\infty. \tag{4.1.17}$$

Thus we have $c = 0$ in (4.1.15), and then by noting (4.1.13) we arrive at the following explicit expression for the perturbed part of the density matrix,

$$\begin{aligned} \rho'(t) &= \frac{1}{i\hbar} e^{-iKt/\hbar} \int_{-\infty}^{t} dt'\, k(t')\, e^{iKt/\hbar} \\ &= \frac{1}{i\hbar} \int_{-\infty}^{t} e^{-iK(t-t')/\hbar} \left[H'(t'), \rho\right] e^{iK(t-t')/\hbar}\, dt'. \end{aligned} \tag{4.1.18}$$

By putting this result into (4.1.8) we finally obtain the linear response

$$\langle A(t) \rangle = \frac{1}{i\hbar} \int_{-\infty}^{t} \mathrm{tr}\left(e^{-iK(t-t')/\hbar} \left[H'(t'), \rho\right] e^{iK(t-t')/\hbar} A\right) dt'. \tag{4.1.19}$$

This is the essence of the Kubo theory.

Let us rewrite the result of (4.1.19) in a simpler form. If the external perturbation has the time dependence of (4.1.2), the linear response to it also will have the same form of time dependence,

$$\langle A(t) \rangle = \langle A(\omega) \rangle e^{-i\omega t}. \tag{4.1.20}$$

If we put this and (4.1.2) into (4.1.19), and put $t - t' = \tau$, we have

$$\langle A(\omega) \rangle = \frac{i}{\hbar} \int_0^\infty e^{i\omega\tau} \operatorname{tr}\left(\rho\,[A_{\rm I}(\tau), B]\right) d\tau, \tag{4.1.21}$$

where we used the cyclic property of the trace, (3.4.8). While $A_{\rm I}(\tau)$ is in the interaction representation defined in (4.1.13), from now on we write it simply as $A(\tau)$. Since H' does not appear in (4.1.21) at all, we may view $A(\tau)$ as the Heisenberg representation with respect to the natural motion described by K. Let us define the thermal expectation in the absence of an external perturbation as

$$\operatorname{tr}(\rho C) = {}_0\langle C \rangle . \tag{4.1.22}$$

Then we can rewrite (4.1.21) in the following final form

$$\langle A(\omega) \rangle = \frac{i}{\hbar} \int_0^\infty d\tau \; e^{i\omega\tau} \, {}_0\langle [A(\tau), B] \rangle . \tag{4.1.23}$$

Note that nowhere does H' appear in the above expression. The linear response is determined from a correlation function of the fluctuations of the relevant quantities in the absence of the external perturbation.

4.1.2 Kubo Formula for the Magnetic Susceptibility

Let us derive an explicit expression for the magnetic susceptibility. When an oscillating magnetic field is applied in the ν-direction as in (4.1.3), the thermal expectation of (4.1.9), magnetization in the μ-direction, is obtained from (4.1.23) as

$$\langle M_\mu(\omega) \rangle = \frac{i}{\hbar} \int_0^\infty d\tau \; e^{i\omega\tau} \, {}_0\langle [M_\mu(\tau), M_\nu] \rangle H_\nu . \tag{4.1.24}$$

Defining the dynamic magnetic susceptibility as

$$\chi_{\mu\nu}(\omega) = \frac{\langle M_\mu(\omega)\rangle}{H_\nu}, \tag{4.1.25}$$

we have

$$\chi_{\mu\nu}(\omega) = \frac{i}{\hbar}\int_0^\infty d\tau\; e^{i\omega\tau}\,{}_0\langle [M_\mu(\tau), M_\nu]\rangle. \tag{4.1.26}$$

This is the Kubo formula for the dynamic magnetic susceptibility.

Let us proceed to obtain the magnetic response to a magnetic field oscillating in space, as well as in time,

$$H_\nu(\boldsymbol{r}, t) = H_\nu(\boldsymbol{q})\, e^{i\boldsymbol{q}\boldsymbol{r}}\, e^{-i\omega t}. \tag{4.1.27}$$

This gives rise to the following perturbation Hamiltonian,

$$H' = -\int M_\nu(\boldsymbol{r})\, H_\nu(\boldsymbol{r}, t)\, d\boldsymbol{r} = -\,M_\nu(-\boldsymbol{q})\, H_\nu(\boldsymbol{q})\, e^{-i\omega t}, \tag{4.1.28}$$

in place of (4.1.3). We defined the spatial Fourier transform of magnetization as (cf.(1.5.29)),

$$M_\nu(\boldsymbol{q}) = \int M_\nu(\boldsymbol{r})\, e^{-i\boldsymbol{q}\boldsymbol{r}}\, d\boldsymbol{r}. \tag{4.1.29}$$

The inverse transformation, then, is given as

$$M_\nu(\boldsymbol{r}) = \frac{1}{V}\sum_{\boldsymbol{q}} M_\nu(\boldsymbol{q})\, e^{i\boldsymbol{q}\boldsymbol{r}}, \tag{4.1.29'}$$

where the sum over $\boldsymbol{q}$ extends to all wave vector satisfying the periodic boundary condition of (1.1.7).

The linear response to such magnetic field will take the following form,

$$\langle M_\mu(\boldsymbol{r}, t)\rangle = \frac{1}{V}\langle M_\mu(\boldsymbol{q}, \omega)\rangle e^{i\boldsymbol{q}\boldsymbol{r}}\, e^{-i\omega t}. \tag{4.1.30}$$

Thus, substituting

$$M_\mu(\omega) \Rightarrow \frac{1}{V} M_\mu(\boldsymbol{q}, \omega)\, e^{i\boldsymbol{q}\boldsymbol{r}},$$

on the left hand side of (4.1.24), and

$$M_\mu(\tau) \Rightarrow M_\mu(\boldsymbol{r}, \tau) = \frac{1}{V} M_\mu(\boldsymbol{q}, \tau)\, e^{i\boldsymbol{q}\boldsymbol{r}},$$

$$M_\nu H_\nu \Rightarrow M_\nu(-\boldsymbol{q}) H_\nu(\boldsymbol{q}),$$

on the right hand side, we obtain

$$\big\langle M_\mu(\boldsymbol{q}, \omega)\big\rangle = \frac{i}{\hbar} \int_0^\infty d\tau\; e^{i\omega\tau}\; {}_0\big\langle [M_\mu(\boldsymbol{q}, \tau), M_\nu(-\boldsymbol{q})]\big\rangle H_\nu(\boldsymbol{q}). \tag{4.1.31}$$

If we define the magnetic susceptibility representing the response to the magnetic field of (4.1.27) as

$$\chi_{\mu\nu}(\boldsymbol{q}, \omega) = \frac{\big\langle M_\mu(\boldsymbol{q},\omega)\big\rangle}{H_\nu(\boldsymbol{q})}, \tag{4.1.32}$$

we obtain

$$\chi_{\mu\nu}(\boldsymbol{q}, \omega) = \frac{i}{\hbar} \int_0^\infty d\tau\, e^{i\omega\tau}\,{}_0\big\langle [M_\mu(\boldsymbol{q}, \tau), M_\nu(-\boldsymbol{q})]\big\rangle. \tag{4.1.33}$$

In (4.1.33), the integrand on the right hand side contains a kind of correlation function of magnetization density. Note that here the magnetization is that of thermal fluctuation around the equilibrium in the absence of external magnetic field, namely,

$${}_0\big\langle M_\mu(\boldsymbol{q}, \tau)\big\rangle = {}_0\big\langle M_\nu(-\boldsymbol{q})\big\rangle = 0. \tag{4.1.34}$$

An important point to note is that in order to satisfy the initial condition of (4.1.16), we set

$$\omega \Rightarrow \omega + i0^+ \tag{4.1.35}$$

in (4.1.2) and, accordingly, in the results such as (4.1.23) and (4.1.33). Here 0^+ stands for a positive infinitesimal.

Note that we can rewrite the results of (4.1.31) and (4.1.32) as

$$\langle M_\mu(\boldsymbol{r}, t)\rangle = \int_V d\boldsymbol{r}' \int_{-\infty}^{t} dt'\, \chi_{\mu\nu}(\boldsymbol{r}-\boldsymbol{r}', t-t')\, H_\nu(\boldsymbol{r}', t')\,. \tag{4.1.36}$$

The meaning of $\chi_{\mu\nu}(\boldsymbol{q}, t)$ may be obvious; $\chi_{\mu\nu}(\boldsymbol{q}, \omega)$ is the Fourier transform of $\chi_{\mu\nu}(\boldsymbol{r}, t)$. Because $\chi_{\mu\nu}(\boldsymbol{r}, t)$ has to be real, we have the following relation for its Fourier transform,

$$\chi_{\mu\nu}(-\boldsymbol{q}, -\omega) = \chi_{\mu\nu}{}^*(\boldsymbol{q}, \omega). \tag{4.1.37}$$

Here * stands for the complex conjugate. Note also that from the assumption of causality we require [4.2, 4.3]

$$\chi_{\mu\nu}(\boldsymbol{r}-\boldsymbol{r}', t-t') = 0 \quad \text{for} \quad t < t'. \tag{4.1.38}$$

4.1.3 The Fluctuation-Dissipation Theorem

We have seen that a linear response function is given in the form

$$\chi_{AB}(\boldsymbol{q}, \omega) = \frac{i}{\hbar}\int_0^{\infty} d\tau\, e^{i\omega\tau}{}_0\langle [A(\boldsymbol{q}, \tau), B(-\boldsymbol{q})]\rangle\,. \tag{4.1.39}$$

We now show that this response function can be related to an ordinary correlation function, ${}_0\langle A(\boldsymbol{q}, \tau)\, B(-\boldsymbol{q})\rangle$.

By extending the interval of integration to $(-\infty, \infty)$ in (4.1.39), we define a function as

$$f_{AB}(\boldsymbol{q}, \omega) = \frac{i}{\hbar}\int_{-\infty}^{\infty} d\tau\, e^{i\omega\tau}{}_0\langle [A(\boldsymbol{q}, \tau), B(-\boldsymbol{q})]\rangle$$

$$= \frac{i}{\hbar}\int_0^{\infty} d\tau\, e^{i\omega\tau}{}_0\langle [A(\boldsymbol{q}, \tau), B(-\boldsymbol{q})]\rangle + \frac{i}{\hbar}\int_0^{\infty} d\tau\, e^{-i\omega\tau}{}_0\langle [A(\boldsymbol{q}), B(-\boldsymbol{q}, \tau)]\rangle\,. \tag{4.1.40}$$

In obtaining the second term in the second line, we changed τ to $-\tau$, and moved the factors $\exp(\pm\, iK\tau/\hbar)$ around $A(\boldsymbol{q})$ to around $B(-\boldsymbol{q})$ by employing the cyclic property of trace. If we note (4.1.39), this function is related to the linear response functions as

$$f_{AB}(\boldsymbol{q},\omega) = \chi_{AB}(\boldsymbol{q},\omega) - \chi_{BA}(-\boldsymbol{q},-\omega)$$

$$= \chi_{AB}(\boldsymbol{q},\omega) - \chi_{BA}{}^*(\boldsymbol{q},\omega)\,, \qquad (4.1.41)$$

where we used the relation

$$\chi_{AB}(-\boldsymbol{q},-\omega) = \chi_{AB}{}^*(\boldsymbol{q},\omega)\,, \qquad (4.1.37')$$

which is a generalization of (4.1.37). Next, let us see how the function $f_{AB}(\boldsymbol{q}, \omega)$ is related to the ordinary correlation function.

If we note the relation

$$\int_{-\infty}^{\infty} d\tau\, e^{i\omega\tau}{}_0\langle A(\tau)B\rangle = e^{\beta\hbar\omega}\int_{-\infty}^{\infty} d\tau\, e^{i\omega\tau}{}_0\langle BA(\tau)\rangle\,, \qquad (4.1.42)$$

which we will prove below, the first line of (4.1.40) is rewritten as

$$f_{AB}(\boldsymbol{q},\omega) = (1-e^{-\beta\hbar\omega})\frac{i}{\hbar}\int_{-\infty}^{\infty} d\tau\, e^{i\omega\tau}{}_0\langle A(\boldsymbol{q},\tau)B(-\boldsymbol{q})\rangle\,. \qquad (4.1.43)$$

From (4.1.41) and (4.1.43) we arrive at the desired relation between the linear response function and the ordinary correlation function,

$$\int_{-\infty}^{\infty} dt\, e^{i\omega t}{}_0\langle A(\boldsymbol{q},t)B(-\boldsymbol{q})\rangle = \frac{i\hbar}{\left(e^{-\beta\hbar\omega}-1\right)}\left[\chi_{AB}(\boldsymbol{q},\omega) - \chi_{BA}{}^*(\boldsymbol{q},\omega)\right]. \qquad (4.1.44)$$

For $A = B$, in particular, we have

$$\int_{-\infty}^{\infty} dt\, e^{i\omega t}{}_0\langle A(\boldsymbol{q},t)A(-\boldsymbol{q})\rangle = \frac{-\hbar}{e^{-\beta\hbar\omega}-1}\,2\,\mathrm{Im}\,\chi_{AA}(\boldsymbol{q},\omega), \qquad (4.1.45)$$

where Im indicates to take the imaginary part. As for the imaginary part of the response function, if a system has the inversion symmetry, from (4.1.37) we have

$$\mathrm{Im}\,\chi_{\mu\nu}(\boldsymbol{q},-\omega) \;=\; \mathrm{Im}\,\chi_{\mu\nu}(-\boldsymbol{q},-\omega) \;=\; -\;\mathrm{Im}\,\chi_{\mu\nu}(\boldsymbol{q},\omega)\,. \tag{4.1.46}$$

Similarly, for the real part we have

$$\mathrm{Re}\,\chi_{\mu\nu}(\boldsymbol{q},-\omega) \;=\; \mathrm{Re}\,\chi_{\mu\nu}(\boldsymbol{q},\omega)\,. \tag{4.1.47}$$

The result of (4.1.44) and (4.1.45) shows how the correlation function of thermal fluctuations is related to linear response functions. Relations such as (4.1.44) or (4.1.45) are called the *fluctuation-dissipation theorem* [4.1–4.3].

Finally, let us prove the relation (4.1.42). We write the Fourier transforms of two correlation functions appearing in (4.1.42) as

$$J(\omega) \;=\; \int_{-\infty}^{\infty} dt\; e^{i\omega t}\,{}_0\langle A(t)B\rangle\,; \quad J'(\omega) \;=\; \int_{-\infty}^{\infty} dt\; e^{i\omega t}\,{}_0\langle BA(t)\rangle\,. \tag{4.1.48}$$

The inverse transforms are then given by

$${}_0\langle A(t)B\rangle \;=\; \frac{1}{2\pi}\int_{-\infty}^{\infty} d\omega\; e^{-i\omega t}J(\omega)\,;$$

$${}_0\langle BA(t)\rangle \;=\; \frac{1}{2\pi}\int_{-\infty}^{\infty} d\omega\; e^{-i\omega t}J'(\omega)\,. \tag{4.1.49}$$

These two correlation functions are related in the following way,

$$\begin{aligned}
{}_0\langle A(t)B\rangle &= \mathrm{tr}\big[e^{-\beta K}e^{iKt/\hbar}A\,e^{-iKt/\hbar}B\big]\,/\Xi \\
&= \mathrm{tr}\big[B\,e^{-\beta K}e^{iKt/\hbar}A\,e^{-iKt/\hbar}e^{\beta K}e^{-\beta K}\big]\,/\Xi \\
&= \mathrm{tr}\big[e^{-\beta K}B\,e^{iK(t+i\hbar\beta)/\hbar}A\,e^{-iK(t+i\hbar\beta)/\hbar}\big]\,/\Xi \\
&= {}_0\big\langle BA(t+i\hbar\beta)\big\rangle\,,
\end{aligned} \tag{4.1.50}$$

where we used the cyclic property of trace, (3.4.8). If we rewrite the first and the last expressions by using the relations of (4.1.48) and (4.1.49), we obtain

$$\frac{1}{2\pi}\int_{-\infty}^{\infty} d\omega\, e^{-i\omega t} J(\omega) = \frac{1}{2\pi}\int_{-\infty}^{\infty} d\omega\, e^{-i\omega(t + i\hbar\beta)} J'(\omega),$$

and, therefore,

$$J(\omega) = e^{\beta\hbar\omega} J'(\omega)\,. \tag{4.1.51}$$

This relation, together with (4.1.48) and (4.1.49), leads to (4.1.42).

Note that the relations of (4.1.44) and (4.1.45) can be rewritten as

$${}_0\langle A(\boldsymbol{q}, t)\, B(-\boldsymbol{q})\rangle = \frac{1}{2\pi}\int_{-\infty}^{\infty} d\omega\, e^{-i\omega t} \frac{i\hbar}{\left(e^{-\beta\hbar\omega} - 1\right)}\left[\chi_{AB}(\boldsymbol{q},\omega) - \chi_{BA}{}^*(\boldsymbol{q},\omega)\right] \tag{4.1.52}$$

$${}_0\langle A(\boldsymbol{q}, t)\, A(-\boldsymbol{q})\rangle = -\frac{1}{\pi}\int_{-\infty}^{\infty} d\omega\, e^{-i\omega t} \frac{\hbar}{e^{-\beta\hbar\omega} - 1}\, \mathrm{Im}\, \chi_{AA}(\boldsymbol{q},\omega)\,, \tag{4.1.53}$$

by taking the inverse Fourier transform.

4.1.4 Neutron Scattering Cross-Section and Magnetic Susceptibility

It had been shown by van Hove [4.4, 1.1] that the differential magnetic scattering cross-section per unit solid angle per unit energy range of (unpolarized) neutrons by electron spins is given in terms of the correlation function of the electron spins,

$$\begin{aligned} S_{\mu,\nu}(\kappa, \omega) &= \left\langle S_\mu(\kappa, \omega)\, S_\nu(-\kappa) \right\rangle \\ &= \frac{1}{2\pi}\int_{-\infty}^{\infty} dt\, e^{i\omega t} \left\langle S_\mu(\kappa, t)\, S_\nu(-\kappa) \right\rangle, \end{aligned} \tag{4.1.54}$$

as

$$\frac{d^2\sigma}{d\Omega\, d\omega} = \left(\frac{g_N e^2}{mc^2}\right)^2 \frac{k'}{k} |f(\kappa)|^2 \sum_{\mu,\nu} (\delta_{\mu,\nu} - \hat{\kappa}_\mu \hat{\kappa}_\nu)\, S_{\mu,\nu}(\kappa, \omega)\,. \tag{4.1.55}$$

Here $S_\mu(\kappa)$ is the Fourier transform of the μ (x, y, or z) component of electron spin density (note that $M_\mu(\kappa) = -2\mu_B S_\mu(\kappa)$; see (4.1.29)), $S_\mu(\kappa, t)$ is its Heisenberg representation, m is the electron mass, $g_N = -1.91$ is the neutron magnetic moment in nuclear magnetons, $\boldsymbol{k}$ and $\boldsymbol{k}' = \boldsymbol{k} - \kappa$ are the initial and final wave vectors of the neutron, $\hat{\kappa}_\mu = \kappa_\mu/\kappa$, and $\hbar\omega = \hbar^2k^2/2M_N - \hbar^2k'^2/2M_N$,
M_N being the neutron mass. Finally $f(\kappa)$ is what corresponds to the atomic form factor in the localized spin system; it comes out from that the wave functions of metallic electrons is not plane waves but Bloch functions. In the simple approximation of neglecting the overlap of neighboring Wannier functions (3.5.4) we have

$$f(\kappa) = \int |w_0^*(\boldsymbol{r})|^2 e^{-i\kappa \boldsymbol{r}} d^3\boldsymbol{r}\,. \tag{4.1.56}$$

Here we note that the spin correlation function in the expression for the neutron scattering cross-section can be rewritten in terms of the imaginary part of magnetic susceptibility by using the fluctuation-dissipation theorem as follows. First, from (4.1.44) we have

$$\left\langle S_\mu(\kappa, \omega) S_\nu(-\kappa) \right\rangle$$
$$= \frac{i\hbar}{e^{-\beta\hbar\omega} - 1} \frac{1}{(g\mu_B)^2} \left[\chi_{\mu\nu}(\kappa, \omega) - \chi_{\nu\mu}^*(\kappa, \omega)\right], \tag{4.1.57}$$

where $g \cong 2$, and $\chi_{\mu,\nu}(\kappa, \omega)$ is defined in (4.1.33). This relation is symmetrized with respect to μ and ν as

$$\frac{1}{2}\left\{\left\langle S_\mu(\kappa, \omega) S_\nu(-\kappa) \right\rangle + \left\langle S_\nu(\kappa, \omega) S_\mu(-\kappa) \right\rangle\right\}$$

$$= \frac{i\hbar}{e^{-\beta\hbar\omega} - 1} \frac{1}{(g\mu_B)^2} \frac{1}{2}\left\{\chi_{\mu\nu}(\kappa, \omega) - \chi_{\nu\mu}^*(\kappa, \omega) + \chi_{\nu\mu}(\kappa, \omega) - \chi_{\mu\nu}^*(\kappa, \omega)\right\}$$

$$= -\frac{1}{e^{-\beta\hbar\omega} - 1} \frac{\hbar}{(g\mu_B)^2} \left\{\mathrm{Im}\,\chi_{\mu\nu}(\kappa, \omega) + \mathrm{Im}\,\chi_{\nu\mu}(\kappa, \omega)\right\}. \tag{4.1.58}$$

By putting (4.1.58) into (4.1.54)–(4.1.55) we arrive at the following result [3.13, 3.14],

$$\frac{d^2\sigma}{d\Omega\, d\omega} = \left(\frac{-1.91\, e}{\hbar c}\right)^2 |f(\kappa)|^2 \frac{k'}{k} \frac{2\hbar}{1 - e^{-\beta\hbar\omega}}$$

$$\times \sum_{\mu,\nu} \left(\delta_{\mu,\nu} - \widehat{\kappa}_\mu \widehat{\kappa}_\nu\right) \operatorname{Im} \chi_{\mu\nu}(\kappa, \omega) \tag{4.1.59}$$

where we inserted the explicit expression for μ_B, (2.1.5).

Thus the neutron scattering cross-section is directly related to the imaginary part of the dynamical susceptibility of the system from which the neutron is magnetically scattered. This result, which I first found, is especially useful in dealing with inelastic scattering since $\operatorname{Im} \chi_{\mu,\nu}(\kappa, \omega)$ represents the very spectral density of the magnetic excitations of the system; see, for instance, 4.4 and 6.5 concerning spin waves.

4.2 LINEAR RESPONSES OF FREE ELECTRONS

The linear response of free electrons to a spatially uniform ($\boldsymbol{q} = 0$), static ($\omega = 0$) magnetic field is the Pauli susceptibility, as we saw in 3.1. In this section we obtain the magnetic response of free electrons to a magnetic field of the form of (4.1.27), changing both in space and time. We also obtain the linear response of free electrons to a charge potential oscillating both in space and time.

In obtaining a linear response, it is not always most convenient to use the Kubo formula. Here we use a method which is more elementary than, but equivalent to the Kubo formula. As for the Kubo formula, we will show how to use it in Chapter 9.

As we saw in the preceding section, if we know the density matrix of (4.1.4) corresponding to the total Hamiltonian of (4.1.1) including that of time dependent external perturbation, the time dependent thermal expectation of a physical quantity A is calculated as follows,

$$\langle A(t) \rangle = \operatorname{tr}(\rho_{\text{tot}}(t) A). \tag{4.2.1}$$

We differentiate both sides of this equation with respect to t to obtain

$$i\hbar \frac{\partial}{\partial t} \langle A(t) \rangle = \operatorname{tr}\left\{\left(i\hbar \frac{\partial}{\partial t} \rho_{\text{tot}}\right) A\right\}. \tag{4.2.2}$$

Since A is an operator in the Schrödinger representation, it does not depend on t. If we note the von Neumann equation of (1.4.10) or (4.1.10), (4.2.2) is rewritten as

$$i\hbar \frac{\partial}{\partial t}\langle A(t)\rangle = \mathrm{tr}\{[H_{\mathrm{tot}}, \rho_{\mathrm{tot}}]A\} = \mathrm{tr}\{\rho_{\mathrm{tot}}[A, H_{\mathrm{tot}}]\}$$

$$= \langle [A, H_{\mathrm{tot}}] \rangle, \tag{4.2.3}$$

where we have used the cyclic property of trace, (3.4.8).

If the external perturbation has the time dependence of the form of (4.1.2), the linear response will also have the time dependence of (4.1.20), as we noted in deriving the Kubo formula. Thus, (4.2.3) can be rewritten as

$$\hbar\omega\langle A(t)\rangle \equiv \hbar\omega\langle A(\omega)\rangle e^{-i\omega t}$$

$$= \langle [A, H]\rangle + \langle [A, H'] \rangle \cong \langle [A, H]\rangle + {}_{\mathrm{o}}\langle [A, H'] \rangle, \tag{4.2.4}$$

within the linear response approximation. We often abbreviate $\langle A(t)\rangle$ by $\langle A\rangle$. Note that the last term in the above equation represents the thermal expectation in the absence of the external perturbation. Note also that here ω is to be understood as $\omega + i0^+$ by the same reason as in Kubo theory (see (4.1.35)). In the remainder of this chapter we obtain various linear responses by applying (4.2.4).

4.2.1 The Magnetic Susceptibility of Free Electrons $\chi^{\mathrm{o}}_{zz}(\boldsymbol{q}, \omega)$: The Lindhard Function

We consider an electron gas enclosed in a box of volume $V = L^3$ as in the preceding chapters. If a magnetic field of the form of (4.1.27) is applied in the direction of the z-axis, according to (4.1.28) the perturbation Hamiltonian is given by

$$H'_{\mathrm{m}} = -M_z(-\boldsymbol{q})H_z(\boldsymbol{q})e^{-i\omega t}, \tag{4.2.5}$$

where $M_z(\boldsymbol{q})$ is the Fourier component of the magnetization which is given explicitly as

$$M_z(\boldsymbol{q}) = -\mu_{\mathrm{B}}\sigma_z(\boldsymbol{q}), \tag{4.2.6}$$

$$\sigma_z(\boldsymbol{q}) = \sum_{\sigma,\sigma'} \int \psi^{+}_{\sigma}(\boldsymbol{r})\sigma_z\psi_{\sigma'}(\boldsymbol{r})e^{-i\boldsymbol{q}\boldsymbol{r}}\,d\boldsymbol{r}$$

$$= \sum_{\boldsymbol{k}} a^{+}_{\boldsymbol{k}+}a_{\boldsymbol{k}+\boldsymbol{q},+} - \sum_{\boldsymbol{k}} a^{+}_{\boldsymbol{k}-}a_{\boldsymbol{k}+\boldsymbol{q},-} = n_+(\boldsymbol{q}) - n_-(\boldsymbol{q}). \tag{4.2.7}$$

The above notation is the same as in Chapters 1 and 3; see (3.1.2), in particular.

Note that in this subsection, since we are neglecting the Coulomb interaction between electrons, H in (4.2.4) is simply the one particle energy of electrons, H_0. Note also that it does not matter whether we use H or K in (4.2.3) and (4.2.4).

In obtaining the thermal expectation of $A = M_z(\boldsymbol{q})$ by using (4.2.4), we need to know $\langle a_{k\sigma}^{+} a_{k+q,\sigma} \rangle$. The equation to be solved, then, is

$$\hbar\omega\langle a_{k\sigma}^{+} a_{k+q,\sigma} \rangle = \langle [a_{k\sigma}^{+} a_{k+q,\sigma}, H_0] \rangle + {}_0\langle [a_{k\sigma}^{+} a_{k+q,\sigma}, H'_m] \rangle. \tag{4.2.8}$$

By using (1.5.5), the commutators involved are calculated as,

$$[a_{k\sigma}^{+} a_{k+q,\sigma}, H_0] = [a_{k\sigma}^{+} a_{k+q,\sigma}, \sum_{l,\sigma'} \varepsilon_l a_{l\sigma'}^{+} a_{l\sigma'}]$$

$$= \left(\varepsilon_{k+q} - \varepsilon_k\right) a_{k\sigma}^{+} a_{k+q,\sigma}, \tag{4.2.9}$$

$$[a_{k\sigma}^{+} a_{k+q,\sigma}, \sum_l a_{l\sigma'}^{+} a_{l-q,\sigma'}] = \left(a_{k\sigma}^{+} a_{k\sigma} - a_{k+q,\sigma}^{+} a_{k+q,\sigma}\right) \delta_{\sigma,\sigma'}, \tag{4.2.10}$$

and, accordingly,

$$[a_{k\sigma}^{+} a_{k+q,\sigma}, H'_m] = \sigma\mu_B H_z(\boldsymbol{q}) e^{-i\omega t}\left(a_{k\sigma}^{+} a_{k\sigma} - a_{k+q,\sigma}^{+} a_{k+q,\sigma}\right). \tag{4.2.11}$$

It is left as an exercise for the reader to derive the results of (4.2.9)–(4.2.11) by using the commutation relations for fermion of (1.5.5).

If we put (4.2.9) and (4.2.11) into the right hand side of (4.2.8), we have

$$\hbar\omega\langle a_{k\sigma}^{+} a_{k+q,\sigma} \rangle = \left(\varepsilon_{k+q} - \varepsilon_k\right)\langle a_{k\sigma}^{+} a_{k+q,\sigma} \rangle$$

$$+ \sigma\mu_B H_z(\boldsymbol{q}) e^{-i\omega t}\left({}_0\langle a_{k\sigma}^{+} a_{k\sigma} \rangle - {}_0\langle a_{k+q,\sigma}^{+} a_{k+q,\sigma} \rangle\right). \tag{4.2.12}$$

Then, if we note that ${}_0\langle a_{k\sigma}^{+} a_{k\sigma} \rangle$ is the Fermi distribution of (1.6.5), it is straightforward to obtain,

$$\langle M_z(\boldsymbol{q}) \rangle = -\mu_B\left[\sum_k \langle a_{k+}^{+} a_{k+q,+} \rangle - \sum_k \langle a_{k-}^{+} a_{k+q,-} \rangle\right]$$

$$= 2F(\boldsymbol{q}, \omega)\mu_B^2 H_z(\boldsymbol{q}) e^{-i\omega t}, \tag{4.2.13}$$

where we introduced the *Lindhard function* as

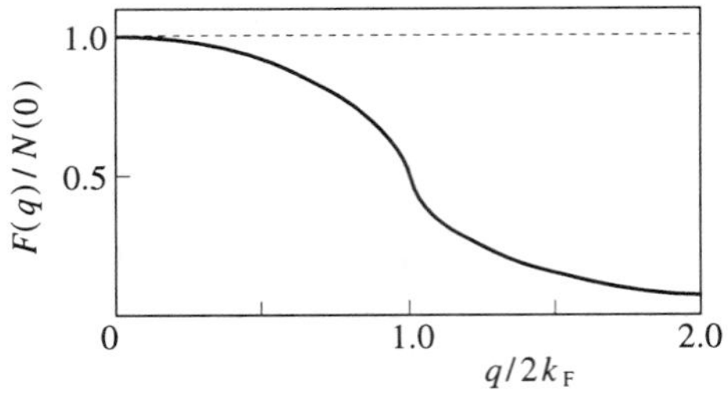

Fig.4.1 The static Lindhard function of a free electron gas at $T = 0$.

$$F(\boldsymbol{q},\,\omega) \;=\; \sum_{\boldsymbol{k}} \frac{f(\varepsilon_{\boldsymbol{k}}) - f(\varepsilon_{\boldsymbol{k+q}})}{\varepsilon_{\boldsymbol{k+q}} - \varepsilon_{\boldsymbol{k}} - \hbar\omega \;-\; i0^{+}}\;, \tag{4.2.14}$$

where we noted the requirement, (4.1.35). The dynamic Lindhard function should always be understood as in (4.2.14) with an infinitesimal imaginary constant in the denominator, although we often omit it. If we let

$$\langle M_z(\boldsymbol{q})\rangle \;=\; \langle M_z(\boldsymbol{q},\,\omega)\rangle e^{-i\omega t}, \tag{4.2.15}$$

as in (4.1.20), we obtain the magnetic susceptibility defined in (4.1.32) as

$$\chi^{\rm o}_{zz}(\boldsymbol{q},\,\omega) \;=\; 2\,\mu_{\rm B}{}^2 F(\boldsymbol{q},\,\omega). \tag{4.2.16}$$

In the paramagnetic state of an electron gas, since the system is isotropic we have

$$\chi^{\rm o}_{xx}(\boldsymbol{q},\,\omega) \;=\; \chi^{\rm o}_{yy}(\boldsymbol{q},\,\omega) \;=\; \chi^{\rm o}_{zz}(\boldsymbol{q},\,\omega). \tag{4.2.17}$$

The Lindhard function plays a central role in describing the behavior of electrons in a metal. It will appear repeatedly hereafter in this book. We summarized in Appendix A the basic properties of the Lindhard function for an electron gas. It is not a simple task to calculate the Lindhard function for $T > 0$, $\boldsymbol{q} \neq 0$ and $\omega \neq 0$ for a real metal.

The most important property of the Lindhard function is

$$\lim_{q\to 0} F(\boldsymbol{q},\,0) \;\equiv\; \lim_{q\to 0} F(\boldsymbol{q}) \;=\; \sum_{\boldsymbol{k}} -\frac{\partial f(\varepsilon_{\boldsymbol{k}})}{\partial \varepsilon_{\boldsymbol{k}}}$$

$$=\; F(0) \;\cong\; N(0), \tag{4.2.18}$$

as we saw in (1.6.14). Thus, fundamentally the Lindhard function $F(0)$ represents the electronic density of states at the Fermi surface. We call $F(\boldsymbol{q})$ the static Lindhard function.

For $\omega = 0$ and $\boldsymbol{q} = 0$, the result of (4.2.16) reduces to

$$\lim_{q \to 0} \chi^{0}_{zz}(\boldsymbol{q}, 0) \cong 2\mu_{\mathrm{B}}^{2} N(0). \tag{4.2.19}$$

This is the Pauli susceptibility of (3.1.8). Here note that since the total magnetization is an extensive quantity proportional to the size of a system, so is the magnetic susceptibility. Thus, in dealing with susceptibility data we normalize it either by volume or by molar weight.

For a free electron gas, the static Lindhard function at $T = 0$ is obtained readily as

$$F(\boldsymbol{q}) = N(0)\left[\frac{1}{2} + \frac{1-x^2}{4x}\ln\left|\frac{1+x}{1-x}\right|\right] \tag{4.2.20}$$

with $x = q/2k_{\mathrm{F}}$. The behavior of the free electron static Lindhard function is illustrated in Fig.4.1. As can be seen from (4.2.20), the Lindhard function has a logarithmically diverging derivative at $q = 2k_{\mathrm{F}}$.

4.2.2 RKKY-Friedel Oscillation

As an application of the result of the preceding subsection let us calculate the magnetization to be induced in an electron gas by a static magnetic field of the delta function type applied at the origin of the coordinates,

$$H_z(\boldsymbol{r}) = \left(\frac{V}{N}\right) H_z\,\delta(\boldsymbol{r}), \tag{4.2.21}$$

N being the total number of atoms. This represents a situation that a magnetic field H_z is applied within one atomic volume, V/N, around the origin. If we note (1.5.25b), this magnetic field is Fourier decomposed as

$$H_z(\boldsymbol{r}) = \sum_{\boldsymbol{q}} H_z(\boldsymbol{q})\, e^{i\boldsymbol{qr}} = \sum_{\boldsymbol{q}} \left(\frac{H_z}{N}\right) e^{i\boldsymbol{qr}}. \tag{4.2.22}$$

The induced magnetization is, then, given as

$$\langle M_z(\boldsymbol{r})\rangle = \frac{1}{V}\sum_{\boldsymbol{q}} \langle M_z(\boldsymbol{q})\rangle e^{i\boldsymbol{qr}}$$

$$= \frac{1}{V}\sum_{\boldsymbol{q}} \chi^{o}_{zz}(\boldsymbol{q},\,0)\,H_z(\boldsymbol{q})\; e^{i\boldsymbol{qr}} \;=\; \frac{2\mu_B^2 H_z}{VN}\sum_{\boldsymbol{q}} F(\boldsymbol{q})\, e^{i\boldsymbol{qr}}. \qquad (4.2.23)$$

If we carry out the summation over $\boldsymbol{q}$ in the final expression of (4.2.23) by putting in (4.2.20), we obtain

$$\frac{1}{V}\sum_{\boldsymbol{q}} F(\boldsymbol{q})\, e^{i\boldsymbol{qr}} \;=\; 6\pi\frac{n}{V}N(0)\,\frac{\sin(2k_F r) - 2k_F r\cos(2k_F r)}{(2k_F r)^4}$$

$$\equiv\; F(\boldsymbol{r}), \qquad (4.2.24)$$

where n is the total number of electrons in the system.

The result of (4.2.24) is illustrated in Fig.4.2, where we assumed the values $k_F = 0.5 \times 10^8$/cm and $H_z = N/(2N(0)\mu_B{}^2)$. A delta function form for the magnetic field produces an oscillating magnetization within metallic electrons which is called the *RKKY oscillation*. This phenomenon was first noted theoretically in connection with the conduction electron spin polarization due to interaction with a nuclear spin by Ruderman and Kittel [4.5], and, later with an atomic magnetic moment by Kasuya [4.6] and Yosida [4.7].

As we will see shortly, an electric charge potential of the delta function type produces also an oscillating electron charge polarization of the form of (4.2.24). Since it was first noted by Friedel [4.8], we call it the *Friedel oscillation*.

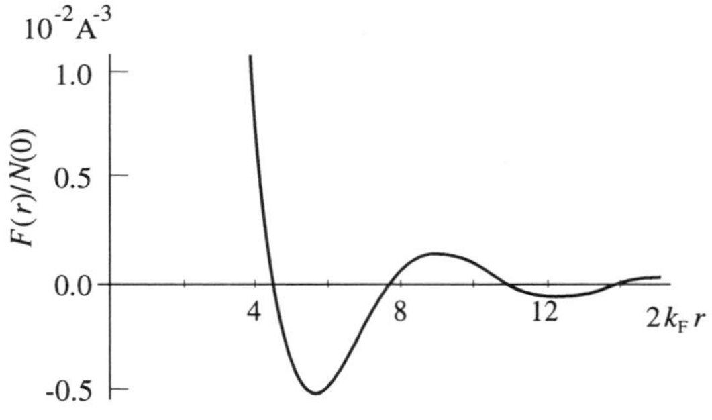

Fig.4.2 The RKKY-Friedel oscillation of the spin or charge densities in a free electron gas. We assume $k_F = 0.5 \times 10^8$/cm throughout in Figs.4.2–4.9.

4.2.3 Electric Charge Susceptibility of Free Electrons: $\chi^{\mathrm{o}}_{\mathrm{ee}}(\boldsymbol{q}, \omega)$

In this subsection we study how metallic electrons respond to an external charge potential. Here we first neglect the Coulomb interaction between electrons, as in the preceding subsection.

If we have an external charge potential of the form of

$$U(\boldsymbol{r}, t) = U(\boldsymbol{q})\, e^{i\boldsymbol{q}\boldsymbol{r}}\, e^{-i\omega t}, \qquad (4.2.25)$$

the corresponding perturbation Hamiltonian is obtained, by the procedure of 1.5 (see (1.5.27)), to be

$$H'_{\mathrm{e}} = -\, e\, U(\boldsymbol{q})\, n(-\boldsymbol{q})\, e^{-i\omega t}, \qquad (4.2.26)$$

where

$$n(\boldsymbol{q}) = n_{+}(\boldsymbol{q}) + n_{-}(\boldsymbol{q}) = \sum_{\boldsymbol{k}} \left(a^{+}_{\boldsymbol{k}+}\, a_{\boldsymbol{k}+\boldsymbol{q},+} + a^{+}_{\boldsymbol{k}-}\, a_{\boldsymbol{k}+\boldsymbol{q},-} \right) \qquad (4.2.27)$$

is the Fourier component of the electron density (see (1.5.29)–(1.5.30)).

The perturbed electron density polarization to be induced by the external perturbation of (4.2.26) will have the following dependence on space and time,

$$\langle n(\boldsymbol{q}) \rangle\, e^{i\boldsymbol{q}\boldsymbol{r}} = \langle n(\boldsymbol{q}, \omega) \rangle\, e^{i\boldsymbol{q}\boldsymbol{r} - i\omega t}. \qquad (4.2.28)$$

We can calculate this response by using the procedure of (4.2.4) in the same way that we calculated the magnetic susceptibility in the preceding subsection. In setting up the equation of motion for the same operator $a^{+}_{\boldsymbol{k}\sigma}\, a_{\boldsymbol{k}+\boldsymbol{q},\sigma}$ the only change to be made in (4.2.8) is to replace H'_{m} by H'_{e} in the last term;

$$\hbar\omega \langle a^{+}_{\boldsymbol{k}\sigma} a_{\boldsymbol{k}+\boldsymbol{q},\sigma} \rangle = \langle [a^{+}_{\boldsymbol{k}\sigma} a_{\boldsymbol{k}+\boldsymbol{q},\sigma}\, , H_{\mathrm{o}}] \rangle + {}_{\mathrm{o}}\langle [a^{+}_{\boldsymbol{k}\sigma} a_{\boldsymbol{k}+\boldsymbol{q},\sigma}\, , H'_{\mathrm{e}}] \rangle . \qquad (4.2.29)$$

The new commutator is calculated, using (4.2.10), to be

$$\left[a^{+}_{\boldsymbol{k}\sigma} a_{\boldsymbol{k}+\boldsymbol{q},\sigma}\, , H'_{\mathrm{e}} \right] = -\, e\, U(\boldsymbol{q})\, e^{-i\omega t} \left(a^{+}_{\boldsymbol{k}\sigma} a_{\boldsymbol{k}\sigma} - a^{+}_{\boldsymbol{k}+\boldsymbol{q},\sigma} a_{\boldsymbol{k}+\boldsymbol{q},\sigma} \right) . \qquad (4.2.30)$$

By putting (4.2.9) and (4.2.30) into (4.2.29), we obtain

$$\hbar\omega \langle a^{+}_{\boldsymbol{k}\sigma} a_{\boldsymbol{k}+\boldsymbol{q},\sigma} \rangle = (\varepsilon_{\boldsymbol{k}+\boldsymbol{q}} - \varepsilon_{\boldsymbol{k}}) \langle a^{+}_{\boldsymbol{k}\sigma} a_{\boldsymbol{k}+\boldsymbol{q},\sigma} \rangle$$

$$-\, e\, U(\boldsymbol{q})\, e^{-i\omega t} \left({}_{\mathrm{o}}\langle n_{\boldsymbol{k}\sigma} \rangle - {}_{\mathrm{o}}\langle n_{\boldsymbol{k}+\boldsymbol{q},\sigma} \rangle \right) , \qquad (4.2.31)$$

and, then,

$$\langle n(\boldsymbol{q})\rangle = \langle n(\boldsymbol{q}, \omega)\rangle e^{-i\omega t} = 2F(\boldsymbol{q}, \omega)\, e\, U(\boldsymbol{q})\, e^{-i\omega t}. \tag{4.2.32}$$

If we define the electron charge susceptibility as

$$\chi_{\mathrm{ee}}(\boldsymbol{q}, \omega) = \frac{-e\langle n(\boldsymbol{q}, \omega)\rangle}{-U(\boldsymbol{q})}, \tag{4.2.33}$$

for the present case of free electrons it is given by

$$\chi^{\mathrm{o}}_{\mathrm{ee}}(\boldsymbol{q}, \omega) = 2e^2 F(\boldsymbol{q}, \omega). \tag{4.2.34}$$

Apart from a constant factor, this result is identical to that of (4.2.16); when the interaction between electrons are neglected charge and spin responses are identical. The linear response of free electrons is proportional to the electronic density of states at the Fermi surface, independent of the nature of the perturbation. As we will see in the remainder of this chapter, (4.2.16) and (4.2.34) are the two most basic results in discussing the linear responses of a metallic electron system. Note, however, that when the Coulomb interaction between electrons are considered, charge and spin responses become immediately different, as we will see shortly.

By using the result of (4.2.34) let us explore how the electron density distribution will be modified when an external static point charge is put into a metallic electron system. If the magnitude of the charge is Ze and its location is at the origin of the coordinates, we have

$$U(\boldsymbol{r}) = \frac{Ze}{r} \tag{4.2.35}$$

for its potential. As for its Fourier transform, from (1.5.23) we have

$$U(\boldsymbol{q}) = \frac{1}{V}\frac{4\pi Ze}{q^2}\ ; \quad U(\boldsymbol{r}) = \sum_{\boldsymbol{q}} U(\boldsymbol{q})\, e^{i\boldsymbol{q}\boldsymbol{r}}. \tag{4.2.36}$$

According to (4.2.33), the response to this charge potential is given by

$$\begin{aligned} -e\langle n(\boldsymbol{r})\rangle &= -\frac{1}{V}\sum_{\boldsymbol{q}} \chi^{\mathrm{o}}_{\mathrm{ee}}(\boldsymbol{q}, 0)\, U(\boldsymbol{q})\, e^{i\boldsymbol{q}\boldsymbol{r}} \\ &= -\frac{1}{V}\sum_{\boldsymbol{q}} 2e^2 F(\boldsymbol{q})\, \frac{4\pi Ze}{Vq^2}\, e^{i\boldsymbol{q}\boldsymbol{r}}. \end{aligned} \tag{4.2.37}$$

We need a numerical calculation to see the spatial behavior of this perturbed part of electron density polarization. Note that when $U(\boldsymbol{q})$ is a constant, independent of $\boldsymbol{q}$, we obtain the Friedel oscillation.

Let us assume $Z > 0$. Electrons will, then, be attracted to the point charge. We can calculate the total charge of such attracted electrons as

$$-e \int_V \langle n(\boldsymbol{r}) \rangle \, d\boldsymbol{r} = -\frac{1}{V} \sum_{q} \int_V d\boldsymbol{r} \, \chi^0_{\rm ee}(\boldsymbol{q}, 0)\, U(\boldsymbol{q})\, e^{i\boldsymbol{q}\boldsymbol{r}}.$$

If here we note (1.5.25a), we have

$$-e \int_V \langle n(\boldsymbol{r}) \rangle \, d\boldsymbol{r} = -\lim_{q \to 0} \chi^0_{\rm ee}(\boldsymbol{q}, 0)\, U(\boldsymbol{q})$$

$$= -2e^2 N(0) \lim_{q \to 0} \frac{4\pi Z e}{V \boldsymbol{q}^2} = -\infty. \tag{4.2.38}$$

The total number of electrons to be attracted to the point charge is not Z but is infinite. This, of course, is an unphysical result. The calculation itself is correct. The infinite result stems from the neglect of the Coulomb interaction between electrons, as we will show in the next section.

4.2.4 The s-d or s-f Exchange Model: Rare Earth Metals and Magnetic Dilute Alloys

In connection with the RKKY spin polarization studied in 4.2.2, let us introduce here the *s-d* or *s-f exchange model* as an important model for describing the magnetic properties of rare earth metals and dilute alloys [4.5, 4.6].

Suppose a localized magnetic moment is immersed in a sea of metallic electrons. One such example may be a Mn atom in a sample of Cu metal. Of the $(3d)^5\,(4s)^2$ electrons of the Mn atom, five $3d$ electrons form a localized moment of the size $S = 5/2$; two $4s$ electrons mingle with the conduction electrons of Cu. For the interaction between the localized moment $\boldsymbol{S}$ and the spin density of the conduction electrons $\sigma(\boldsymbol{r})$, we may assume the form of

$$H_{\rm sd} = -\int J(\boldsymbol{r} - \boldsymbol{R})\, \boldsymbol{S}(\boldsymbol{R})\, \sigma(\boldsymbol{r}) \, d\boldsymbol{r}\, d\boldsymbol{R}, \tag{4.2.39}$$

where, if the site of the localized moment is $\boldsymbol{R}_i$,

$$S(\boldsymbol{R}) = \boldsymbol{S}_i\,\delta(\boldsymbol{R} - \boldsymbol{R}_i). \tag{4.2.40}$$

This interaction is called as the *s-d* (or, *s-f*) *exchange interaction.*

In the simplest case, the exchange interaction between the localized moment and the conduction electron spin can be taken to be

$$J(\boldsymbol{r} - \boldsymbol{R}) = J\frac{V}{N}\,\delta(\boldsymbol{r} - \boldsymbol{R}). \tag{4.2.41}$$

This expression is understood similarly to (4.2.21); the interaction takes place when the conduction electron is within one atomic distance from the localized spin. J thus defined has a value of the order of 1 eV. If we put (4.2.40) and (4.2.41) into (4.2.39), we have

$$H_{\mathrm{sd}} = -J\frac{V}{N}\,\boldsymbol{S}_i\,\boldsymbol{\sigma}(\boldsymbol{R}_i). \tag{4.2.42}$$

Equation (4.2.42) can be alternatively written as

$$H_{\mathrm{sd}} = -\int \{-\mu_{\mathrm{B}}\,\boldsymbol{\sigma}(\boldsymbol{r})\}\,\boldsymbol{H}_{\mathrm{d}}(\boldsymbol{r})\,d\boldsymbol{r}, \tag{4.2.43}$$

with the effective delta function type magnetic field due to the localized spin,

$$\boldsymbol{H}_{\mathrm{d}}(\boldsymbol{r}) = -\frac{V}{N}\frac{1}{\mu_{\mathrm{B}}}\,J\boldsymbol{S}_i\,\delta(\boldsymbol{r} - \boldsymbol{R}_i). \tag{4.2.44}$$

Thus, an RKKY spin polarization of the conduction electrons is induced around the localized spin $\boldsymbol{S}_i$ at the site $\boldsymbol{R}_i$.

Suppose there is another localized spin $\boldsymbol{S}_j$ at a site $\boldsymbol{R}_j$. The RKKY spin polarization produced by $\boldsymbol{S}_i$ will extend to the site $\boldsymbol{R}_j$ and interact with $\boldsymbol{S}_j$ again by the *s-d* exchange interaction, as illustrated in Fig.4.3. The energy of such an interaction between the localized moments $\boldsymbol{S}_i$ and $\boldsymbol{S}_j$ is obtained in the following way.

If we put the spin polarization of the conduction electrons produced by $\boldsymbol{S}_i$ equal to $1/2\langle\sigma_i(\boldsymbol{r})\rangle$, then from (4.2.23), (4.2.24) and (4.2.44) we have

$$\langle\sigma_i(\boldsymbol{r})\rangle = 2\left(\frac{J}{N}\right)\boldsymbol{S}_i\,F(\boldsymbol{r} - \boldsymbol{R}_i). \tag{4.2.45}$$

The interaction energy between this $\langle\sigma_i(\boldsymbol{r})\rangle$ and $\boldsymbol{S}_j$ is found using (4.2.42) to be

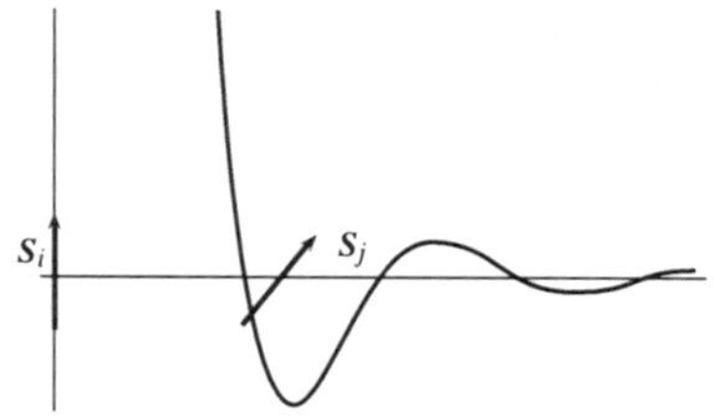

Fig.4.3 The mechanism of the RKKY interaction between a pair of localized spins imbedded in the sea of metallic electrons.

$$H_{\mathrm{RKKY}} = -\frac{V}{N} J \langle \sigma_i(\boldsymbol{R}_j) \rangle \boldsymbol{S}_j$$

$$= -2\left(\frac{J}{N}\right)^2 V F(\boldsymbol{R}_j - \boldsymbol{R}_i) \boldsymbol{S}_i \boldsymbol{S}_j. \qquad (4.2.46)$$

We call this interaction between a pair of localized spins mediated by conduction electrons the *RKKY interaction*. In addition to describing systems such as CuMn dilute alloys, the RKKY interaction is the principal interaction mechanism between the localized 4*f* spins in a rare earth metal. The size of the interaction energy between a pair of spins separated by a distance R is estimated to be

$$(J/N)^2 V |F(R)| = O\left[(J^2/\varepsilon_F)/(k_F R)^3\right]. \qquad (4.2.47)$$

For the total interaction energy of the 4*f* localized spins in a rare earth metal, we have to sum over all the pairs of spins. If we recall (4.2.24) and (2.5.4), we can write down such summation as

$$H_{\mathrm{RKKY}} = -2 \sum_{i>j} \sum_{\boldsymbol{q}} \left(\frac{J}{N}\right)^2 F(\boldsymbol{q}) e^{i\boldsymbol{q}(\boldsymbol{R}_j - \boldsymbol{R}_i)} \boldsymbol{S}_i \boldsymbol{S}_j$$

$$= -\sum_{\boldsymbol{q}} \left(\frac{J}{N}\right)^2 F(\boldsymbol{q}) S(\boldsymbol{q}) S(-\boldsymbol{q}). \qquad (4.2.48)$$

Note that if we put

$$J(\boldsymbol{q}) = \left(\frac{J}{N}\right)^2 F(\boldsymbol{q}) \qquad (4.2.49)$$

in (4.2.48), it takes the form of (2.5.19). For a free electron band for which the Lindhard function behaves as given in Fig.4.1, it is for $\boldsymbol{q} = 0$ that $J(\boldsymbol{q})$ takes the maximum value. In a real rare-earth metal, however, $F(\boldsymbol{q})$ and, therefore, $J(\boldsymbol{q})$ can attain their maximum value at $\boldsymbol{q} \neq 0$. In this case we can have a spin ordering such as (*d*) or (*e*) of Fig.2.2, as we discussed in 2.6. For a review and references, see Ref. [4.8].

Concerning localized magnetic state in a metal, there is the Anderson model [4.10]. With the Anderson model one can discuss under what condition an impurity atom in a metal can form a localized moment. It was shown, then, that the interaction between such localized moments mediated by conduction electrons is also of RKKY form [4.10, 4.11]. Such a similarity between the s-d exchange model and the Anderson model is now well known from the Schrieffer-Wolf transformation [4.12].

The *s-d* exchange Hamiltonian of (4.2.39) or (4.2.42) is rewritten by representing the spin density operator of conduction electrons in the second quantized form as

$$\begin{aligned} H_{\rm sd} &= -\frac{J}{N}\left[S_{iz}\,\sigma_z(\boldsymbol{R}_i) + 2\{S_{i+}\,\sigma_-(\boldsymbol{R}_i) + S_{i-}\,\sigma_+(\boldsymbol{R}_i)\}\right] \\ &= -\frac{J}{N}\sum_{\boldsymbol{k},\boldsymbol{q}} e^{i\boldsymbol{q}\boldsymbol{R}_i}\Big[S_{iz}\,(a^+_{\boldsymbol{k}+}a_{\boldsymbol{k}+\boldsymbol{q},+} - a^+_{\boldsymbol{k},-}a_{\boldsymbol{k}+\boldsymbol{q},-}) \\ &\qquad + 2S_{i+}a^+_{\boldsymbol{k}-}a_{\boldsymbol{k}+\boldsymbol{q},+} + 2S_{i-}a^+_{\boldsymbol{k}+}a_{\boldsymbol{k}+\boldsymbol{q},-}\Big], \end{aligned} \tag{4.2.50}$$

here $\sigma_\pm$ and $S_{i\pm}$ are the spin operators introduced in (1.1.23) and (2.5.15). When there is more than one localized spins in the system we have to sum over the localized spin sites *i*.

The *s-d* exchange interaction had been known also as a principal mechanism of the electrical resistance of a magnetic dilute alloy [4.13]. Quite unexpectedly, however, it was found that if one calculates the scattering amplitude of a metallic electron up to the 2nd order in the *s-d* interaction the electrical resistance diverges as log T at low temperatures [4.14]. With this result, which is called the *Kondo effect,* Kondo succeeded in explaining the phenomenon of the resistance minimum in a magnetic dilute alloy.

Almost simultaneously, I found by using a canonical transformation that the *s-d* exchange interaction gives rise to an effective interaction between conduction electrons in a dilute alloy [4.15]; the procedure is quite analogous to that which Bardeen and Pines [4.16] used to derive the phonon mediated attractive electron-electron interaction of Fröhlich [4.17]. The result was first met with strong skepticism but soon confirmed [4.18, 4.19]. It was then pointed out by Heeger [4.20] that the electron scattering due to the effective electron-electron interaction leads immediately to the Kondo result. These aspects of the Kondo effect were emphasized recently also by Mattis [2.8]. The higher order electron scattering amplitudes due to the *s-d* exchange interaction turned out the more divergent than the 2nd order one at low temperatures. Considerable effort was made to resolve this problem. For a review of recent progress and references see, for instance, Ref. [4.21].

4.2.5 The Effect of Magnetic Field on the Orbital Motion of Electrons: Landau Diamagnetic Susceptibility

It is not only the spin of an electron on which magnetic field has an effect. The orbital motion of the electron also is affected by magnetic field. We discuss this subject in this subsection.

If we represent a magnetic field by a vector potential $\boldsymbol{A}(\boldsymbol{r})$, for the Hamiltonian of a free electron system we have

$$H = \int \psi^{+}(x)\, \frac{1}{2m} \left\{ \boldsymbol{p} + \frac{e}{c}\boldsymbol{A}(\boldsymbol{r}) \right\}^2 \psi(x)\, dx$$

$$= H_o + H_1 + H_2 \tag{4.2.51}$$

in place of (1.5.14), where, H_o, H_1, and H_2 are, respectively, terms in the zeroth, first, and second orders in $\boldsymbol{A}$. Since we are interested in the linear response to $\boldsymbol{A}$, it suffices to consider up to H_1, which is given as

$$H_1 = -\frac{ie\hbar}{2mc} \int \psi^{+}(x) \left\{ \boldsymbol{A}\nabla + \nabla \boldsymbol{A} \right\} \psi(x)\, dx \,. \tag{4.2.52}$$

We assume the vector potential is of the following form,

$$\boldsymbol{A}(\boldsymbol{r}) = \boldsymbol{A}(\boldsymbol{q})\, e^{i\boldsymbol{q}\boldsymbol{r}} \,. \tag{4.2.53}$$

Then, from (4.2.52) we have

$$H_1 = \frac{e\hbar}{mc} \sum_{\boldsymbol{k},\sigma} \left(\boldsymbol{A}(\boldsymbol{q}) \cdot \boldsymbol{k} \right) a^{+}_{\boldsymbol{k}\sigma} a_{\boldsymbol{k}-\boldsymbol{q},\sigma} \,, \tag{4.2.54}$$

where we use the gauge such that $\boldsymbol{q}\boldsymbol{A}(\boldsymbol{q}) = 0$, and, therefore, $\boldsymbol{A}(\boldsymbol{q})(\boldsymbol{k} + \boldsymbol{q}/2) = \boldsymbol{A}(\boldsymbol{q})\boldsymbol{k}$.

As the response to the perturbation first we consider the electric current density $\boldsymbol{J}(\boldsymbol{r})$. The magnetization density is then obtained through the following relation,

$$J(r) = c\nabla \times M(r) . \tag{4.2.55}$$

The current density is given by (see, for instance, Ref. [1.5] p.268)

$$\begin{aligned} J(r) &= -\frac{e}{2m}\left\{ \sum_\sigma \psi_\sigma^+(r)\left(\frac{\hbar}{i}\nabla + \frac{eA(r)}{c}\right)\psi_\sigma(r) \right. \\ &\qquad \left. + \left[\left(\frac{\hbar}{i}\nabla + \frac{eA(r)}{c}\right)\psi_\sigma(r)\right]^+ \psi_\sigma(r) \right\} \\ &= \frac{ie\hbar}{2m}\sum_\sigma \left[\psi_\sigma^+(r)\nabla\psi_\sigma(r) - \left(\nabla\psi_\sigma(r)\right)^+\psi_\sigma(r)\right] - \frac{e^2}{mc}A(r)\sum_\sigma \psi_\sigma^+(r)\,\psi_\sigma(r) \\ &= J_{\rm p}(r) + J_{\rm d}(r) , \end{aligned} \tag{4.2.56}$$

where $J_{\rm p}$ and $J_{\rm d}$ are called, respectively, the paramagnetic, and diamagnetic currents density; we will see later why we call them so.

The paramagnetic current is explicitly written as

$$J_{\rm p}(r) = \frac{1}{V}J_{\rm p}(q)\,e^{iqr} \;\; ; \;\; J_{\rm p}(q) = -\frac{e\hbar}{m}\sum_{k,\sigma}\left(k + \frac{q}{2}\right)a_{k\sigma}^+ a_{k+q,\sigma} . \tag{4.2.57}$$

As for the diamagnetic current density,

$$J_{\rm d}(r) = \frac{1}{V}J_{\rm d}(q)\,e^{iqr} \;\; ; \;\; J_{\rm d}(q) = -\frac{e^2}{mc}\,nA(q) , \tag{4.2.58}$$

where n is the total number of electrons.

Let us calculate the paramagnetic current, $\langle J_{\rm p}(r)\rangle$, responding to the magnetic field by using the procedure of (4.2.4). We put

$$J_{\rm p}(q) = -\frac{2e\hbar}{m}\sum_k J_k(q) \;\; ; \;\; J_k(q) = \left(k + \frac{q}{2}\right)a_k^+ a_{k+q} , \tag{4.2.59}$$

where by summing over the spin states of electrons beforehand we dropped the spin subscripts from the electron creation and annihilation operators. We have following equation for $J_{\rm p}(q)$:

$$\hbar\omega\langle J_k(q)\rangle = \langle [J_k(q), H_{\rm o}]\rangle + {}_{\rm o}\langle [J_k(q), H_1]\rangle , \tag{4.2.60}$$

where we assumed the vector potential is time dependent with the factor $e^{-i\omega t}$.

If we note (4.2.9)–(4.2.10), the commutators in (4.2.60) are readily obtained as

$$[\boldsymbol{J}_{\boldsymbol{k}}(\boldsymbol{q}),\, H_{\mathrm{o}}] \;=\; (\varepsilon_{\boldsymbol{k}+\boldsymbol{q}} - \varepsilon_{\boldsymbol{k}})\boldsymbol{J}_{\boldsymbol{k}}(\boldsymbol{q}), \tag{4.2.61}$$

$$\begin{aligned}
[\boldsymbol{J}_{\boldsymbol{k}}(\boldsymbol{q}),\, H_1] &= \frac{e\hbar}{mc}\sum_{\boldsymbol{l}} \Big(\boldsymbol{A}(\boldsymbol{q})\cdot\boldsymbol{l}\Big)\Big(\boldsymbol{k} + \frac{\boldsymbol{q}}{2}\Big)\big[a_{\boldsymbol{k}}^{+} a_{\boldsymbol{k}+\boldsymbol{q}},\, a_{\boldsymbol{l}}^{+} a_{\boldsymbol{l}-\boldsymbol{q}}\big] \\
&= \frac{e\hbar}{mc}\Big(\boldsymbol{A}(\boldsymbol{q})\cdot\boldsymbol{k}\Big)\Big(\boldsymbol{k} + \frac{\boldsymbol{q}}{2}\Big)\big(a_{\boldsymbol{k}}^{+} a_{\boldsymbol{k}} - a_{\boldsymbol{k}+\boldsymbol{q}}^{+} a_{\boldsymbol{k}+\boldsymbol{q}}\big).
\end{aligned} \tag{4.2.62}$$

Thus, (4.2.60) leads to

$$\begin{aligned}
\hbar\omega\langle\, \boldsymbol{J}_{\boldsymbol{k}}(\boldsymbol{q})\,\rangle &= (\varepsilon_{\boldsymbol{k}+\boldsymbol{q}} - \varepsilon_{\boldsymbol{k}})\langle\, \boldsymbol{J}_{\boldsymbol{k}}(\boldsymbol{q})\,\rangle \\
&\quad + \frac{e\hbar}{mc}\Big(\boldsymbol{A}(\boldsymbol{q})\cdot\boldsymbol{k}\Big)\Big(\boldsymbol{k} + \frac{\boldsymbol{q}}{2}\Big)\Big(f(\varepsilon_{\boldsymbol{k}}) - f(\varepsilon_{\boldsymbol{k}+\boldsymbol{q}})\Big)
\end{aligned} \tag{4.2.63}$$

and, then,

$$\langle\, \boldsymbol{J}_{\mathrm{p}}(\boldsymbol{q},\omega)\,\rangle = \frac{2e^2\hbar^2}{m^2c}\sum_{\boldsymbol{k}} \Big(\boldsymbol{A}(\boldsymbol{q})\cdot\boldsymbol{k}\Big)\Big(\boldsymbol{k} + \frac{\boldsymbol{q}}{2}\Big)\frac{f(\varepsilon_{\boldsymbol{k}+\boldsymbol{q}}) - f(\varepsilon_{\boldsymbol{k}})}{\varepsilon_{\boldsymbol{k}} - \varepsilon_{\boldsymbol{k}+\boldsymbol{q}} + \hbar\omega}. \tag{4.2.64}$$

Let us take $\boldsymbol{q}$ in the direction of z axis. Then, by noting the relation, $\boldsymbol{q}\boldsymbol{A}(\boldsymbol{q}) = 0$, we assume $\boldsymbol{A}(\boldsymbol{q})$ is in the direction of x-axis. The current induced, then, will be in the direction of x-axis; current in other directions cancels out as can be easily checked from (4.2.64).

Thus, we have $\langle \boldsymbol{J}_{\mathrm{p}}(\boldsymbol{q})\rangle = \langle \boldsymbol{J}_{\mathrm{p}}(\boldsymbol{q},\, 0)\rangle = (\langle J_{\mathrm{p}x}(\boldsymbol{q})\rangle,\, 0,\, 0)$, for $\boldsymbol{A}(\boldsymbol{q}) = (A(\boldsymbol{q}),\, 0,\, 0)$,

$$\langle\, J_{\mathrm{p}x}(\boldsymbol{q},\, 0)\,\rangle = \frac{2e^2\hbar^2}{m^2c}A(\boldsymbol{q})\sum_{\boldsymbol{k}} k_x^2\,\frac{f(\varepsilon_{\boldsymbol{k}+\boldsymbol{q}}) - f(\varepsilon_{\boldsymbol{k}})}{\varepsilon_{\boldsymbol{k}} - \varepsilon_{\boldsymbol{k}+\boldsymbol{q}}}. \tag{4.2.65}$$

This current is in the same direction of $\boldsymbol{A}(\boldsymbol{q})$. Thus it is called paramagnetic current.

The corresponding diamagnetic current, (4.2.58), is in the same direction, $\langle \boldsymbol{J}_{\mathrm{d}}(\boldsymbol{q},\, 0)\rangle = (\langle J_{\mathrm{d}x}(\boldsymbol{q})\rangle,\, 0,\, 0)$, and given as

$$\langle\, J_{\mathrm{d}x}(\boldsymbol{q})\,\rangle = -\frac{e^2 n}{mc}A(\boldsymbol{q}), \tag{4.2.66}$$

where n is the total number of electrons. This current flows in the direction opposite to that of $\boldsymbol{A}$. Thus it is called diamagnetic current.

The total current, $\langle \boldsymbol{J}(\boldsymbol{q})\rangle = (\langle J_x(\boldsymbol{q})\rangle,\, 0,\, 0)$, is obtained from (4.2.65) and (4.2.66) as

$$\langle J_x(\boldsymbol{q}) \rangle = -\frac{e^2 n}{mc} A(\boldsymbol{q}) \left[1 - \frac{2\hbar^2}{mn} \sum_{\boldsymbol{k}} k_x^2 \frac{f(\varepsilon_{\boldsymbol{k}+\boldsymbol{q}}) - f(\varepsilon_{\boldsymbol{k}})}{\varepsilon_{\boldsymbol{k}} - \varepsilon_{\boldsymbol{k}+\boldsymbol{q}}} \right]$$

$$\equiv -\frac{e^2 n}{mc} A(\boldsymbol{q}) L(\boldsymbol{q}) . \tag{4.2.67}$$

Unlike the cases of charge and spin responses, the above result is not given in terms of the Lindhard function. However, we can see that it is again the excitation processes near the Fermi surface that determines the paramagnetic current response.

An important observation on the result (4.2.67) is that

$$\lim_{q \to 0} \langle J_x(\boldsymbol{q}) \rangle = 0. \tag{4.2.68}$$

This can be proved from

$$\lim_{q \to 0} \frac{2\hbar^2}{mn} \sum_{\boldsymbol{k}} k_x^2 \frac{f(\varepsilon_{\boldsymbol{k}+\boldsymbol{q}}) - f(\varepsilon_{\boldsymbol{k}})}{\varepsilon_{\boldsymbol{k}} - \varepsilon_{\boldsymbol{k}+\boldsymbol{q}}}$$

$$= \frac{4}{3n} \sum_{\boldsymbol{k}} \varepsilon_{\boldsymbol{k}} \left(-\frac{\partial f(\varepsilon_{\boldsymbol{k}})}{\partial \varepsilon_{\boldsymbol{k}}} \right) = 1, \tag{4.2.68'}$$

in (4.2.67). This result is understood that the spatially uniform part of the paramagnetic current and the diamagnetic current cancel each other.

The above observation is important in conjuction with superconductivity. In the superconducting state there is a gap in the electron excitation spectrum near the Fermi surface. Then, the quantity corresponding to that which was estimated in (4.2.68') becomes zero. This leaves the diamagnetic current, (4.2.66), entirely intact. The diamagnetic current is nothing other than the superconducting current that is responsible for the Meissner effect [1.2]. Now we understood why we divided the current into the paramagnetic and diamagnetic components.

Let us proceed to calculate the magnetization by using (4.2.55). First we note that the magnetic field is obtained from the vector potential as

$$\boldsymbol{H}(\boldsymbol{q})\, e^{i\boldsymbol{q}\boldsymbol{r}} = \nabla \times \left(\boldsymbol{A}(\boldsymbol{q})\, e^{i\boldsymbol{q}\boldsymbol{r}} \right) = i\boldsymbol{q} \times \boldsymbol{A}(\boldsymbol{q})\, e^{i\boldsymbol{q}\boldsymbol{r}}. \tag{4.2.69}$$

Then if we introduce the diamagnetic susceptibility $\chi_{\rm dia}(\boldsymbol{q})$ as the total magnetic response through the orbital motion of electrons, to which both the paramagnetic and diamagnetic currents contribute, (4.2.55) is rewritten as

$$-\frac{e^2 n}{mc^2} A(\boldsymbol{q}) L(\boldsymbol{q}) e^{i\boldsymbol{qr}} = \chi_{\rm dia}(\boldsymbol{q})\, i\boldsymbol{q} \times \big(i\boldsymbol{q} \times \boldsymbol{A}(\boldsymbol{q})\big) e^{i\boldsymbol{qr}}. \tag{4.2.70}$$

If we recall that $\boldsymbol{q}$ and $\boldsymbol{A}(\boldsymbol{q})$ are, respectively, in the directions of z- and x-axes, from (4.2.70) we obtain

$$\chi_{\rm dia}(\boldsymbol{q}) = -\frac{e^2 n}{mc^2} \frac{L(\boldsymbol{q})}{q^2}. \tag{4.2.71}$$

Although we need a numerical calculation to estimate $\chi_{\rm dia}(\boldsymbol{q})$ for $\boldsymbol{q} \neq 0$, the value for $\boldsymbol{q} \to 0$ can be analytically obtained. If we put

$$f(\varepsilon_{\boldsymbol{k}+\boldsymbol{q}}) - f(\varepsilon_{\boldsymbol{k}}) = f'(\varepsilon_{\boldsymbol{k}})(\varepsilon_{\boldsymbol{k}+\boldsymbol{q}} - \varepsilon_{\boldsymbol{k}}) + \frac{1}{2} f''(\varepsilon_{\boldsymbol{k}})(\varepsilon_{\boldsymbol{k}+\boldsymbol{q}} - \varepsilon_{\boldsymbol{k}})^2 + \frac{1}{6} f'''(\varepsilon_{\boldsymbol{k}})(\varepsilon_{\boldsymbol{k}+\boldsymbol{q}} - \varepsilon_{\boldsymbol{k}})^3$$

into (4.2.67) and note (4.2.68), we obtain

$$\begin{aligned}\lim_{q\to 0} \frac{L(\boldsymbol{q})}{q^2} &= \lim_{q\to 0} \frac{2\hbar^2}{q^2 mn} \left[\frac{1}{2} \frac{\hbar^2}{2m} \sum_{\boldsymbol{k}} k_x^2 f''(\varepsilon_{\boldsymbol{k}})(2\boldsymbol{kq} + \boldsymbol{q}^2) + \frac{1}{6}\left(\frac{\hbar^2}{2m}\right)^2 \sum_{\boldsymbol{k}} k_x^2 f'''(\varepsilon_{\boldsymbol{k}})(2\boldsymbol{kq} + \boldsymbol{q}^2)^2 \right] \\ &= \frac{2}{n}\left(\frac{\hbar^2}{2m}\right)^2 \sum_{\boldsymbol{k}} k_x^2 f''(\varepsilon_{\boldsymbol{k}}) + \frac{2}{3n}\left(\frac{\hbar^2}{2m}\right)^3 4 \sum_{\boldsymbol{k}} k_x^2 k_z^2 f'''(\varepsilon_{\boldsymbol{k}}),\end{aligned} \tag{4.2.72}$$

where we noted that $\boldsymbol{q}$ is in the direction of z-axis and considered symmetry in summing over $\boldsymbol{k}$. The necessary integrals are readily done:

$$\sum_{\boldsymbol{k}} k_x^2 f''(\varepsilon_{\boldsymbol{k}}) = \frac{1}{3}\left(\frac{2m}{\hbar^2}\right) \frac{\partial}{\partial \varepsilon} \big(\varepsilon N(\varepsilon)\big)\Big|_{\varepsilon=\varepsilon_{\rm F}} = \frac{m}{\hbar^2} N(\varepsilon_{\rm F}), \tag{4.2.73}$$

$$\sum_{\boldsymbol{k}} k_x^2\, k_z^2 f'''(\varepsilon_{\boldsymbol{k}}) = -\frac{1}{15}\left(\frac{2m}{\hbar^2}\right)^2 \frac{\partial^2}{\partial \varepsilon^2} \big(\varepsilon^2 N(\varepsilon)\big)\Big|_{\varepsilon=\varepsilon_{\rm F}} = -\frac{1}{4}\left(\frac{2m}{\hbar^2}\right)^2 N(\varepsilon_{\rm F}), \tag{4.2.74}$$

where we assumed $(k_{\rm B}T/\varepsilon_{\rm F})^2 << 1$.

Thus we have

$$\lim_{q\to 0} \frac{L(\boldsymbol{q})}{q^2} = \frac{\hbar^2}{2mn} N(0)\left[1 - \frac{2}{3}\right] = \frac{\hbar^2}{6mn} N(0), \tag{4.2.75}$$

and, finally, from (4.2.71), the Landau diamagnetic susceptibility,

$$\chi_{\text{Landau}} = \lim_{q\to 0} \chi_{\text{dia}}(\boldsymbol{q}) = -\frac{2}{3} \mu_{\text{B}}^2 N(0)$$

$$= -\frac{1}{3} \chi_{\text{P}}, \tag{4.2.76}$$

where we noted (2.1.5), and χ_{P} is the Pauli spin susceptibility given in (3.1.8); for Landau's original procedure of deriving this result, see Peierls [4.22] (see also Enz [1.6]).

4.3 THE EFFECT OF THE INTERACTION BETWEEN ELECTRONS VARIOUS LINEAR RESPONSES OF METALLIC ELECTRONS

In the preceding section we discussed the spin and charge responses of metallic electrons, but without considering the effects of the interaction between electrons. There, regarding the charge response, we encountered the difficulty with the screening charge sum rule in 4.2.3. As for the spin response, we earlier found in Chapter 3 that it is essential to take into account the effect of the exchange interaction between electrons. In this section we present a simple, systematic approach to include the effect of the electron-electron interaction in various linear responses.

4.3.1 Generalized Mean Field Approximation of the Coulomb Interaction between Electrons

In 4.2 we calculated the linear responses, $\chi^0_{zz}(\boldsymbol{q}, \omega)$ and $\chi^0_{ee}(\boldsymbol{q}, \omega)$, by using the prescription of (4.2.4). To include the effect of the Coulomb interaction between electrons on the calculation of these responses, we have to add a new term,

$$\left\langle [a^+_{\boldsymbol{k}\sigma} a_{\boldsymbol{k}+\boldsymbol{q},\sigma}, H_{\text{C}}] \right\rangle, \tag{4.3.1}$$

on the right hand sides of (4.2.8) and (4.2.29), respectively. With this term, however, it is no longer feasible to carry out the calculation of the linear responses exactly. We, thus, introduce the mean field approximation to this term.

As we already saw in 3.2, within the mean field approximation H_{C} of (1.5.24) reduces to

$$H_{\mathrm{C,m}} = \sum_{k,l,\kappa,\sigma,\sigma'}{}' \mathrm{v}(\kappa) \left\langle a_{l\sigma'}^{+} a_{l+\kappa,\sigma'} \right\rangle a_{k\sigma}^{+} a_{k-\kappa,\sigma}$$

$$- \sum_{k,l,\kappa,\sigma,\sigma'}{}' \mathrm{v}(\kappa) \left\langle a_{k\sigma}^{+} a_{l+\kappa,\sigma'} \right\rangle a_{l\sigma'}^{+} a_{k-\kappa,\sigma} . \qquad (4.3.2)$$

The first and second terms on the right hand side represent, respectively, the Hartree, or the direct Coulomb repulsion, and the Fock, or the exchange interaction, terms.

Regarding the expectation values appearing in (4.3.2), in Chapter 3 we imposed the requirement of (3.2.5) reflecting the spatial homogeneity of the system. However, if a system is under an external perturbation breaking spatial homogeneity, such as (4.2.5), (3.2.5) is no longer adequate. When an external field has a spatial dependence characterized by a wave number $\boldsymbol{q}$, we should require

$$\left\langle a_{l\sigma}^{+} a_{l+\kappa,\sigma} \right\rangle \neq 0, \quad \text{and,} \quad \left\langle n_{\sigma}(\kappa) \right\rangle \neq 0, \quad \text{for} \quad \kappa = 0 \quad \text{or} \quad \boldsymbol{q}, \qquad (4.3.3)$$

in place of (3.2.5). With (4.3.3) in (4.3.2), we arrive at the following generalized mean field approximation result,

$$H_{\mathrm{C,m}} = \mathrm{v}(\boldsymbol{q}) \left\langle n(\boldsymbol{q}) \right\rangle \sum_{k,\sigma} a_{k\sigma}^{+} a_{k-q,\sigma}$$

$$- \widetilde{V}(0) \sum_{k,\sigma} n_{\sigma}\, a_{k\sigma}^{+} a_{k\sigma} - \widetilde{V}(\boldsymbol{q}) \sum_{k,\sigma} \left\langle n_{\sigma}(\boldsymbol{q}) \right\rangle a_{k\sigma}^{+} a_{k-q,\sigma} . \qquad (4.3.4)$$

The meaning of the first term on the right hand side is clear; it represents the Coulomb potential due to the electron density polarization $\langle n(\boldsymbol{q})\rangle$ caused by the external field. The second and third terms represent the exchange interaction effect. As for the second term, we already encountered it earlier in (3.3.2); it represents the spatially uniform part of the exchange potential.

$\widetilde{V}(\boldsymbol{q})$ in the third term is introduced as follows: If we retain only those contributions with $\sigma = \sigma'$, $\boldsymbol{l}+\kappa = \boldsymbol{k}+\boldsymbol{q}$ in the expectation appearing in the second term on the right hand side of (4.3.2), it reduces to

$$- \sum_{k,l,\sigma} \mathrm{v}(\boldsymbol{k}+\boldsymbol{q}-\boldsymbol{l}) \left\langle a^+_{k\sigma} a_{k+q,\sigma} \right\rangle a^+_{l\sigma} a_{l-q,\sigma}$$

$$\cong - \widetilde{V}(\boldsymbol{q}) \sum_{k,l,\sigma} \left\langle a^+_{k\sigma} a_{k+q,\sigma} \right\rangle a^+_{l\sigma} a_{l-q,\sigma}$$

$$= - \widetilde{V}(\boldsymbol{q}) \sum_{l,\sigma} \langle n_\sigma(\boldsymbol{q}) \rangle \, a^+_{l\sigma} a_{l-q,\sigma} \,. \tag{4.3.5}$$

We introduced $\widetilde{V}(\boldsymbol{q})$ by averaging the role of $\mathrm{v}(\boldsymbol{k}+\boldsymbol{q}-\boldsymbol{l})$, as we earlier did in rewriting (3.3.1) as (3.3.2) by introducing $\widetilde{V}(0) = \widetilde{V}$. $\widetilde{V}(\boldsymbol{q})$ can be understood as the coupling constant of the exchange potential associated with the spin density component $\langle n_\sigma(\boldsymbol{q}) \rangle$.

The result of (4.3.4) may be understood as the result of a mean field approximation on the following effective Hamiltonian

$$H_\mathrm{C} = \frac{1}{2} \sum_{\kappa}{}' \mathrm{v}(\kappa) n(\kappa) n(-\kappa) - \frac{1}{2} \sum_{\kappa,\sigma} \widetilde{V}(\kappa) n_\sigma(\kappa) n_\sigma(-\kappa). \tag{4.3.6}$$

The prime on the summation indicates that $\kappa = 0$ is to be excluded from the sum. Note, however, that (4.3.6) is not an exact result.

In rewriting the total Hamiltonian including the one particle energy part H_o in the mean field approximation, it is convenient to set

$$\varepsilon_{k\sigma} = \varepsilon_k - \widetilde{V} \langle n_\sigma \rangle, \tag{4.3.7}$$

as we previously did in (3.3.4) (note that $\widetilde{V}(0) = \widetilde{V}$) we call the second term on the right hand side the exchange self-energy of an electron. Thus, we have

$$H = H_\mathrm{o} + H_\mathrm{C} \cong H_\mathrm{o} + H_{\mathrm{C,m}}$$

$$= \sum_{k,\sigma} \varepsilon_{k\sigma} a^+_{k\sigma} a_{k\sigma} + \mathrm{v}(\boldsymbol{q}) \langle n(\boldsymbol{q}) \rangle \sum_{k,\sigma} a^+_{k\sigma} a_{k-q,\sigma}$$

$$- \widetilde{V}(\boldsymbol{q}) \sum_{k,\sigma} \langle n_\sigma(\boldsymbol{q}) \rangle a^+_{k\sigma} a_{k-q,\sigma} \,. \tag{4.3.8}$$

Note that the system is subjected to an external field with the spatial dependence proportional to $\exp(i\boldsymbol{q}\boldsymbol{r})$ with $\boldsymbol{q} \neq 0$. In 3.3 we discussed the case of $\boldsymbol{q} = 0$ and, found that only the first term in the last expression of (4.3.8) appeared in the result.

With the above preparation let us now start to discuss the effect of the Coulomb interaction on various linear responses of metallic electrons. Here we follow Refs. [4.23–4.25], but an essentially equivalent result can also be found in Rajagopal et al. [4.26].

4.3.2 Magnetic Susceptibility of Interacting Electrons: $\chi_{zz}(\boldsymbol{q}, \omega)$

Suppose an external magnetic field of the form of (4.1.27) is applied in the direction of the z-axis. If we are to obtain the magnetization induced in a system of interacting electrons by the procedure of 4.2.1 within the mean field approximation, we have only to add $\langle [a_{\boldsymbol{k}\sigma}^{+} a_{\boldsymbol{k}+\boldsymbol{q},\sigma}, H_{\mathrm{C}}] \rangle \cong {}_{0}\langle [a_{\boldsymbol{k}\sigma}^{+} a_{\boldsymbol{k}+\boldsymbol{q},\sigma}, H_{\mathrm{Cm}}] \rangle$ to the right hand side of (4.2.8). Note that $\langle n_{\sigma}(\boldsymbol{q}) \rangle$ in (4.2.8) is already proportional to the external field.

Let us approximate H_{C} by $H_{\mathrm{C,m}}$, as in (4.3.8). Then, from the commutation relations such as (4.2.11), we obtain

$$\begin{aligned} \hbar\omega \langle a_{\boldsymbol{k}\sigma}^{+} a_{\boldsymbol{k}+\boldsymbol{q},\sigma} \rangle &= (\varepsilon_{\boldsymbol{k}+\boldsymbol{q},\sigma} - \varepsilon_{\boldsymbol{k}\sigma}) \langle a_{\boldsymbol{k}\sigma}^{+} a_{\boldsymbol{k}+\boldsymbol{q},\sigma} \rangle \\ &+ \left[\mathrm{v}(\boldsymbol{q}) \langle n(\boldsymbol{q}) \rangle - \widetilde{V}(\boldsymbol{q}) \langle n_{\sigma}(\boldsymbol{q}) \rangle + \sigma \mu_{\mathrm{B}} H_z(\boldsymbol{q}) e^{-i\omega t} \right] \\ &\times \left\{ {}_{0}\langle a_{\boldsymbol{k}\sigma}^{+} a_{\boldsymbol{k}\sigma} \rangle - {}_{0}\langle a_{\boldsymbol{k}+\boldsymbol{q},\sigma}^{+} a_{\boldsymbol{k}+\boldsymbol{q},\sigma} \rangle \right\}, \end{aligned} \tag{4.3.9}$$

corresponding to (4.2.12) for the case of the free electrons. This equation leads to

$$\begin{aligned} \langle n_{\sigma}(\boldsymbol{q}) \rangle = \langle n_{\sigma}(\boldsymbol{q}, \omega) \rangle e^{-i\omega t} &= F_{\sigma}(\boldsymbol{q}, \omega) \left[- \sigma \mu_{\mathrm{B}} H_z(\boldsymbol{q}) e^{-i\omega t} \right. \\ &\left. - \mathrm{v}(\boldsymbol{q}) \left(\langle n_{+}(\boldsymbol{q}) \rangle + \langle n_{-}(\boldsymbol{q}) \rangle \right) + \widetilde{V}(\boldsymbol{q}) \langle n_{\sigma}(\boldsymbol{q}) \rangle \right], \end{aligned} \tag{4.3.10}$$

as (4.2.12) led to (4.2.13), where

$$F_{\sigma}(\boldsymbol{q}, \omega) = \sum_{\boldsymbol{k}} \frac{f(\varepsilon_{\boldsymbol{k}\sigma}) - f(\varepsilon_{\boldsymbol{k}+\boldsymbol{q},\sigma})}{\varepsilon_{\boldsymbol{k}+\boldsymbol{q},\sigma} - \varepsilon_{\boldsymbol{k}\sigma} - \hbar\omega} \tag{4.3.11}$$

is the Lindhard function with an explicit spin dependence. Note that the Fermi distribution

$${}_{0}\langle a_{\boldsymbol{k}\sigma}^{+} a_{\boldsymbol{k}\sigma} \rangle = \frac{1}{e^{\beta(\varepsilon_{\boldsymbol{k}\sigma} - \mu)} + 1} = f_{\sigma}(\varepsilon_{\boldsymbol{k}}) \tag{4.3.12}$$

is spin dependent in the ferromagnetic state as we already saw in (3.4.18)–(3.4.19). In the paramagnetic state we have

$$F_+(\boldsymbol{q}, \omega) = F_-(\boldsymbol{q}, \omega) = F(\boldsymbol{q}, \omega). \tag{4.3.13}$$

The meaning of the result of (4.3.10) is easily understood. When there is no interaction between electrons, we have only the first term in the bracket on the right hand side, and it reduces to that of (4.2.13). The second and the third terms in the bracket, respectively, represent the Coulomb and the exchange potentials due to the polarized electron density. Note the difference in signs of these potentials; while the repulsive Coulomb potential tends to reduce the electron density response, the attractive exchange potential tends to increase it.

By solving (4.3.10) for $\langle n_+(\boldsymbol{q}) \rangle$ and $\langle n_-(\boldsymbol{q}) \rangle$ we obtain

$$\langle n_\sigma(\boldsymbol{q}) \rangle = -\frac{\widetilde{F}_\sigma(\boldsymbol{q}, \omega) + 2\mathrm{v}(\boldsymbol{q})\widetilde{F}_+(\boldsymbol{q}, \omega)\widetilde{F}_-(\boldsymbol{q}, \omega)}{1 + \mathrm{v}(\boldsymbol{q})[\widetilde{F}_+(\boldsymbol{q}, \omega) + \widetilde{F}_-(\boldsymbol{q}, \omega)]}\, \sigma\mu_{\mathrm{B}} H_z(\boldsymbol{q})\, e^{-i\omega t}, \tag{4.3.14}$$

where we have introduced the exchange enhanced Lindhard function as

$$\widetilde{F}_\sigma(\boldsymbol{q}, \omega) = \frac{F_\sigma(\boldsymbol{q}, \omega)}{1 - \widetilde{V}(\boldsymbol{q}) F_\sigma(\boldsymbol{q}, \omega)}\ . \tag{4.3.15}$$

From (4.3.14), the magnetic susceptibility is obtained to be

$$\chi_{zz}(\boldsymbol{q}, \omega) = \mu_{\mathrm{B}}{}^2 \frac{\widetilde{F}_+(\boldsymbol{q}, \omega) + \widetilde{F}_-(\boldsymbol{q}, \omega) + 4\mathrm{v}(\boldsymbol{q})\widetilde{F}_+(\boldsymbol{q}, \omega)\widetilde{F}_-(\boldsymbol{q}, \omega)}{1 + \mathrm{v}(\boldsymbol{q})[\widetilde{F}_+(\boldsymbol{q}, \omega) + \widetilde{F}_-(\boldsymbol{q}, \omega)]}. \tag{4.3.16}$$

For the paramagnetic state the above result reduces to

$$\begin{aligned} \chi_{zz}(\boldsymbol{q}, \omega) &= 2\mu_{\mathrm{B}}{}^2 \frac{F(\boldsymbol{q}, \omega)}{1 - \widetilde{V}(\boldsymbol{q}) F(\boldsymbol{q}, \omega)} \\ &\equiv 2\mu_{\mathrm{B}}{}^2 \widetilde{F}(\boldsymbol{q}, \omega)\,. \end{aligned} \tag{4.3.17}$$

This expression reduces to the Stoner susceptibility (3.3.8) for $\omega = 0$ and $\boldsymbol{q} = 0$. Thus, (4.3.17) is understood to be the extension of the Stoner susceptibility to the case of $\boldsymbol{q} \neq 0$ and $\omega \neq 0$.

Exercise: Confirm that in the ferromagnetic state, for $\omega = 0$ and $\boldsymbol{q} = 0$ (4.3.16) reduces to the high field magnetic susceptibility of (3.4.24). Note that

$$F_\sigma(\boldsymbol{q}, 0) = \int d\varepsilon\, N(\varepsilon)\left(-\frac{\partial f_\sigma(\varepsilon)}{\partial \varepsilon}\right)$$

$$= -\int d\varepsilon\, N_\sigma(\varepsilon)\frac{\partial f(\varepsilon)}{\partial \varepsilon} = F_\sigma(0) = N_\sigma(0). \qquad (4.3.18)$$

Note that in the ferromagnetic state of a system, the magnetic response is no longer isotropic; the relation of (4.2.17) does not hold. We will discuss χ_{xx} and χ_{yy} for the ferromagnetic state in 4.4.1.

The effect of the interaction between electrons is most clearly seen in the result of (4.3.17); the magnetic susceptibility is enhanced by the exchange effect from (4.2.16) to (4.3.17). As an application of (4.3.17), let us explore the spatial behavior of the electron spin polarization induced by a delta function type magnetic field [4.27]. What is to be done is simply to replace $\chi^{\mathrm{o}}_{zz}(\boldsymbol{q}, 0)$ by $\chi_{zz}(\boldsymbol{q}, 0)$ of (4.3.17) in (4.2.23). In this case we need a numerical procedure.

In Fig.4.4 we give an example of such results, where we assume an electron gas with $k_{\mathrm{F}} = 0.5 \times 10^8$/cm, the same value used to obtain Fig.4.2, and put $\widetilde{V}(\boldsymbol{q}) = \widetilde{V}$, independent of $\boldsymbol{q}$. Corresponding to (4.2.23)–(4.2.24) we put $\langle M_z(\boldsymbol{r})\rangle = 2\, m^2 (H_z/N)\widetilde{F}(\boldsymbol{r})$. Note that the case of $\overline{V} = 0$ in the figure is identical to the result of Fig.4.2.

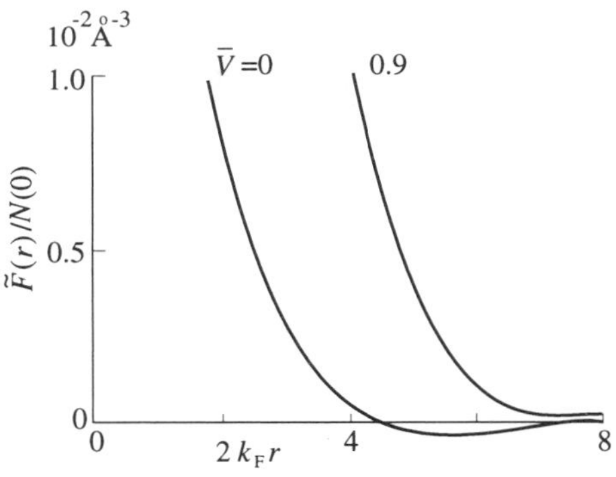

Fig.4.4 The exchange enhanced RKKY-Friedel oscillation. The curve for $\overline{V} = 0$ is identical to given in Fig.4.2.

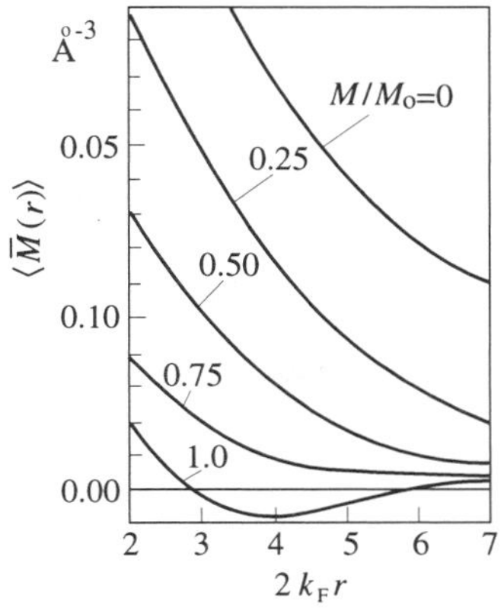

Fig.4.5 Spin polarization of a ferromagnetic electron gas due to a nega-that tive δ-function type magnetic field of (4.2.21) for different relative magnetizations. The size of δ function type magnetic field is $|H_z| = N[\mu_{\mathrm{B}}N(0)]^{-1}$.

As can be seen from Fig.4.4, the exchange interaction significantly affects the spatial behavior of the spin polarization; the magnitude of the polarization is enhanced and its spatial oscillation is drastically reduced.

Another interesting, but less well-known, problem is the spatial behavior of the response of ferromagnetic metals, with spin-split electron energy bands, to a delta function form of applied magnetic field [4.28]. This can be studied by putting $\chi_{zz}(\boldsymbol{q}, 0)$ of (4.3.16) in place of $\chi^0_{zz}(\boldsymbol{q}, 0)$ in (4.2.23). In Fig.4.5 we present the results of numerical calculations of $\langle \overline{M}(\boldsymbol{r})\rangle = \langle n_+(\boldsymbol{r})\rangle - \langle n_-(\boldsymbol{r})\rangle$, for an electron gas with $k_F = 0.5\times 10^8$/cm, the same value used in Fig.4.4, for different values of magnetization, M/M_0, of the system, M_0 being the maximum possible magnetization. The magnitude of the negative magnetic field in (4.2.21) is such that $\mu_B|H_z| = N/N(0)$. Different magnetizations, M/M_0, are realized by changing the value of $\overline{V} = \widetilde{V}N(0)$ within the Stoner model at $T = 0$; we assume $n_+ > n_-$ for the ferromagnetic state, as in Fig.3.5(a), and put $\widetilde{V}(\boldsymbol{q}) = \widetilde{V}$ independent of $\boldsymbol{q}$.

According to the results of Fig.4.5, the behavior of the spin polarization induced by a delta function type magnetic field changes quite dramatically with the magnetization of the system. When the system is fully magnetized, the case of $M/M_0 = 1$, the spin polarization shows the typical RKKY oscillation with the smallest amplitude. For the cases of smaller M/M_0, we have electron spin polarizations with larger amplitudes and slower oscillation.

Note that we actually observe such a tendency in PdFe dilute alloys, for instance. Although pure Pd metal is not ferromagnetic, when we put a small concentration of Fe in it, it becomes ferromagnetic at low temperatures. We may explain such behavior by using the *s-d* exchange model of 4.2.4; we associate a localized moment with each Fe atom. With this picture we expect that the spin splitting of the conduction electron bands of Pd will increase with increasing Fe concentration. Then, the conduction electron polarizations per Fe impurity of the host Pd metal will change with increasing Fe concentration; with increasing Fe concentration, the behavior of the spin polarizations around each Fe atom will change from the case of $M/M_0 = 0$ toward $M/M_0 = 1$ in Fig.4.5 [4.28]. Such a systematic trend was actually observed by Low et al. [4.29].

4.3.3 Charge Response to Magnetic Field: $\chi_{em}(\boldsymbol{q}, \omega)$

From (4.3.14) we have

$$\langle n_+(\boldsymbol{q})\rangle + \langle n_-(\boldsymbol{q})\rangle = -\frac{\widetilde{F}_+(\boldsymbol{q},\omega) - \widetilde{F}_-(\boldsymbol{q},\omega)}{1 + v(\boldsymbol{q})[\widetilde{F}_+(\boldsymbol{q},\omega) + \widetilde{F}_-(\boldsymbol{q},\omega)]}\mu_B H(\boldsymbol{q})e^{-i\omega t}. \quad (4.3.19)$$

This result tells us that in the ferromagnetic state of a metal, where $\widetilde{F}_+(\boldsymbol{q},\omega)\neq \widetilde{F}_-(\boldsymbol{q},\omega)$ magnetic field induces a polarization of the electron number density. While the responses of ± spin electrons to the magnetic field are proportional to

$N_\pm(0)$, with opposite signs for ± spin electrons, in the ferromagnetic state we have $N_+(0) - N_-(0) \neq 0$.

Let us define the electron charge response to a magnetic field as

$$\chi_{\rm em}(\boldsymbol{q}, \omega) = \frac{-e(\langle n_+(\boldsymbol{q}, \omega)\rangle + \langle n_-(\boldsymbol{q}, \omega)\rangle)}{H(\boldsymbol{q})} . \qquad (4.3.20)$$

Then (4.3.19) gives

$$\chi_{\rm em}(\boldsymbol{q}, \omega) = e\,\mu_{\rm B} \frac{\widetilde{F}_+(\boldsymbol{q}, \omega) - \widetilde{F}_-(\boldsymbol{q}, \omega)}{1 + {\rm v}(\boldsymbol{q})[\,\widetilde{F}_+(\boldsymbol{q}, \omega) + \widetilde{F}_-(\boldsymbol{q}, \omega)\,]} . \qquad (4.3.21)$$

In the paramagnetic state we have, of course, $\chi_{\rm em} = 0$. Note also that

$$\lim_{\boldsymbol{q}\to 0} \chi_{\rm em}(\boldsymbol{q}, 0) = 0. \qquad (4.3.22)$$

Thus, unless the magnetic field is of the form of $H(\boldsymbol{q}) \propto 1/q^2$, or more diverging for $\boldsymbol{q} \to 0$ the induced electron density polarization vanishes when it is integrated over the entire volume of a system.

4.3.4 Charge Susceptibility of an Interacting Electron Gas, $\chi_{\rm ee}(\boldsymbol{q}, \omega)$, and the Screening Constant, $\varepsilon_\sigma(\boldsymbol{q}, \omega)$

In 4.2.4 we derived the charge susceptibility for the non-interacting electron gas, $\chi^0_{\rm ee}(\boldsymbol{q}, \omega)$, and noted the difficulty with it of (4.2.38). Here we discuss the charge susceptibility including the effect of the Coulomb interaction between electrons. Also we discuss the charge susceptibility in the ferromagnetic state of a metal.

According to the prescription of (4.2.3), in order to include the effect of the electron interaction in the electron density response, we have to add the term

$$\langle [a^+_{\boldsymbol{k}\sigma}\, a_{\boldsymbol{k}+\boldsymbol{q},\sigma}\,,\, H_{\rm C,m}] \rangle$$

to the right hand side of (4.2.29) or (4.2.31). Then we obtain,

$$\langle n_\sigma(\boldsymbol{q})\rangle = F_\sigma(\boldsymbol{q}) \big[e\,U(\boldsymbol{q})\, e^{-i\omega t}$$
$$- {\rm v}(\boldsymbol{q})\,(\langle n_+(\boldsymbol{q})\rangle + \langle n_-(\boldsymbol{q})\rangle) + \widetilde{V}(\boldsymbol{q})\langle n_\sigma(\boldsymbol{q})\rangle \big], \qquad (4.3.23)$$

by the same procedure used to derive (4.3.10) for the magnetic response. The role of the interaction between electrons becomes clear if we compare this result with (4.2.32). While the repulsive Coulomb potential acts to oppose the electron polarization, the attractive exchange potential enhances it, as we already noted in (4.3.10).

From (4.3.23) we obtain

$$\langle n_\sigma(\boldsymbol{q}) \rangle = \langle n_\sigma(\boldsymbol{q}, \omega) \rangle e^{-i\omega t}$$

$$= \frac{\widetilde{F}_\sigma(\boldsymbol{q}, \omega)}{1 + \mathrm{v}(\boldsymbol{q}) [\widetilde{F}_+(\boldsymbol{q}, \omega) + \widetilde{F}_-(\boldsymbol{q}, \omega)]} e\, U(\boldsymbol{q}) e^{-i\omega t}. \tag{4.3.24}$$

The charge susceptibility defined in (4.2.33) is, then, given as [4.23]

$$\chi_{\mathrm{ee}}(\boldsymbol{q}, \omega) = e^2 \frac{\widetilde{F}_+(\boldsymbol{q}, \omega) + \widetilde{F}_-(\boldsymbol{q}, \omega)}{1 + \mathrm{v}(\boldsymbol{q}) [\widetilde{F}_+(\boldsymbol{q}, \omega) + \widetilde{F}_-(\boldsymbol{q}, \omega)]} . \tag{4.3.25}$$

This is valid for both the ferromagnetic and the paramagnetic states. In the paramagnetic state it reduces to

$$\chi_{\mathrm{ee}}(\boldsymbol{q}, \omega) = e^2 \frac{2\widetilde{F}(\boldsymbol{q}, \omega)}{1 + 2\mathrm{v}(\boldsymbol{q}) \widetilde{F}(\boldsymbol{q}, \omega)} \tag{4.3.26}$$

$$= e^2 \frac{\chi_{zz}(\boldsymbol{q}, \omega)/\mu_{\mathrm{B}}{}^2}{1 + \mathrm{v}(\boldsymbol{q}) \chi_{zz}(\boldsymbol{q}, \omega)/\mu_{\mathrm{B}}{}^2} \tag{4.3.26'}$$

$$= e^2 \frac{2F(\boldsymbol{q}, \omega)}{1 + [2\mathrm{v}(\boldsymbol{q}) - \widetilde{V}(\boldsymbol{q})] F(\boldsymbol{q}, \omega)}, \tag{4.3.26''}$$

where $\chi_{zz}(\boldsymbol{q}, \omega)$ is the spin susceptibility given by (4.3.17). The result in the form of (4.3.26") was obtained by Hubbard [4.30]. However, the direct and close relation between the charge and spin susceptibilities given in (4.3.26') was not noticed before our work [4.23].

The results of (4.3.25) and (4.3.26) are very useful in various problems. We will base our entire discussion on phonons in a metal on these results. Here we examine some of their fundamental aspects.

Let us first see how the difficulty of (4.2.38) is resolved with (4.3.25). If we put $\chi_{\mathrm{ee}}(\boldsymbol{q}, 0)$ of (4.3.25) in place of $\chi^0_{\mathrm{ee}}(\boldsymbol{q}, 0)$ into (4.2.38), we have

$$-e\int\langle n(\boldsymbol{r})\rangle d\boldsymbol{r} = -\lim_{q\to 0} e^2 \frac{\widetilde{F}_+(\boldsymbol{q},0)+\widetilde{F}_-(\boldsymbol{q},0)}{1+\frac{4\pi e^2}{Vq^2}\left[\widetilde{F}_+(\boldsymbol{q},0)+\widetilde{F}_-(\boldsymbol{q},0)\right]}\cdot\frac{4\pi Ze}{Vq^2}$$

$$= -Ze. \tag{4.3.27}$$

The total number of electrons polarized to screen an external point charge *Ze* put into a metal is now correctly *Z*. Note that this result is valid for the ferromagnetic state as well as for the paramagnetic state.

The electron charge response is often discussed in terms of the *dielectric constant,* or, the *screening constant.* Let us see how the result of (4.3.25) can be related to the screening constant.

First, in the paramagnetic state, we can rewrite (4.3.24) as

$$\langle n(\boldsymbol{q},\omega)\rangle = \frac{2\widetilde{F}(\boldsymbol{q},\omega)}{1+2\mathrm{v}(\boldsymbol{q})\widetilde{F}(\boldsymbol{q},\omega)}\, eU(\boldsymbol{q}) \equiv 2F(\boldsymbol{q},\omega)\frac{eU(\boldsymbol{q})}{\varepsilon(\boldsymbol{q},\omega)}, \tag{4.3.28}$$

by introducing the screening constant as

$$\varepsilon(\boldsymbol{q},\omega) = 1 + \left(2\mathrm{v}(\boldsymbol{q}) - \widetilde{V}(\boldsymbol{q})\right)F(\boldsymbol{q},\omega). \tag{4.3.29}$$

The Coulomb interaction between electrons serves to screen the charge potential and to modify it from $U(\boldsymbol{q})$ to $U(\boldsymbol{q})/\varepsilon(\boldsymbol{q},\omega)$; the response to the *screened potential* is, then, given by that of free electrons without interaction between them.

When we neglect the effect of the exchange interaction in (4.3.29), we obtain the well known result,

$$\varepsilon_0(\boldsymbol{q},\omega) = 1 + 2\mathrm{v}(\boldsymbol{q})F(\boldsymbol{q},\omega). \tag{4.3.30}$$

The *Thomas-Fermi screening constant* is obtained by putting $\omega = 0$ and, then, $F(\boldsymbol{q},0) \cong N(0)$ so that

$$\varepsilon_{\mathrm{TF}}(\boldsymbol{q}) = 1 + 2\mathrm{v}(\boldsymbol{q})N(0). \tag{4.3.31}$$

The screening constant of the form of (4.3.29) including the exchange effect was first noted by Hubbard [4.30].

When we take into account the role of the exchange interaction, the screening constant depends upon the particular charges to be screened; the screening constant of a pair of external charges, such as ions, is different from that of an electron and an external charge. We will discuss this in 4.3.7.

Let us proceed to discuss the screening constant in the ferromagnetic state of a metal. Starting from (4.3.24) we obtain the relations corresponding to (4.3.28) and (4.3.29) of the paramagnetic state,

$$\langle n_\sigma(\boldsymbol{q},\omega)\rangle = F_\sigma(\boldsymbol{q},\omega)\,\frac{e\,U(\boldsymbol{q})}{\varepsilon_\sigma(\boldsymbol{q},\omega)}\,, \tag{4.3.32}$$

and the spin dependent screening constant [4.31],

$$\varepsilon_\sigma(\boldsymbol{q},\omega) = \left[1-\widetilde{V}(\boldsymbol{q})F_\sigma(\boldsymbol{q},\omega)\right]\left\{1+\mathrm{v}(\boldsymbol{q})\left[\widetilde{F}_+(\boldsymbol{q},\omega)+\widetilde{F}_-(\boldsymbol{q},\omega)\right]\right\}. \tag{4.3.33}$$

In the ferromagnetic state we have $\varepsilon_+(\boldsymbol{q},\omega) \neq \varepsilon_-(\boldsymbol{q},\omega)$. This implies that while the original charge potential $U(\boldsymbol{q})$ is independent of electron spin, the screened charge potential, $U(\boldsymbol{q})/\varepsilon_\sigma(\boldsymbol{q},\omega)$, is spin dependent.

In we show how the static screening constant $\varepsilon_\sigma(\boldsymbol{q},0) \equiv \varepsilon_\sigma(\boldsymbol{q})$ behaves for different spins as a function of the magnetization of the electrons. In carrying out this numerical calculation we used the same model and method as in Fig.4.5.

As we can anticipate from the expression of (4.3.33), we find $\varepsilon_\sigma(\boldsymbol{q})$ becomes negative in small $\boldsymbol{q}$ region for one of the spins; in the present case it is for – spin electrons that the screening constant becomes negative. This implies that if a positive point charge is brought into such an electron system in the ferromagnetic state, electrons with – spins are repelled from it, while electrons with + spins are attracted to it.

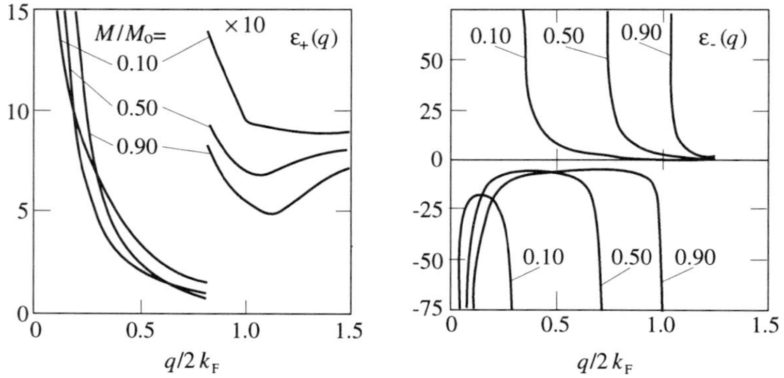

Fig.4.6 Spin dependent screening constant of a ferromagnetic electron gas for different relative magnetizations of the system. We assume $n_+ > n_-$ as illustrated in Fig.3.5.

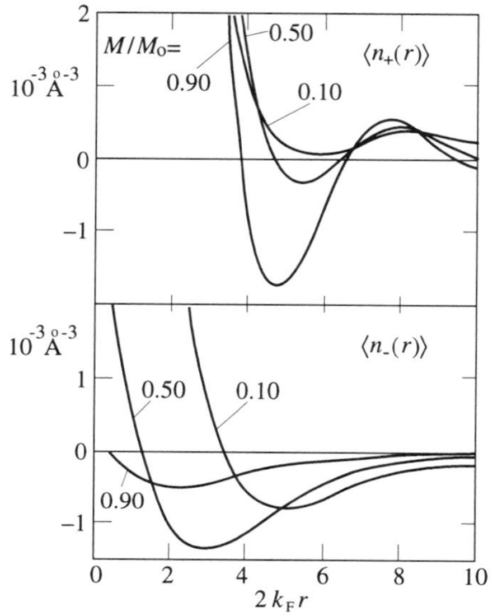

Fig.4.7 The electron density polarization for each spin of a ferromagnetic electron gas produced by a unit positive point charge for different magnetizations of the system.

In order to examine the situation more closely, in Fig.4.7 we show how the electrons of ± spins are polarized when a unit positive point charge is brought into a metal in the ferromagnetic state. We used the same model and method in obtaining the results in Fig.4.7 as were used in Figs.4.5 and 4.6; the point charge is located at $r = 0$. While the amplitude of $n_+(\boldsymbol{r})$ is dominantly positive at small r, that of $n_-(\boldsymbol{r})$ is negative in the same region; for a most illustrative situation, see the case of $M/M_0 = 0.9$.

For a more direct illustration, let us calculate separately the total number of ± spin electrons, $\langle n_\pm \rangle$, attracted to a unit positive point charge embedded in a metal in the ferromagnetic state. First, from (4.3.24) we have

$$\langle n_\pm \rangle = \int d\boldsymbol{r} \,\langle n_\pm(\boldsymbol{r}) \rangle = \lim_{\boldsymbol{q}\to 0} \frac{\widetilde{F}_\pm(\boldsymbol{q},0)}{1 + \dfrac{4\pi e^2}{V\boldsymbol{q}^2}[\widetilde{F}_+(\boldsymbol{q},0) + \widetilde{F}_-(\boldsymbol{q},0)]} \cdot \frac{4\pi e^2}{V\boldsymbol{q}^2}$$

$$= \frac{N_\pm(0)\,[1 - \widetilde{V} N_\mp(0)\,]}{N_+(0) + N_-(0) - 2\widetilde{V}\, N_+(0)\, N_-(0)} \,. \qquad (4.3.34)$$

Then, the numerical result of Fig.4.8 is obtained by using the same model and method as were used in Figs.4.5–4.7.

As we anticipated, $\langle n_- \rangle < 0$, while $\langle n_+ \rangle > 0$. And, the magnitudes of both of $\langle n_+ \rangle$ and $\langle n_- \rangle$ can be much larger than unity, while the condition of

$$\langle n_+ \rangle + \langle n_- \rangle = 1 \tag{4.3.35}$$

is satisfied.

4.3.5 Magnetic Response to Charge Potential: $\chi_{\rm me}(\boldsymbol{q}, \omega)$

According to the result of the preceding subsection, in the ferromagnetic state of a metal we have $\langle n_+(\boldsymbol{r}) \rangle \neq \langle n_-(\boldsymbol{r}) \rangle$ for the response to a charge potential. This implies that a charge potential can produce a magnetization. Let us define the magnetic response to a charge potential as

$$\chi_{\rm me}(\boldsymbol{q}, \omega) = \frac{-\mu_{\rm B}\big(\langle n_+(\boldsymbol{q}, \omega) \rangle - \langle n_-(\boldsymbol{q}, \omega) \rangle \big)}{-U(\boldsymbol{q})}. \tag{4.3.36}$$

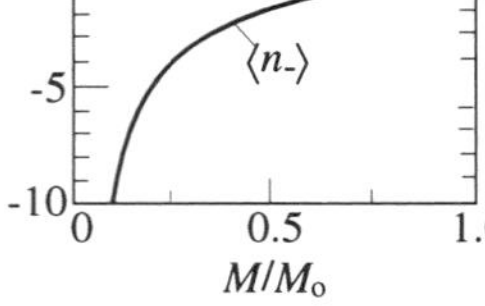

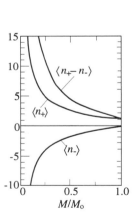

Fig.4.8 Total screening charge of each spin around a unit positive charge as the function of the magnetization of an electron gas. This result is derived from those of Fig.4.7.

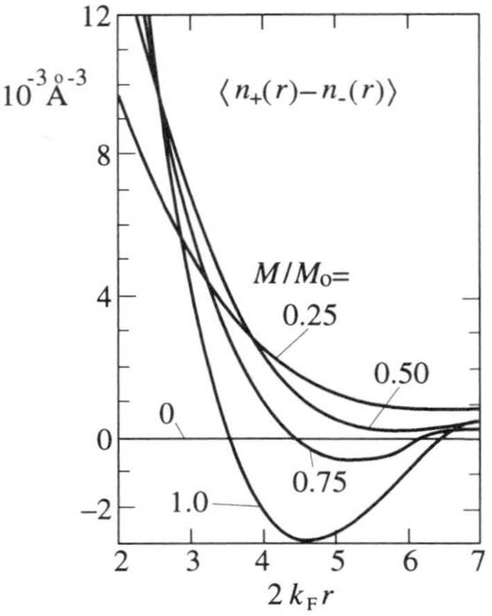

Fig.4.9 Spin polarization of a ferromagnetic electron gas due to a unit point charge as a function of M/M_0.

From (4.3.24), or, (4.3.32) and (4.3.33) we find, then, that

$$\chi_{\rm me}(\boldsymbol{q}, \omega) = \chi_{\rm em}(\boldsymbol{q}, \omega), \tag{4.3.37}$$

where $\chi_{\rm em}(\boldsymbol{q}, \omega)$ is given in (4.3.21).

In Fig. 4.9 we show how a positive unit point charge put into a metal in the ferromagnetic state induces a magnetization around it. While

$$-\mu_{\rm B}\langle \overline{M}(\boldsymbol{r})\rangle = -\frac{1}{V}\sum_{\boldsymbol{q}} \chi_{\rm me}(\boldsymbol{q}, 0)\frac{4\pi e}{Vq^2}\, e^{i\boldsymbol{q}\boldsymbol{r}}, \tag{4.3.38}$$

in Fig.4.8 we calculated $\langle \overline{M}(\boldsymbol{r})\rangle = \langle n_+(\boldsymbol{r})\rangle - \langle n_-(\boldsymbol{r})\rangle$ by using the same model and method as in Figs.4.5–4.7, for different values of the magnetization, M/M_0, of the electron gas. The total magnetization induced by the unit positive point charge is obtained by integrating the both sides of (4.3.38) over the volume,

$$-\mu_{\rm B}\int \langle \overline{M}(\boldsymbol{r})\rangle d\boldsymbol{r} = -\mu_{\rm B}\big(\langle n_+\rangle - \langle n_-\rangle\big) = -\lim_{q\to 0}\chi_{\rm me}(\boldsymbol{q}, 0)\frac{4\pi e}{Vq^2}$$

$$\cong -\mu_{\rm B}\frac{N_+(0) - N_-(0)}{N_+(0) + N_-(0) - 2\widetilde{V}N_+(0)N_-(0)}. \tag{4.3.39}$$

The numerical result for $(\langle n_+\rangle - \langle n_-\rangle)$ from this expression is given in Fig.4.8.

As can be seen from Fig.4.8, in the ferromagnetic state of a metal with a small magnetization, $M/M_0 \ll 1$, a unit positive point charge can induce an electron density polarization such that

$$\langle n_+\rangle \gg 1, \quad \langle n_-\rangle < 0 \quad \text{with} \quad |n_-| \gg 1, \tag{4.3.40}$$

and therefore,

$$\langle n_+\rangle - \langle n_-\rangle \gg 1. \tag{4.3.41}$$

The magnitude of the magnetic moment an electron can have is $1\mu_{\rm B}$. However, in the ferromagnetic state a unit positive point charge put into a metal can produce a magnetization much larger than $1\mu_{\rm B}$. This rather surprising possibility does not yet seem to be widely known. The result with $\widetilde{V} = 0$ in (4.3.39) had been obtained earlier by Friedel [4.8]; see also Gomes and Campbell [4.32]. For this case without the exchange effect, however, we have always $|\langle n_+\rangle - \langle n_-\rangle|/n = |N_+(0) - N_-(0)|/(N_+(0) + N_-(0)) \le 1$ in place of

(4.3.41). The role of the exchange effect on the charge response or screening constant is really dramatic.

4.3.6 Role of the Exchange Interaction in the Charge Response

In the preceding subsection we encountered the interesting possibility that the spin-dependent screening of a unit positive charge in a ferromagnetic metal can result in an induced magnetic moment in much excess of $1\mu_B$. At the base of this possibility is the exchange interaction between electrons. Recall that the energies of the direct Coulomb repulsion and the exchange interaction due to an electron polarization $\langle n(\boldsymbol{q})\rangle = \langle n_+(\boldsymbol{q})\rangle + \langle n_-(\boldsymbol{q})\rangle$ are given as $1/2\,\mathrm{v}(\boldsymbol{q})\langle n(\boldsymbol{q})\rangle^2$ and $-1/2\,\widetilde{V}(\boldsymbol{q})[\langle n_+(\boldsymbol{q})\rangle^2 + \langle n_-(\boldsymbol{q})\rangle^2]$, respectively. Thus, the exchange energy favors spin-polarized screening over screening which is spin unpolarized.

Note that the screening charge sum rule of (4.3.27) is independent of the exchange interaction between electrons. The exchange effect does not change the total number of polarized electrons; its effect is to change the spatial behavior of the electron density polarization of each spin, both in the ferromagnetic and the paramagnetic states.

In discussing the effect of the exchange interaction on the charge response in the paramagnetic state, let us first discuss how $\widetilde{V}(\boldsymbol{q})$ depends on $\boldsymbol{q}$. Since $\widetilde{V}(\boldsymbol{q})$ was introduced by an averaging process such as in (4.3.5), its dependence on $\boldsymbol{q}$ will not be as sensitive as that of $\mathrm{v}(\boldsymbol{q})$. Further, in a more advanced treatment, as we will see in Chapter 8, the effective exchange interaction turns out to be screened; this also tends to make $\widetilde{V}(\boldsymbol{q})$ less dependent on $\boldsymbol{q}$. Thus, we might set

$$\widetilde{V}(\boldsymbol{q}) = \frac{4\pi e^2}{V(\kappa_o^2 + \boldsymbol{q}^2)}, \tag{4.3.42}$$

with $\kappa_o \cong k_F$ [4.30].

With this form for $\widetilde{V}(\boldsymbol{q})$ we can examine $\chi_{ee}(\boldsymbol{q}, \omega)$ given by (4.3.26"). For small q the electron density response is not much affected by the exchange effect. It is for $q \cong k_F$ where $\widetilde{V}(\boldsymbol{q}) \cong \mathrm{v}(\boldsymbol{q})$ that $\chi_{ee}(\boldsymbol{q}, \omega)$ is significantly enhanced by the exchange effect. Thus, if we put a point charge in a metal, it is in the region within the distance $r \leq 1/k_F$ from the point charge that the electron polarization is exchange enhanced.

Finally, note that with the Thomas-Fermi screening constant of (4.3.31) we obtain the screened Coulomb interaction in the form of (4.3.42),

$$\frac{\mathrm{v}(\boldsymbol{q})}{\varepsilon_{TF}(\boldsymbol{q})} = \frac{4\pi e^2}{V(\kappa_{TF}^2 + \boldsymbol{q}^2)}, \tag{4.3.43}$$

with

$$\kappa_{\mathrm{TF}} = \sqrt{8\pi e^2 N(0)/V} \ . \tag{4.3.44}$$

If we use for $N(0)$ its value for the free electrons, (1.3.5), we find

$$\kappa_{\mathrm{TF}} = k_{\mathrm{F}} \left(\frac{2e^2 k_{\mathrm{F}}}{\pi \hbar^2 k_{\mathrm{F}}^2 /2m} \right)^{1/2} \cong O(k_{\mathrm{F}}). \tag{4.3.45}$$

As is well known, (4.3.43) leads to the following screened Coulomb potential

$$\sum_q \frac{4\pi e^2}{V(\kappa_{\mathrm{TF}}^2 + q^2)} e^{iqr} = \frac{e^2}{r} \exp(-\kappa_{\mathrm{TF}} r) \ . \tag{4.3.46}$$

4.3.7 The Effect of the Exchange Interaction on the Screening Constants; Two Kinds of Screening Constants

One of the most fundamental questions in the physics of metals is: What happens when we put an external point charge into a metal? The most basic answer to this question is the screening of the point charge potential with the Thomas-Fermi screening constant of (4.3.31) and the results of (4.3.43)–(4.3.46). The next step in the progress of our understanding is to obtain the screening constant of (4.3.30). The third stage of our progress is to obtain the Hubbard screening constant of (4.3.29) which includes the effect of the exchange interaction between electrons; note that this exchange effect is often called the *local field correction* in the literature. And, then, we extend these ideas to include the ferromagnetic case and obtained the results given in (4.3.32) and (4.3.33).

Note that when the effect of the exchange interaction between electrons is included the screening constant is dependent upon the charges to be screened. The screening constants of (4.3.29) and (4.3.33) are for the screening of the interaction between an external charge and electrons. As we will show next, the screening constant of the interaction between a pair of external charges in a metal is quite different from (4.3.29) or (4.3.33); by an *external charge* we mean a *non-electronic charge*. Let us see this in the following.

A *Screening of the interaction between a pair of external point charges*

Suppose first that an external point charge $Z_1 e$ is put at $\boldsymbol{R}_1$ in a metal. Then we place another external point charge $Z_2 e$ at $\boldsymbol{R}_2$, and ask what interaction potential it will feel due to the presence of the first point charge $Z_1 e$. If we write this interaction energy as $W(\boldsymbol{R}_2 - \boldsymbol{R}_1)$, it is given as

$$W(\boldsymbol{R}_2 - \boldsymbol{R}_1) = W^{\circ}(\boldsymbol{R}_2 - \boldsymbol{R}_1) - Z_2\, e^2 \int \frac{\langle n_1(\boldsymbol{r})\rangle}{|\boldsymbol{R}_2 - \boldsymbol{r}|}\, d\boldsymbol{r}, \tag{4.3.47}$$

where

$$W^{\circ}(\boldsymbol{R}_2 - \boldsymbol{R}_1) = \frac{Z_2\, Z_1\, e^2}{|\boldsymbol{R}_2 - \boldsymbol{R}_1|} \tag{4.3.48}$$

is the direct Coulomb interaction energy between the external point charges, and $\langle n_1(\boldsymbol{r})\rangle$ is the electron density polarization at $\boldsymbol{r}$ induced by the first point charge $Z_1 e$ at $\boldsymbol{R}_1$. From (4.2.33), the Fourier component of $\langle n_1(\boldsymbol{r}) \rangle$ is given in terms of the electron density response of (4.3.25) as

$$e \langle n_1 (\boldsymbol{q}) \rangle = \chi_{\mathrm{ee}}(\boldsymbol{q}, 0)\, U_1^{\circ}(\boldsymbol{q}), \tag{4.3.49}$$

where

$$U_1^{\circ}(\boldsymbol{q}) = \frac{4\pi Z_1 e}{Vq^2}\, e^{-i\boldsymbol{q}\boldsymbol{R}_1} \tag{4.3.50}$$

is the Fourier transform of the potential, $U^{\circ}(\boldsymbol{r}-\boldsymbol{R}_1) = Z_1 e \,/\, |\, \boldsymbol{r}-\boldsymbol{R}_1\, |$, due to the first point charge. Finally, by Fourier transforming the both sides of (4.3.47) and putting (4.3.25) into it we obtain

$$\begin{aligned} W(\boldsymbol{q}) &= W^{\circ}(\boldsymbol{q}) \left[1 - \frac{4\pi}{Vq^2}\, \chi_{\mathrm{ee}}(\boldsymbol{q}, 0) \right] \\ &= \frac{W^{\circ}(\boldsymbol{q})}{1 + \mathrm{v}(\boldsymbol{q}) \left[\widetilde{F}_{+}(\boldsymbol{q}, 0) + \widetilde{F}_{-}(\boldsymbol{q}, 0) \right]}, \end{aligned} \tag{4.3.51}$$

where

$$W^{\circ}(\boldsymbol{q}) = \frac{4\pi Z_2 Z_1 e^2}{Vq^2}. \tag{4.3.52}$$

Note that we arrive at the same result by pursuing what potential the point charge $Z_1 e$ will feel due to the presence of the point charge $Z_2 e$; $W(\boldsymbol{R}_1 - \boldsymbol{R}_2) = W(\boldsymbol{R}_2 - \boldsymbol{R}_1)$.

From (4.3.51) it is natural to identify the screening constant of the interaction between a pair of external charges in a metal as

$$\varepsilon_e(\boldsymbol{q},\,\omega) \;=\; 1 + \mathrm{v}(\boldsymbol{q})\big[\widetilde{F}_+(\boldsymbol{q},\,\omega) + \widetilde{F}_-(\boldsymbol{q},\,\omega)\,\big]\,. \tag{4.3.53}$$

In the paramagnetic states it reduces to

$$\begin{aligned}\varepsilon_e(\boldsymbol{q},\,\omega) \;&=\; 1 + 2\mathrm{v}(\boldsymbol{q})\frac{F(\boldsymbol{q},\,\omega)}{1-\widetilde{V}(\boldsymbol{q})\,F(\boldsymbol{q},\,\omega)} \\ &=\; 1 + \mathrm{v}(\boldsymbol{q})\,\chi_{zz}(\boldsymbol{q},\omega)\,/\mu_B{}^2,\end{aligned} \tag{4.3.54}$$

where $\chi_{zz}(\boldsymbol{q},\,\omega)$ is the exchange enhanced spin susceptibility of (4.3.17).

These results are quite different from the corresponding ones of (4.3.33) and (4.3.29) for the screening constants of the interaction between an external charge and electrons in a metal. In the case of the paramagnetic state, while $\varepsilon_e(\boldsymbol{q},\,\omega)$ of (4.3.54) is enhanced from $\varepsilon_0(\boldsymbol{q},\omega)$ of (4.3.30), $\varepsilon(\boldsymbol{q},\,\omega)$ of (4.3.29) is reduced from $\varepsilon_0(\boldsymbol{q},\,\omega)$ by the exchange effect. Let us next discuss how this difference comes out.

B *Screening of the interaction between an external point charge and electrons*

Suppose an external point charge Ze is put at $\boldsymbol{R}$ in a metal. Then the interaction energy between this point charge and the *test* electron density $n_\sigma(\boldsymbol{r})$ may be expressed in terms of the unscreened Coulomb potential $U^{\mathrm{o}}(\boldsymbol{r}-\boldsymbol{R})$ of the external charge as in the right hand side of the following equation,

$$\begin{aligned}-e\int\langle\, n_\sigma(\boldsymbol{r})\,\rangle\, U_{\mathrm{sc}\sigma}(\boldsymbol{r}-\boldsymbol{R})\,d\boldsymbol{r} \;&\equiv\; -e\int\langle\, n_\sigma(\boldsymbol{r})\,\rangle\, U^{\mathrm{o}}(\boldsymbol{r}-\boldsymbol{R})\,d\boldsymbol{r} \\ &+ \int \mathrm{v}\,(\boldsymbol{r}-\boldsymbol{r}')\,\langle\; n_\sigma(\boldsymbol{r})\,n\,(\boldsymbol{r}')\,\rangle\, d\boldsymbol{r}\,d\boldsymbol{r}'\;,\end{aligned} \tag{4.3.55}$$

where $\mathrm{v}\,(\boldsymbol{r}-\boldsymbol{r}')$ is the Coulomb interaction between electrons, and $\langle n(\boldsymbol{r}')\rangle$ is the electron density induced by the external point charge; $U_{\mathrm{sc}\sigma}(\boldsymbol{r}-\boldsymbol{R})$ is the screened potential of the point charge to be obtained. This equation is Fourier transformed into the following form,

$$\begin{aligned}-e\sum_{\boldsymbol{q}}\langle\, n_\sigma(-\boldsymbol{q})\,\rangle\, U_{\mathrm{sc}\sigma}(\boldsymbol{q})\; e^{-i\boldsymbol{q}\boldsymbol{R}} \;&=\; -e\sum_{\boldsymbol{q}}\langle\, n_\sigma(-\boldsymbol{q})\,\rangle\, U^{\mathrm{o}}(\boldsymbol{q})\; e^{-i\boldsymbol{q}\boldsymbol{R}} \\ &+ \sum_{\boldsymbol{q}} \mathrm{v}(\boldsymbol{q})\langle\, n_\sigma(-\boldsymbol{q})\; n(\boldsymbol{q})\,\rangle\,.\end{aligned} \tag{4.3.55'}$$

If we apply the mean field approximation as described in 4.3.1 on the second term in the right hand side of (4.3.55'), we have

$$\mathrm{v}(\boldsymbol{q})\,\langle n_\sigma(-\boldsymbol{q})n(\boldsymbol{q})\rangle \;\cong\; \mathrm{v}(\boldsymbol{q})\,\langle n(\boldsymbol{q})\rangle\langle n_\sigma(-\boldsymbol{q})\rangle - \widetilde{V}(\boldsymbol{q})\,\langle n_\sigma(\boldsymbol{q})\rangle\langle n_\sigma(-\boldsymbol{q})\rangle\,. \tag{4.3.56}$$

Here the thermal expectations $\langle n(\boldsymbol{q})\rangle$ and $\langle n_\sigma(\boldsymbol{q})\rangle$ are to be taken by including the effect of the external point charge Ze. These thermal expectations are then given similarly to (4.3.49) and (4.3.24). By putting such explicit expressions into (4.3.55') and eliminating the common factor, we finally obtain

$$\begin{aligned}-e\,\langle n_\sigma(-\boldsymbol{q})\rangle\,U_{\mathrm{sc}\sigma}(\boldsymbol{q}) \;&=\; -e\langle n_\sigma(-\boldsymbol{q})\rangle U^{\mathrm{o}}(\boldsymbol{q})\\ &+\left[\mathrm{v}(\boldsymbol{q})\frac{\widetilde{F}_{+}(\boldsymbol{q},0)+\widetilde{F}_{-}(\boldsymbol{q},0)}{1+\mathrm{v}(\boldsymbol{q})\left\{\widetilde{F}_{+}(\boldsymbol{q},0)+\widetilde{F}_{-}(\boldsymbol{q},0)\right\}}\right.\\ &\left.-\widetilde{V}(\boldsymbol{q})\frac{\widetilde{F}_{\sigma}(\boldsymbol{q},0)}{1+\mathrm{v}(\boldsymbol{q})\left\{\widetilde{F}_{+}(\boldsymbol{q},0)+\widetilde{F}_{-}(\boldsymbol{q},0)\right\}}\right] e\,\langle n_\sigma(-\boldsymbol{q})\rangle\,U^{\mathrm{o}}(\boldsymbol{q})\,.\end{aligned} \tag{4.3.57}$$

This result can be rewritten as

$$U_{\mathrm{sc}\sigma}(\boldsymbol{q}) \;=\; \frac{U^{\mathrm{o}}(\boldsymbol{q})}{\varepsilon_\sigma(\boldsymbol{q},0)}\,, \tag{4.3.58}$$

with the spin dependent screening constant ε_σ given in (4.3.33); the potential due to the external charge is modified by screening from $U^{\mathrm{o}}(\boldsymbol{q})$ to $U_{\mathrm{sc}\sigma}(\boldsymbol{r})$; in the ferromagnetic state the screened potential is spin dependent.

If we neglect the second term on the right hand side of (4.3.56), then the second term in the bracket on the right hand side of (4.3.57) will be eliminated and we will end up with the screening constant $\varepsilon_{\mathrm{e}}(\boldsymbol{q},\,\omega)$ of (4.3.53). In (4.3.55), while $n_\sigma(\boldsymbol{r})$ is the test electron charge density, $n(\boldsymbol{r}')$ is the density of the electrons polarized to screen the external charge. Then, the second term on the right hand side of (4.3.56) represents the exchange interaction between the test and the screening electron densities. When the test charge is not electronic, there can be no such exchange interaction between the test charge and the screening electrons. This is the physical origin of the difference between the two kinds of screening constants.

Finally, how about the screening of the Coulomb interaction between electrons? We will discuss this subject in Chapter 10 (see (10.2.6)). We will find there still another form of screening constant.

4.3.8 Electrical Resistance due to Non-magnetic Impurities in a Ferromagnetic Metal

As an illustrative application of the result of (4.3.33) on the screening of the interaction between an electron and an external charge, let us discuss the electrical resistance of a ferromagnetic metal due to non-magnetic impurities [4.33].

While the electrical conductivity σ of a metal consists of contributions from $\pm$ spin electrons with

$$\sigma = \sigma_+ + \sigma_- , \tag{4.3.59}$$

in the ferromagnetic state with the magnetization M the conductivities of $\pm$ spin electrons are, in general, different with

$$\sigma_\pm(M) = \frac{e^2 n_\pm \tau_\pm(\varepsilon_{F\pm})}{m^*} , \tag{4.3.60}$$

where m^* is the effective mass of an electron and $\tau_\pm$ is the relaxation time of $\pm$ spin electrons. In the paramagnetic state where $n_+ = n_- = n/2$ and $\tau_+ = \tau_- = \tau$, we obtain the familiar expression

$$\sigma(0) = \frac{e^2 n\tau}{m^*} . \tag{4.3.61}$$

The spin dependent relaxation time of an electron due to impurities in the ferromagnetic state is obtained following the standard procedure [4.34] as

$$\frac{1}{\tau_\pm} = N_i \frac{\pi}{\hbar} N_\pm(\varepsilon_{F\pm}) \int_0^\pi \left| U_{sc\pm}(2k_{F\pm}\sin\frac{\theta}{2}) \right|^2 (1-\cos\theta)\, d\theta , \tag{4.3.62}$$

where θ is the electron scattering angle, N_i is the number of impurities, and

$$U_{sc\pm}(\boldsymbol{q}) = \frac{U(\boldsymbol{q})}{\varepsilon_\pm(\boldsymbol{q}, 0)} \tag{4.3.63}$$

is the screened impurity potential with the screening constant of (4.3.33); if we represent the impurity atom by a point charge, $U(\boldsymbol{q})$ is given as in (4.2.36). It was earlier noted that in the ferromagnetic state of a metal the electron relaxation time becomes spin dependent through the spin dependence of the factor $N_\pm(\varepsilon_{F\pm})$ in (4.3.62). Now, from the above result, we find that the screened impurity potential also has a spin dependence originating with the screening constant $\varepsilon_{\pm}(\boldsymbol{q}, 0)$ in (4.3.63).

For a parabolic electron energy band we have

$$\sigma_{\pm}(M) = \sigma_0 \frac{(1 \pm M/M_0)^2}{\Theta_{\pm}(M)} \tag{4.3.64}$$

with

$$\sigma_0 = \frac{4n\varepsilon_F^2}{\pi Z^2 e^2 \hbar k_F N_i}, \tag{4.3.65}$$

and

$$\Theta_{\pm}(M) = \int_0^{\pi} d\theta\,(1 - \cos\theta)\,\sin\theta$$

$$\times \left\{ \varepsilon_{\pm}\left[2k_F(1 \pm M/M_0)^{1/3} \sin^2(\theta/2) \right] \right\}^{-2}. \tag{4.3.66}$$

Let us carry out a numerical calculation on (4.3.64)–(4.3.66) for an electron gas with $k_F = 10^8$/cm as in previous examples. For the effective exchange interaction we assume the form of $\widetilde{V}(\boldsymbol{q}) = \widetilde{V}(0)/\left[1 + Cq^2/k_F\right]$ and consider the cases of $C = 0$ and 1. Thus we obtain the results of Fig.4.10, where we put $\sigma_+(0) = \sigma_-(0) \equiv \sigma(0)$. We also give the corresponding electrical resistance, $\rho(M) = 1/\sigma$ in Fig.4.11, where we put $\rho_0 = 1/\sigma_0$.

The results of Fig.4.10 show the possibility of observing $\sigma_- > \sigma_+$ although $n_+ > n_-$ and $N_+(0) > N_-(0)$ in the ferromagnetic state of a metal. It is because we have $\tau_+ < \tau_-$, and this effect overrides that of $n_+ > n_-$ in (4.3.60), or (4.3.64). Correspondingly, the electrical resistance behaves as in Fig.4.11. The broken line in Fig.4.11, which is almost independent of the magnetization, is the electrical resistance calculated with the RPA screening constant obtained by putting $\widetilde{V}(\boldsymbol{q}) = 0$ in (4.3.33). The difference in the behaviors of the resistance is striking. In a variety of ferromagnetic alloys a large increase and/or a minimum in the electrical resistance has been observed as the temperature is lowered below T_C. See, for instance, Refs. [4.35–4.39].

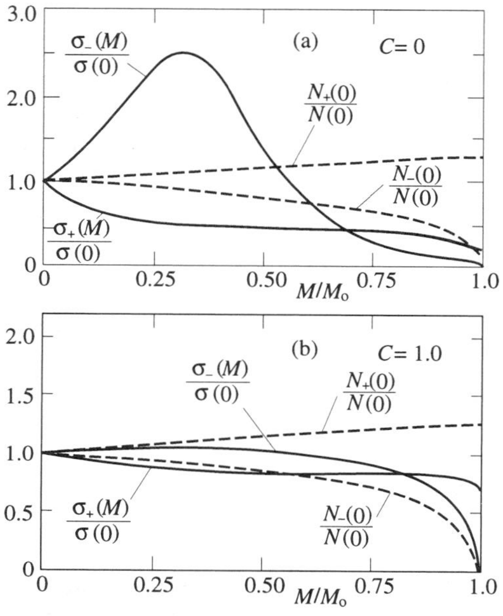

Fig.4.10 The conductivity of the majority "+" spin and minority "−" spin electrons as a function of the relative magnetization, M/M_o, of the conduction electrons, $\sigma_+(0) = \sigma_-(0) = \sigma(0)$. The density of states at the Fermi surface of each spin band is shown by a broken line.

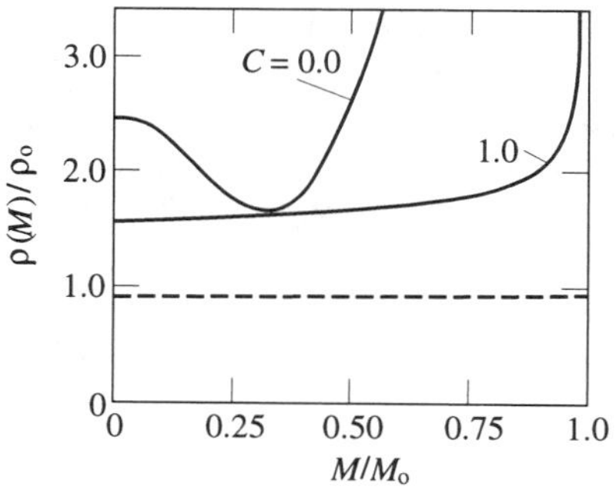

Fig.4.11 Resistivity as a function of magnetization for two values of the range of the exchange interaction. The broken line is the resistivity using the RPA screening constant which does not completely include the exchange effect.

Other interesting observations which may be related to the results of Fig.4.10 are the recent spin polarized photo-emission [4.40] and tunneling [4.41] experiments on ferromagnetic metals. In both experiments the relative numbers of ± spin electrons which emerge from states near the Fermi surface, either by photo-emission or tunneling, are opposite to that expected based on the bulk densities of states at the Fermi surface of ± spin electrons. In both of these experiments, the electrons inside a ferromagnetic metal must travel to the surface before either being emitted, or tunneled. Therefore, the electron mean free path is involved and it seems possible to understand the experiments of spin polarized photo-emission and tunneling by a mechanism discussed in the above.

More recently, in studying the magneto-resistive behavior of n–type $Cd_{1-x}Mn_xSe$, a magnetic semiconductor, Shapira et al. [4.42] noted the usefulness of the findings of this subsection; the Thomas-Fermi or RPA screening of the ionized impurity potential, which gives rise to the band splitting dependence of electrical resistance given by the broken line in Fig.4.11, was found to lead a magnetic field dependence of resistivity which is much smaller than the observed one.

A situation similar to the above is also anticipated for the electrical resistance due to phonons. A model numerical calculation was recently carried out which shows that it is possible to have $\sigma_+ < \sigma_-$ although $n_+ > n_-$ [4.43].

4.3.9 The Effect of Electron Correlation on Linear Responses

In this section we have been systematically discussing the effect of the Coulomb interaction on the magnetic and charge responses of a metallic electron gas. Thus we have found, for instance, the relation of (4.3.26') which will play a central role in the next chapter in relating the phonon frequency of a metal to its magnetic properties. Note, however, that in (4.3.26') both $\chi_{ee}(\boldsymbol{q},\omega)$ and $\chi_{zz}(\boldsymbol{q},\omega)$ are the results of the mean field approximation; such a simple relation between χ_{ee} and χ_{zz} breaks down if we go beyond the mean field approximation; see, for instance, Ref. [4.44]. Very pleasingly, however, $\chi_{ee}(\boldsymbol{q},\omega)$ turns out to be exchange enhanced even more than is implied by the relation (4.3.26'), as we will see below.

The fundamental equations of (4.3.10) and (4.3.23) can be combined to give

$$\langle n_\sigma(\boldsymbol{q},\omega)\rangle = F_\sigma(\boldsymbol{q},\omega)\Big[-\sigma\mu_B H_z(\boldsymbol{q}) + eU(\boldsymbol{q}) - V_{\sigma,\sigma}(\boldsymbol{q})\langle n_\sigma(\boldsymbol{q},\omega)\rangle - V_{\sigma,-\sigma}(\boldsymbol{q})\langle n_{-\sigma}(\boldsymbol{q},\omega)\rangle\Big], \tag{4.3.67}$$

with

$$V_{\sigma,\sigma}(\boldsymbol{q}) \;\cong\; \mathrm{v}(\boldsymbol{q}) - \widetilde{V}(\boldsymbol{q}), \tag{4.3.68a}$$

$$V_{\sigma,-\sigma}(\boldsymbol{q}) \;\cong\; \mathrm{v}(\boldsymbol{q}), \tag{4.3.68b}$$

in the mean field approximation. The origin of the difference between $V_{\sigma,\sigma}(\boldsymbol{q})$ and $V_{\sigma,-\sigma}$ is the fact that a pair of electrons with parallel spins tend to avoid each other owing to the Pauli principle. This effect is absent in the case of a pair of electrons with antiparallel spins. However, even without the Pauli principle, a pair of electrons will tend to stay apart in order to reduce the Coulomb repulsion energy. Such an effect is called the *correlation effect.* The correlation effect is present for a pair of electrons with spins either parallel or antiparallel. However, in the case of a parallel spin electron pair, since they are already kept apart by the Pauli principle's requirements, the additional effect of correlation is much less significant than for the case of an antiparallel spin electron pair. If we represent the effect on the Coulomb energy of the correlation as $-\Delta V_{\sigma,\sigma}$ and $-\Delta V_{\sigma,-\sigma}$, with positive $\Delta V_{\sigma,\sigma}$ and $\Delta V_{\sigma,-\sigma}$, we have $\Delta V_{\sigma,\sigma} \ll \Delta V_{\sigma,-\sigma}$, and, therefore

$$V_{\sigma,\sigma}(\boldsymbol{q}) \;=\; \mathrm{v}(\boldsymbol{q}) - \widetilde{V}(\boldsymbol{q}) - \Delta V_{\sigma,\sigma}(\boldsymbol{q}), \tag{4.3.69a}$$

$$V_{\sigma,-\sigma}(\boldsymbol{q}) \;=\; \mathrm{v}(\boldsymbol{q}) \;-\; \Delta V_{\sigma,-\sigma}(\boldsymbol{q}). \tag{4.3.69b}$$

With (4.3.69) in (4.3.67), in the paramagnetic state, where $V_{++} = V_{--}$, we have

$$\begin{aligned}\chi_{zz}(\boldsymbol{q},\omega) &= 2\mu_B^2 \frac{F(\boldsymbol{q},\omega)}{1 - \widetilde{V}(\boldsymbol{q})\, F(\boldsymbol{q},\omega)} \\ &= 2\mu_B^2 \frac{F(\boldsymbol{q},\omega)}{1 - V_m(\boldsymbol{q})\, F(\boldsymbol{q},\omega)}, \end{aligned}\tag{4.3.70}$$

with

$$V_m(\boldsymbol{q}) \cong \widetilde{V}(\boldsymbol{q}) \quad , \tag{4.3.71}$$

and

$$\chi_{ee}(\boldsymbol{q},\omega) \;=\; 2e^2 \frac{F(\boldsymbol{q},\omega)}{1 + \{V_{+-}(\boldsymbol{q}) + V_{++}(\boldsymbol{q})\}\, F(\boldsymbol{q},\omega)}$$

$$\chi_{ee}(\boldsymbol{q},\omega) \;=\; 2e^2 \frac{F(\boldsymbol{q},\omega)}{1 + \{V_{+-}(\boldsymbol{q}) + V_{++}(\boldsymbol{q})\}\, F(\boldsymbol{q},\omega)}$$

$$= 2e^2 \frac{\dfrac{F(\boldsymbol{q}, \omega)}{1 - V_{\mathrm{e}}(\boldsymbol{q}) F(\boldsymbol{q}, \omega)}}{1 + 2\mathrm{v}(\boldsymbol{q}) \dfrac{F(\boldsymbol{q}, \omega)}{1 - V_{\mathrm{e}}(\boldsymbol{q}) F(\boldsymbol{q}, \omega)}}, \qquad (4.3.72)$$

with

$$V_{\mathrm{e}}(\boldsymbol{q}) \cong \widetilde{V}(\boldsymbol{q}) + \Delta V_{+-}(\boldsymbol{q}) > V_{\mathrm{m}}(\boldsymbol{q}). \qquad (4.3.73)$$

These results can be compared with those of (4.3.26) and (4.3.17). When the correlation effects are taken into account, certainly the exchange enhancement of the magnetic response is reduced from the mean field approximation result. Note, however, that the charge susceptibility is exchange enhanced even more strongly than it is anticipated from the relation between the charge and the magnetic susceptibilities within the mean field approximation. This conclusion may be more directly expressed by rewriting (4.3.72) in terms of the magnetic susceptibility which includes correlation effects, (4.3.70), as

$$\bar{\chi}_{\mathrm{ee}}(\boldsymbol{q}, \omega) = \frac{\bar{\chi}_{zz}(\boldsymbol{q}, \omega)}{1 + \left\{\mathrm{v}(\boldsymbol{q}) - \frac{1}{2}\Delta V_{+-}(\boldsymbol{q})\right\} \bar{\chi}_{zz}(\boldsymbol{q}, \omega)}. \qquad (4.3.72')$$

We use throughout this book the conventions,

$$\bar{\chi}_{\mathrm{ee}}(\boldsymbol{q}, \omega) = \chi_{\mathrm{ee}}(\boldsymbol{q}, \omega) / e^2, \quad \bar{\chi}_{zz}(\boldsymbol{q}, \omega) = \chi_{zz}(\boldsymbol{q}, \omega) / \mu_{\mathrm{B}}^2, \text{ etc.} \qquad (4.3.74)$$

4.4 LINEAR RESPONSE AND COLLECTIVE EXCITATION

We saw that the divergence of a static response function is associated with a phase transition. $\chi_{zz}(0, 0) = \infty$, for instance, implies a ferromagnetic transition. More generally, $\chi_{zz}(\boldsymbol{q}, 0) = \infty$ for some $\boldsymbol{q} \neq 0$ implies the spin density wave (SDW) [4.45]. $\chi_{\mathrm{ee}}(\boldsymbol{q}, 0) = \infty$ is associated with a charge density wave (CDW) transition [4.46] (Strictly speaking, however, this statement is not correct as we will see in 5.3.3).

What, then, is implied by the divergence of a dynamic response function, $\chi(\boldsymbol{q}, \omega) = \infty$? It implies that without any external field there can be an oscillation with a frequency ω and a wave number $\boldsymbol{q}$ of some physical quantity such as spin or charge densities in a system. It is an excitation mode of the system with an energy $\hbar\omega$ and a wave number $\boldsymbol{q}$. Thus, the spectrum of the electronic plasma oscillation of a metal is derived from the poles of $\chi_{\mathrm{ee}}(\boldsymbol{q}, \omega)$. In this section we discuss the spin wave excitations of an itinerant electron ferromagnet with such an approach [3.13, 3.14].

4.4.1 Transverse Magnetic Susceptibility: $\chi_{-+}(\boldsymbol{q},\omega)$

According to the result illustrated in Fig.2.4, a spin wave in a localized spin ferromagnet is a propagation of the precession of spins in clockwise direction in the x-y plane; we assume spins are orderd in the direction of the z-axis in the ground state. In order to induce such a motion of spins we need a magnetic field of the form

$$\boldsymbol{H} = -H\left[\hat{\boldsymbol{x}}\cos(\omega t+\boldsymbol{qr}) - \hat{\boldsymbol{y}}\sin(\omega t+\boldsymbol{qr})\right], \tag{4.4.1}$$

where $\hat{\boldsymbol{x}}$, and $\hat{\boldsymbol{y}}$ are the unit vectors in the directions of the x-and y-axes, respectively, and $H>0$. When this magnetic field is applied to an electron system it attains the following Zeeman energy:

$$H' = \mu_{\rm B}H\left[\sigma_+(-\boldsymbol{q})\,e^{i\omega t} + \sigma_-(\boldsymbol{q})\,e^{-i\omega t}\right], \tag{4.4.2}$$

where, as we already showed in (4.2.50),

$$\sigma_+(-\boldsymbol{q}) = \sum_{\boldsymbol{k}} a^+_{\boldsymbol{k}+}a_{\boldsymbol{k}-\boldsymbol{q},-}\,; \quad \sigma_-(\boldsymbol{q}) = \sum_{\boldsymbol{k}} a^+_{\boldsymbol{k}-}a_{\boldsymbol{k}+\boldsymbol{q},+}. \tag{4.4.3}$$

Since the two terms on the right hand side of (4.4.2) are Hermite conjugate to each other, it suffices to consider one of them; we retain the first one,

$$H' = \mu_{\rm B}H\sigma_+(-\boldsymbol{q})\,e^{i\omega t}. \tag{4.4.4}$$

The quantity $\langle\sigma_-(\boldsymbol{q})\rangle$ responds to such an external perturbation with the time dependence of $\exp(i\omega t)$, as we will see below. Let us obtain this response by using the prescription of (4.2.4).

We need to pursue the following equation:

$$-\hbar\omega\left\langle a^+_{\boldsymbol{k}-}a_{\boldsymbol{k}+\boldsymbol{q},+}\right\rangle = \left\langle\left[a^+_{\boldsymbol{k}-}a_{\boldsymbol{k}+\boldsymbol{q},+}\,,\,H_{\rm o}\right]\right\rangle + \left\langle\left[a^+_{\boldsymbol{k}-}a_{\boldsymbol{k}+\boldsymbol{q},+}\,,\,H_{\rm C}\right]\right\rangle$$

$$+\,{}_{\rm o}\!\left\langle\left[a^+_{\boldsymbol{k}-}a_{\boldsymbol{k}+\boldsymbol{q},+}\,,\,H'\right]\right\rangle, \tag{4.4.5}$$

where

$$\left[a^+_{\boldsymbol{k}-}a_{\boldsymbol{k}+\boldsymbol{q},+}\,,\,H_{\rm o}\right] = \left(\varepsilon_{\boldsymbol{k}+\boldsymbol{q}} - \varepsilon_{\boldsymbol{k}}\right)a^+_{\boldsymbol{k}-}a_{\boldsymbol{k}+\boldsymbol{q},+}, \tag{4.4.6}$$

$$\left[a^+_{\boldsymbol{k}-}a_{\boldsymbol{k}+\boldsymbol{q},+}\,,\,H'\right] = \mu_{\rm B}He^{i\omega t}\left[a^+_{\boldsymbol{k}-}a_{\boldsymbol{k}-} - a^+_{\boldsymbol{k}+\boldsymbol{q},+}a_{\boldsymbol{k}+\boldsymbol{q},+}\right]. \tag{4.4.7}$$

As for the commutator involving H_C, we first rewrite H_C within the mean field approximation. Noting that the system is under the external magnetic perturbation of (4.4.4), we have

$$H_{\mathrm{C,m}} = -\widetilde{V}(0)\sum_{\boldsymbol{k},\sigma} n_\sigma a^+_{\boldsymbol{k}\sigma} a_{\boldsymbol{k}\sigma} - \widetilde{V}(\boldsymbol{q})\langle\sigma_-(\boldsymbol{q})\rangle\sigma_+(-\boldsymbol{q}). \tag{4.4.8}$$

The first term on the right hand side can be incorporated into the one particle energy as in (4.3.7). As for the second term, since it is of the same structure as that of (4.4.4), the commutator involving it can be calculated similarly to (4.4.7). Thus, (4.4.5) is reduced to

$$\begin{aligned}&\left(\varepsilon_{\boldsymbol{k}+\boldsymbol{q},+} - \varepsilon_{\boldsymbol{k}-} + \hbar\omega\right)\left\langle a^+_{\boldsymbol{k}-}\, a_{\boldsymbol{k}+\boldsymbol{q},+}\right\rangle \\ &\qquad = \left(f(\varepsilon_{\boldsymbol{k}-}) - f(\varepsilon_{\boldsymbol{k}+\boldsymbol{q},+})\right)\left[-\mu_{\mathrm{B}}H\, e^{i\omega t} + \widetilde{V}(\boldsymbol{q})\left\langle\sigma_-(\boldsymbol{q})\right\rangle\right].\end{aligned} \tag{4.4.9}$$

From this equation we obtain

$$\left\langle\sigma_-(\boldsymbol{q})\right\rangle = \sum_{\boldsymbol{k}}\left\langle a^+_{\boldsymbol{k}-}\, a_{\boldsymbol{k}+\boldsymbol{q},+}\right\rangle = -\frac{F_{-+}(\boldsymbol{q},-\omega)}{1-\widetilde{V}(\boldsymbol{q})F_{-+}(\boldsymbol{q},-\omega)}\,\mu_{\mathrm{B}}H\, e^{i\omega t}, \tag{4.4.10}$$

where we introduced the transverse Lindhard function

$$F_{-+}(\boldsymbol{q},\omega) = \sum_{\boldsymbol{k}}\frac{f(\varepsilon_{\boldsymbol{k}-}) - f(\varepsilon_{\boldsymbol{k}+\boldsymbol{q},+})}{\varepsilon_{\boldsymbol{k}+\boldsymbol{q},+} - \varepsilon_{\boldsymbol{k}-} - \hbar\omega}. \tag{4.4.11}$$

In discussing a linear response so far we have been assuming the time dependence of an external field as $\exp(i\omega t)$. According to such convention the time dependence in (4.4.4) is to be associated with a negative frequency, $-\omega$. Since we define the transverse Lindhard function for a conventional positive frequency as in (4.4.11), the frequency arguments of the Lindhard functions appearing in (4.4.10) have negative sign.

From (4.4.10) and (4.4.11), the transverse spin susceptibility is defined, and obtained as

$$\chi_{-+}(\boldsymbol{q},\omega) = \frac{-\mu_{\mathrm{B}}\langle\sigma_-(\boldsymbol{q})\rangle}{H} = \mu_{\mathrm{B}}^2\,\frac{F_{-+}(\boldsymbol{q},\omega)}{1-\widetilde{V}(\boldsymbol{q})F_{-+}(\boldsymbol{q},\omega)}. \tag{4.4.12}$$

Then, the result of (4.4.10) may be rewritten as

$$-\mu_B \langle \sigma_-(\boldsymbol{q}) \rangle = \chi_{-+}(\boldsymbol{q}, -\omega)\, H\, e^{i\omega t}. \tag{4.4.10'}$$

$\chi_{+-}(\boldsymbol{q}, \omega)$ can be similarly defined, and obtained simply by replacing the spin subscripts $\pm$ by $\mp$ in (4.4.11) and (4.4.12). In the paramagnetic state we have

$$\chi_{-+}(\boldsymbol{q}, \omega) = \chi_{+-}(\boldsymbol{q}, \omega) = 1/2\, \chi_{zz}(\boldsymbol{q}, \omega). \tag{4.4.13}$$

4.4.2 Spin Waves from the Poles of the Transverse Spin Susceptibility

From our discussion in 4.4.1, we expect the condition $\chi_{-+}(\boldsymbol{q}, -\omega) = \infty$ to give the spectrum of the spin waves. Let us show how such expectation is actually realized [3.14].

As we can see from (4.4.2), in place of the poles of $\chi_{-+}(\boldsymbol{q}, -\omega)$ we may examine those of $\chi_{+-}(\boldsymbol{q}, \omega)$. Then from

$$\chi_{+-}(\boldsymbol{q}, \omega) = \infty,$$

we have

$$1 = \widetilde{V}(\boldsymbol{q}) \sum_{\boldsymbol{k}} \frac{f(\varepsilon_{\boldsymbol{k}+}) - f(\varepsilon_{\boldsymbol{k}+\boldsymbol{q},-})}{\varepsilon_{\boldsymbol{k}+\boldsymbol{q},-} - \varepsilon_{\boldsymbol{k}+} - \hbar\omega}. \tag{4.4.14}$$

Our goal is to obtain the relation between ω and $\boldsymbol{q}$ such that (4.4.14) is satisfied. If we put $\boldsymbol{q} = 0$ in (4.4.14), we have

$$1 = \widetilde{V} \frac{n_+ - n_-}{\widetilde{V}\left(n_+ - n_-\right) - \hbar\omega}. \tag{4.4.15}$$

Thus, we find that $\omega = 0$ for $\boldsymbol{q} = 0$. Note that since the right hand side of (4.4.14) is invariant to the inversion $\boldsymbol{q} \to -\boldsymbol{q}$, ω should be an even function of $\boldsymbol{q}$. Thus, for a small $\boldsymbol{q}$ we will have

$$\hbar\omega_{\mathrm{sw}}(\boldsymbol{q}) = Dq^2. \tag{4.4.16}$$

Let us derive an explicit expression for the exchange stiffness, D.

We assume that the system is in the ferromagnetic state, as illustrated in Fig.3.5(a), and put the magnitude of the spin splitting of the bands as

$$\Delta = \varepsilon_{\boldsymbol{k}-} - \varepsilon_{\boldsymbol{k}+} = \widetilde{V}(n_+ - n_-). \tag{4.4.17}$$

Since for $\boldsymbol{q} \to 0$ we have

$$\hbar\omega / \Delta \ll 1, \qquad \left|\varepsilon_{\boldsymbol{k}+\boldsymbol{q},-} - \varepsilon_{\boldsymbol{k}+}\right| / \Delta \ll 1, \tag{4.4.18}$$

the right hand side of (4.4.14) can be expanded in the following way,

$$\begin{aligned} 1 &= \widetilde{V} \sum_{\boldsymbol{k}} \left[f(\varepsilon_{\boldsymbol{k}+}) - f(\varepsilon_{\boldsymbol{k}+\boldsymbol{q},-}) \right] \frac{1}{\Delta} \left[1 + \frac{1}{\Delta}\left(\varepsilon_{\boldsymbol{k}+\boldsymbol{q}} - \varepsilon_{\boldsymbol{k}}\right) - \frac{\hbar\omega}{\Delta} \right]^{-1} \\ &= \widetilde{V} \sum_{\boldsymbol{k}} \left[f(\varepsilon_{\boldsymbol{k}+}) - f(\varepsilon_{\boldsymbol{k}+\boldsymbol{q},-}) \right] \frac{1}{\Delta} \left[1 - \left\{ \frac{1}{\Delta}\left(\varepsilon_{\boldsymbol{k}+\boldsymbol{q}} - \varepsilon_{\boldsymbol{k}}\right) - \frac{\hbar\omega}{\Delta} \right\} \right. \\ &\qquad \left. + \left\{ \frac{1}{\Delta}\left(\varepsilon_{\boldsymbol{k}+\boldsymbol{q}} - \varepsilon_{\boldsymbol{k}}\right) - \frac{\hbar\omega}{\Delta} \right\}^2 + \cdots \right], \end{aligned} \tag{4.4.19}$$

where we put $\widetilde{V}(\boldsymbol{q}) = \widetilde{V}$. From this equation we arrive at

$$\begin{aligned} \hbar\omega &= \frac{\widetilde{V}}{\Delta} \sum_{\boldsymbol{k}} \left(f(\varepsilon_{\boldsymbol{k}+}) + f(\varepsilon_{\boldsymbol{k}-}) \right) \frac{1}{2} (\boldsymbol{q}\nabla_{\boldsymbol{k}})^2 \varepsilon_{\boldsymbol{k}} \\ &\quad - \left(\frac{1}{\Delta}\right)^2 \widetilde{V} \sum_{\boldsymbol{k}} \left(f(\varepsilon_{\boldsymbol{k}+}) - f(\varepsilon_{\boldsymbol{k}-}) \right) (\boldsymbol{q}\nabla_{\boldsymbol{k}}\varepsilon_{\boldsymbol{k}})^2 , \end{aligned} \tag{4.4.20}$$

where $\nabla_{\boldsymbol{k}} = (\partial/\partial k_x, \partial/\partial k_y, \partial/\partial k_z)$.

For the parabolic energy dispersion of electrons, $\varepsilon_k = \hbar^2 k^2/2m$, D of (4.4.16) is explicitly obtained to be

$$D = \frac{\hbar^2}{2m}\left(\frac{n}{n_+ - n_-}\right)\left[1 - \frac{4}{5}\,\frac{1}{\widetilde{V}(n_+ - n_-)}\,\frac{n_+\varepsilon_{F+} - n_-\varepsilon_{F-}}{n}\right], \tag{4.4.21}$$

where $\varepsilon_{F\pm} = \hbar^2 k_{F\pm}^2 / 2m$ is the Fermi energy of ± spin electrons measured from the bottom of each spin band. Note we have $D \propto (n_+ - n_-)$ for $(n_+ - n_-) \to 0$.

As we already mentioned the existence of spin waves in an itinerant electron ferromagnet was also discussed by Herring and others [3.9–3.12] with the method of 2.5.1 analogously to the case of localized spin system. More discussion will be given on spin wave in later chapters. In 6.6.2, for instance, we show that the long standing dogma that an excitation of a spin wave causes a decrease of magnetization by $2\mu_B$ should be drastically modified.

4.4.3 Stoner Excitations and Spin Waves

It is not only the spin wave excitations that satisfies the condition of (4.4.14). Let us see this. For given $\boldsymbol{k}$ and $\boldsymbol{q}$, the denominator of the corresponding term on the right hand side of (4.4.14) vanishes for an ω such that

$$\hbar\omega = \varepsilon_{k+q,-} - \varepsilon_{k+} > 0. \tag{4.4.22}$$

Thus, if

$$f(\varepsilon_{k+}) - f(\varepsilon_{k+q,-}) > 0, \tag{4.4.23}$$

(4.4.14) will be satisfied with an ω close to that of (4.4.22). For both (4.4.22) and (4.4.23) to be valid, we require

$$\varepsilon_{k+} < \mu, \quad \varepsilon_{k+q,-} > \mu, \tag{4.4.24}$$

where μ is the chemical potential. Thus, this excitation is to remove a + spin electron from inside of the Fermi sphere and put it at outside of the Fermi surface with a reversal of its spin. We call this the *Stoner excitation*.

For the parabolic energy dispersion of electron, (4.4.22) is rewritten as

$$\hbar\omega = \frac{\hbar^2}{m}\left(\boldsymbol{k}\cdot\boldsymbol{q} + \frac{1}{2}q^2\right) + \tilde{V}(n_+ - n_-). \tag{4.4.25}$$

Here, for a given $\boldsymbol{q}$, $\boldsymbol{k}\cdot\boldsymbol{q}$ can take the following range of values

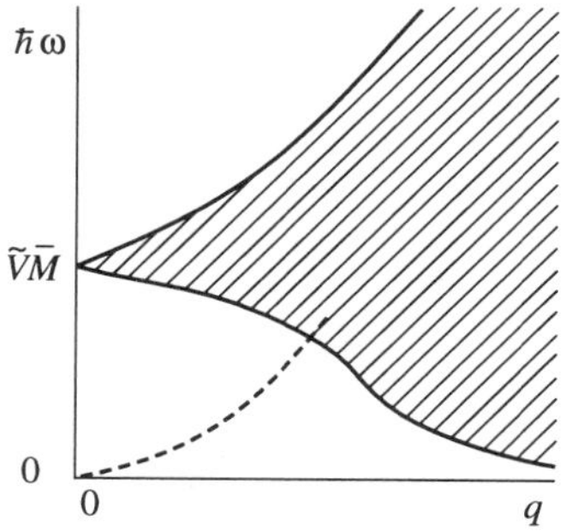

Fig.4.12 The spin wave dispersion (the broken line), and the Stoner excitations (the shaded area).

$$- k_{\mathrm{F}+}q \; < \; \boldsymbol{k} \cdot \boldsymbol{q} \; < \; k_{\mathrm{F}+}q. \tag{4.4.26}$$

Thus, unlike the case of the spin wave, the Stoner excitation with a wave number q can have its energy in a wide range as shown by the shaded region in Fig.4.12.

In Fig.4.12, the spin wave dispersion is shown by a broken line. At a large value of q, the broken line merges into the shaded region; the spin wave mingles with the Stoner excitations. What will occur at this merging point ? Continuing efforts have been made by means of neutron diffraction to clarify such problems. For a recent review and references, see, for instance, Cook et al. [4.47] and Mook [4.48].

PROBLEMS FOR CHAPTER 4

4.1 Confirm the two basic commutation relations of (4.2.9) and (4.2.10).

4.2 Derive the result of (4.2.20) for the static Lindhard function of an electron gas at $T = 0$ by noting

$$\sum_k \frac{f(\varepsilon_k) - f(\varepsilon_{k-q})}{\varepsilon_{k+q} - \varepsilon_k} = 2\sum_k \frac{f(\varepsilon_k)}{\varepsilon_{k+q} - \varepsilon_k} .$$

c.f. Appendix A.

4.3 Derive the RKKY oscillation of (4.2.24). See Ref. [4.4].

4.4 Integrate (4.2.23) over the entire volume of the system and discuss the physical implication of the result.

4.5 Derive the *s-d* exchange Hamiltonian of (4.2.50).

4.6 Obtain the high field magnetic susceptibility of (3.4.24) from (4.3.16).

4.7 Obtain $\chi_{zz}(\boldsymbol{q}, \omega)$ of the ferromagnetic state for the Hubbard model of (3.5.17) with the mean field approximation. Confirm that the result is reproduced from (4.3.16) by replacing every $v(\boldsymbol{q})$ and $V(\boldsymbol{q})$ appearing in it with I/N. Then, show that the Hubbard model result fails to reproduce the high field magnetic susceptibility of (3.4.24).

4.8 Obtain the result (4.3.46) for the Thomas-Fermi screened potential.

4.9 Obtain the electron plasma frequency, $\omega_{\rm plo} = [4\pi\, n\, e^2 / V\, m]^{1/2}$, from the pole of the charge susceptibility (4.3.26), without the exchange effect ($\tilde{V}(\boldsymbol{q}) = 0$). Discuss the effects of the exchange interaction, and, then, the ferromagnetic spin splitting on the plasma frequency.

4.10 Obtain the electron charge susceptibility and screening constants in the Hubbard model within the mean field approximation. As in the problem 4.6, the Hubbard model results will be reproduced from the long range Coulomb interaction model of the text by replacing every $v(\boldsymbol{q})$ and $\tilde{V}(\boldsymbol{q})$ in the latter with I/N. Show that such Hubbard model result can not properly yield the electron plasma oscillation.

4.11 Derive (4.4.2) from (4.4.1).

4.12 From (4.4.21) show that the exchange stiffness constant vanishes as $D \propto (n_+ - n_-)$ for $(n_+ - n_-) \to 0$.

4.13 Derive the spin wave spectrum of a ferromagnetic electron gas with the method given in 2.5.1. See Refs. [3.9–3.12].

Chapter 5

ELECTRON-PHONON INTERACTION: THE EFFECT OF MAGNETISM ON LATTICE VIBRATIONS

Based on the results of Chapter 4, in Chapters 5–7 we discuss how closely the phonon and elasticity properties and the magnetic properties are related in a metal. We start our discussion by exploring how directly the magnetic property of a metal is reflected in its phonon spectrum and elastic property. A more thorough discussion of this subject based on the results of this and the next chapters will be given in Chapter 7.

5.1 HOW THE PHONON AND ELASTICITY PROPERTIES OF A METAL ARE DETERMINED BY ITS MAGNETIC PROPERTIES

A metal is a system in which ions form a periodic lattice and electrons are mobile. In dealing with the behavior of electrons in a metal, for simplicity we often neglect the effect of the motion of its ionic lattice. We have been making such an approximation in this book. In discussing the magnetic susceptibility of a metal, for instance, we did not take into account the possible effect of the interaction between electrons and lattice vibrations. We know, however, that there are cases in which the interaction between electrons and lattice vibrations plays an important role. An example is that of the electrical resistance of a metal; in a pure metal, the origin of electrical resistance is the scattering of electrons by the lattice vibrations. Another example is found in the superconductivity of a metal; the attraction between electrons to form a Cooper pair is caused by the interaction of electrons with lattice vibrations. The elastic and thermal properties of a metal are more directly related to the lattice vibrational properties.

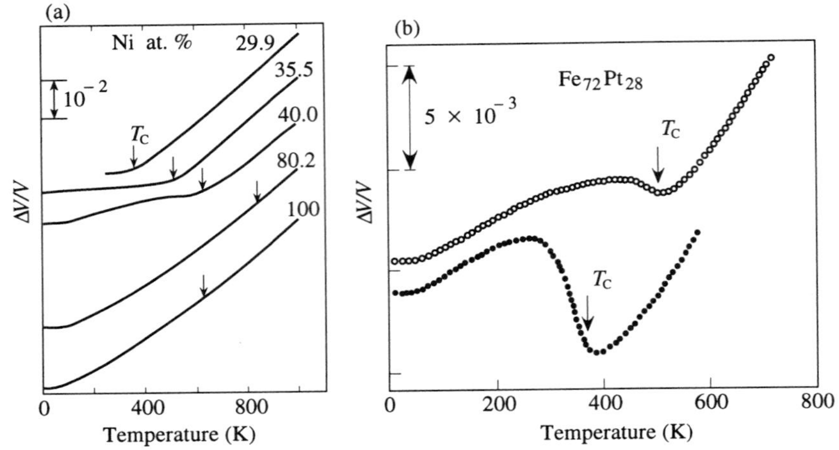

Fig.5.1 The temperature dependence of volume for FeNi alloys [5.1], (a) and $Fe_{72}Pt_{28}$ alloy [5.2], (b). The arrows show the locations of the Curie temperatures. In (b), the data with the higher T_C are for an ordered alloy and those with the lower T_C are for a disordered one.

In this chapter we discuss how the lattice vibrational properties of a metal are related to its magnetic properties. We ask how the magnetic properties are reflected in the elastic properties in a metal.

Our answer to this question is rather straightforward. While the lattice vibrational spectrum is determined by the interaction between ions, in a metal the bare ion-ion interaction is screened by conduction electrons. The screening constant, then, sensitively reflects the magnetic properties, as we saw in Chapter 4.

From this view of the close relation between the magnetic properties and the lattice vibrational spectrum, we can discuss various problems under a new light. For example, lattice vibrations are involved in structural phase transitions through the mechanism of phonon softening. Thus, a magnetic property can be a trigger of a structural phase transition in a metal.

It is well known that some ferromagnetic metals and alloys show unusual volume and elasticity behaviors. As shown in Fig.5.1(a), some FeNi alloys, for instance, show very small or zero thermal expansion below the Curie points T_C; we call such a material an *Invar*. Note that the volume behavior of $Fe_{72}Pt_{28}$ alloy is even more anomalous as shown in Fig.5.1(b); in a wide temperature region below T_C, thermal expansion becomes negative.

As shown in Fig.5.2 the elastic constant also show anomalous behavior in ferromagnetic metals and alloys. As in the case of pure Ni shown in Fig.5.2(c), however, not all metallic ferromagnets are anomalous in their elastic properties.

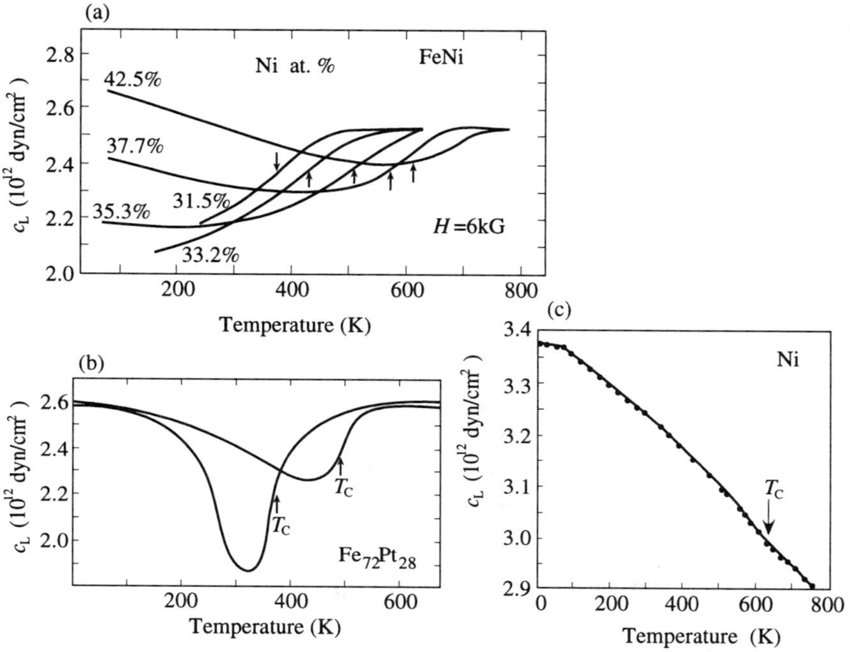

Fig.5.2 The temperature dependence of the elastic constant $c_L = 1/2(c_{11} + c_{12} + 2c_{44})$ of FeNi alloys [5.3], (a); $Fe_{72}Pt_{28}$ alloys [5.4], (b); Ni [5.5], (c). The arrows show the Curie points. In (b), the curve with the higher T_C is for an ordered alloy and that with the lower T_C is for a disordered one.

(As we will discuss in Chapter 7, even in Ni elasticity and volume behaviors become anomalous near T_C.) As to the origin of such volume and elasticity properties of ferromagnetic metals it is very controversial as of now. We will give our current view on this problem in Chapter 7. The content of this chapter is preparatory for such later discussion.

We may raise a question opposite to that of this chapter: What effect will the electron interaction with the lattice vibration have on the magnetic properties of a metal? This question has been repeatedly asked, and the currently prevailing answer is predominantly negative. In Chapter 6 we will give an answer opposite to the prevalent one. The results of this chapter will become also the basis of that discussion.

Throughout in Chapters 5–7, the central role is played by the following result for the screening of phonon frequency in a metal [5.6],

$$\omega_{\boldsymbol{q}}^2 = \Omega_{\boldsymbol{q}}^2 - |g(\boldsymbol{q})|^2 \,\overline{\chi}_{\rm ee}(\boldsymbol{q}, \omega_{\boldsymbol{q}}), \qquad (5.1.1)$$

where ω_q and Ω_q are, respectively, the screened and the bare phonon frequencies, $g(\boldsymbol{q})$ is the electron-phonon interaction constant (see (5.2.38)), and we put

$$\bar{\chi}_{\mathrm{ee}}(\boldsymbol{q}, \omega) = \chi_{\mathrm{ee}}(\boldsymbol{q}, \omega)/e^2, \tag{5.1.2}$$

$\chi_{\mathrm{ee}}(\boldsymbol{q}, \omega)$ being the electron charge susceptibility of (4.3.25), namely,

$$\bar{\chi}_{\mathrm{ee}}(\boldsymbol{q}, \omega) = \frac{\widetilde{F}_+(\boldsymbol{q}, \omega) + \widetilde{F}_-(\boldsymbol{q}, \omega)}{1 + \mathrm{v}(\boldsymbol{q})\left[\widetilde{F}_+(\boldsymbol{q}, \omega) + \widetilde{F}_-(\boldsymbol{q}, \omega)\right]}. \tag{5.1.3}$$

In (5.1.1), the second term on the right hand side represents the effect of the screening by electrons, and the magnetic properties of the metal enter through χ_{ee}.

In this chapter first we derive this fundamental result of (5.1.1) with (5.1.3), and then discuss above mentioned problems based on it.

5.2 THE ELECTRON-PHONON INTERACTION

5.2.1 Lattice Vibrations

In this subsection we summarize the basics of lattice vibrations in a solid. For a more thorough introduction, see, for instance, Refs. [5.7–5.11].

Let be designated the location of ions by $\boldsymbol{R}_i$, ($i = 1, 2, \cdots, N$) and the mass by M_{I}. For simplicity we assume the solid consists of a single kind of ions. If we let the momentum of an ion be $\boldsymbol{P}_i = M_I\dot{\boldsymbol{R}}_i$ ($\dot{\boldsymbol{R}}_i = d\boldsymbol{R}_i/dt$), the energy of the ions is given by

$$H_{\mathrm{ph}} = \sum_{i=1}^{N} \frac{1}{2M_{\mathrm{I}}}\boldsymbol{P}_i^2 + \frac{1}{2}\sum_{i \neq j} W(\boldsymbol{R}_i - \boldsymbol{R}_j), \tag{5.2.1}$$

where $W(\boldsymbol{R}_i - \boldsymbol{R}_j)$ is the interaction potential of ions. Referring to the equilibrium position $\boldsymbol{R}_i^{\circ}$ of ions, the ionic position is written as

$$\boldsymbol{R}_i = \boldsymbol{R}_i^{\circ} + \boldsymbol{u}_i, \tag{5.2.2}$$

with the ionic displacement $\boldsymbol{u}_i$. The interaction potential between ions may then be expanded in terms of $\boldsymbol{u}_i$ as

$$\frac{1}{2}\sum_{i \neq j} W(\boldsymbol{R}_i - \boldsymbol{R}_j) = \frac{1}{2}\sum_{i \neq j} \Big[W(\boldsymbol{R}_i^\circ - \boldsymbol{R}_j^\circ)$$

$$+ \Big\{ (\boldsymbol{u}_i \nabla_i)(\boldsymbol{u}_i \nabla_i) + (\boldsymbol{u}_i \nabla_i)(\boldsymbol{u}_j \nabla_j) \Big\} W(\boldsymbol{R}_i^\circ - \boldsymbol{R}_j^\circ) \Big]$$

$$\equiv \frac{1}{2}\sum_{i \neq j} W(\boldsymbol{R}_i^\circ - \boldsymbol{R}_j^\circ) + \frac{1}{2}\sum_{i,\,j} \boldsymbol{u}_i A_{ij} \boldsymbol{u}_j, \tag{5.2.3}$$

where ∇_i is the nabla with respect to $\boldsymbol{R}_i^\circ$. We retained terms only up to the second order in the displacements. The reason why no terms linear in $\boldsymbol{u}_i$ appear is that at the equilibrium positions we have $\nabla_i W(\boldsymbol{R}_i^\circ - \boldsymbol{R}_j^\circ) = 0$. Note that each of A_{ij} is a tensor with components such as $A_{ix,\,jy}$.

With the ion-ion interaction of (5.2.3) the equation of motion for ions are obtained as

$$M_{\mathrm{I}} \ddot{\boldsymbol{u}}_i = -\sum_j A_{ij} \boldsymbol{u}_j, \tag{5.2.4}$$

with the following obvious relation

$$A_{ij} = A_{ji}\,. \tag{5.2.5}$$

Although (5.2.4) appears to be a Newtonian equation, it is derived quantum mechanically by using the following commutation relations

$$[\, R_{i\mu}\,, P_{j\nu} \,] = [\, u_{i\mu}\,, P_{j\nu} \,] = i\hbar\, \delta_{i\mu,\, j\nu}\,,$$

$$[\, R_{i\mu}\,, R_{j\nu} \,] = [\, u_{i\mu}\,, u_{j\nu} \,] = [\, P_{i\mu}\,, P_{j\nu} \,] = 0, \tag{5.2.6}$$

μ, ν being x, y or z.

In solving (5.2.4), we put

$$\boldsymbol{u}_i = \frac{1}{(NM_{\mathrm{I}})^{1/2}} \sum_{\boldsymbol{q},\lambda}^{\mathrm{1BZ}} \varepsilon_{\boldsymbol{q}\lambda}\, Q_{\boldsymbol{q}\lambda}\, e^{i\boldsymbol{q}\boldsymbol{R}_i^\circ}, \tag{5.2.7}$$

by introducing the *normal coordinates*,

$$Q_{\boldsymbol{q}\lambda} \propto \exp(-i\Omega_{\boldsymbol{q}\lambda}\, t\,), \tag{5.2.8}$$

which represent the amplitude of a plane wave of ions with the wave vector $\boldsymbol{q}$ and the polarization $\varepsilon_{\boldsymbol{q}\lambda}$. The polarization vector $\varepsilon_{\boldsymbol{q}\lambda}$ represents the unit vector in the direction of the ionic oscillation. In the present case, for each $\boldsymbol{q}$ we have one longitudinal mode ($\varepsilon_{q\lambda} // q$) corresponding to, say, $\lambda = 1$, and two transverse modes ($\varepsilon_{q\lambda} \perp q$) corresponding to $\lambda = 2$, and 3. Since q is given as in (2.5.5) and confined to the 1st Brillouin zone, the total number of lattice vibration modes is $3N$.

The inverse transformation of (5.2.7) is given by

$$Q_{q\lambda} = \left(\frac{M_{\mathrm{I}}}{N}\right)^{1/2} \sum_{i=1}^{N} \boldsymbol{u}_i\, \varepsilon_{q\lambda}\, e^{-i\boldsymbol{q}\boldsymbol{R}_i^{\circ}}. \tag{5.2.9}$$

If we put $\boldsymbol{u}(\boldsymbol{q}) = (1/NM_{\mathrm{I}})^{1/2}\Sigma_{\lambda}\varepsilon_{q\lambda}\, Q_{q\lambda}$, (5.2.9) constitutes the ordinary Fourier transform of $\boldsymbol{u}_i$. Corresponding to (5.2.7) and (5.2.9) we define

$$\boldsymbol{P}_i = \left(\frac{M_{\mathrm{I}}}{N}\right)^{1/2} \sum_{q,\lambda}^{\mathrm{1BZ}} P_{q\lambda}\, \varepsilon_{q\lambda}\, e^{i\boldsymbol{q}\boldsymbol{R}_i^{\circ}}, \tag{5.2.10}$$

$$P_{q\lambda} = \frac{1}{(NM_{\mathrm{I}})^{1/2}} \sum_{i=1}^{N} \boldsymbol{P}_i\, \varepsilon_{q\lambda}\, e^{-i\boldsymbol{q}\boldsymbol{R}_i^{\circ}}, \tag{5.2.11}$$

for ionic momentum. Then, we have the following simple commutation relations,

$$[\, Q_{q\lambda}, P_{q'\lambda'} \,] = i\hbar\delta_{q,q'}\, \delta_{\lambda\lambda'}; \quad [\, Q_{q\lambda}, Q_{q'\lambda'} \,] = [\, P_{q\lambda}, P_{q'\lambda'} \,] = 0. \tag{5.2.12}$$

The ion-ion potential is Fourier decomposed as

$$W(\boldsymbol{R}) = \sum_{\boldsymbol{k}} W(\boldsymbol{k})\, e^{i\boldsymbol{k}\boldsymbol{R}}, \tag{5.2.13}$$

$$W(\boldsymbol{k}) = \frac{1}{V}\int W(\boldsymbol{R})\, e^{-i\boldsymbol{k}\boldsymbol{R}}\, d\boldsymbol{R}. \tag{5.2.14}$$

Since here $\boldsymbol{R}$ is a continuous variable, $\boldsymbol{k}$ is not restricted to the 1st Brillouin zone. By putting (5.2.13) into (5.2.3) we have

$$A_{ij} = \sum_{\boldsymbol{k}} \boldsymbol{k}{*}\boldsymbol{k}\; W(\boldsymbol{k})\, e^{i\boldsymbol{k}(\boldsymbol{R}_i^{\circ}-\boldsymbol{R}_j^{\circ})} \quad \text{for } i \neq j, \tag{5.2.15}$$

$$A_{ii} = -\sum_{\boldsymbol{k}} \sum_{j(\neq i)} \boldsymbol{k}{*}\boldsymbol{k}\; W(\boldsymbol{k})\, e^{i\boldsymbol{k}(\boldsymbol{R}_i^{\circ}-\boldsymbol{R}_j^{\circ})}, \tag{5.2.16}$$

where we put

$$\boldsymbol{k}{*}\boldsymbol{k} \;\equiv\; \begin{pmatrix} k_x \\ k_y \\ k_z \end{pmatrix} (k_x,\, k_y,\, k_z)\,. \tag{5.2.17}$$

$\boldsymbol{k}{*}\boldsymbol{k}$ is a matrix whose xy component is $k_x k_y$. Thus, we have, for instance,

$$A_{ix,\,jy} \;=\; \sum_{\boldsymbol{k}} k_x k_y \; W(\boldsymbol{k})\, e^{i\boldsymbol{k}(\boldsymbol{R}_i^{\circ} - \boldsymbol{R}_j^{\circ})}.$$

With (5.2.7) and (5.2.15)–(5.2.16), the equation of motion of (5.2.4) is rewritten as

$$M_{\rm I}\sum_{\boldsymbol{q},\lambda}^{\rm 1BZ} \Omega_{\boldsymbol{q}\lambda}^2 \varepsilon_{\boldsymbol{q}\lambda}\, e^{i\boldsymbol{q}\boldsymbol{R}_i^{\circ}}\, Q_{\boldsymbol{q}\lambda} \;=\; \sum_{j(\neq i)} \sum_{\boldsymbol{k}} \boldsymbol{k}{*}\boldsymbol{k}\; W(\boldsymbol{k})\, e^{i\boldsymbol{k}(\boldsymbol{R}_i^{\circ} - \boldsymbol{R}_j^{\circ})} \sum_{\boldsymbol{q},\lambda}^{\rm 1BZ} Q_{\boldsymbol{q}\lambda}\, \varepsilon_{\boldsymbol{q}\lambda}\, e^{i\boldsymbol{q}\boldsymbol{R}_j^{\circ}}$$

$$-\sum_{j(\neq i)} \sum_{\boldsymbol{k}} \boldsymbol{k}{*}\boldsymbol{k}\; W(\boldsymbol{k})\, e^{i\boldsymbol{k}(\boldsymbol{R}_i^{\circ} - \boldsymbol{R}_j^{\circ})} \sum_{\boldsymbol{q},\lambda}^{\rm 1BZ} Q_{\boldsymbol{q}\lambda}\, \varepsilon_{\boldsymbol{q}\lambda}\, e^{i\boldsymbol{q}\boldsymbol{R}_i^{\circ}}. \tag{5.2.18}$$

We carry out the summation over j and use the relation (2.5.11) to obtain

$$\Omega_{\boldsymbol{q}\lambda}^2\, \varepsilon_{\boldsymbol{q}\lambda} \;=\; (N/M_{\rm I}) \sum_{K_n} \Big[(\boldsymbol{q} + \boldsymbol{K}_n) * (\boldsymbol{q} + \boldsymbol{K}_n)\; W(\boldsymbol{q} + \boldsymbol{K}_n)$$

$$- \boldsymbol{K}_n{*}\boldsymbol{K}_n W(\boldsymbol{K}_n) \Big]\, \varepsilon_{\boldsymbol{q}\lambda}\,, \tag{5.2.19}$$

by equating the coefficients of $Q_{\boldsymbol{q}\lambda}$ in the both sides of the equation. This is the final form of the equation to determine the phonon frequency (eigenvalue) $\Omega_{\boldsymbol{q}\lambda}$ and the corresponding phonon polarization (eigenvector) $\varepsilon_{\boldsymbol{q}\lambda}$ for each $\boldsymbol{q}$, for given ion-ion interaction $W(\boldsymbol{k})$ and crystal structure ($\boldsymbol{K}_n$, the reciprocal lattice vector).

Since the matrix multiplying to $\varepsilon_{\boldsymbol{q}\lambda}$ on the right hand side of (5.2.19), which is called the *dynamical matrix*, is hermitian, its eigenvalues $\Omega_{\boldsymbol{q}\lambda}^2$ are real, and its eigenvectors are orthogonal,

$$\varepsilon_{\boldsymbol{q}\lambda}^{+}\, \varepsilon_{\boldsymbol{q}\lambda'} \;=\; \delta_{\lambda,\,\lambda'}\,. \tag{5.2.20}$$

Then, multiplying the both sides of (5.2.19) by $\varepsilon_{\boldsymbol{q}\lambda}^{+}$ and summing over λ, we obtain the following frequency sum rule,

$$M_{\mathrm{I}} \sum_{\lambda} \Omega_{q\lambda}^{2} = NW(\boldsymbol{q})\,q^{2} + N \sum_{K_n \neq 0} \left[W(\boldsymbol{q}+\boldsymbol{K}_n)(\boldsymbol{q}+\boldsymbol{K}_n)^2 - W(\boldsymbol{K}_n)\boldsymbol{K}_n^2 \right], \tag{5.2.21}$$

where we used the relation $\sum_{\lambda} \varepsilon_{q\lambda}^{+}(\boldsymbol{k}*\boldsymbol{k})\varepsilon_{q\lambda} = \boldsymbol{k}^2$, which can be easily confirmed by using (5.2.17) and (5.2.20).

In terms of the above normal coordinates, the Hamiltonian of an ionic system, which was originally given as in (5.2.1), can be rewritten as

$$H_{\mathrm{ph}} = \frac{1}{2} \sum_{q,\lambda}^{1\mathrm{BZ}} \left(P_{q\lambda}^{+} P_{q\lambda} + \Omega_{q\lambda}^{2} Q_{q\lambda}^{+} Q_{q\lambda} \right), \tag{5.2.22}$$

where we took note of the relations

$$P_{q\lambda}^{+} = P_{-q\lambda}, \qquad Q_{q\lambda}^{+} = Q_{-q\lambda}. \tag{5.2.23}$$

This Hamiltonian represents a collection of harmonic oscillators.

If we let

$$Q_{q\lambda} = \left(\frac{\hbar}{2\Omega_{q\lambda}} \right)^{1/2} \left(b_{q\lambda} + b_{-q\lambda}^{+} \right),$$

$$P_{q\lambda} = i \left(\frac{\hbar \Omega_{q\lambda}}{2} \right)^{1/2} \left(b_{q\lambda}^{+} - b_{-q\lambda} \right), \tag{5.2.24}$$

(5.2.22) can be further rewritten as

$$H_{\mathrm{ph}} = \sum_{q,\lambda} \hbar \Omega_{q\lambda} \left(b_{q\lambda}^{+} b_{q\lambda} + \frac{1}{2} \right). \tag{5.2.25}$$

As to the new operators $b_{q\lambda}$ and $b_{q\lambda}^{+}$, we may use (5.2.12) to obtain the following boson commutation relations (see (2.5.30)),

$$[\, b_{q\lambda}, b_{q'\lambda'}^{+} \,] = \delta_{q\lambda,\, q'\lambda'}, \qquad [\, b_{q\lambda}, b_{q'\lambda'} \,] = [\, b_{q\lambda}^{+}, b_{q'\lambda'}^{+} \,] = 0. \tag{5.2.26}$$

Thus $b_{q\lambda}^{+}$ is the creation operator of a Bose particle characterized by $(\boldsymbol{q}, \lambda)$ and energy $\hbar\Omega_{q\lambda}$. We call this particle, or quasi particle, a *phonon*. In (5.2.25), $b_{q\lambda}^{+} b_{q\lambda}$ is the number operator of phonons; $\hbar\Omega_{q\lambda}/2$ represents the contribution of the zero point oscillations of ions.

The above is a general formulation to deal with lattice vibrations in a solid. We can obtain the phonon spectrum from (5.2.19) if the crystal structure $(\boldsymbol{K}_n)$

and the ion-ion interaction ($W(\boldsymbol{k})$) are given. In this book, however, we will not actually carry out such a program for a real crystal. Since our primary interest lies elsewhere, we simplify our approach by neglecting the crystal structure of a solid; we assume that the spatial distribution of ions is like that of a liquid. The *jellium model* of metal, which assumes a uniform distribution of positive ionic charges, embodies such a simplification. In this book we extensively use this jellium model with some extension.

If we assume a uniform distribution of ions, we have

$$\sum_j e^{i\boldsymbol{q}\boldsymbol{R}_j^{\circ}} = N\delta_{\boldsymbol{q},0}, \tag{5.2.27}$$

in place of (2.5.11). Then, (5.2.19) and (5.2.21) are, respectively, reduced to

$$M_{\rm I}\,\Omega_{\boldsymbol{q}\lambda}^2\,\varepsilon_{\boldsymbol{q}\lambda} = N\boldsymbol{q}*\boldsymbol{q}\,W(\boldsymbol{q})\,\varepsilon_{\boldsymbol{q}\lambda}, \tag{5.2.19'}$$

$$M_{\rm I}\sum_\lambda \Omega_{\boldsymbol{q}\lambda}^2 = N\,W(\boldsymbol{q})\,q^2. \tag{5.2.21'}$$

In the jellium model the summation over the reciprocal lattice vectors is eliminated.

Let us first see that in such a system the frequency of transverse lattice vibration becomes zero. Consider (5.2.19') for a transverse mode with $\varepsilon_{q{\rm t}} \perp \boldsymbol{q}$. If we multiply the both sides of the equation with $\varepsilon_{q{\rm t}}^{+}$ from left, we obtain

$$\Omega_{q{\rm t}}^2 = 0. \tag{5.2.28}$$

Then, the frequency of a longitudinal phonon is obtained from (5.2.21') as

$$\Omega_{q{\rm l}}^2 = \left(\frac{N}{M_{\rm I}}\right) W(\boldsymbol{q})\,q^2. \tag{5.2.29}$$

Now, let us assume that each ion is a point charge Ze, with the mass $M_{\rm I}$. Then, the ion-ion interaction, and its Fourier transform are given, respectively, as

$$W(\boldsymbol{R}_i - \boldsymbol{R}_j) = \frac{Z^2e^2}{|\boldsymbol{R}_i - \boldsymbol{R}_j|}; \quad W(\boldsymbol{q}) = \frac{4\pi Z^2 e^2}{Vq^2}. \tag{5.2.30}$$

If these ions are uniformly distributed, as in the jellium model, the frequency of a longitudinal phonon is found from (5.2.29) to be

$$\Omega_{q\mathrm{l}}^2 = \frac{4\pi Z^2 e^2 N}{M_\mathrm{I} V} \equiv \Omega_{\mathrm{pl}}^2. \tag{5.2.31}$$

We recognize this to be the *ionic plasma frequency*.

In dealing with the lattice vibration of a metal, it is essential to take into account the role of the conduction electrons. The positive ionic charges are neutralized by the electrons in the metal: $NZ = n$, n being the total number of electrons. Due to the presence of these electrons the vibrational frequency of the ions becomes drastically different from the ionic plasma frequency of (5.2.31). It is one of the central themes of this book to see how the phonon frequency is modified by the electron screening of the ions in a metal. In the next subsection we will begin our discussion of this subject.

5.2.2 The Electron-Phonon Interaction and Phonon Frequency

Let the interaction potential between an ion at $\boldsymbol{R}_i$ and an electron at $\boldsymbol{r}$ be $U(\boldsymbol{r}-\boldsymbol{R}_i)$. The interaction energy of an electron with N ions of a system is, then, given as

$$\sum_{i=1}^{N} U(\boldsymbol{r}-\boldsymbol{R}_i) = \sum_{i=1}^{N} U(\boldsymbol{r}-\boldsymbol{R}_i^{\circ}) - \sum_{i=1}^{N} \boldsymbol{u}_i \nabla_{\boldsymbol{r}} U(\boldsymbol{r}-\boldsymbol{R}_i^{\circ}), \tag{5.2.32}$$

where we used the relation $\nabla_{\boldsymbol{R}_i} = -\nabla_{\boldsymbol{r}}$, and (5.2.2). The first term on the right hand side is the periodic potential which goes into the determination of the electron energy bands; (in (3.5.1) we wrote the periodic potential simply as $U(\boldsymbol{r})$). It is the second term on the right hand side of (5.2.32) that gives rise to the interaction between electrons and lattice vibrations, the *electron-phonon interaction* (EPI).

Let us represent this EPI in the second quantized form with respect to electrons. We use the Bloch functions of (3.5.2) as the basis in carrying out the procedure of 1.5;

$$\begin{aligned} H_{\text{el-ph}} &= -\sum_i \boldsymbol{u}_i \int d\boldsymbol{r}\, \psi^{+}(\boldsymbol{r})\, \nabla_{\boldsymbol{r}} U(\boldsymbol{r}-\boldsymbol{R}_i^{\circ})\, \psi(\boldsymbol{r}) \\ &= -\sum_i \boldsymbol{u}_i \sum_{\boldsymbol{k},\boldsymbol{k}',\sigma}^{\mathrm{1BZ}} a_{\boldsymbol{k}\sigma}^{+} a_{\boldsymbol{k}'\sigma} \int d\boldsymbol{r}\, \varphi_{\boldsymbol{k}}^{*}(\boldsymbol{r})\, \nabla_{\boldsymbol{r}} U(\boldsymbol{r})\, \varphi_{\boldsymbol{k}'}(\boldsymbol{r}) e^{-i(\boldsymbol{k}-\boldsymbol{k}')\boldsymbol{R}_i^{\circ}}, \end{aligned} \tag{5.2.33}$$

where we used the property of (3.5.3), namely,

$$\varphi_{\boldsymbol{k}}(\boldsymbol{r}+\boldsymbol{R}_i^{\circ}) = e^{i\boldsymbol{k}\boldsymbol{R}_i^{\circ}} \varphi_{\boldsymbol{k}}(\boldsymbol{r}). \tag{5.2.34}$$

By putting (5.2.7) into (5.2.33), we obtain

$$H_{\text{el-ph}} = \sum_{\boldsymbol{k},\boldsymbol{q},\sigma,\lambda}^{1\text{BZ}} g_{\boldsymbol{k}\lambda}(\boldsymbol{q})\, a^{+}_{\boldsymbol{k}\sigma}\, a_{\boldsymbol{k}-\boldsymbol{q}\sigma}\, Q_{\boldsymbol{q}\lambda}, \tag{5.2.35}$$

$$g_{\boldsymbol{k}\lambda}(\boldsymbol{q}) = -\left(\frac{N}{M_{\text{I}}}\right)^{1/2} \varepsilon_{\boldsymbol{q}\lambda} \int d\boldsymbol{r}\; \varphi^{*}_{\boldsymbol{k}}(\boldsymbol{r})\, \nabla_{\boldsymbol{r}} U(\boldsymbol{r})\, \varphi_{\boldsymbol{k}-\boldsymbol{q}}(\boldsymbol{r}). \tag{5.2.36}$$

Note that both $\boldsymbol{k}$ and $\boldsymbol{q}$ are limited to the 1st Brillouin zone. If $\boldsymbol{k}-\boldsymbol{q}$ is outside of the 1st Brillouin zone, it is understood that it is to be brought back within the 1st Brillouin zone by adding an appropriate reciprocal lattice vector. In a real metal, the conduction electron band usually is comprised of more than one band. In such a case the electron-phonon interaction constant, (5.2.36), can have inter-band matrix elements.

When we neglect the periodic structure of ionic lattice, the Bloch functions reduce to plane waves. Then $g_{\boldsymbol{k}}(\boldsymbol{q})$ becomes independent of $\boldsymbol{k}$; $g_{\boldsymbol{k}}(\boldsymbol{q}) = g(\boldsymbol{q})$, and (5.2.35) and (5.2.36) reduce, respectively, to

$$H_{\text{el-ph}} = \sum_{\boldsymbol{q},\boldsymbol{k},\sigma} g(\boldsymbol{q})\, a^{+}_{\boldsymbol{k}\sigma}\, a_{\boldsymbol{k}-\boldsymbol{q},\sigma}\, Q_{\boldsymbol{q}} = \sum_{\boldsymbol{q}} g(\boldsymbol{q})\, n(-\boldsymbol{q})\, Q_{\boldsymbol{q}}, \tag{5.2.37}$$

$$g(\boldsymbol{q}) = -\left(\frac{N}{M_{\text{I}}}\right)^{1/2} \left(\frac{\boldsymbol{q}}{q}\right) \frac{1}{V} \int d\boldsymbol{r}\; e^{-i\boldsymbol{q}\boldsymbol{r}}\, \nabla_{\boldsymbol{r}} U(\boldsymbol{r}). \tag{5.2.38}$$

In this case we have only longitudinal phonons and hence we have eliminated the subscript λ.

If an ion is approximated by a point charge Ze at $\boldsymbol{R}_i$, we have

$$U(\boldsymbol{r}-\boldsymbol{R}_i) = -\frac{Ze^2}{|\boldsymbol{r}-\boldsymbol{R}_i|} = -\sum_{\boldsymbol{q}} \frac{4\pi Ze^2}{Vq^2}\, e^{i\boldsymbol{q}(\boldsymbol{r}-\boldsymbol{R}_i)}. \tag{5.2.39}$$

As previously mentioned, the jellium model of a metal assumes that in the equilibrium state the point charge ions are distributed uniformly. For the jellium model, by putting (5.2.39) into (5.2.38) we obtain

$$g(\boldsymbol{q}) = i\left(\frac{N}{M_{\text{I}}}\right)^{1/2} \frac{1}{V}\, \frac{4\pi Ze^2}{q}. \tag{5.2.38'}$$

If we use (5.2.24), (5.2.37) can be rewritten in the following form,

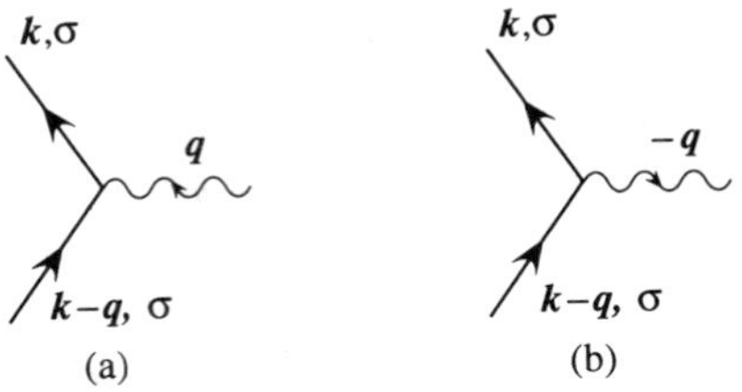

Fig.5.3 The electron-phonon interaction of (5.2.40). The propagation of a phonon is represented by a wavy line while that of an electron is represented by a solid line.

$$H_{\text{el-ph}} = \sum_{q,k,\sigma} \alpha(\boldsymbol{q})\, a^{+}_{k\sigma}\, a_{k-q,\sigma}\,(\, b_{q} + b^{+}_{-q}\,), \tag{5.2.40}$$

$$\alpha(\boldsymbol{q}) = \left(\frac{\hbar}{2\Omega_q}\right)^{1/2} g(\boldsymbol{q}). \tag{5.2.41}$$

The electron-phonon interaction (EPI) of (5.2.40) can be represented by the diagrams given in Fig.5.3; the solid and wavy lines represent electrons and phonons, respectively; an electron is scattered either by absorbing a phonon $\boldsymbol{q}$, (a), or by emitting a phonon $-\boldsymbol{q}$, (b). Also note that (5.2.37) tells us that the electron-phonon interaction is the interaction between the ionic displacement, Q_q, and the electron density polarization, $n(-\boldsymbol{q})$.

Finally, note that even in the jellium model, the summation over $\boldsymbol{q}$ should be restricted to the 1st Brillouin zone; there can not be more than 3N different phonons in a solid consisting of N ions. Thus, the terms of the Hamiltonian which include the phonon variables will henceforth be taken to be

$$H_{\text{ph}} = \frac{1}{2}\sum_{q}^{1\text{BZ}} \left(\, P^{+}_{q}P_{q} + \Omega_q^{2}\, Q^{+}_{q}Q_{q}\right), \tag{5.2.22'}$$

or

$$H_{\text{ph}} = \sum_{q}^{1\text{BZ}} \hbar\Omega_q \left(\, b^{+}_{q}b_{q} + \frac{1}{2}\right), \tag{5.2.25'}$$

and the EPI of (5.2.37) or (5.2.40). Using these terms we will first seek to determine how the bare phonon frequencies, Ω_q, are affected by the screening of the bare ion-ion interactions by the EPI.

If we note the commutation relation of (5.2.12), we obtain the Heisenberg equation of motion for Q_q as

$$\frac{\partial^2}{\partial t^2} Q_q + \Omega_q^2 Q_q = -g(-q)n(q). \qquad (5.2.42)$$

The right hand side of the equation represents the effect of EPI. If the phonon frequency is modified by this effect of EPI from Ω_q to ω_q, we will have

$$Q_q \propto \exp(-i\omega_q t), \qquad (5.2.43)$$

in place of (5.2.8). We seek to obtain an expression for ω_q.

The equation of motion of (5.2.42) is an exact one for the present model. Note that $n(q)$ in it is an operator. If we are to treat (5.2.42) exactly, we have to set up another equation of motion for $n(q)$, and such a chain of equations will further continue. It is generally not feasible to treat such a chain of equations exactly to obtain ω_q. Thus, we introduce an approximation in the following way.

In the present model, if ions are in the equilibrium positions, the electron density distribution is uniform: $\langle n(q) \rangle = 0$ for $q \neq 0$. When ions are displaced from the equilibrium positions, now we can have $\langle n(q) \rangle \neq 0$. In (5.2.42), we approximate the right hand side by replacing the operator $n(q)$ by its thermal average $\langle n(q) \rangle$ in the presence of displacement of ionic positions from the equilibrium one, and, then, treat $\langle n(q) \rangle$ within the linear response approximation. $\langle n(q) \rangle$ is the electron density polarization to screen the displaced ions.

With (5.2.43), we note that the EPI of (5.2.37) has the form of the charge potential perturbation of (4.2.26); the following correspondence holds,

$$-eU(q)\, e^{-i\omega t} \leftrightarrow g(q)\, Q_q.$$

Thus, in terms of the electron charge susceptibility defined in (4.2.33) and (5.1.2) we have

$$\langle n(q) \rangle = \langle n(q, \omega_q) \rangle e^{-i\omega_q t} = -\overline{\chi}_{ee}(q, \omega_q)\, g(q)\, Q_q. \qquad (5.2.44)$$

We put this in the place of $n(q)$ in (5.2.42) and take note of (5.2.43), to obtain

$$(-\omega_q^2 + \Omega_q^2) Q_q = |g(q)|^2 \overline{\chi}_{ee}(q, \omega_q)\, Q_q.$$

This immediately gives the result of (5.1.1) for the screened phonon frequencies.

5.3 EFFECT OF MAGNETISM ON THE LATTICE VIBRATION OF METAL

5.3.1 Phonons in the Paramagnetic State

For the paramagnetic state, if we use (4.3.26), (5.1.1) is reduced to

$$\omega_q^2 = \Omega_q^2 - |g(\boldsymbol{q})|^2 \frac{\bar{\chi}_{zz}(\boldsymbol{q},\omega_q)}{1 + \mathrm{v}(\boldsymbol{q})\,\bar{\chi}_{zz}(\boldsymbol{q},\omega_q)} \tag{5.3.1}$$

$$= \left[\Omega_q^2 - \frac{|g(\boldsymbol{q})|^2}{\mathrm{v}(\boldsymbol{q})}\right] + \frac{|g(\boldsymbol{q})|^2/\,\mathrm{v}(\boldsymbol{q})}{1 + \mathrm{v}(\boldsymbol{q})\,\bar{\chi}_{zz}(\boldsymbol{q},\omega_q)} \tag{5.3.1'}$$

$$= \Omega_q^2 - |g(\boldsymbol{q})|^2 \frac{2F(\boldsymbol{q},\omega_q)}{1 + \left[2\mathrm{v}(\boldsymbol{q}) - \widetilde{V}(\boldsymbol{q})\right]F(\boldsymbol{q},\omega_q)}\ , \tag{5.3.1''}$$

where, in a manner similar to that used in (5.1.2), we let

$$\bar{\chi}_{zz}(\boldsymbol{q},\omega) = \chi_{zz}(\boldsymbol{q},\omega)\,/\,\mu_B^2\ . \tag{5.3.2}$$

Note that in the paramagnetic state, we have

$$\bar{\chi}_{zz}(\boldsymbol{q},\omega) = \frac{2F(\boldsymbol{q},\omega)}{1 - \widetilde{V}(\boldsymbol{q})\,F(\boldsymbol{q},\omega)}\ . \tag{5.3.3}$$

See (4.3.17).

According to (5.3.1) or (5.3.1') the frequencies of phonons in the paramagnetic state of a metal are directly related to the magnetic susceptibility of the metal. This is a very interesting result. The larger is the magnetic susceptibility, the lower the phonon frequency becomes. We may call this observation the *magnetic softening of phonons*.

The expression of (5.3.1") for the screened phonon frequencies in the paramagnetic state had been derived earlier [5.12, 5.13]; it corresponds to using the Hubbard screening constant $\varepsilon(\boldsymbol{q},\omega)$ of (4.3.29), in place of $\varepsilon_0(\boldsymbol{q},\omega)$ of (4.3.30), in the screening of the EPI. However, it was only after I obtained the result in the form of (5.3.1) or (5.3.1') that I began to appreciate the deep involvement of magnetism in determining the phonon spectrum of a metal [5.6].

What does the first term on the right hand side of (5.3.1') represent ? If ions are really point charges, as in the pure jellium model, the bare phonon frequency is given by the ionic plasma frequency of (5.2.31),

$$\Omega_q^{\ 2} \ = \ \Omega_{\rm pl}^{\ 2}. \tag{5.3.4}$$

Then, from (5.2.38') we note that

$$|\, g(\boldsymbol{q})\,|^2/\ {\rm v}(\boldsymbol{q}) \ = \ \Omega_{\rm pl}^{\ 2}. \tag{5.3.5}$$

Thus, for the pure jellium model, we have

$$\Omega_q^{\ 2} \ - \ |\, g(\boldsymbol{q})\,|^2/\ {\rm v}(\boldsymbol{q}) \ = \ 0, \tag{5.3.6}$$

and, therefore

$$\omega_q^2 \ = \ \frac{\Omega_{\rm pl}^2}{1 + {\rm v}(\boldsymbol{q})\,\overline{\chi}_{zz}(\boldsymbol{q},\ \omega_q)} \ = \ \frac{\Omega_{\rm pl}^2}{\varepsilon_{\rm e}(\boldsymbol{q},\ \omega_q)}\ , \tag{5.3.7}$$

where $\varepsilon_{\rm e}(\boldsymbol{q},\ \omega)$ is the screening constant of (4.3.54). Note that this screening constant is the one which is applicable to the screening of the interaction between a pair of external charges. The screened phonon frequency is obtained by screening the bare ion-ion interaction by $\varepsilon_{\rm e}(\boldsymbol{q},\ \omega)$.

If in (5.3.7) we use $\varepsilon_{\rm o}(\boldsymbol{q},\ \omega)$ of (4.3.30) in place of $\varepsilon_{\rm e}(\boldsymbol{q},\ \omega)$ by neglecting the exchange effect on the screening, we have the well-known result of

$$\omega_q^{\ 2} \ = \ \frac{\Omega_{\rm pl}^2}{1 + 2{\rm v}(\boldsymbol{q})\,F(\boldsymbol{q},\ \omega_q)}\,. \tag{5.3.8}$$

This result corresponds to that obtained by using the Pauli susceptibility of (4.2.16) for the magnetic susceptibility in (5.3.7).

In a real metal, however, ions are not point charges; the interactions between ions, and between an ion and an electron are much more complicated than that of the jellium model. Thus, the relations of (5.3.4)–(5.3.6) do not hold generally in real metals. For $q \rightarrow 0$, however, we expect (5.3.6) to hold even in a real metal, since at large inter-ion, inter-electron and ion-electron distances the interactions between them will be predominantly Coulombic. Thus, for a small q, such as $q \ll 1/a$, a being the distance between neighboring ions, we may introduce a non-jellium correction in the form

$$\Omega_q^{\ 2} - |\, g(\boldsymbol{q})\,|^2/\ {\rm v}(\boldsymbol{q}) \ = \ \xi\, s_{\rm o}^2 q^2 \tag{5.3.9}$$

in (5.3.1'), where

$$s_o = \Omega_{pl}/(8\pi e^2 N(0)/V)^{1/2} \tag{5.3.10}$$

is the sound velocity corresponding to the phonon dispersion of (5.3.8), as we will see shortly. We call s_o the *Bohm-Staver sound velocity* [5.14]. The parameter ξ is considered to be independent of $\boldsymbol{q}$ for small q.

The meaning of the parameter ξ may become clearer if we note that the energy due to the non-Coulombic direct repulsive interaction between ion cores can be represented as $\xi N M_I s_o^2/2$, as we will see in Chapter 7 (see (7.2.24)). Then, ξ is considered to take the magnitude of $O(1)$ with positive sign. In 5.3.2, however, we will show that ξ or $\Omega_q^2 - |g(\boldsymbol{q})|^2/\mathrm{v}(\boldsymbol{q})$ in (5.3.9) can effectively take a negative value.

The parameter ξ plays a very important role. However, in this book we treat it simply as a phenomenological parameter.

In calculating the phonon frequency $\omega_{\boldsymbol{q}}$ from (5.1.1) or (5.3.1), note that $\omega_{\boldsymbol{q}}$ appears also in the response function on the right hand side of the equation. Because of this, solving for $\omega_{\boldsymbol{q}}$ from these equations is a rather complicated task.

Generally, however, the motion of ions is much slower than that of electrons. If we compare the Fermi velocity of an electron with the sound velocity of (5.3.10), we have

$$s_o / \mathrm{v_F} \cong (m / M_I)^{1/2} << 1. \tag{5.3.11}$$

Thus, for electrons near the Fermi surface the ionic motion is slow and will appear to be almost static. We may approximate (5.1.1) and (5.3.1), then, as

$$\omega_{\boldsymbol{q}}^2 = \Omega_{\boldsymbol{q}}^2 - |g(\boldsymbol{q})|^2 \overline{\chi}_{ee}(\boldsymbol{q}, 0) , \tag{5.3.12}$$

$$\omega_{\boldsymbol{q}}^2 = \left(\Omega_{\boldsymbol{q}}^2 - \frac{|g(\boldsymbol{q})|^2}{\mathrm{v}(\boldsymbol{q})} \right) + \frac{|g(\boldsymbol{q})|^2 / \mathrm{v}(\boldsymbol{q})}{1 + \mathrm{v}(\boldsymbol{q}) \overline{\chi}_{zz}(\boldsymbol{q}, 0)} . \tag{5.3.13}$$

We call the simplification of the electron response contained in (5.3.12) or (5.3.13) the *adiabatic approximation.* Note that there are situations in which we can not use the adiabatic approximation. In the next section we will see that the attenuation of sound is outside of the adiabatic approximation.

Finally, we note that (5.3.1') and (5.3.13) can be approximately rewritten in the following form,

$$\omega_{\boldsymbol{q}}^2 = \left(\Omega_{\boldsymbol{q}}^2 - \Omega_{pl}^2 \right) + \frac{\Omega_{pl}^2}{1 + \mathrm{v}(\boldsymbol{q}) \overline{\chi}_{zz}(\boldsymbol{q}, \omega_{\boldsymbol{q}})} \tag{5.3.14}$$

$$\cong \left(\Omega_q^{2} - \Omega_{\rm pl}^{2}\right) + \frac{\Omega_{\rm pl}^{2}}{1 + {\rm v}(\boldsymbol{q})\,\overline{\chi}_{zz}(\boldsymbol{q}, 0)} . \qquad (5.3.14')$$

5.3.2 The Effect of Electron Correlation on Phonon Frequency

The result of 5.3.1 is based on (4.3.26). If we consider the effect of electron correlation as we treated it in 4.3.9, and use (4.3.72') in place of (4.3.26) in (5.1.1), we have

$$\omega_q^2 = \Omega_q^{2} - |\,g(\boldsymbol{q})\,|^2 \frac{\overline{\chi}_{zz}(\boldsymbol{q})}{1 + \left\{{\rm v}(\boldsymbol{q}) - \Delta V_{+-}(\boldsymbol{q})\right\}\overline{\chi}_{zz}(\boldsymbol{q})} . \qquad (5.3.15)$$

with $\overline{\chi}_{zz}$ given by (4.3.70). Thus, with $\Delta V_{+-} > 0$, the effect of electron correlations enhances the dependence of phonon frequency on the magnetic susceptibility by replacing ${\rm v}(\boldsymbol{q})$ in (5.3.1) by $({\rm v}(\boldsymbol{q}) - \Delta V_{+-}(\boldsymbol{q}))$. Compare (5.3.15) with (5.3.14).

Note that for ${\rm v}(\boldsymbol{q})\,\overline{\chi}_{zz} \gg 1$, in which case we are most interested, (5.3.15) can be approximated as

$$\omega_q^2 = \left[\Omega_q^{2} - \Omega_{\rm pl}^{2} - \Omega_{\rm pl}^{2}\frac{\Delta V_{+-}(\boldsymbol{q})}{{\rm v}(\boldsymbol{q})}\right] + \frac{\Omega_{\rm pl}^{2}}{1 + {\rm v}(\boldsymbol{q})\,\overline{\chi}_{zz}(\boldsymbol{q})} , \qquad (5.3.16)$$

where we used (5.3.5). Compare this result with (5.3.14): The effect of electron correlation is to change the first term on the right hand side of (5.3.14) as in (5.3.16). This change is rather important. As we discussed below (5.3.9), we expect that $\Omega_q{}^2 - \Omega_{\rm pl}{}^2 > 0$, and, therefore, $\xi > 0$ in (5.3.9). However, in (5.3.16) we may now conceive of a situation to have

$$\Omega_q^{2} - \Omega_{\rm pl}^{2} - \Omega_{\rm pl}^{2}\frac{\Delta V_{+-}(\boldsymbol{q})}{{\rm v}(\boldsymbol{q})} < 0 . \qquad (5.3.17)$$

Thus, if we newly define ξ as

$$\Omega_q^{2} - \Omega_{\rm pl}^{2} - \Omega_{\rm pl}^{2}\frac{\Delta V_{+-}(\boldsymbol{q})}{{\rm v}(\boldsymbol{q})} \equiv \xi\, s_0^2 q^2 \qquad (5.3.18)$$

for small q, this ξ can have a negative value. Note that since ΔV_{+-} is short-ranged, $\Delta V_{+-}(\boldsymbol{q})$ would not much depend on $\boldsymbol{q}$, unlike ${\rm v}(\boldsymbol{q})$.

In the following part of this book, we use (5.3.13) or (5.3.14), but not (5.3.15), for the phonon frequency in the paramagnetic state. Similarly, we use (5.3.28) for the phonon frequency in the ferromagnetic state. The quantity

$(\Omega_q^2 - \Omega_{\rm pl}^2)$ or the parameter ξ appearing there, however, should always be understood as in (5.3.18), with the possibility of becoming negative.

5.3.3 The Charge Density Wave (CDW) and The Spin Density Wave (SDW)

In 3.3 we saw that the divergence of the static uniform Stoner magnetic susceptibility of a metal, $\chi_{zz}(0, 0) = \infty$, implies a ferromagnetic phase transition of the metal. Similarly, if we have

$$\chi_{zz}(\boldsymbol{q}, 0) = \infty , \tag{5.3.19}$$

for some $\boldsymbol{q} \neq 0$, it means to have

$$\langle \sigma_z(\boldsymbol{q}) \rangle \neq 0 \tag{5.3.20}$$

without any external field; we have in a metal a spontaneous static, spatially oscillating spin density of the form of $\langle \sigma_z(\boldsymbol{r}) \rangle = \langle \sigma_z(\boldsymbol{q}) \rangle \exp(i\boldsymbol{qr})$. We call it the *spin density wave* (SDW) [4.45], and the condition for its occurrence is given by (4.3.17) as

$$1 - \widetilde{V}(\boldsymbol{q})F(\boldsymbol{q}) \leq 0. \tag{5.3.21}$$

For $\boldsymbol{q} = 0$, this condition reduces to the Stoner condition of (3.3.10).

Next, what will happen if we have

$$\chi_{\rm ee}(\boldsymbol{q}, 0) = \infty, \tag{5.3.22}$$

for some $\boldsymbol{q} \neq 0$? We will have

$$\langle n(\boldsymbol{q}) \rangle \neq 0, \tag{5.3.23}$$

without any external charge potential. Such a non-uniform electron charge density distribution is called a *charge density wave* (CDW) [4.46].

From (4.3.26"), the condition of (5.3.22) is explicitly given by

$$1 + \left[2{\rm v}(\boldsymbol{q}) - \widetilde{V}(\boldsymbol{q})\right] F(\boldsymbol{q}) \leq 0. \tag{5.3.24}$$

If we compare the above condition with that for the SDW given in (5.3.21), however, we find it not possible to satisfy the CDW condition before the SDW condition, since $\mathrm{v}(\boldsymbol{q}) > 0$. This approach leads us to the erroneous conclusion that it is impossible to observe a CDW within the present mean field approximation.

The situation is the same even if we use the response function which includes the effect of electron correlation: In that case, according to (4.3.70) and (4.3.72) we obtain the condition

$$1 - \widetilde{V}_{\mathrm{m}} F(\boldsymbol{q}) = 1 - \left\{ \widetilde{V}(\boldsymbol{q}) - \Delta V_{+-}(\boldsymbol{q}) \right\} F(\boldsymbol{q}) \leq 0 \tag{5.3.25}$$

for the occurrence of a SDW, and the condition

$$1 + \left\{ 2\mathrm{v}(\boldsymbol{q}) - V_{\mathrm{e}}(\boldsymbol{q}) \right\} F(\boldsymbol{q}) = 1 + \left\{ 2\mathrm{v}(\boldsymbol{q}) - \widetilde{V}(\boldsymbol{q}) - \Delta V_{+-}(\boldsymbol{q}) \right\} F(\boldsymbol{q}) \leq 0 \tag{5.3.26}$$

for the occurrence of a CDW. Since it is expected that $\mathrm{v}(\boldsymbol{q}) > \Delta V_{+-}$, as well as $\mathrm{v}(\boldsymbol{q}) > \widetilde{V}(\boldsymbol{q})$, it is not possible to satisfy the condition (5.3.26) before the condition (5.3.25).

To set matters right, we need to note that (5.3.22) is not the correct condition for a CDW. Rather, we should require

$$\omega_{\boldsymbol{q}} = 0, \tag{5.3.27}$$

the softening of a phonon. This condition implies that ions may be displaced to form a static ionic charge density wave without any cost in energy; the electron density distribution is, of course, to be modulated accordingly. Thus, the CDW is essentially equivalent to the *Peierls transition* first noted for a linear metallic system [4.19]. Such a change in the ionic configuration is nothing other than a structural phase transition. In a metal, a structural phase transition will always accompany a CDW.

Now, by adopting (5.3.27) as the condition for the occurrence of a CDW, we find that the sign of the quantity $\Omega_{\boldsymbol{q}}^2 - |g(\boldsymbol{q})|^2/\mathrm{v}(\boldsymbol{q})\{1 + \Delta V_{+-}(\boldsymbol{q})/\mathrm{v}(\boldsymbol{q})\}$, or ξ in (5.3.18) determines whether the SDW or the CDW will occur in a metal for $\chi_{zz}(\boldsymbol{q}) \to \infty$.

(i) $\Omega_{\boldsymbol{q}}^2 - \{ |g(\boldsymbol{q})|^2 / \mathrm{v}(\boldsymbol{q}) \} \{ 1 + \Delta V_{+-}(\boldsymbol{q})/\mathrm{v}(\boldsymbol{q}) \} > 0.$

A SDW occurs when $\chi_{zz}(\boldsymbol{q}, 0) = \infty$; the CDW condition of (5.3.27) can not be realized.

(ii) $\Omega_{\boldsymbol{q}}^2 - \{ |g(\boldsymbol{q})|^2 / \mathrm{v}(\boldsymbol{q}) \} \{ 1 + \Delta V_{+-}(\boldsymbol{q})/\mathrm{v}(\boldsymbol{q}) \} < 0.$

We have $\omega_{\boldsymbol{q}} = 0$ before $\chi_{zz}(\boldsymbol{q}, 0)$ diverges; a CDW occurs before a possible SDW.

(iii) $\Omega_q^2 - \{|g(\boldsymbol{q})|^2 / \mathrm{v}(\boldsymbol{q})\}\{1 + \Delta V_{+-}(\boldsymbol{q})/\mathrm{v}(\boldsymbol{q})\} = 0.$

A SDW and a CDW will occur simultaneously when $\chi_{zz}(\boldsymbol{q}, 0) = \infty$.

Recently the CDW and the SDW are most intensively discussed in connection with the quasi one-dimensional organic metals. For a recent review see [5.15]. The reason why the one-dimensionality has been regarded as important in this problem is that in a one-dimensional electron system the static Lindhard function diverges for $q = 2k_F$ at $T = 0$ (*problem* 5.4.). In this case $\chi^0_{ee}(2k_F, 0) = 2e^2F(2k_F, 0) = \infty$, which implies a CDW if we adopt the condition of (5.3.22) with the charge susceptibility (4.2.34) for free electrons.

We are now finding, however, that such a conclusion is not justified. There will be competition between the CDW and the SDW, and it is the sign of the quantity $\Omega_q^2 - |g(\boldsymbol{q})|^2/\mathrm{v}(\boldsymbol{q})\{1 + \Delta V_{+-}(\boldsymbol{q})/\mathrm{v}(\boldsymbol{q})\}$ that determines which one will actually occur. This idea, however, does not seem to be widely appreciated at this time. Note, here, that in order to realize the above condition (ii) for a CDW to occur, the favorable situation is that with strong electron correlation; the ratio $\Delta V_{+-}(\boldsymbol{q})/\mathrm{v}(\boldsymbol{q})$ can be viewed to represent the size of electron correlation effect.

Here we should note that the mean field type magnetic susceptibility, of either (5.3.3) or (4.3.70), which comes in the expression for phonon frequency, (5.3.1), does not satisfactorily represent actual observations; the Stoner susceptibility fails to reproduce the widely observed Curie-Weiss like temperature dependence, for instance, as we noted at the end of Chapter 3. Thus, within the scope of the discussion of this chapter we can not identify the mean field susceptibility which appears in (5.3.13) directly with the observed magnetic susceptibility of a metal.

In Chapter 7, however, we will indicate that the magnetic susceptibility which should be entered into (5.3.1) is actually the observed, Curie-Weiss like one, rather than the Stoner one. Note also that the mean field magnetic susceptibility is still a good measure of the magnetic tendency of a metal. Thus, our conclusion on the close relation between phonon spectrum and magnetism remains valid.

5.3.4 Phonons in the Ferromagnetic State

For the ferromagnetic state, (5.1.1) is rewritten as [5.6]

$$\omega_q^2 = \left[\Omega_q^2 - \frac{|g(\boldsymbol{q})|^2}{\mathrm{v}(\boldsymbol{q})}\right] + \frac{|g(\boldsymbol{q})|^2/\,\mathrm{v}(\boldsymbol{q})}{1 + \mathrm{v}(\boldsymbol{q})\left[\widetilde{F}_+(\boldsymbol{q}, \omega_q) + \widetilde{F}_-(\boldsymbol{q}, \omega_q)\right]} \tag{5.3.28}$$

$$\cong \left[\Omega_q^{\,2} - \frac{|g(\boldsymbol{q})|^2}{\mathrm{v}(\boldsymbol{q})}\right] + \frac{\Omega_{\mathrm{pl}}^{\,2}}{1 + \mathrm{v}(\boldsymbol{q})\left[\widetilde{F}_+(\boldsymbol{q},0) + \widetilde{F}_-(\boldsymbol{q},0)\right]}. \tag{5.3.29}$$

The phonon frequency changes with magnetization as $\widetilde{F}_{\pm}(\boldsymbol{q},0)$ change with the spin splitting of the electron energy bands. As we will see later, the dependence of the phonon frequency on magnetization can be quite large; the relative size of the change of phonon frequency can be of order unity. Thus it seems to be more possible to have a CDW, namely, a structural phase transition in the ferromagnetic state of a metal than in the paramagnetic state.

5.4 SOUND PROPAGATION AND MAGNETISM IN METALS

5.4.1 Sound Propagation in Metals

In this chapter our discussion has thus far centered on phonons. However, to observe phonons directly we have to employ neutron diffraction. Much simpler to perform are experiments dealing with the velocity of sound, and yet they can yield very useful information. In fact, it is on studies of the velocity of sound that most of the important observations shown in Fig.5.2, concerning the relation between elasticity and magnetism, were made. Note that, the sound velocity, s, and the elastic constant (bulk modulus), B, are related by $B \propto s^2$ (see (7.2.14')).

In this subsection we present a general formulation on the velocity and attenuation of sound propagation in a metal. Then in the following two subsections we will discuss sound propagation in the paramagnetic and the ferromagnetic states.

Let us note that the phonon frequency is a complex quantity of the following form,

$$\omega_q = \bar{\omega}_q - i\gamma_q , \tag{5.4.1}$$

with real, positive quantities $\bar{\omega}_q$ and γ_q. The time dependence of the phonon normal coordinate is, then, a damped oscillation,

$$Q_q \propto e^{-i\bar{\omega}_q t}\, e^{-\gamma_q t}. \tag{5.4.2}$$

Let us now consider how the imaginary part of phonon frequency may be obtained. (For a general discussion on the following procedure, see, for instance, Ref. [5.16]. see also 9.2)

In the discussion of the Kubo theory of linear response, we noted that we have to take the frequency as (4.1.35) in the response function, corresponding to the initial condition (4.1.17) on the density matrix. Then, for a given real frequency ω the electron charge response function must be complex,

$$\overline{\chi}_{ee}(\boldsymbol{q}, \omega) = \frac{\widetilde{F}_+(\boldsymbol{q}, \omega + i0^+) + \widetilde{F}_-(\boldsymbol{q}, \omega + i0^+)}{1 + v(\boldsymbol{q})\left[\widetilde{F}_+(\boldsymbol{q}, \omega + i0^+) + \widetilde{F}_-(\boldsymbol{q}, \omega + i0^+)\right]}$$

$$\equiv \mathrm{Re}\,\overline{\chi}_{ee}(\boldsymbol{q}, \omega) + i\,\mathrm{Im}\,\overline{\chi}_{ee}(\boldsymbol{q}, \omega). \tag{5.4.3}$$

With this, (5.1.1) is rewritten as

$$\omega_{\boldsymbol{q}}^2 = \Omega_{\boldsymbol{q}}^2 - |g(\boldsymbol{q})|^2\,\mathrm{Re}\,\overline{\chi}_{ee}(\boldsymbol{q}, \overline{\omega}_{\boldsymbol{q}}) - i|g(\boldsymbol{q})|^2\,\mathrm{Im}\,\overline{\chi}_{ee}(\boldsymbol{q}, \overline{\omega}_{\boldsymbol{q}}). \tag{5.4.4}$$

Both Re $\overline{\chi}_{ee}(\boldsymbol{q}, \omega)$ and Im $\overline{\chi}_{ee}(\boldsymbol{q}, \omega)$ are continuous functions of ω.

Now, since $\mathrm{Im}\overline{\chi}_{ee}(\boldsymbol{q}, \omega) \neq 0$, $\omega_{\boldsymbol{q}}$ can not be a real quantity in (5.4.4); $\omega_{\boldsymbol{q}}$ should take the form of (5.4.1). We anticipate, however, that

$$0 < \gamma_{\boldsymbol{q}} \ll \overline{\omega}_{\boldsymbol{q}}, \tag{5.4.5}$$

as we will confirm shortly. Then, in rewriting (5.4.4), we may approximate $\mathrm{Re}\,\overline{\chi}_{ee}(\boldsymbol{q}, \overline{\omega}_{\boldsymbol{q}} - i\gamma_{\boldsymbol{q}}) \cong \mathrm{Re}\,\overline{\chi}_{ee}(\boldsymbol{q}, \overline{\omega}_{\boldsymbol{q}})$ etc. to obtain

$$(\overline{\omega}_{\boldsymbol{q}} - i\gamma_{\boldsymbol{q}})^2 = \Omega_{\boldsymbol{q}}^2 - |g(\boldsymbol{q})|^2\,\mathrm{Re}\,\overline{\chi}_{ee}(\boldsymbol{q}, \overline{\omega}_{\boldsymbol{q}}) - i\,|g(\boldsymbol{q})|^2\,\mathrm{Im}\,\overline{\chi}_{ee}(\boldsymbol{q}, \overline{\omega}_{\boldsymbol{q}}). \tag{5.4.6}$$

By separating this equation into its real and imaginary parts we arrive at

$$\overline{\omega}_{\boldsymbol{q}}^2 - \gamma_{\boldsymbol{q}}^2 = \Omega_{\boldsymbol{q}}^2 - |g(\boldsymbol{q})|^2\,\mathrm{Re}\,\overline{\chi}_{ee}(\boldsymbol{q}, \overline{\omega}_{\boldsymbol{q}}) \tag{5.4.7}$$

$$\cong \overline{\omega}_{\boldsymbol{q}}^{\,2}, \tag{5.4.7'}$$

$$2\gamma_{\boldsymbol{q}}\,\overline{\omega}_{\boldsymbol{q}} = |g(\boldsymbol{q})|^2\,\mathrm{Im}\,\overline{\chi}_{ee}(\boldsymbol{q}, \overline{\omega}_{\boldsymbol{q}}). \tag{5.4.8}$$

These are the basic equations for determining $\overline{\omega}_{\boldsymbol{q}}$ and $\gamma_{\boldsymbol{q}}$.

If we put $\gamma_{\boldsymbol{q}} = 0$ in (5.4.7), it reduces to the familiar result. In order to understand the implication of (5.4.8) for $\gamma_{\boldsymbol{q}}$ let us consider the simplest case in which χ_{ee}^{o} of (4.2.34) is used for χ_{ee}:

$$\omega_{\boldsymbol{q}}^2 = \Omega_{\boldsymbol{q}}^2 - |g(\boldsymbol{q})|^2\,\overline{\chi}_{ee}^{o}(\boldsymbol{q}, \omega_{\boldsymbol{q}})$$

$$= \Omega_{\boldsymbol{q}}^2 - 2\,|g(\boldsymbol{q})|^2\,F(\boldsymbol{q}, \omega_{\boldsymbol{q}}). \tag{5.4.9}$$

In the complex Lindhard function

$$F(\mathbf{q}, \omega + i0^+) = \operatorname{Re} F(\mathbf{q}, \omega + i0^+) + i \operatorname{Im} F(\mathbf{q}, \omega + i0^+)$$

$$\equiv R(\mathbf{q}, \omega) + i\, I(\mathbf{q}, \omega), \tag{5.4.10}$$

the imaginary part is obtained to be

$$I(\mathbf{q}, \omega) = \operatorname{Im} \sum_{k} \frac{f(\varepsilon_k) - f(\varepsilon_{k+q})}{\varepsilon_{k+q} - \varepsilon_k - \hbar\omega - i\,0^+}$$

$$= \pi \sum_{k} [\, f(\varepsilon_k) - f(\varepsilon_{k+q})\,]\, \delta(\, \varepsilon_{k+q} - \varepsilon_k - \hbar\omega\,). \tag{5.4.11}$$

We used the following well-known relation

$$\frac{1}{x \pm i\,0^+} = \mathrm{P}\frac{1}{x} \mp i\pi\delta(x), \tag{5.4.12}$$

where P signifies that the principal value of the integral is to be obtained.

Substituting (5.4.11) into (5.4.8), with $\overline{\chi}_{\mathrm{ee}}$ replaced by $\overline{\chi}^{\mathrm{o}}_{\mathrm{ee}}$ and changing $g(\mathbf{q})$ into $\alpha(\mathbf{q})$ by (5.2.41), we obtain

$$2\gamma_q = \frac{2\pi}{\hbar} |\, \alpha(\mathbf{q})\,|^2 \left(\frac{\Omega_q}{\overline{\omega}_q} \right) \Bigg[2 \sum_{k} f(\varepsilon_k)\, [\, 1 - f(\varepsilon_{k+q})\,] \delta(\varepsilon_{k+q} - (\varepsilon_k + \hbar\omega_q)\,)$$

$$- 2 \sum_{k} f(\varepsilon_{k+q})\, [\, 1 - f(\varepsilon_k)\,]\, \delta((\varepsilon_k + \hbar\omega_q\,) - \varepsilon_{k+q}) \Bigg]. \tag{5.4.13}$$

This result is reminiscent of Fermi's golden rule for transition probabilities. The first and the second terms on the right hand side correspond to the processes given in Figs. 5.4(a) and (b) respectively. Thus, decay or damping rate γ_q of a phonon with wave vector $\mathbf{q}$ is determined to be the difference between these probabilities of phonon annihilation and creation.

The result of (5.4.13), however, goes beyond that of the simple golden rule. At first, the phonons involved are not the bare ones with frequency Ω_q, but the screened ones with frequency ω_q. In the second, (5.4.13) has the factor $\Omega_q / \overline{\omega}_q$ which does not appear in the golden rule result. Finally, (5.4.13) is valid for $T \neq 0$.

Note that the result of (5.4.13) was obtained for the simplest possible case of (5.4.9). In the succeeding subsections we shall deal with the more complicated case of (5.1.1)–(5.1.3). The result we will thus obtain will contain much richer physics than is inherent in the golden rule approach.

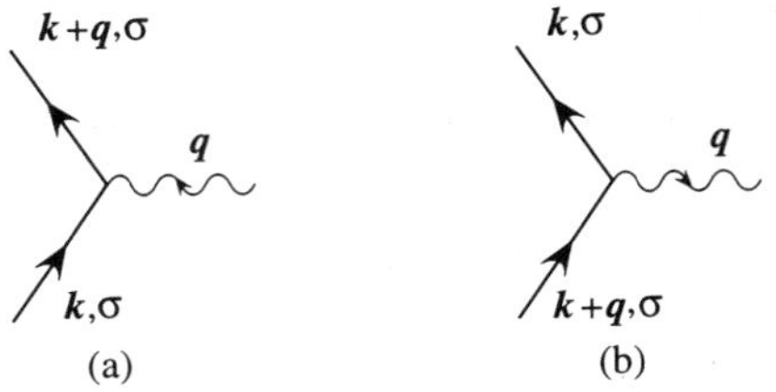

Fig.5.4 Absorption, (a), and emission, (b), by electrons of a phonon with wave number $\boldsymbol{q}$.

As a preparation for our discussion, let us confirm in the Appendix A that the imaginary part of the Lindhard function obtained from (5.4.11) is

$$I(\boldsymbol{q},\,\omega) \;=\; \frac{\pi}{2}N(0)\,\frac{\omega}{\mathrm{v_F}q}\,, \tag{5.4.14}$$

for $k_{\mathrm{B}}T \ll \varepsilon_{\mathrm{F}}$, $q/k_{\mathrm{F}} \ll 1$ and

$$|\omega| \;\ll\; \mathrm{v_F}q\,, \tag{5.4.15}$$

where $\mathrm{v_F}$ is the Fermi velocity of electron. The sound velocity s is related to the phonon frequency by

$$\lim_{q\to 0}\,\bar{\omega}_{\boldsymbol{q}} \;=\; sq\,. \tag{5.4.16}$$

The condition (5.4.15) is equivalent to the following quite plausible requirement

$$s \;\ll\; \mathrm{v_F}. \tag{5.4.15'}$$

Also, since $R(0,\,0) \cong N(0)$, the condition (5.4.15) leads to

$$|\,I(\boldsymbol{q},\,\omega)\,| \;\ll\; R(\boldsymbol{q},\,\omega). \tag{5.4.17}$$

As for the attenuation of sound, often the spatial energy attenuation constant, $\alpha_{\boldsymbol{q}}$, is used in place of $\gamma_{\boldsymbol{q}}$, the temporal one. Since the propagation distance x is given as $x = st$, we have $\exp(-\gamma_{\boldsymbol{q}}\,t) = \exp(-\,[\gamma_{\boldsymbol{q}}/s\,]\,x)$, and, then,

$$\alpha_{\boldsymbol{q}} \;=\; 2\gamma_{\boldsymbol{q}}\,/\,s. \tag{5.4.18}$$

From the result of (5.3.8), it is straightforward to obtain

$$\overline{\omega}_{q} = s_{o} q, \tag{5.4.19}$$

where s_o is the Bohm-Staver sound velocity of (5.3.10). For the free electron energy dispersion, we have

$$s_{o} = \frac{1}{\sqrt{3}} \left(\frac{mZ}{M_{I}} \right)^{1/2} v_{F}, \tag{5.4.20}$$

from (1.3.5) and (5.2.31). Note that this simple result succeeded in reproducing observed sound velocity in wide range of nonmagnetic metals [5.8]. The corresponding attenuation constant is obtained as

$$\gamma_{qo} = \frac{\pi}{4 v_{F} q} \overline{\omega}_{q}^{2}; \quad \alpha_{qo} = \frac{\pi}{2} \overline{\omega}_{q} / v_{F}. \tag{5.4.21}$$

In the following two subsections, we will show how drastically sound propagation is affected by the exchange interaction between electrons and the spin splitting of the electron energy bands. Here note that the above and succeeding results are for the case in which the mean free path of the electron is much larger than the wave length of the phonon. Such a situation is realized in either a metal of high purity or for high frequency sound waves, or both. It is known that when these conditions are not met the result is quite different. For a review and references, see Kittel [1.1].

5.4.2 Sound Propagation in the Ferromagnetic State

We first discuss the velocity and attenuation of sound in the ferromagnetic state of a metal. Our discussion is based on the mean field approximation result of (5.1.1)–(5.1.3). Remember that in our model, which neglects the crystal structure of the ions, we have only longitudinal acoustic phonons. Correspondingly, the elastic constant which we obtain coincides with the bulk modulus.

We apply the prescription of (5.4.7) and (5.4.8) to (5.3.28). If we put

$$2N(0) \Big/ \sum_{\sigma} \frac{F_{\sigma}(\boldsymbol{q}, \omega_{q} + i0^{+})}{1 - \widetilde{V}(\boldsymbol{q}) F_{\sigma}(\boldsymbol{q}, \omega_{q} + i0^{+})} = \boldsymbol{R} - i\boldsymbol{I}, \tag{5.4.22}$$

for $\boldsymbol{q} \to 0$, we have

$$R = 2N(0) / \sum_{\sigma} \widetilde{F}_{\sigma}(0), \tag{5.4.23}$$

$$I = 2N(0) \sum_{\sigma} \frac{I_{\sigma}}{\left(1 - \widetilde{V} F_{\sigma}(0)\right)^2} \frac{1}{\left(\sum_{\sigma} \widetilde{F}_{\sigma}(0)\right)^2}, \tag{5.4.24}$$

where we put

$$F_{\sigma}(\boldsymbol{q}, \omega + i0^{+}) = R_{\sigma}(\boldsymbol{q}, \omega) + i I_{\sigma}(\boldsymbol{q}, \omega), \tag{5.4.25}$$

as in (5.4.10), and

$$R_{\sigma}(\boldsymbol{q}, 0) = F_{\sigma}(\boldsymbol{q}, 0) \equiv F_{\sigma}(\boldsymbol{q}). \tag{5.4.26}$$

Then, by noting $I_{\sigma} = (\pi/2) N_{\sigma}(0)(s / \mathrm{v}_{\mathrm{F}\sigma})$, where $\mathrm{v}_{\mathrm{F}\sigma}$ is the Fermi velocity of σ spin electrons (see Appendix A), we have [5.6]

$$\left(\frac{s}{s_{\mathrm{o}}}\right)^2 = \xi + 2N(0) / \sum_{\sigma} \widetilde{F}_{\sigma}(0), \tag{5.4.27}$$

$$\frac{\alpha_{\omega}}{\alpha_{\omega\mathrm{o}}} = 2\left(\frac{s_{\mathrm{o}}}{s}\right)^2 N(0) \sum_{\sigma} \frac{\mathrm{v}_{\mathrm{F}}}{\mathrm{v}_{\mathrm{F}\sigma}} \frac{F_{\sigma}(0)}{\left(1 - \widetilde{V} F_{\sigma}(0)\right)^2} \frac{1}{\left[\sum_{\sigma} \widetilde{F}_{\sigma}(0)\right]^2}, \tag{5.4.28}$$

where (see(4.3.15))

$$\begin{aligned} \widetilde{F}_{\pm}(0) &= \frac{F_{\pm}(0)}{1 - \widetilde{V} F_{\pm}(0)} \\ &\cong \frac{N_{\pm}(0)}{1 - \widetilde{V} N_{\pm}(0)} \equiv \widetilde{N}_{\pm}(0), \end{aligned} \tag{5.4.29}$$

and α_{ω} and $\alpha_{\omega\mathrm{o}}$ are damping constants for a sound wave with the common frequency ω, corresponding to α_{q} and $\alpha_{q\mathrm{o}}$; in an experiment, what is given is the frequency, not the wave length.

According to (5.4.27)–(5.4.28), it is the electronic density of states at the Fermi surface that determines the velocity and attenuation constant. This conclusion seems to be very physical. Thus, although it is for the case of the paparabolic energy dispersion of free electrons that the above result is actually derived, in this book we assume its validity beyond such a limitation.

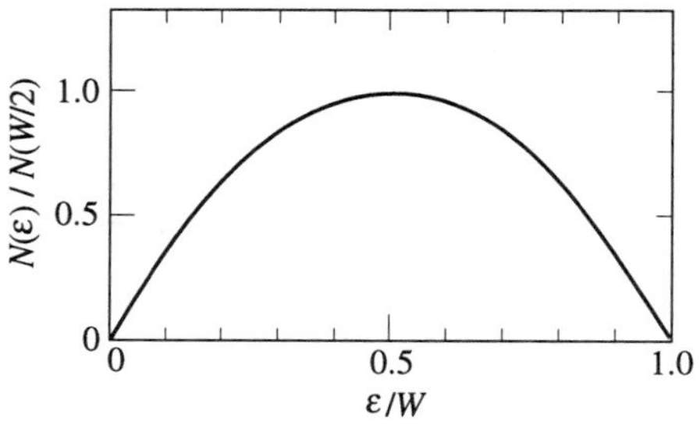

Fig.5.5 The model electronic density of states given by (5.4.31).

Note also that from a given $N_\sigma(\varepsilon)$, we can not uniquely determine $v_{F\sigma}$ in (5.4.28). Here we simply assume the relation,

$$N_\pm(0)/v_{F\pm} = N(0)/v_F, \tag{5.4.30}$$

which is valid for the parabolic energy dispersion of free electrons.

Thus we carry out a numerical calculation on (5.4.27) and (5.4.28) by using a model electronic density of states which is different from that of the parabolic energy dispersion.

First, let us assume the following form of model electronic density of states,

$$N(\varepsilon) = \frac{6N}{W^3}\,\varepsilon(W-\varepsilon)\,, \tag{5.4.31}$$

which is illustrated in Fig.5.5; N and W are the total number of atoms in the system and the width of the band, respectively. This band can accommodate up to one electron per spin and atom. We frequently use this simple model electronic density of states in this book.

If the total number of electrons, or, the value of ε_F/W, ε_F being the Fermi energy in the spin unsplit state, and the magnitude of $\widetilde{V}$ are given, the temperature dependence of the magnetization is determined by (3.6.1) within the mean field theory. Here, however, we first employ the simplified procedure used in obtaining the result of Fig.4.5 etc.; we produce the spin splitting of the bands by changing the value of $\widetilde{V}$ while keeping the temperature at $T = 0$. In this way we obtain the result of Figs.5.6 and 5.7, respectively, for (5.4.27) and (5.4.28), for the model electronic density of states of (5.4.31). Note that since $N(\varepsilon) = N(W-\varepsilon)$, we have the same result for $\varepsilon_F/W = x$ and $1-x$.

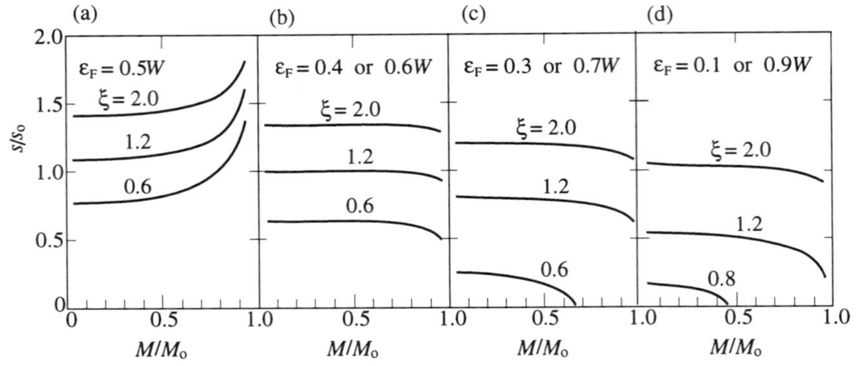

Fig.5.6 The magnetization dependence of sound velocity at absolute zero of temperature calculated from (5.4.27) for different locations of Fermi energy in the model electronic density of states of Fig.5.5. M_o is the maximum magnetization for each case.

According to the results of Fig.5.6, (i) the elastic constant can either increase or decrease with magnetization depending upon the location of ε_F, and (ii) the relative size of the change in the elastic constant due to magnetization can be of the order of a good fraction of unity. These results seem to represent quite well what is actually observed in FeNi alloys. According to the experimental observations shown in Fig.5.2, the characteristic temperature dependence of elastic constant of FeNi alloys changes quite drastically with alloy composition. If we assume a rigid band picture, different alloy compositions correspond to different locations of ε_F in the band. Thus, if we are to understand the observation on Ni of Fig.5.2(c), which is very similar to the case of Fig.5.6(a), with $N(\varepsilon)$ of Fig.5.5, ε_F is required to be located very near a peak of $N(\varepsilon)$ in Ni. On the other hand, to observe an elastic softening with increasing magnetization, ε_F is required to be located far from the peak of $N(\varepsilon)$.

Comparing the behavior of s/s_0 in Fig.5.6 to that of $\alpha_\omega/\alpha_{\omega 0}$ in Fig.5.7, we find that $\alpha_\omega/\alpha_{\omega 0}$ is considerably more sensitive than s/s_0 to changes in both the magnetization and the location of the Fermi energy, ε_F. It would be interesting to have experimental results with which we can compare these theoretical predictions.

The numerical results given in Figs.5.6 and 5.7 were obtained by changing the values of $\tilde{V}N(0) = \tilde{V}$, but not temperature, to produce different magnetizations; the temperature was kept at $T = 0$. Now in Figs.5.8 and 5.9 we present the results for (s/s_0) and $\alpha_\omega/\alpha_{\omega 0}$ which were obtained by actually changing temperature for given constant values of $\tilde{V}N(0)$. We find the results are delicately different from those of Figs.5.6 and 5.7.

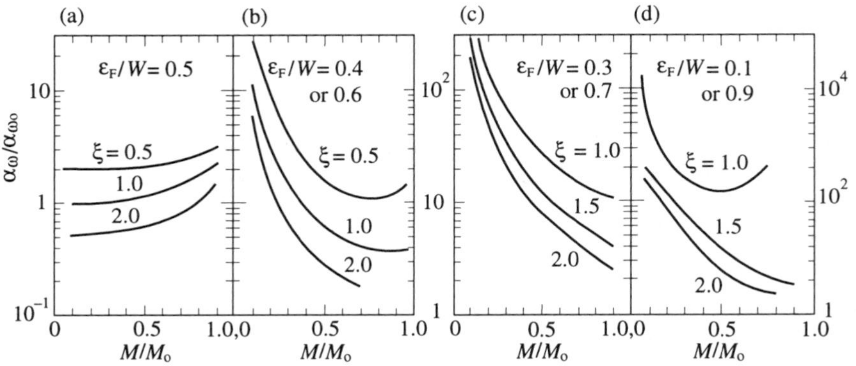

Fig.5.7 The magnetization dependence of the attenuation constant of sound at absolute zero temperature calculated from (5.4.28) for different occupations of the band of Fig.5.5.

If we compare Fig.5.6 with Fig.5.8, for instance, we note that while the case of $\varepsilon_F/W = 0.5$ are similar, the other cases are different; the elastic softening with decreasing temperature is observed only for $\varepsilon_F/W = 0.4$ in Fig.5.8 while in Fig.5.6 it is observed in all the cases except that of $\varepsilon_F/W = 0.5$.

Thus, the result of (5.4.27) and (5.4.28) for sound velocity and attenuation requires careful handling. In this respect note that in a more self-consistent treatment, $\widetilde{V}$ should be replaced by an effective exchange interaction which includes the effects of EPI and spin fluctuations and, therefore, changes with temperature, as we will see in Chapters 6 and 7. Thus, the result of Fig.5

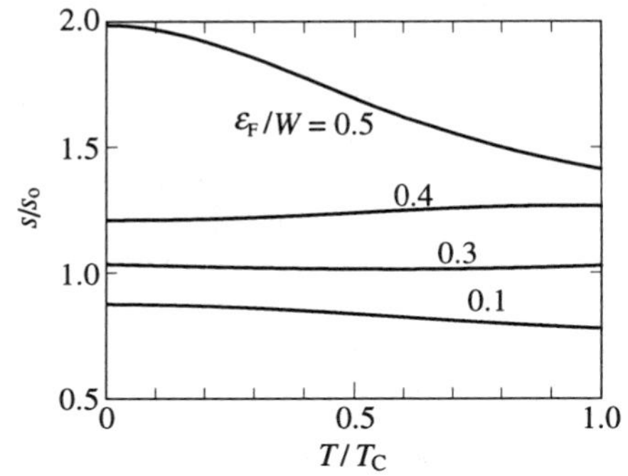

Fig.5.8 The temperature dependence of sound velocity calculated from (5.4.27) for different locations of the Fermi energy in the spin unsplit state in density of states of Fig.5.5. We assumed that W = 1 eV, $\overline{V}$ = 1.3, and ξ = 2.0. Note the abrupt change at $T = T_C$ for ε_F/W = 0.1 and 0.4.

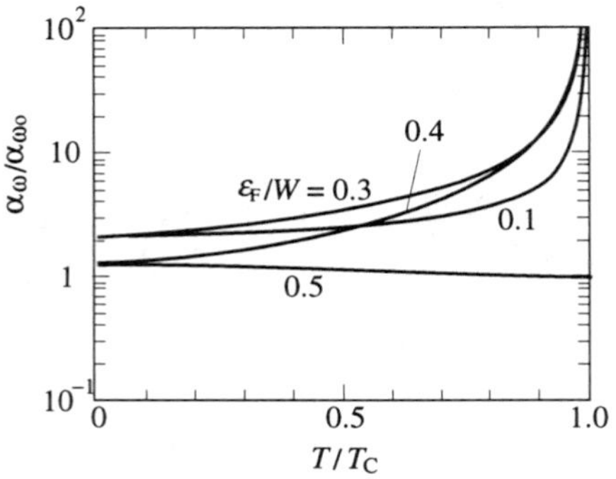

Fig.5.9 The temperature dependence of sound attenuation calculated from (5.4.28) similarly to Fig.5.8.

.8 may not necessarily be closer to the reality than that of Fig.5.6.

At this stage we should content ourselves by reconfirming that the result of (5.4.27) can in principle account for the experimental observations such as given in Fig.5.2, both in sizes and varieties. It is important to note that the size of the magnetization dependence of the screening of phonon frequency as embodied in (5.4.27) is *not too large* compared with experiments. In Chapter 6 we will discuss the effect of EPI on the magnetic properties of a metal based on the very same (5.1.1)–(5.1.3), or (5.4.27) as in this chapter, and find it to be much larger than generally assumed. The result and conclusion of this subsection, then, can be taken as a strong support for such conclusion of Chapter 7.

Now, according to Fig.5.2(b), in $Fe_{72}Pt_{28}$ the magnetization dependence of elastic constant is qualitatively different from that of FeNi alloys; as we lower temperature below T_C, the elastic constant first decreases, and, then, after reaching the minimum it begins to increase. In order to see whether it is at all possible to obtain such a behavior of elastic constant from the result of (5.4.27), we tried a numerical calculation for another model electronic density of states given in Fig.5.10 [5.17]. Note that in the result of Fig.5.11, which was obtained similarly to that of Fig.5.6, $(s/s_o)^2$ can become negative since we put $\xi = 0$. The observed behavior of $Fe_{72}Pt_{28}$ given in Fig.5.2(b) seems to be qualitatively reproduced in the case (d) of Fig.5.11; the positions of ε_F corresponding to this case are indicated by arrows in Fig.5.10. This findng provides a convincing support for the possibility of understanding even the quite anomalous elastic behavior in the ferromagnetic state of $Fe_{72}Pt_{28}$ with the result of (5.4.27).

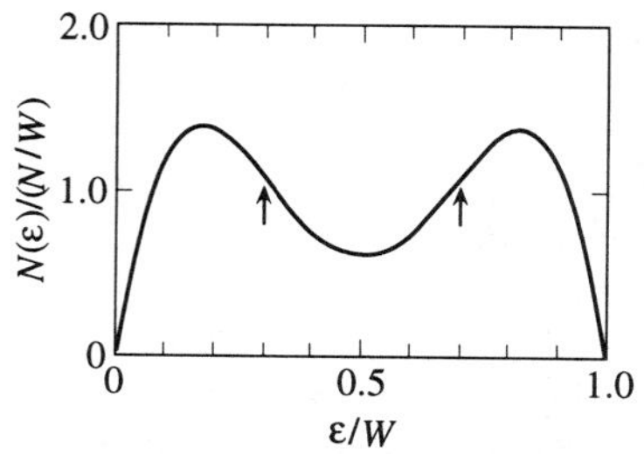

Fig.5.10 The model electronic density of states used in the calculation given in Fig.5.11. The arrows indicate the locations of the Fermi levels corresponding to the case (d) in Fig.5.11.

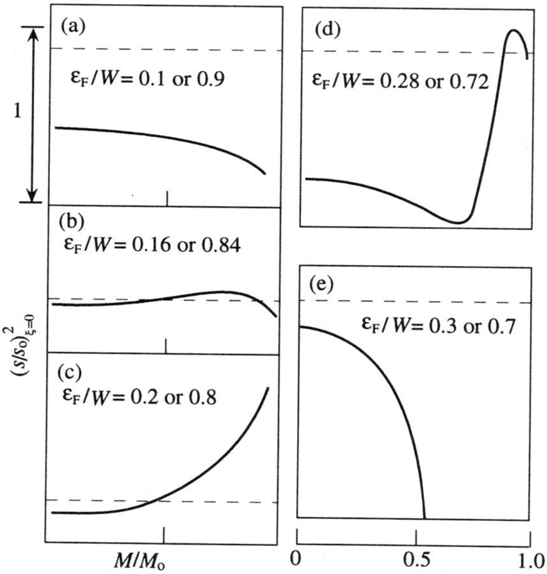

Fig.5.11 The magnetization dependence of sound velocity for the model electronic density of states of Fig.5.10 calculated in the same way as in Fig.5.6. We put $\xi = 0$ and the origin of the ordinate is shown by a horizontal broken line in each case.

Our final observation is concerned with the behavior near T_C. According to the result of this subsection, the sound velocity or the elastic constant changes discontinuously at $T = T_C$. As can be seen from (5.4.32) below, when the temperature approaches T_C from above we have $(s/s_o)^2 = \xi$; thus we have $(s/s_o)^2 = 0$ for $\xi = 0$. According to the result of Fig.5.11, for instance, however, for $M \to 0$, $(s/s_0)^2$ for $\xi = 0$ remains negative. Thus, $(s/s_0)^2$ changes discontinuously at $T = T_C$. Such behavior is also seen in Fig.5.8. If we treat the temperature region around T_C more carefully by abandoning the adiabatic approximation, the change in $(s/s_0)^2$ around T_C turns out to be a continuous one; however, the result that $(s/s_0)^2$ changes rather abruptly in the vicinity below T_C remains unchanged [5.18].

Matter of fact such phenomena have long been known as the *ΔE effect* and the origin of such behavior has been attributed to the reorientation of magnetic domains [5.19]. It now seems gradually to be recognized that the *ΔE* effect is not such domain effect [5.20]; it is rather a single domain phenomena such as we observe in Fig.5.8. We will discuss this subject again in Chapter 7.

Note that recently such observations are actually made not only on itinerant ferromagnets but also on antiferromagnet [5.21].

As to the behavior of sound attenuation, the result of Figs.5.7 and 5.9 shows that, except the cases of $\varepsilon_F/W = 0.5$, it rapidly increases as temperature approaches T_C from below. Thus, the attenuation constant would have a peak at immediately below T_C [5.18].

Note that our discussion on the elastic behavior of itinerant electron ferromagnet will continue in Chapter 7; there we will find our result in this subsection should be modified, particularly in the temperature region immediately below T_C.

5.4.3 Sound Propagation in the Paramagnetic State

For the paramagnetic state, in which $F_+(0) = F_-(0) = F(0)$, the results of (5.4.27) and (5.4.28) reduce, respectively, to

$$(s/s_0)^2 = \xi + [\,1 - \widetilde{V}F(0)\,] = \xi + \chi_P/\chi_S\,, \tag{5.4.32}$$

$$\alpha_\omega/\alpha_{\omega 0} = (s_0/s)^2 = 1/[\xi + \chi_P/\chi_S]\,, \tag{5.4.33}$$

where χ_P and χ_S are, respectively, the Pauli and the Stoner spin susceptibilities.

Concerning sound velocity for $T > T_C$, however, the result of (5.4.32) seems to fail to reproduce the observed various temperature dependence such as given in Fig.5.2. While in Ni sound velocity decreases with increasing temperature, in the result given in Fig.5.8, sound velocity for $T > T_C$ always increases with increasing temperature, independent of the location of ε_F in the model electronic density of states of Fig.5.5. Such result is natural since on the right hand side of (5.4.32), χ_S decreases with increasing temperature.

What is, then, wrong with (5.4.32)? This problem will be discussed in Chapter 7. Let us reiterate that result of this chapter serves as the basis of such discussion there.

5.5 THE EFFECT OF A MAGNETIC FIELD ON THE ELASTIC PROPERTIES OF A METAL

According to Fig.5.12, in FeNi alloys a magnetic field of 6 kG can produce a change in the sound velocity by ~1% both above and below T_C [5.3]. This is a particularly surprising observation. If we put the sound velocity in a magnetic field equal to

$$s(H) = s + \Delta s(H), \tag{5.5.1}$$

we anticipate

$$\left|\frac{\Delta s(H)}{s}\right| = \begin{cases} O\left(\dfrac{\mu_B H}{\varepsilon_F}\right)^2 & \text{for } T > T_C, \quad (5.5.2) \\[2ex] O\left(\dfrac{\mu_B H}{\varepsilon_F}\right) & \text{for } T < T_C, \quad (5.5.3) \end{cases}$$

to the lowest orders, considering the effect of the Zeeman splitting of the electron energy bands and the symmetry concerning the sign of the magnetic field; we do not consider the effect of the magnetic field on the orbital motion of electron. We will explicitly derive these results shortly. If we assume $\varepsilon_F \cong 1$ eV and $H = 10$ kG, then $\mu_B H/\varepsilon_F \cong 10^{-4}$. Thus, to obtain $|\Delta s(H)/s| \cong 1$ %, as observed in FeNi alloys, we require enhancement factors of $\sim 10^6$ for $T > T_C$, and $\sim 10^2$ for $T < T_C$. The required enhancement factor for $T > T_C$ is particularly large.

Another system in which the magnetic field effect on elasticity has been carefully studied is the A15 compounds. In V_3Si, for instance, a magnetic field of ~80 kG led to $|\Delta s(H)/s| \cong 1$ % [5.22]. To produce a result from (5.5.2) in agreement with such an observation, we are required to assume $\varepsilon_F \cong W < 10^{-2}$ eV [5.23], W being the electron energy band width. This was actually used to corroborate the linear chain model for the A15 compounds [5.24, 5.25]. However, detailed band calculations have shown that a conduction electron band with such a small width is unsuggestible; see, for instance, Klein et al. [5.26].

If $W > 0.1$ eV, as various band calculations suggest, we need an enhancement factor of $\sim 10^2$ to account for the observation.

In this section, after a general formulation of the subject in 5.5.1, we will show that large enhancement factors for the magnetic field effect are obtainable from the results of the preceding sections of this chapter, both for $T > T_C$, and $T < T_C$, respectively, in 5.5.2 and 5.5.3. We will also discuss the effect of a magnetic field on sound attenuation, and will predict that the magnetic field effect on the attenuation of sound can be even more enhanced. These results regarding the effect of a magnetic field on the elastic properties of a metal provi-

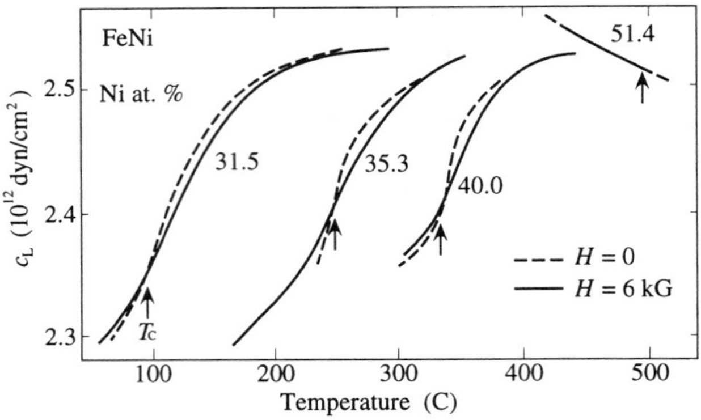

Fig.5.12 The magnetic field dependence of the elastic constant c_L of FeNi alloys [5.3].

de the basis of our discussion in the next chapter on the effect of EPI on the magnetic properties of a metal. How the sound velocity changes with a magnetic field is directly related to how the magnetic susceptibility and magnetization are affected by EPI in a metal. This relation makes it very important to study the magnetic field dependence of sound velocity in a metal both experimentally and theoretically [5.27].

5.5.1 Magnetic Field Effect on Sound Propagation: General Formulation

In the expression for phonon frequency of (5.1.1), where will the effect of a magnetic field enter? It is in the electron charge susceptibility χ_{ee}. Ω_q will not be affected by magnetic fields of available magnitudes. The magnetic field acts on the spin of conduction electrons and χ_{ee} thereby is modified. Although the orbital motion of electrons is also affected by the magnetic field we neglect such effect in our discussion.

So far in this book, we assumed mostly the situation of $n_+ > n_-$, or $\overline{M} = (n_+ - n_-) > 0$ for the ferromagnetic state of a metal; (see Fig.3.5(a)); this corresponds to assuming for the physical magnetization that $M = -\mu_B \overline{M} < 0$. Then, in order to increase the magnetization, we have to apply a magnetic field in the negative direction of z-axis. This situation may cause confusion concerning the sign of magnetic field. Thus, in dealing with the effect of an external magnetic field let us assume exclusively the situation of $M > 0$, or, $n_- > n_+$, as in Fig.3.5(b). Then a positive magnetic field causes to increase the spin splitting of the electron energy bands and the physical magnetization, M.

If we apply a magnetic field H in the z-direction, corresponding to (5.4.27) we obtain

$$\left(\frac{s(H)}{s_0}\right)^2 = \xi + \frac{2N(0)}{\dfrac{F_+(0;H)}{1-\widetilde{V}F_+(0;H)} + \dfrac{F_-(0;H)}{1-\widetilde{V}F_-(0;H)}}, \tag{5.5.4}$$

where $F_\sigma(\boldsymbol{q}; H)$ is the static Lindhard function in the presence of a magnetic field H; $F_\sigma(\boldsymbol{q}; 0) = F_\sigma(\boldsymbol{q})$. For $H = 0$, (5.5.4) reduces either to (5.4.32) or (5.4.27), respectively, for the paramagnetic or the ferromagnetic state.

In exploring $F_\sigma(\boldsymbol{q}; H)$ we put the spin splitting of the bands induced by the external field equal to

$$\begin{aligned} 2\eta &= \widetilde{V}\Delta M/\mu_B + 2\mu_B H \\ &= -\widetilde{V}\Delta\overline{M} + 2\mu_B H, \end{aligned} \tag{5.5.5}$$

where ΔM and $\Delta\overline{M}$ are, respectively, the changes in M and $\overline{M}$ due to the external field. Note that the electron system may be in the ferromagnetic state. In that case, 2η represents the *additional spin splitting* of the bands due to the external magnetic field. The electron energy under the magnetic field is then written as

$$\varepsilon_{k\sigma}(\eta) = \varepsilon_{k\sigma} + \sigma\eta. \tag{5.5.6}$$

With such a spin splitting of the bands we have the Fermi distribution of electrons in the form

$$f(\varepsilon_{k\sigma}(\eta)) = \frac{1}{1+\exp\{\beta((\varepsilon_{k\sigma}-\mu)+(\sigma\eta-\Delta\mu))\}} \equiv f_\sigma(\varepsilon_k; \eta), \tag{5.5.7}$$

where $\Delta\mu$ is the change in the chemical potential due to the magnetic field. The Lindhard function for $\boldsymbol{q} = 0$ under a magnetic field is then given by

$$F_\sigma(0; H) = \int d\varepsilon\, N(\varepsilon)\left(-\frac{\partial f_\sigma(\varepsilon;\eta)}{\partial\varepsilon}\right). \tag{5.5.8}$$

Expanding the right hand side in terms of $\sigma\eta - \Delta\mu$, we obtain

$$\begin{aligned} F_\sigma(0; H) &\equiv F_\sigma(0; \eta) \\ &= F_\sigma(0) + F_\sigma'(0)(\sigma\eta - \Delta\mu) + \frac{1}{2!}F_\sigma''(0)(\sigma\eta - \Delta\mu)^2 + \cdots, \end{aligned} \tag{5.5.9}$$

where

$$F_\sigma^{(n)}(0) \equiv -\int d\varepsilon\, N(\varepsilon) \left. \frac{\partial^{(n+1)} f_\sigma(\varepsilon;\eta)}{\partial \varepsilon^{(n+1)}} \right|_{\eta=0}. \tag{5.5.10}$$

The change in the chemical potential, $\Delta\mu$, appearing in (5.5.9) is obtained by requiring the conservation of electron number,

$$\sum_{\boldsymbol{k},\sigma} f(\varepsilon_{\boldsymbol{k}\sigma}(\eta))$$

$$= \sum_{\boldsymbol{k},\sigma} \left[f(\varepsilon_{\boldsymbol{k}\sigma}) + f'(\varepsilon_{\boldsymbol{k}\sigma})(\sigma\eta - \Delta\mu) + \frac{1}{2!} f''(\varepsilon_{\boldsymbol{k}\sigma})(\sigma\eta - \Delta\mu)^2 + \cdots \right]$$

Retaining up to $O(\eta^2)$, we have

$$\Delta\mu = \frac{\Pi}{P}\eta + \frac{2}{P^3}\left[F_+^2(0)F_-'(0) + F_-^2(0)F_+'(0)\right]\eta^2 + \cdots, \tag{5.5.11}$$

where we put

$$P = \sum_\sigma F_\sigma(0); \quad \Pi = \sum_\sigma \sigma F_\sigma(0). \tag{5.5.12}$$

By putting the above result on $\Delta\mu$ into (5.5.9) we obtain

$$F_\sigma(0;\eta) = F_\sigma(0)\left[1 + b_{1\sigma}\left(\frac{\eta}{W}\right) + b_{2\sigma}\left(\frac{\eta}{W}\right)^2 + \cdots \right], \tag{5.5.9'}$$

$$b_{1\sigma} = 2\sigma \frac{F_\sigma'(0)F_{-\sigma}(0)}{F_\sigma(0)P} W, \tag{5.5.13}$$

$$b_{2\sigma} = \left[\frac{F_\sigma''(0)}{2P}\left\{ 3\frac{F_{-\sigma}(0)}{F_\sigma(0)} - 1 \right\} - 2\frac{F_\sigma'(0)}{F_\sigma(0)} \frac{\sum_{\sigma'} F_{\sigma'}^2(0)F_{-\sigma'}'(0)}{P^3} \right] W^2, \tag{5.5.14}$$

where W is the width of the electron energy band.

Then if we put as

$$s(\eta) = s + \Delta s(\eta), \tag{5.5.15}$$

from (5.5.4) we obtain

$$\frac{\Delta s(\eta)}{s} = f_1\left(\frac{\eta}{W}\right) + f_2\left(\frac{\eta}{W}\right)^2 + \cdots , \tag{5.5.16}$$

$$f_1 = -\frac{1}{2}\left(\frac{s_0}{s}\right)^2 \frac{2N(0)}{\widetilde{P}^2} \sum_\sigma D_\sigma^2 F_\sigma(0)\, b_{1\sigma} , \tag{5.5.17}$$

$$f_2 = -\frac{1}{2}\left(\frac{s_0}{s}\right)^2 \frac{2N(0)}{\widetilde{P}}\left[\frac{1}{\widetilde{P}}\sum_\sigma D_\sigma^2 F_\sigma(0)\left\{\widetilde{V} F_\sigma(0)\, b_{1\sigma}^2 + b_{2\sigma}\right\}\right.$$

$$\left. -2\frac{1}{\widetilde{P}^2}\left\{\sum_\sigma D_\sigma^2 F_\sigma(0)\, b_{1\sigma}\right\}^2\right] , \tag{5.5.18}$$

where we put

$$\widetilde{P} = \sum_\sigma \widetilde{F}_\sigma(0); \qquad D_\sigma = 1/[1 - \widetilde{V}F_\sigma(0)]. \tag{5.5.19}$$

Starting from (5.4.28) we can similarly pursue how the attenuation constant of sound wave depends on the (additional) spin splitting of the electron energy bands due to an external magnetic field. If we put as

$$\alpha_\omega(\eta) = \alpha_\omega + \Delta\alpha_\omega(\eta) , \tag{5.5.20}$$

we obtain

$$\frac{\Delta\alpha_\omega(\eta)}{\alpha_\omega} = g_1 \frac{\eta}{W} + g_2\left(\frac{\eta}{W}\right)^2 + \cdots , \tag{5.5.21}$$

$$g_1 = 2\left[\left(\frac{s}{s_0}\right)^2 \frac{\widetilde{P}}{N(0)} - 1\right] f_1 + 2\sum_\sigma D_\sigma \widetilde{V} F_\sigma(0)\, b_{1\sigma} . \tag{5.5.22}$$

As for g_2, since its expression is too complicated we give it only for the paramagnetic state,

$$g_2 = 2\left[\left(\frac{s}{s_0}\right)^2 \frac{\widetilde{P}}{N(0)} - 1\right] f_2 + 2\widetilde{V}\,\widetilde{F}(0) \sum_\sigma \left\{b_{2\sigma} - \widetilde{V}\,\widetilde{F}(0)\, b_{1\sigma}^2\right\} . \tag{5.5.23}$$

Based on the above results we proceed to discuss the effect of an external magnetic field on the velocity and attenuation of sound wave for the paramagnetic and ferromagnetic state of a metal.

5.5.2 Magnetic Field Effect on Sound Propagation in the Paramagnetic State of a Metal

A *Magnetic field effect on sound velocity for $T > T_C$*

In the paramagnetic state, from (5.5.17) and (5.5.13) we find $f_1 = 0$. Thus, we have

$$\frac{\Delta s(\eta)}{s} \equiv -\frac{1}{2}\left(\frac{s_0}{s}\right)^2 K\left(\frac{\eta}{W}\right)^2, \tag{5.5.24}$$

$$K = \left[\frac{1}{2}\left\{\frac{F''(0)}{F(0)} - \left(\frac{F'(0)}{F(0)}\right)^2\right\} + \widetilde{V}F(0)\,D_0\left(\frac{F'(0)}{F(0)}\right)^2\right]W^2, \tag{5.5.25}$$

where

$$D_0 = \frac{1}{1-\widetilde{V}F(0)} \tag{5.5.26}$$

is the Stoner exchange enhancement factor. If we use the relation,

$$\eta = \left(1 + \frac{\widetilde{V}}{2\mu_B{}^2}\chi_s\right)\mu_B H = D_0\mu_B H, \tag{5.5.27}$$

χ_S being the Stoner susceptibility, we arrive at [5.28]

$$\frac{\Delta s(H)}{s} = k\left(\frac{\mu_B H}{W}\right)^2, \qquad k = -\frac{1}{2}\left(\frac{s_0}{s}\right)^2 D_0{}^2 K. \tag{5.5.28}$$

This result is certainly of the form of (5.5.2). If k is of the order of unity, this magnetic field effect is far too small to account for actual observations in FeNi alloys and A15 compounds. However, as can be seen in (5.5.28), k is explicitly proportional to $D_0{}^2$; further K contains a term proportional to D_0. Thus, we have $k \propto D_0{}^3$.

Note that for $(k_B T/\varepsilon_F)^2 \ll 1$, in (5.5.25) we can use the following approximations,

$$F'(0) = \int N'(\varepsilon) f'(\varepsilon)\, d\varepsilon \cong -N'(0),$$

$$F''(0) = -\int N''(\varepsilon) f'(\varepsilon)\, d\varepsilon \cong N''(0) \tag{5.5.29}$$

If we are given the electronic density of states, the location of the Fermi energy, and the values of $\widetilde{V}$ or $\overline{V}$ and ξ, we can estimate k from (5.5.25) and (5.5.29). In Fig.5.13 we give the result of such numerical calculation for the electronic density of states of Fig.5.5. We put $\xi = 1$, and $W = 1$ eV, and assumed $(k_B T/\varepsilon_F)^2 \ll 1$.

According to the results of Fig.5.13, the sign and magnitude of k depend very sensitively on ε_F/W and $\overline{V}$, and the magnitude can be very large. For two values of $\varepsilon_F/W \cong 0.2$ or 0.8 and $\overline{V} = 0.9$, so that $D_0 = 10$, we obtain $k \cong -10^4$; if $\overline{V}$ is closer to unity, the magnitude of k can become even larger. Thus, it seems possible to explain the observed large magnetic field effect of Fig.5.12 in FeNi alloys on the basis of this theory. Note also that the obtained value of k has right sign.

According to Fig.5.13, if ε_F is located near the peak in the electronic density of states, $\varepsilon_F/W \cong 0.4$, or 0.6, for instance, for the same value of $\overline{V} = 0.9$ as above, we find $k \cong -10^2$, smaller than the above case by two orders of magnitude; thus we can understand why in Ni the effect of magnetic field on the elastic constant is much smaller than in FeNi alloys. The observed result on the

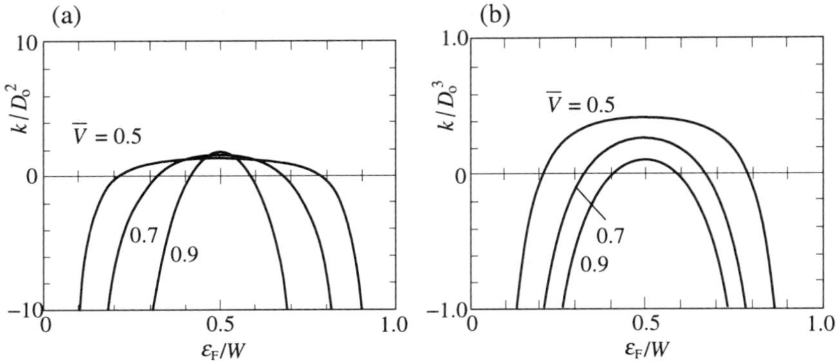

Fig.5.13 The magnetic field dependence of sound velocity in the paramagnetic state, as defined in (5.5.25)−(5.5.29), for the model electronic density of states of Fig.5.5. Here $\xi = 1$, and D_0 is the Stoner factor of (5.5.26). The left (a) and the right (b) figures differ only in the scales of ordinates.

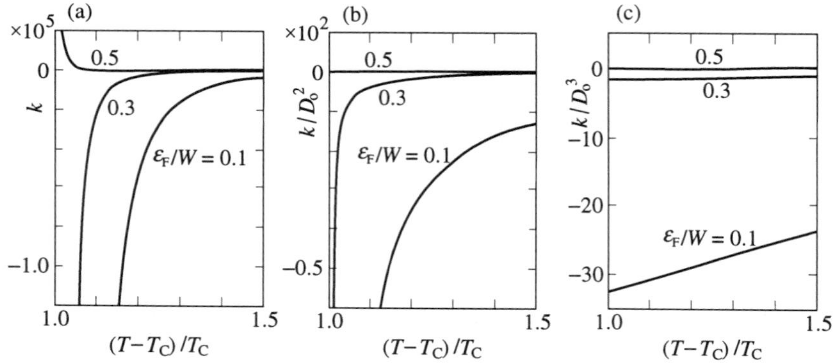

Fig.5.14 The temperature dependence of the magnetic field dependence of sound velocity as defined in (5.5.25)–(5.5.29) for $T > T_C$ of an itinerant electron ferromagnet, for different locations of the Fermi energy in the model electronic density of states of Fig.5.5. We assumed $\overline{V} = 1.3$ and $\xi = 2$. To see that k is proportional to $D_0{}^3$, we showed the results in three different ways, in k, $k/D_0{}^2$, and $k/D_0{}^3$ respectively, in (a), (b), and (c).

A15 compound may also be similarly understood without requiring a much too small value of W if we choose a moderate size of $\overline{V}$.

Note that the quantity k changes with temperature. Particularly in a ferromagnet, where $\overline{V} > 1$, the temperature dependence is large, as shown in the numerical example given Fig.5.14. In this example, we used the same model electronic density of states as in Fig.5.13 and changed the location of the Fermi energy in it as before. We assumed $\overline{V} = 1.3$ and $\xi = 2$. Note that $k \propto D_0{}^3$.

As we will see in Chapter 6, the quantity $k/D_0{}^2$ or K is directly related to the effective exchange interaction between electrons due to EPI. It is this relation that makes it important to study the magnetic field dependence of sound velocity in the paramagnetic state of a metal.

B *Magnetic field effect on sound attenuation for* $T > T_C$

If we write the attenuation constant in the presence of an external magnetic field $\boldsymbol{H}$ as

$$\alpha_\omega(H) = \alpha_\omega + \Delta\alpha_\omega(H) \tag{5.5.30}$$

correspondingly to (5.5.20), from the relation (5.5.21)–(5.5.23) we obtain [5.28]

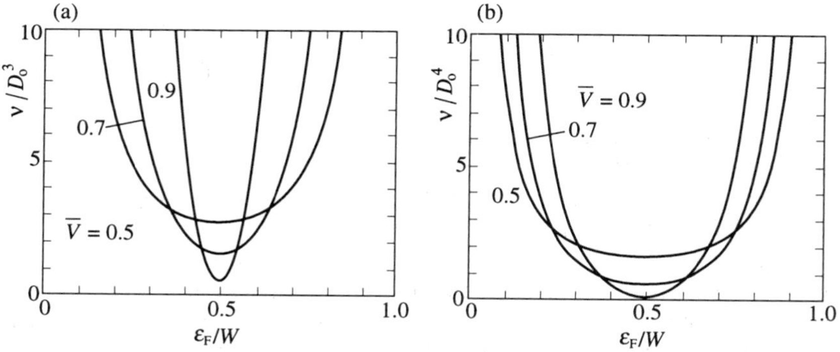

Fig.5.15 The magnetic field dependence of the attenuation of sound in the paramagnetic state, as defined in (5.5.31), for the model electronic density of Fig.5.5 with W = 1 eV.

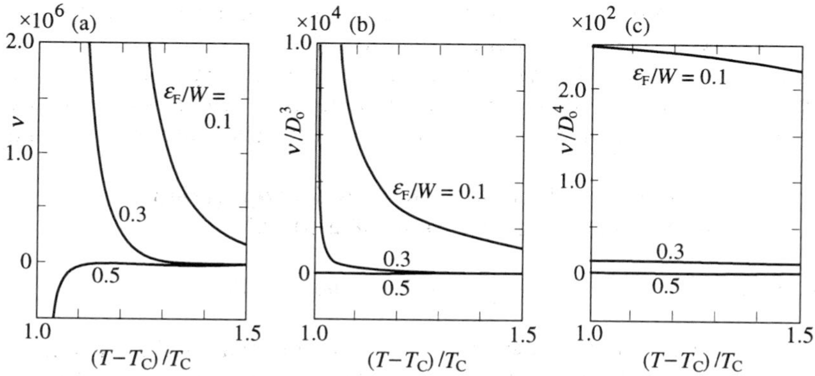

Fig.5.16 The temperature dependence of the shift of sound attenuation due to magnetic filed as defined in (5.5.31)–(5.5.32) for $T > T_C$ of an itinerant electron ferromagnet, for different locations of the Fermi energy in the model electronic density of states of Fig.5.5. We assumed $\overline{V} = 1.3$ and $\xi = 2$. To demonstrate that v is proportional to $D_0{}^4$, we presented the results in v, $v/D_0{}^3$, and $v/D_0{}^4$, respectively, in (a), (b), and (c).

$$\Delta\alpha_\omega(H)/\alpha_\omega = v\,(\mu_B H / W)^2,$$

$$v = D_0{}^2 K\left[\left(\frac{s_0}{s}\right)^2 - 2\right] + G, \qquad (5.5.31)$$

where K is given in (5.5.25), and

$$G = \left[\overline{V} D_0^{\,3} \left\{ \frac{F''(0)}{F(0)} - \left(\frac{F'(0)}{F(0)} \right)^2 \right\} + 3\overline{V}^{\,2} D_0^{\,4} \left(\frac{F'(0)}{F(0)} \right)^2 \right] W^2. \quad (5.5.32)$$

Note that the magnetic field effect on the attenuation constant is exchange enhanced by D_0^4. (In [5.28] and [3.23] there is an error in the expression for ν ; it should be corrected as in (5.5.31). Correspondingly, Fig.7.3 of [3.23] should be corrected as Fig.5.15.)

In Fig.5.15 we present the result of a numerical calculation of ν, obtained in a manner similar to those shown in Fig.5.13, using the model electronic density of states of Fig.5.5. We find that the value of ν sensitively depends on the value of $\overline{V}$ and the location of the Fermi energy in the model electronic density, similarly to the case of k.

The temperature dependence of ν for $T > T_C$ of a ferromagnet, $\overline{V} > 1$, is shown in Fig.5.16. This results is obtained similarly to that of Fig.5.14 for k, with the same model electronic density of states of Fig.5.5.

We have found that in the paramagnetic state of a metal the effects of a magnetic field on the velocity and attenuation of sound is exchange enhanced by the factor of D_0^3 and D_0^4, respectively. Recall that the response of electrons to a magnetic field, namely, the spin susceptibility is exchange enhanced only by D_0. In this sense, phonons in metals are very magnetically responsive entities.

5.5.3 Magnetic Field Effect on Sound Propagation in the Ferromagnetic State of a Metal

A *Magnetic field effect on sound velocity for $T < T_C$*

Differing from the case of the paramagnetic state, in the ferromagnetic state the magnetic field effect begins from the first order in η or H. Thus, here we retain only up to the first order terms.

Thus, from (5.5.16) and (5.5.17) we obtain

$$\frac{\Delta s(\eta)}{s} = f_1\!\left(\frac{\eta}{W}\right) = \left(\frac{s_0}{s}\right)^2 Y(\varepsilon_F, \overline{V})\, \frac{4\, F_+(0)\, F_-(0)}{F_+(0) + F_-(0)}\, \eta, \quad (5.5.33)$$

where $(s/s_0)^2$ is given in (5.4.27), and we put

$$Y(\varepsilon_F, \overline{V}) = -\frac{1}{2} N(0) \sum_{\sigma} \sigma \frac{F_\sigma'(0)/F_\sigma(0)}{\{1 - \widetilde{V} F_\sigma(0)\}^2} \Big/ \left[\sum_{\sigma} \frac{F_\sigma(0)}{1 - \widetilde{V} F_\sigma(0)} \right]^2. \quad (5.5.34)$$

In the ferromagnetic state, η is related to the external magnetic field, H, as

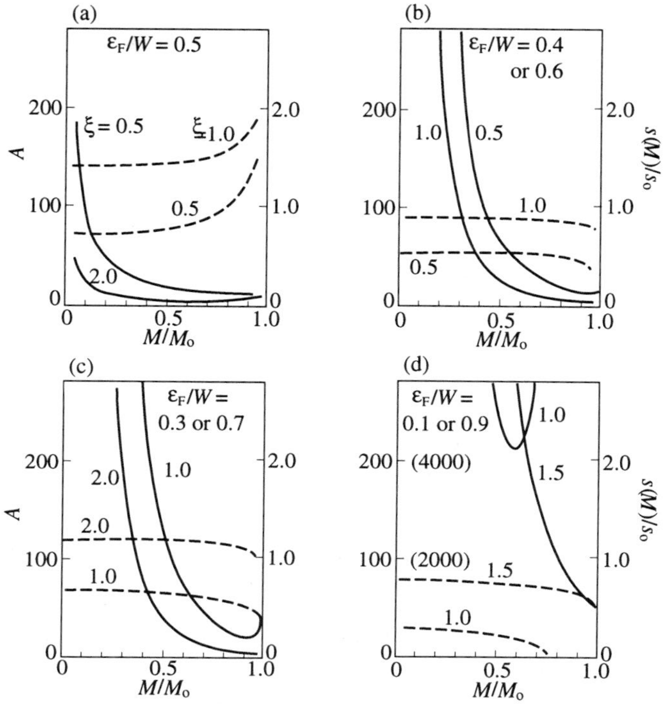

Fig.5.17 The magnetic field dependence of sound velocity in the ferromagnetic state of a metal as defined in (5.5.36)–(5.5.37) for the model electronic density of states of Fig.5.5 with $W = 1$ eV. In (d) the scale shown with parentheses is used for the case of $\xi = 1$. For comparison, the result of Fig.5.6 on the magnetization dependence of the sound velocity is shown by broken lines.

$$\eta = (1 + \frac{1}{2}\widetilde{V}\overline{\chi}_{\rm hf})\mu_{\rm B}H = \frac{F_+(0) + F_-(0)}{4F_+(0)\, F_-(0)}\overline{\chi}_{\rm hf}\mu_{\rm B}H, \qquad (5.5.35)$$

where $\chi_{\rm hf} = \mu_{\rm B}{}^2\overline{\chi}_{\rm hf}$ is the high field magnetic susceptibility given in (3.4.24). Thus, we obtain finally [5.29]

$$\frac{\Delta s(H)}{s} = \frac{A}{W}\,\mu_{\rm B}H, \qquad (5.5.36)$$

$$\frac{A}{W} = \frac{F_+(0) + F_-(0)}{4F_+(0)\, F_-(0)}\overline{\chi}_{\rm hf}\frac{f_1}{W} = (s_{\rm o}/s)^2\, Y\, \overline{\chi}_{\rm hf}. \qquad (5.5.37)$$

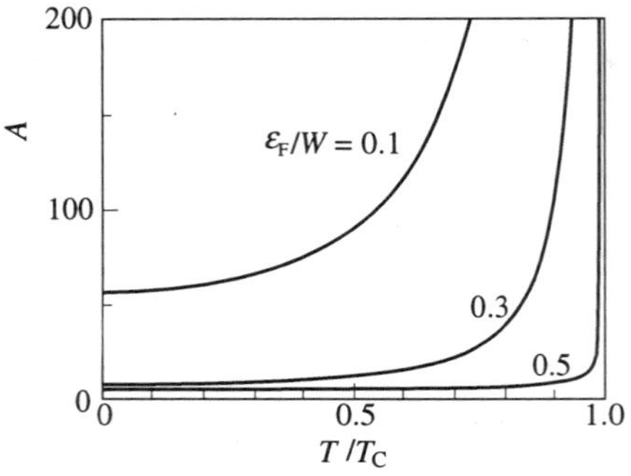

Fig.5.18 The temperature dependence of the magnetic field depend-ence of sound velocity in the ferromagnetic state of a metal cal-culated from (5.5.37) for the model electronic density of states of Fig.5.5. We assumed $\overline{V} = 1.1$ and $\xi = 2$.

In order to see how large the factor A can be, in Fig.5.17 we present a numerical example with the model electronic density of states of Fig.5.5. The procedure is the same as for Fig.5.6; we put $W = 1$ eV, and the magnetization is changed by changing the value of $\overline{V}$ while keeping the temperature at $T = 0$.

In Fig.5.18 we show also the temperature dependence of A which is calculated from (5.5.37) for the same model electronic density of states of Fig.5.5. We assumed $\overline{V} = 1.1$ and $\xi = 2$.

From Fig.5.17 and 5.18 we find that it is quite possible to have $A \cong 10^2$ as required to account for the observations in FeNi alloys. It is also possible to have a much smaller value for A; for $\varepsilon_F/W \cong 0.5$ and $\xi > 1$, as we have been assuming for Ni within the model electronic density of Fig.5.5, $A \cong O(1)$. For $\varepsilon_F/W \cong 0.1$ or 0.9, A becomes very large, and in the case of $\xi = 1$ the scale of theordinate is given within parentheses. Note that the divergence of A for $M \to 0$ is caused by that of χ_{hf} at $M = 0$; a more careful treatment is required in this region.

B *Magnetic field effect on sound attenuation for* $T < T_C$

Retaining up to first order terms in (5.5.21) and putting as (5.5.30) we have

$$\frac{\Delta\alpha_\omega(H)}{\alpha_\omega} = g_1\left[1 + \frac{1}{2}\widetilde{V}\overline{\chi}_{hf}\right]\frac{\mu_B H}{W} \equiv \lambda\,\frac{\mu_B H}{W}, \tag{5.5.38}$$

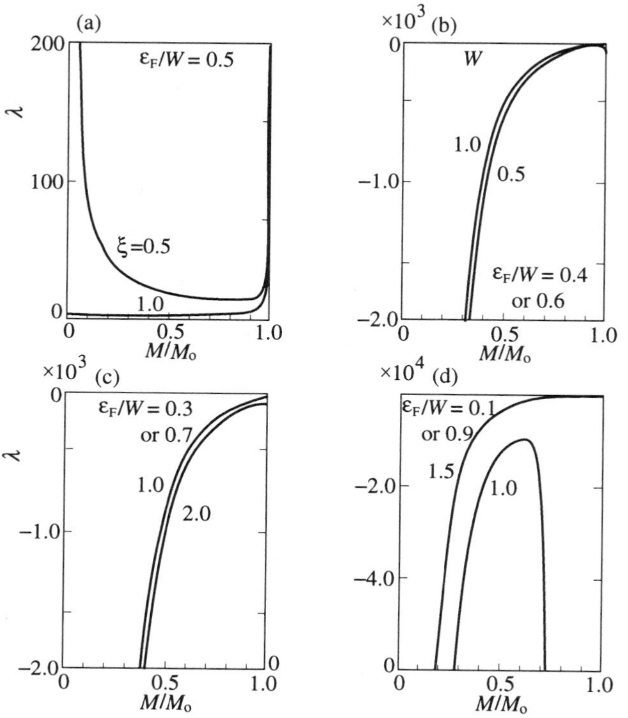

Fig.5.19 The magnetization dependence of sound attenuation calculated from (5.5.39). We used the model electronic density of states of Fig.5.5 with $W = 1$ eV and assumed $\xi = \underline{2}$. The change in magnetization is produced by changing the value of $\overline{V}$ while keeping the temperature at $T = 0$.

$$\frac{\lambda}{W} = 2\left[\left(\frac{s}{s_o}\right)^2 \frac{\widetilde{P}(0)}{N(0)} - 1\right]\frac{A}{W} + \frac{\widetilde{V}\overline{\chi}_{\rm hf}}{\sum_\sigma D_\sigma^2} \sum_\sigma \sigma D_\sigma^3 \frac{F_\sigma'(0)}{F_\sigma(0)} . \qquad (5.5.39)$$

In Fig.5.19 we present the result of numerical calculation on (5.5.39) which is carried out similarly to Fig.5.17 with the same model electronic density of states of Fig.5.5; here, magnetization is changed by changing the value of $\overline{V}$, while keeping the temperature at $T = 0$.

The temperature dependence of λ is also calculated from (5.5.39) in Fig.5.20 by using the same model electronic density of states. We assumed $\overline{V} = 1.1$ and $\xi = 2$. Thus, we have found that in the ferromagnetic state too, phonons can be far more sensitive to an external magnetic field than is generally

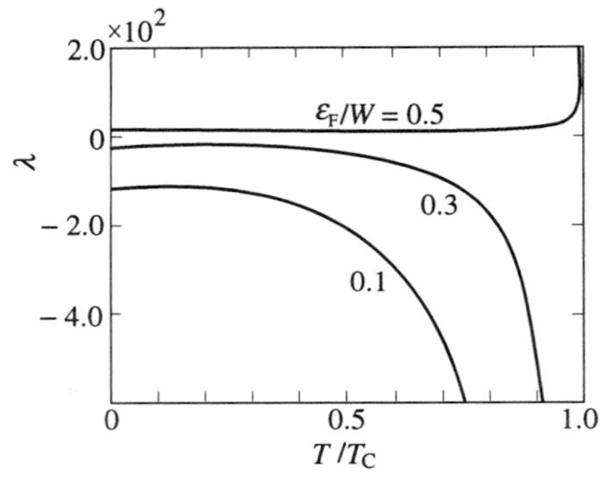

Fig.5.20 The temperature dependence of the effect of magnetic field on sound attenuation for of an itinerant electron ferromagnet calculated from (5.5.39) with the model electronic density of states of Fig.5.5 with $W = 1$ eV. We assumed $\bar{V} = 1.3$ and $\xi = 2$.

assumed. This fact again indicates the possible involvement of phonons in an important way in shaping the magnetic properties of a metal. As we will see in Chapter 6, the quantity A determines the effect of EPI on the spontaneous magnetization of a ferromagnetic metal; $|A| \cong 10^2$ implies that the magnetization of the metal is affected by EPI as much as $\cong 1\ \mu_B$ /atom.

Marked anomalous volume and elasticity behaviors are observed also in various itinerant antiferromagnetic (SDW) systems; see, for instance, Refs. [5.30,5.31]. It is an interesting challenge to extend the discussion for ferromagnetic systems of this chapter, and succeeding Chapters 6 and 7, to antiferromagnetic or SDW systems. Such attempts are being made [5.32, 5.33].

Finally note, as we have found in connection with the temperature dependence of elastic constant in the paramagnetic state, that the result of (5.4.27), on which the entire discussion of this section also is based, is not a fully satisfactory one. In Chapter 7 we will discuss how to improve in an essential way the result of this chapter.

PROBLEMS FOR CHAPTER 5

5.1 Derive the result (5.2.22) starting from (5.2.1).

5.2 Derive (5.2.19) from (5.2.18).

5.3 Derive the boson commutation relation of (5.2.26) from that of (5.2.12). Also derive the result of (5.2.25) from (5.2.22).

5.4 Show that the static Lindhard function at $T = 0$ of 1- and 2- dimensional electron gases are given, respectively, as

$$F(\boldsymbol{q}) = \frac{Lm}{\pi\hbar^2 k_F}\left(\frac{k_F}{q}\right)\ln\left|\frac{2k_F + q}{2k_F - q}\right|,$$

and

$$F(\boldsymbol{q}) = \begin{cases} \dfrac{Sm}{2\pi\hbar^2} = N(0) & \text{for} \quad q/2k_F \leq 1, \\ N(0)\left(1 - \sqrt{1 - \dfrac{1}{x^2}}\right) & \text{for} \quad q/2k_F \equiv x > 1, \end{cases}$$

where L and S are, respectively, the length and area of the 1- and 2-dimensional systems. See problem 4.2.

5.5 Estimate the magnitude of the Bohm-Staver sound velocity from (5.4.20) for simple metals and compare with experiments. See Pines [5.8].

5.6 Derive the result of (5.5.11) for the change in chemical potential due to an additional spin splitting of the electron energy bands in a metal.

Chapter 6

THE ROLE OF THE ELECTRON-PHONON INTERACTION IN METALLIC MAGNETISM

In the preceding chapter we saw how the phonon and elastic properties of a metal are dictated by its magnetic property through the electron-phonon interaction (EPI). In this chapter we discuss the opposite problem; we explore how phonons are involved in determining the magnetic properties of a metal through EPI. In the course of such endeavor we also review the basics of itinerant electron magnetism in general, with some new results.

6.1 WHY CAN THE EFFECT OF EPI BE IMPORTANT IN MAGNETISM OF A METAL?

As we discussed in 3.4, the magnetic susceptibility of a system can be obtained if we know the free energy of the system, $F(M)$, as the function of magnetization, M. Also, for $T < T_C$ of a ferromagnetic system, the spontaneous magnetization is determined by minimizing $F(M)$ with respect to M.

There are different models and approximations for obtaining $F(M)$ of a metal. However, in discussing the magnetic properties of a metal, in most cases it is exclusively the contribution of electrons that is considered. The Stoner model we discussed in Chapter 3 is such an example. As illustrated in the upper part of Fig.6.1 the Stoner model pursues within the mean field approximation how the energy, or free energy changes with spin splitting of the electron energy bands.

In principle, however, the free energy of a metal contains the contribution from the lattice vibrations, or phonons, F_{ph}, in addition to that of the electrons, F_{el},

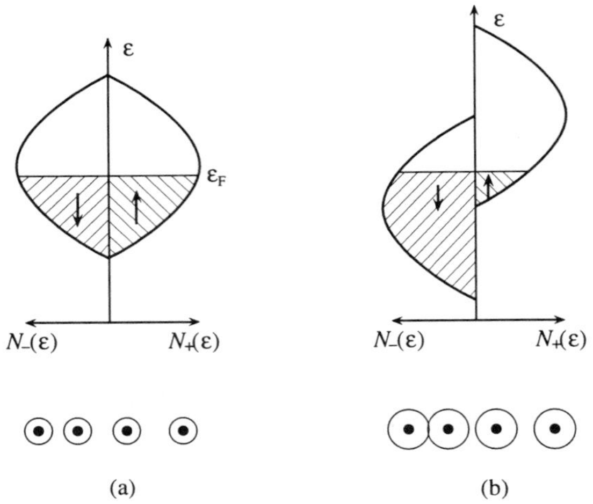

Fig.6.1 Electrons (top) and ions (bottom) in the paramagnetic state (a), and in the spin split states (b) of a metal. We emphasized the difference in the screening of ions, central dots, by the different radii of screening electron distributions.

$$F = F_{\text{el}} + F_{\text{ph}} . \tag{6.1.1}$$

The phonon free energy of a system is determined from the phonon frequency spectrum of the system; see (6.2.3) or (6.1.4). From now on we denote the phonon frequency as $\omega_{\text{ph}}(\boldsymbol{q})$ in place of $\omega_{\boldsymbol{q}}$. If F_{ph} of a metal is to be determined from its screened phonon frequency, F_{ph} should depend on the magnetization since, as we saw in the preceding chapter, $\omega_{\text{ph}}(\boldsymbol{q})$ changes with the spin splitting of the energy band of the electrons which screen the ion-ion interaction. In Chapter 10 we will explicitly derive this result on the phonon free energy.

The most important point to note here is that, quite surprisingly, the magnetization dependent part of F_{ph} can be as large as that of F_{el}. Thus, in discussing magnetism of a metal, it becomes necessary to consider how the free energy of phonons will change with magnetization, as symbolically illustrated in the lower part of Fig.6.1.

When I raise such possible role of phonons to a colleague, the commonest reaction I receive is, "Yes, I know. First, the excitation of phonons in a material implies raising its temperature; of course, temperature does have effects on its magnetic properties. Second, the excitation of phonons also causes thermal expansion; a change in volume of a material would of course affect the magnetic properties of the material." However, what I have been pointing out is entirely different from such a simple expectation. *The phonon effect is present even at $T = 0$ and at a constant volume!*

Before a more detailed discussion in the next section, let us here briefly demonstrate how important a role the phonons can play in determining the paramagnetic spin susceptibility of a metal. For simplicity, let us discuss the problem at $T = 0$. In this case it suffices to consider the energies E_{el} and E_{ph} in place of the corresponding free energies F_{el} and F_{ph}. Let us assume the electron energy bands are slightly spin split as in Fig.3.3, producing a small relative magnetization. Within the mean field approximation, the change in E_{el} due to this magnetization is given, as we already saw in (3.3.11)–(3.3.14), as

$$\Delta E_{el}(M) \;=\; N(0)\,[1 - \widetilde{V}N(0)]\,\eta'^{\,2}, \tag{6.1.2}$$

where η' is half the spin splitting of the bands and other notations are the same as before. Note that, in the present Landau procedure, the spin splitting is a variational one, as emphasized in 3.4.1; we put the prime on η in order to distinguish it from the spin splitting in the equilibrium state which is given in (5.5.5). From Fig.3.3 it is evident that

$$\eta' \;=\; \frac{\overline{M}}{2N(0)} \tag{6.1.3}$$

to the lowest order in $\overline{M} = (n_- - n_+)$ (*Problem* 6.1). Thus, for a small magnetization, $\overline{M}n \ll 1$, we have

$$\Delta E_{el}(M) \;=\; \frac{1}{4N(0)}\left(1 - \widetilde{V}N(0)\right)\overline{M}^{\,2}$$

$$= O\!\left[\, n\,\varepsilon_F\left(1 - \widetilde{V}N(0)\right)\left(\overline{M}/n\right)^2 \right], \tag{6.1.2'}$$

where we noted that $N(0) = O(n/\varepsilon_F)$, n being the total number of electrons.

As for the phonon energy at $T = 0$, it is given by the zero point oscillation contribution,

$$E_{ph} \;=\; \frac{1}{2}\sum_{\boldsymbol{q}} \hbar\,\omega_{ph}(\boldsymbol{q};M)\,. \tag{6.1.4}$$

If we assume that the relation of (5.4.19) holds for all $\boldsymbol{q}$ within the 1st Brillouin zone,

$$\omega_{ph}(\boldsymbol{q};M) \;=\; s(M)\,q, \tag{6.1.5}$$

which is called the *Debye approximation*, (6.1.4) is rewritten as

$$E_{\mathrm{ph}}(M) = \frac{1}{2}\sum_{q} \hbar s(M)\, q = O\left[N\hbar\omega_{\mathrm{D}}\left(\frac{s(M)}{s}\right)\right], \tag{6.1.4'}$$

where s and $s(M) = s + \Delta s(M)$ are, respectively, the sound velocity of a metal without and with magnetization, and N is the total number of atoms. According to the result of 5.5.1, $|\Delta s(M)/s| \propto (\eta'/\varepsilon_{\mathrm{F}})^2$ for a small magnetization, and $|\Delta s(M)/s|$ can be of the order of unity for the full spin splitting of the bands. Thus we may put

$$\left|\Delta s(M)\,/\,s\right| \cong (\,\overline{M}/n\,)^2. \tag{6.1.6}$$

Substituting this into (6.1.4'), we have

$$\left|\,\Delta E_{\mathrm{ph}}(M)\right| = O\big(N\hbar\omega_{\mathrm{D}}\,(\,\overline{M}/n\,)^2\big) \tag{6.1.7}$$

for the change in the phonon energy due to the magnetization.

From the procedure of (3.4.4) with $\Delta E_{\mathrm{el}}(M)$ of (6.1.2'), we obtain the Stoner susceptibility. What would be the size of the contribution of $\Delta E_{\mathrm{ph}}(M)$ to the spin susceptibility of a metal? From (6.1.7) we have

$$\left|\frac{\partial^2 \Delta E_{\mathrm{ph}}(M)\,/\,\partial M^2}{\partial^2 \Delta E_{\mathrm{el}}(M)\,/\,\partial M^2}\right| \cong \left|\frac{\Delta E_{\mathrm{ph}}(M)}{\Delta E_{\mathrm{el}}(M)}\right| \cong O\left(\frac{\hbar\omega_{\mathrm{D}}}{\varepsilon_{\mathrm{F}}\left|1 - \widetilde{V}N(0)\right|}\right), \tag{6.1.8}$$

where we noted $n \cong N$.

In a metal where the exchange interaction between electrons is weak, $\widetilde{V}N(0) \ll 1$, we obtain

$$\left|\frac{\Delta E_{\mathrm{ph}}(M)}{\Delta E_{\mathrm{el}}(M)}\right| = O\left(\frac{\hbar\omega_{\mathrm{D}}}{\varepsilon_{\mathrm{F}}}\right) \cong 10^{-2}. \tag{6.1.9}$$

In this case the effect of EPI on the magnetic susceptibility (at $T = 0$) is very small. It is this situation which has been invoked to justify the claim of the smallness of the effect of EPI. For a review, see Herring [3.10].

In the case of a ferromagnetic metal, however, $\widetilde{V}N(0) \cong 1$ in (6.1.8), and the relative size of the EPI effect on the spin susceptibility can be much larger than that of (6.1.9). Here, whereas we have a fairly clear idea on the size of $\hbar\omega_{\mathrm{D}}$, it is not possible to estimate unambiguously the magnitude of $\widetilde{V}N(0)$, or, $\varepsilon_{\mathrm{F}}\,|1-\widetilde{V}N(0)|$.

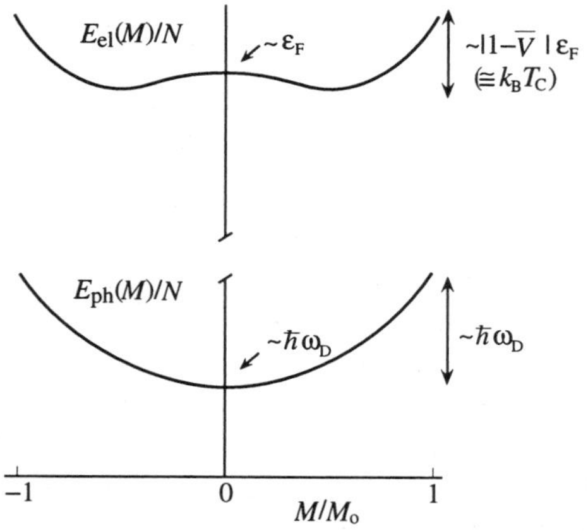

Fig.6.2 Magnetization dependence of the energy of electrons, $E_{el}(M)$, and the zero point oscillation energy of phonons, $E_{ph}(M)$ while $E_{el} \gg E_{ph}$, $\Delta E_{el} \cong \Delta E_{ph}$. In the figure, an abbreviation: $\overline{V} = \widetilde{V}N(0)$ is used.

In this respect, we make an important finding, as we will see in the next section, that with the effect of EPI the Curie temperature of an itinerant electron ferromagnet is given by

$$k_BT_C = O[\,\varepsilon_F(\widetilde{V}N(0) - 1)\,]\,, \tag{6.1.10}$$

with $\widetilde{V}N(0) > 1$ (see (6.2.28)). According to (6.1.2') or (3.3.11)–(3.3.14), the right hand side of (6.1.10) gives the size of the energy per electron to be gained by fully spin splitting the electron energy bands. The Curie temperature of a Heisenberg or Ising spin model is given by such an energy. For itinerant electron systems, however, the Curie temperature has not previously been obtained in this form; see, for instance, the Stoner theory result of (3.3.19).

With (6.1.10) in (6.1.8) we have

$$\left|\Delta E_{ph}(M)\,/\,\Delta E_{el}(M)\right| = O\left(\theta_D/T_C\right), \tag{6.1.11}$$

where θ_D is the phonon Debye temperature defined as

$$\hbar\omega_D = k_B\theta_D\,. \tag{6.1.12}$$

Generally, both θ_D and T_C are of the order of 10^2 K. Thus for a ferromagnetic metal we have

$$\left| \Delta E_{\rm ph}(M) \,/\, \Delta E_{\rm el}(M) \right| \cong O(1), \tag{6.1.13}$$

although

$$\left| E_{\rm ph} \,/\, E_{\rm el} \right| = O\left(\hbar\omega_{\rm D}/\varepsilon_{\rm F}\right) \cong 10^{-2}\,. \tag{6.1.14}$$

We illustrate the situation in Fig.6.2. This is a surprising finding: The role of phonons can be as important as that of electrons in determining the magnetic susceptibility of a ferromagnetic metal.

In arriving at the above conclusion, it was crucial to note the relation of (6.1.10). This new result for T_C of an itinerant electron ferromagnet is to be obtained in the next section by treating the effect of the EPI in a fully temperature dependent way. The size of the EPI effect increases with increasing temperature, since the number of thermally excited phonons increases with increasing temperature.

The importance of the EPI effect is not confined only to ferromagnetic metals. Even in a metal where $\widetilde{V}N(0) < 1$ and, thus, (6.1.9) holds, the EPI effect plays an important role in determining the temperature dependence of the spin susceptibility. We will see this in 6.2–6.3.

What is the effect of EPI on the spontaneous magnetization below T_C of a ferromagnetic metal? From our above discussion and conclusion illustrated in Fig.6.2, it is evident that the location of the minimum of the free energy can be appreciably shifted by the effect of $E_{\rm ph}(M)$ or $F_{\rm ph}(M)$. We will find that the spontaneous magnetization of an itinerant electron ferromagnet can be affected by EPI as much as $\sim 1\ \mu_B$/atom. We will discuss this subject in 6.6–6.7. In the course of the discussion we make two interesting findings. The first is that phonon in a ferromagnet, either localized or itinerant, carries magnetization; we demonstrate that the size of magnetization carried by a phonon can be $\sim 1\ \mu_B$ in an itinerant electron ferromagnet. The second is that the effect of EPI on magnetization manifests itself in the spin splitting of the electron energy bands; we demonstrate that the relative sizes of both the spin wave and EPI contributions can be $O(1)$. This subject is discussed in 6.8. In 6.9 we formulate non-perturbatively the roles of spin wave and phonon in the temperature dependence of magnetization and in 6.10 we discuss the effect of EPI (and spin wave or/and spin fluctuation) on the high field susceptibility. If phonons play a significant role in magnetism of a metal, there may be an isotope effect. In 6.11 we discuss where such an isotope effect can possibly be found.

6.2 THE EFFECT OF EPI ON THE MAGNETIC SUSCEPTIBILITY OF A METAL: PHNON MECHANISM OF THE CURIE-WEISS SUSCEBPTIBILITY

If we apply the procedure of (3.4.4) to (6.1.1), the magnetic susceptibility of a metal is given by

$$\frac{1}{\chi} = \left.\frac{\partial^2 F_{\rm el}(M)}{\partial M^2}\right|_{M=0} + \left.\frac{\partial^2 F_{\rm ph}(M)}{\partial M^2}\right|_{M=0} . \qquad (6.2.1)$$

As for $F_{\rm el}$, we may divide it into two parts to be obtained by the mean field approximation, $F_{\rm el,m}$, and the remaining correlation contribution, $\Delta F_{\rm el,c}$, as

$$F_{\rm el}(M) = F_{\rm el,m} + \Delta F_{\rm el,c} . \qquad (6.2.2)$$

In this chapter, we will consider primarily the mean field part. As for the contribution of $\Delta F_{\rm el,c}$, we will discuss the effect of spin fluctuation based on a result to be obtained later in Chapter 10. Note that within the mean field approximation we earlier derived the result of (3.4.22) from the explicit expression for $F_{\rm el,m}$ of (3.4.10). Our major goal, then, is to pursue the contribution of the phonons, the second term on the right hand side of (6.2.1), to the susceptibility.

6.2.1 Phonon Free Energy and The Effective Exchange Interaction due to EPI

If the phonon spectrum $\omega_{\rm ph}(\boldsymbol{q};M)$ of a system is given, the phonon free energy of the system is obtained as

$$F_{\rm ph} = -k_{\rm B}T \ln {\rm tr}\left(\exp\left[-\frac{1}{k_{\rm B}T}\sum_{\boldsymbol{q}} \hbar\omega_{\rm ph}(\boldsymbol{q};M)\left(b_{\boldsymbol{q}}^{+}b_{\boldsymbol{q}} + \frac{1}{2}\right)\right]\right), \qquad (6.2.3)$$

where $b_{\boldsymbol{q}}^{+}$ and $b_{\boldsymbol{q}}$ are, respectively, the creation and annihilation operators of a phonon with wave vector $\boldsymbol{q}$ as defined in (5.2.24). By actually carrying out the trace operation in (6.2.3), as in (2.5.36), we obtain

$$F_{\rm ph} = \frac{1}{2}\sum_{\boldsymbol{q}} \hbar\omega_{\rm ph}(\boldsymbol{q};M) + k_{\rm B}T\sum_{\boldsymbol{q}} \ln\left[1-\exp\left(-\hbar\omega_{\rm ph}(\boldsymbol{q};M)/k_{\rm B}T\right)\right]. \qquad (6.2.4)$$

As we noted earlier, this phonon free energy changes with magnetization since the phonon frequency depends on the spin splitting of the energy bands of electrons:

$$F_{\rm ph} = F_{\rm ph}[\omega_{\rm ph}(\boldsymbol{q};M)] \,. \tag{6.2.5}$$

If we put

$$J_{\rm ph} = -2\left.\frac{\partial^2 F_{\rm ph}(\overline{M})}{\partial \overline{M}^2}\right|_{\overline{M}=0} = -2\mu_{\rm B}{}^2 \left.\frac{\partial^2 F_{\rm ph}(M)}{\partial M^2}\right|_{M=0}$$

$$= -2\hbar\sum_{\boldsymbol{q}}\left(\frac{1}{2} + n(\hbar\omega_{\rm ph})\right)\left.\frac{\partial^2 \omega_{\rm ph}(\boldsymbol{q};\overline{M})}{\partial \overline{M}^2}\right|_{\overline{M}=0} , \tag{6.2.6}$$

$n(\varepsilon)$ being the Bose distribution function with $\mu = 0$ of (2.5.37), from (6.2.1) we obtain (see (3.4.22)),

$$\frac{\mu_{\rm B}{}^2}{\chi} = \frac{1}{2}\left[\frac{1}{\displaystyle\int N(\varepsilon)\left(-\frac{\partial f(\varepsilon)}{\partial \varepsilon}\right)d\varepsilon} - \widetilde{V}\right] - \frac{1}{2}J_{\rm ph} \tag{6.2.7}$$

or

$$\chi = \frac{2\mu_{\rm B}^2 F(0)}{1 - (\widetilde{V} + J_{\rm ph})F(0)} , \tag{6.2.8}$$

where $-\mu_{\rm B}\overline{M} = M$ (see (3.4.11)). We call $J_{\rm ph}$ the *effective exchange interaction due to EPI.*

We can estimate $J_{\rm ph}$ numerically from (6.2.6) without any further approximation; $\omega_{\rm ph}(M)$ is obtained from (5.3.28) for a given system. In this book, however, we often use the Debye approximation of (6.1.5). In this approximation the phonon free energy of (6.2.4) in the spin split state of a metal is given by

$$F_{\rm ph}(M) = \frac{3}{8}N\hbar\omega_{\rm D}\frac{s(M)}{s}$$

$$+ Nk_{\rm B}T\left[\ln\left\{1 - \exp\left(-\frac{\hbar s(M)\,q_{\rm D}}{k_{\rm B}T}\right)\right\} - \frac{1}{3}D\left(\frac{\hbar s(M)\,q_{\rm D}}{k_{\rm B}T}\right)\right], \tag{6.2.9}$$

where

$$D(x) = \frac{3}{x^3}\int_0^x dz \frac{z^3}{e^z - 1} \tag{6.2.10}$$

is the Debye function, $\omega_D = \omega_D(0) = sq_D$, and q_D is the Debye wave number defined as

$$Vq_D^3/6\pi^2 = N, \tag{6.2.11}$$

N and V being, respectively, the total number of atoms and volume of the system.

Note that here we consider only the longitudinal acoustic phonons since we assume jellium-like ions in our model. In the ordinary Debye approximation result, the right hand side of (6.2.9) is multiplied by a factor 3, reflecting the presence of two transverse phonon modes.

As we already noted in the preceding section, the magnetization in (6.2.1) is not that of an equilibrium state. It is a variational one which is related to η' as in (6.1.3). If we use η' in place of M, and put $s(\eta') = s + \Delta s(\eta')$, we have

$$\frac{\Delta s(\eta')}{s} = -\frac{1}{2}\left(\frac{s_0}{s}\right)^2 K\left(\frac{\eta'}{W}\right)^2, \tag{6.2.12}$$

corresponding to (5.5.24), where K is the same as that of (5.5.25). The only difference between (5.5.21) and (6.2.12) is the difference between η and η'.

Within the Debye approximation, (6.2.6) is rewritten as

$$J_{ph} = -2\frac{\partial F_{ph}}{\partial \omega_D(M)}\frac{\partial^2 \omega_D(M)}{\partial \overline{M}^2}\bigg|_{M=0} - 2\frac{\partial^2 F_{ph}}{\partial \omega_D(M)^2}\left(\frac{\partial \omega_D(M)}{\partial \overline{M}}\right)^2\bigg|_{M=0}$$

$$= -2\frac{\partial F_{ph}}{\partial \omega_D}\frac{\omega_D}{(2F(0))^2}\frac{\partial^2}{\partial \eta'^2}\left(\frac{\Delta s(\eta')}{s}\right)\bigg|_{\eta'=0}. \tag{6.2.13}$$

where we noted $\Delta\overline{M} = 2F(0)\eta'$ (*Problem* 6.1). The second term on the right hand side of the first equality vanishes since $(\partial\omega_D(M)/\partial M)_{M=0} = 0$ for the paramagnetic state. Then by noting (6.2.12) we obtain [6.1]

$$J_{ph}F(0) = \left(\frac{s_0}{s}\right)^2 LP\left(\frac{T}{\theta_D}\right)\frac{\hbar\omega_D}{W}, \tag{6.2.14}$$

$$L = \frac{N}{2WN(0)} K$$

$$= \frac{NW}{2F(0)} \left[\frac{1}{2} \left\{ \frac{F''(0)}{F(0)} - \left(\frac{F'(0)}{F(0)} \right)^2 \right\} + \frac{\widetilde{V}F(0)}{1 - \widetilde{V}F(0)} \left(\frac{F'(0)}{F(0)} \right)^2 \right], \quad (6.2.15)$$

where W, the electron energy band width, is introduced to make the quantity L dimensionless, and the function $P(T/\theta_D)$ is related to the phonon free energy in the Debye approximation, (6.2.9), as

$$P\left(\frac{T}{\theta_D}\right) = \frac{1}{N\hbar} \frac{\partial}{\partial \omega_D} F_{ph} = \frac{3}{8} + \frac{T}{\theta_D} D\left(\frac{\theta_D}{T}\right). \quad (6.2.16)$$

We see that $|L| = O(1)$; note $N(0) = O(N/W)$, $|N'(0)| = O(N/W^2)$ etc. Thus, if we neglect the temperature dependence through $P(T/\theta_D)$ and assume $P(T/\theta_D) \cong O(1)$, we have $|J_{ph}F(0)| = O(\hbar\omega_D/W) \cong 10^{-2}$ in (6.2.8). Such is the hitherto prevailing view [3.10]. Thus, previously the EPI was thought to be able to play a significant role only in situations in which the value of $\widetilde{V}F(0) \cong \widetilde{V}N(0)$ is only slightly smaller than unity by an amount of the order of 10^{-2}. $ZrZn_2$ was proposed to be such a case by Hopfield [6.2]; the Stoner condition for ferromagnetism can then be realized through the effect of the EPI. It was assumed that $J_{ph} > 0$, without any actual justification; it was not known how to calculate the sign and magnitude of J_{ph} beyond the above order of magnitude estimation.

Now we can quantitatively calculate the value J_{ph} of a metal if the electronic density of states of the metal, the location of ε_F in it, and the value of $\overline{V} = \widetilde{V}N(0)$ are given. L can be either positive or negative. As for the function $P(T/\theta_D)$ which determines the temperature dependence of J_{ph}, it is easy to see

$$P\left(\frac{T}{\theta_D}\right) = \begin{cases} \dfrac{3}{8} + \dfrac{\pi^4}{5}\left(\dfrac{T}{\theta_D}\right)^4 & \text{for } T \ll \theta_D, \quad (6.2.17) \\[2ex] \dfrac{T}{\theta_D} & \text{for } T \gg \theta_D. \quad (6.2.18) \end{cases}$$

In Fig.6.3 we plot this function. Note that the high temperature behavior of (6.2.18) is valid at temperatures as low as $T \cong \theta_D/3$.

By putting these results on J_{ph} into (6.2.8), we obtain

$$\chi = \frac{2\mu_B^2 F(0)}{\left(1 - \widetilde{V}F(0)\right) - \left(\frac{s_0}{s}\right)^2 L \frac{k_B\theta_D}{W} P\left(\frac{T}{\theta_D}\right)} \quad (6.2.19)$$

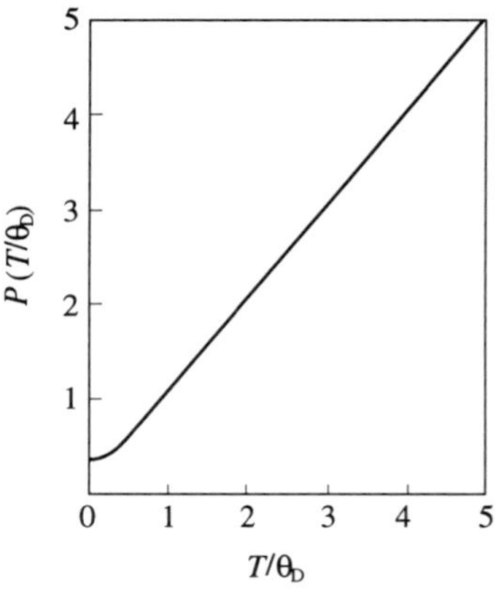

Fig.6.3 Temperature dependence of function $P(T/\theta_D)$ defined by (6.2.16).

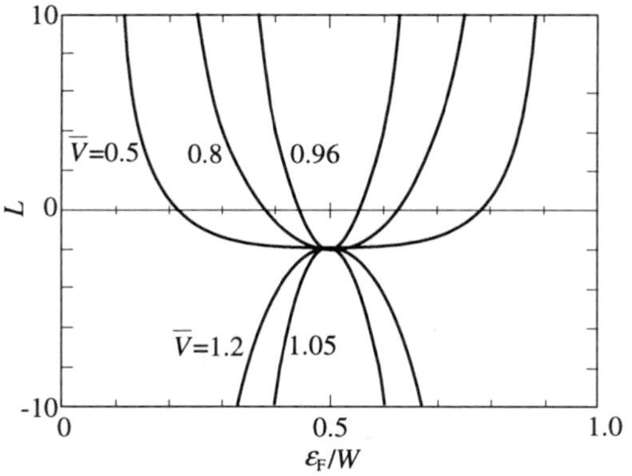

Fig.6.4 Dependence of $L(\bar{V}, \varepsilon_F/W)$ defined by (6.2.15) at $T = 0$ on ε_F/W with the electronic density of states of Fig.5.5 and for different values of $\bar{V} = \tilde{V}N(0)$.

$$
\cong \begin{cases} \dfrac{2\mu_B^2 F(0)}{1 - \tilde{V}F(0) - \frac{3}{8}\left(\frac{s_0}{s}\right)^2 L \frac{k_B\theta_D}{W}} & \text{for } T \ll \theta_D\,(k_B\theta_D/W)^{1/2}, \quad (6.2.20) \\[2ex] \dfrac{2\mu_B^2 F(0)}{\left[1 - \tilde{V}F(0) - \frac{3}{8}\left(\frac{s_0}{s}\right)^2 L \frac{k_B\theta_D}{W}\right] - \frac{\pi^4}{5}\left(\frac{s_0}{s}\right)^2 L \frac{k_B\theta_D}{W}\left(\frac{T}{\theta_D}\right)^4} \\ \qquad \text{for } \theta_D\,(k_B\theta_D/W)^{1/2} < T \ll \theta_D, \quad (6.2.21) \\[2ex] \dfrac{2\mu_B^2 F(0)}{\left(1 - \tilde{V}F(0)\right) - \left(\frac{s_0}{s}\right)^2 L \frac{k_B T}{W}} & \text{for } T \geq \dfrac{\theta_D}{3}. \quad (6.2.22) \end{cases}
$$

The upper temperature bound of the validity of (6.2.20) comes from the requirement that the temperature dependence of $\tilde{V}F(0)$ dominates over that of $P(T/\theta_D)$, namely, $(k_B T/W)^2 > (k_B\theta_D/W)(T/\theta_D)^4$. Since $(k_B\theta_D/W)^{1/2} \cong 10^{-1}$ it is in a narrow low temperature region that the temperature dependence of (6.2.20) or (6.2.21) can be observed.

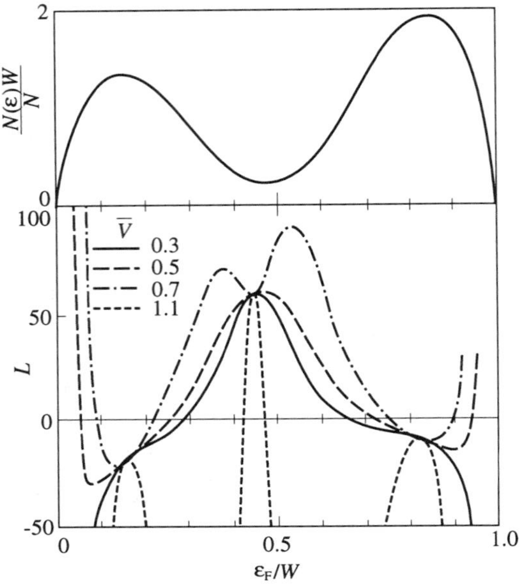

Fig.6.5 Result of numerical calculation of $L(\bar{V}\ \varepsilon_F/W)$ is shown in the lower figure for different values of ε_F/W in the electronic density of states given above for different values of $\bar{V}$. We assumed $k_BT/\varepsilon_F \ll 1$.

The result of (6.2.22) shows us how the Curie-Weiss like temperature dependence can be produced by the effect of EPI in the magnetic susceptibility of a metal; it is not necessary to have localized moments.

The sign and size of J_{ph} is determined by the quantity L defined in (6.2.15). In Fig.6.4 we present an example of a numerical calculation of L for the model electronic density of Fig.5.5. L for $(k_BT/\varepsilon_F) << 1$ is calculated for different values of $\bar{V} = \tilde{V}N(0)$ as the function of the location of ε_F in the electronic density of states. We assumed $W = 1$ eV; the final result for χ is independent of the value of W. We note that for a given $N(\varepsilon)$, L can take a value in quite a wide range; depending upon the values of $\bar{V}$ and ε_F/W, L can be either positive or negative, and the magnitude of L can be much larger than unity.

Another similar numerical example is given in Fig.6.5 (bottom) for the model electronic density of states given in the figure (top). The behavior of L is more complex than in Fig.6.4 reflecting the increased structure in the electronic density of states. We note that $L > 0$ when ε_F is located near the bottom of the valley in $N(\varepsilon)$, and $L < 0$ when ε_F is located near the peaks of $N(\varepsilon)$. Such a relation seems to hold in actual observations, as we will see later.

Note that the quantity L can be experimentally determined by measuring the magnetic field dependence of sound velocity in the paramagnetic state, as indicated in the first line of (6.2.15). As discussed in 5.5, in order to account for some experiments we require the magnitude of the quantity k defined in (5.5.28) to be as large as $\sim 10^6$. Then, according to (5.5.28) and the first line of (6.2.15), we require $|L| = |k/D_o^2| = 10^6/D_o^2$; even with $D_o = 10^2$, we have $|L| = O(10^2)$. Thus, the numerical results on L given in Figs. 6.4 and 6.5 are not unrealistic ones. When $|L| = O(10^2)$, we have $O(|J_{ph}|) = O(\widetilde{V})$. As we will see in 6.4, however, a much smaller EPI effect with $|L| \sim O(1)$ is sufficient to make it vitally important.

6.2.2 Various Temperature Dependence of the Magnetic Susceptibility Caused by EPI

According to (6.2.22), in the temperature region $T \gtrsim \theta_D/3$, the temperature dependence of χ becomes characteristically different depending upon the value of $\overline{V}$ and the sign of L, as illustrated in Fig.6.6. We examine several cases more in detail below.

A $L > 0$ ($J_{ph} > 0$ *EPI effect*)

In this case, $\overline{V} < 1$ is required to make the magnetic susceptibility of (6.2.22) positive. In this temperature region we can put $F(0) = N(0)$ since the temperature dependence of χ is predominantly determined by that of $P(T/\theta_D)$. Thus, for this case (6.2.22) is rewritten as

$$\chi = C/\left(T_a - T\right), \tag{6.2.23}$$

$$C = 2\mu_B^2 N(0)(s/s_o)^2 W/(k_B L), \tag{6.2.24}$$

$$T_a = \left(1 - \overline{V}\right)(s/s_o)^2 W/(k_B L). \tag{6.2.25}$$

Note that both C and T_a are independent of the value of W; W cancels out in W/L. As illustrated in Fig.6.6(a), in this case the magnetic susceptibility increases with increasing temperature.

B $L < 0$ ($J_{ph} < 0$ *EPI effect) and* $\overline{V} > 1$

In this case, (6.2.22) is rewritten in the form of the Curie-Weiss law;

$$\chi = C/(T - T_C), \tag{6.2.26}$$

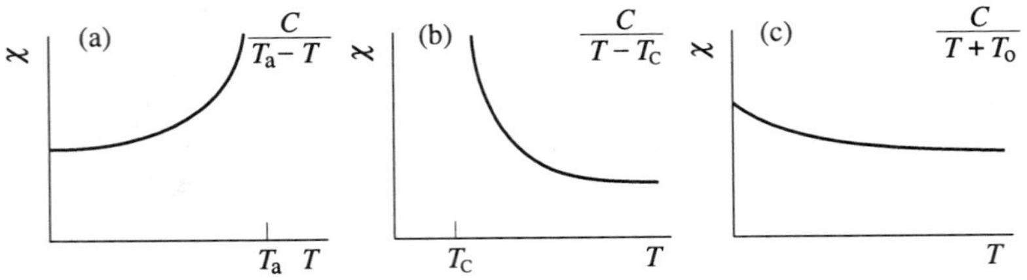

Fig.6.6 Various characteristic temperature dependence of spin susceptibility depending upon the signs of L and $(1-\overline{V})$. (a): $L > 0$ and $\overline{V} < 1$. (b): $L < 0$ and $\overline{V} > 1$. (c): $L < 0$ and $\overline{V} < 1$.

$$C = 2\mu_B^2 N(0)(s/s_o)^2\, W/(k_B|L|)\,, \tag{6.2.27}$$

$$T_C = \left(\overline{V}-1\right)(s/s_o)^2\, W/(k_B|L|)\,. \tag{6.2.28}$$

Note that the Curie constant given in (6.2.27) is of the right order of magnitude, $C = O(N\mu_B^2/k_B)$. However, C has nothing to do with localized moments. As for the Curie temperature T_C, recall that in the discussion of 6.1 it was crucial to have it in the form of (6.1.10); the result of (6.2.28) has this structure.

One long noted difficulty with the Stoner theory is that it generally gives too high a T_C. This difficulty seems to be resolved in our result of (6.2.28). Assume, for instance, $\overline{V} = 1.01$. Then, according to (3.3.18) or (3.3.19), we have $k_B T_C\,(\text{Stoner}) = O(\varepsilon_F(\overline{V}-1)^{1/2}) = O(\varepsilon_F/10)$. However, with the phonon effect we have, with $|L| = O(1)$, $k_B T_C\,(\text{EPI}) = O(\varepsilon_F(\overline{V}-1)) = O(\varepsilon_F/10^2)$.

If the Curie temperature T_C is to be determined by the spin fluctuation effect, it is given as $k_B T_C(\text{sf}) = O(\varepsilon_F(\overline{V}-1)^{3/4})$ [3.18]. Thus we have

$$T_C\,(\text{EPI}) < T_C(\text{sf}) < T_C\,(\text{Stoner}).$$

Note the important role played by L in determining the magnitudes of C and T_C. Since a larger $|L|$ leads to smaller C and T_C, we will have a weak ferromagnet when $|L|$ is large.

From (6.2.27) and (6.2.28) we obtain an interesting simple relation,

$$\frac{C}{T_C} = 2\mu_B^2 \frac{N(0)}{\overline{V}-1}\,. \tag{6.2.29}$$

C and T_C are determined from experiment. Then, if we have $N(0)$, say, from a band calculation, we can determine the value of $\overline{V}$. Note, however, that (6.2.29) doesn't consider the possible effect of spin fluctuations.

C $L < 0$ ($J_{\rm ph} < 0$ *EPI effect*) *and* $\overline{V} < 1$

In this case we have

$$\chi = C/(T_o + T)\,, \tag{6.2.30}$$

$$T_o = \left(1 - \overline{V}\right)(s/s_o)^2\, W/(k_B |L|)\,, \tag{6.2.31}$$

where C is the same as in (6.2.27). The magnetic susceptibility increases with decreasing temperature but does not diverge, as illustrated in Fig.6.6 (c).

In 6.4 we will see how actual experimental results can be successfully analyzed by taking into account the effect of EPI.

6.3 THE EFFECT OF EPI VS. SPIN FLUCTUATION ON THE SPIN SUSCEPTIBILITY

Although our interest is focused on the effect of EPI, here we briefly discuss the effect of spin fluctuation on the spin susceptibility as a part of the electron correlation effect in (6.2.2). Spin fluctuation has been predominantly discussed in connection with the Hubbard model [6.3-6.5]. Historically, the spin fluctuations in an itinerant electron ferromagnet was first discussed with the electron gas model [3.14]. Here it is imperative to use the electron gas model, i.e. the jellium-like model, in order to discuss phonons and spin fluctuations on the same footing.

As will be shown in Chapter 10, see (10.1.31), the spin fluctuation contribution to the free energy of the electron gas model is given as

$$\Delta F_{\rm sf} = -\frac{3}{4\pi}\sum_{\kappa}\int_{-\infty}^{\infty} d\omega\, n(\omega)\left[\int_0^1 dg\, \widetilde{V}(\kappa)\left\{\overline{\chi}_{+-}(\kappa,\omega)_g + \overline{\chi}_{-+}(\kappa,\omega)_g\right\} + \widetilde{V}(\kappa)\left\{F_{+-}(\kappa,\omega) + F_{-+}(\kappa,\omega)\right\}\right], \tag{6.3.1}$$

where $\overline{\chi}_{\sigma,-\sigma}(\kappa,\omega)_g$ is the transverse spin susceptibility with $\mu_B^2 = 1$ of (4.4.12); the subscript g dictates to replace every $\widetilde{V}(\kappa)$, in $\overline{\chi}_{\sigma,-\sigma}$ by $g\,\widetilde{V}(\kappa)$. This result reduces to the well-known result in the Hubbard model [6.3–6.5] if we simply note the correspondence of (3.5.22), namely, $\widetilde{V}(\kappa) = I/N$.

The effect of the spin fluctuation on the spin susceptibility, as first discussed by Béal-Monod et al. [6.6], is to add an additional term, $-1/2J_{sf}$, on the right hand side of (6.2.7), where, corresponding to (6.2.6), we put

$$J_{sf} = -2 \left. \frac{\partial^2 \Delta F_{sf}}{\partial^2 \overline{M}^2} \right|_{\overline{M}=0} . \tag{6.3.2}$$

Then, in place of (6.2.8) we have

$$\chi = \frac{2\mu_B^2 F(0)}{1 - (\widetilde{V} + J_{ph} + J_{sf})\, F(0)} . \tag{6.3.3}$$

Recently an interesting numerical calculation was carried out by Tanaka and Shiina [6.7] to compare the effect of J_{ph} and J_{sf} in (6.3.3). Note that such a calculation becomes possible if we use the jellium model. According to their results, the magnitude of the temperature variations of $J_{ph}F(0)$ and $J_{sf}F(0)$ are generally of the same size. However, their sizes depend very sensitively on the value of $\overline{V}$. When $\overline{V}$ is small the variations of $J_{ph}F(0)$ and $J_{sf}F(0)$ are comparable to that of $\widetilde{V}F(0)$, but when $\overline{V}$ is close to 1, the former two become much larger than the latter one.

An important difference between the effects of phonons and spin fluctuations is that the temperature dependence of $J_{ph}F(0)$ is linear in T, but that of $J_{sf}\,F(0)$ is proportional to T^2 [6.7], as is that of $\widetilde{V}F(0)$. Another marked difference is that J_{ph} can be either positive or negative, while J_{sf} is always negative.

As for the spin fluctuation effect, great efforts have been made to go beyond the above simple result; see, for instance, Refs. [3.17–3.21, 6.8–6.10]. Let us briefly describe one such endeavor.

Note that the spin fluctuation effect of (6.3.1) and, therefore, (6.3.2) is given in terms of the spin susceptibility which does not include the spin fluctuation effect. What was done by Moriya and Kawabata [3.19, 3.20] is to replace the spin susceptibility appearing in (6.3.1) by

$$\overline{\chi}_{-+}(\kappa,\omega) = \frac{F_{-+}(\kappa,\omega)}{\left[\, 1 - \widetilde{V}(\kappa)\, F_{-+}(\kappa,\omega) \,\right] + \lambda} , \tag{6.3.4}$$

where the spin fluctuation effect on the susceptibility is represented by a unknown parameter λ. Then a numerical procedure was taken to determine λ self-consistently within this scheme. Thus they derived the Curie-Weiss like temperature dependence of the spin susceptibility.

Note, however, that such a treatment entirely neglects the role of phonons which, as we have seen, can be as important as that of spin fluctuations. If the role of the EPI is properly taken into account, our outlook on the effect of spin fluctuations would be significantly modified.

6.4 ANALYSIS OF OBSERVED MAGNETIC SUSCEPTIBILITY BY INCLUDING THE EPI EFFECT

In 6.2 it was shown that the effect of EPI can produce three characteristically different temperature dependence in χ, the cases (a), (b), and (c), in Fig.6.6. In this section we show that actual observations belonging to each of such cases can indeed be reproduced if we take into account the effect of EPI [6.11]; here we neglect the possible effect of spin fluctuations.

A *Rh*

As shown in Fig.6.7, in Rh, χ increases with increasing temperature [6.12]. With the EPI effect, such a temperature dependence can be understood if we have $L > 0$. According to the model calculation of Fig.6.5, if ε_F is located near the bottom of a valley of an electronic density of states, we have $L > 0$. The band calculation results of Moruzzi et al. [6.13] given in Fig.6.8(a) show that in Rh, indeed ε_F is located near a minimum in the density of states.

In reproducing the experimental observation on χ with the EPI, the first task is to calculate L. For Rh, we fit a polynomial as in Fig.6.8(b) to the electronic density of states calculated by Moruzzi et al. of Fig.6.8(a) near the Fermi surface. With $N(0)$ as given by the band calculation, we choose $\overline{V}$ as in Table 6.1 so as to reproduce the magnetic susceptibility at $T = 0$. As to the contribution of the possible temperature independent orbital susceptibility, χ_{orb}, our value for Rh is very close to that of Shimizu [6.14]. We put $(s_o/s)^2 = 1$ throughout this section since it will not be much different from unity. We put $W = 1$ eV, again throughout this section; the final result is independent of W. For the temperature dependence in $F(0)$ (see (4.2.18)), in this section we use the approximation of (1.6.14); as for $P(T/\theta_D)$, we use its exact form and do not use the approximations of (6.2.17) and (6.2.18).

In this way we obtained the solid line in Fig.6.7; our result reproduces the observation quite well. For comparison, by the broken line we included the result of Shimizu [6.14] which is based on the same electronic density of states. There are other, earlier and better reproductions of the observation, such as that of Misawa and Kanematsu [6.15], but they seem to be of more phenomenological nature.

Another quite successful analysis with the EPI effect was carried out on Cr for $T > T_N$ [6.11] where the magnetic susceptibility also increases with increasing temperature [6.16].

Rh	$N(0)$ (states/eV atom spin)	$\overline{V}$	θ_D (K)	α/T_F^2	L	$J_{ph}N(0)$	χ_{orb} at T=0 (10^{-4}emu/mol)
Rh	0.68[a]	0.440	480[b]	-6.79×10^{-8}	2.10	0.032	0.17
Ni	1.90[c]	1.215	373[d]	4.10×10^{-8}	−3.50	−0.012	0
Pt	0.90[e]	0.720	230[b]	1.23×10^{-7}	−2.15	−0.016	0

a: Ref. [6.13]. b: From table 1 of chapter 5 of Ref. [2.3]. c: See text.
d: Ref. [6.17]. e: Ref. [6.18].

Table 6.1 Parameters used in the calculation of the magnetic susceptibility of Rh, Ni and Pt.

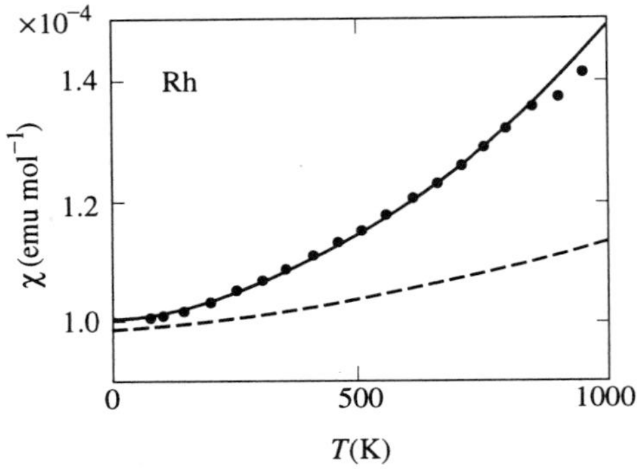

Fig.6.7 Observed magnetic susceptibility of Rh by Kojima et al. [6.12] and its reproduction with the EPI effect (solid line). The analysis of Shimizu [6.14] is shown by the broken line.

B *Ni*

In Ni, for which $\overline{V} > 1$, the calculated electronic density of states is very steep near ε_F. See, for instance, Moruzzi et al. [6.13]. It is difficult to determine L unambiguously for such a case. Thus we tentatively use the electronic density of states of Fig.6.9(a) which was deduced from an electron spectroscopy for chemical analysis (ESCA) experiment by Baer et al. [6.19]. In such an $N(\varepsilon)$ derived from photoemission experiment, any possible fine structures are smoothed away; also the absolute magnitude of $N(0)$, that is the scale of the ordinate of Fig.6.9 can not be determined from the photoemission data alone.

The value of $N(0)$ in Table 6.1 is chosen to be slightly smaller than that of Moruzzi et al. [6.13] which is considered to be slightly too large in view of the

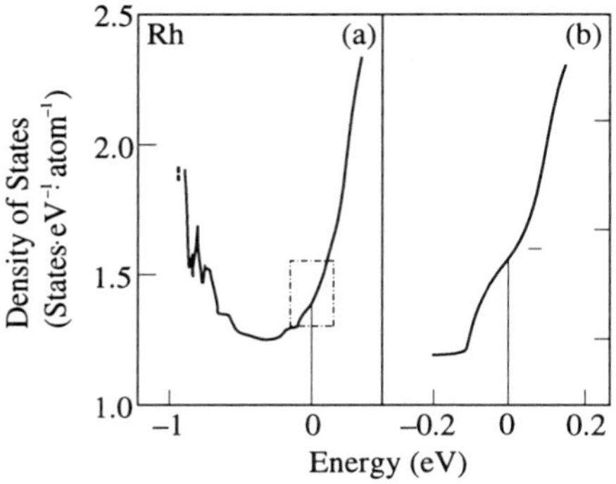

Fig.6.8 The electronic density of states for Rh calculated by Moruzzi et al. [6.13], (a), is approximated by a polynomial near the Fermi level as in (b). The energy is measured from ε_F.

observed electronic specific heats. We proceed as in the case of Rh by fitting such electronic density of states of Fig.6.9(a) by a polynomial as in Fig.6.9(b). Values for other parameters used to obtain the solid line of Fig.6.10 are given in Table 6.1. The agreement of our result with the experiment by Arajs and Colvin [6.20] is quite satisfactory over wide temperature range. For a recent alternative analysis of the magnetic susceptibility of Ni by employing spin fluctuation effect, see Hirooka [6.21].

A few remarks may be in order here. The electronic density of states of Fig.6.9 is without any fine structures. What is shown by the broken line in Fig.6.10 is the Stoner susceptibility calculated for this electronic density of states by choosing the value of $\overline{V}$ so as to reproduce the observed T_C; for such an $N(\varepsilon)$, the Stoner susceptibility can have only a very weak temperature dependence. It is interesting that if the effect of EPI is taken into account, the broad and smooth $N(\varepsilon)$ of Fig.6.9 is sufficient to reproduce the observed temperature dependence in χ of Ni. Note, in addition, that the magnitude of L which we require is very small in view of the wide range of possible values, as can be seen from Fig.6.5.

C *Pt*

As an example of the case of Fig.6.6(c), or (6.2.30) let us take up Pt [6.12]. By using $N(\varepsilon)$ of Fig.6.11(a) calculated by Fradin et al. [6.18] we proceed as before. The values of parameters used to obtain the solid line in Fig.6.12 are given in Table 6.1. As can be seen in Fig.6.12 we succeed in reproducing the observation for quite a wide temperature range.

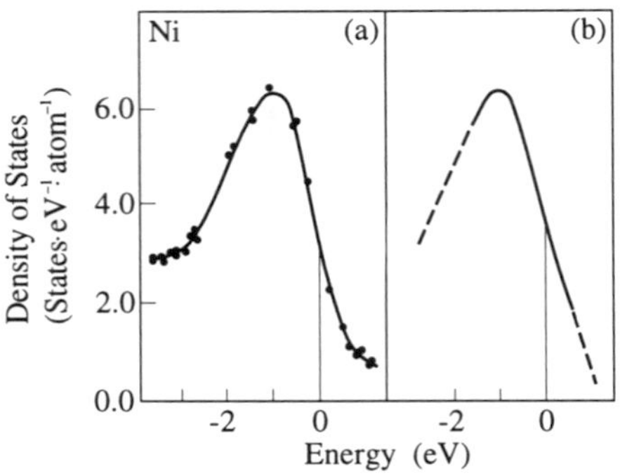

Fig.6.9 The electronic density of states for Ni as determined with an ESCA by Baer et ai. [6.19], (a), is approximated by a polynomial near the Fermi surface as in (b).

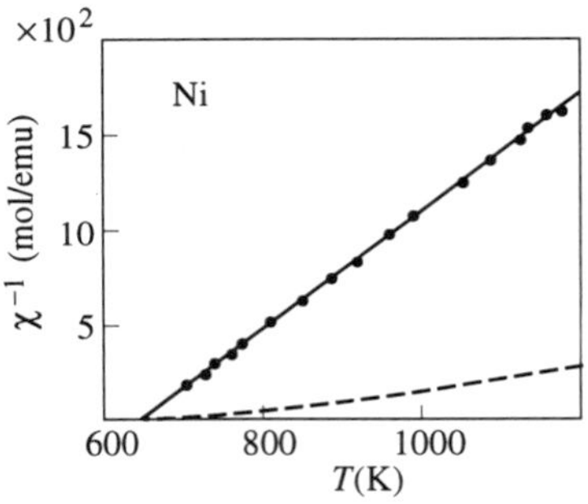

Fig.6.10 Observed magnetic susceptibility of Ni [6.20] and its reproduction with al. the EPI effect (solid line). The Stoner susceptibility for the same electronic density of states is given by the broken line.

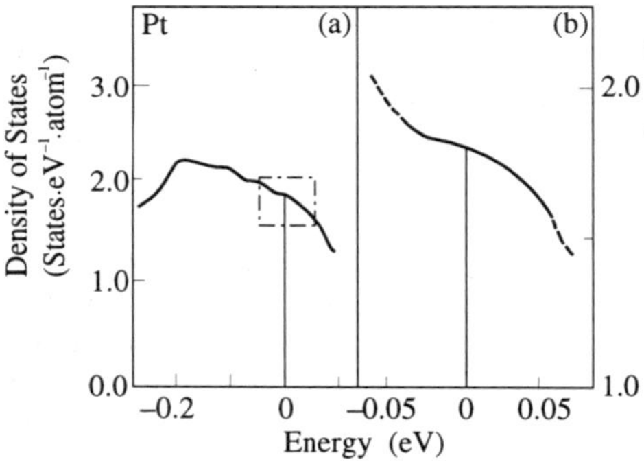

Fig.6.11 The electronic density of states for Pt of Fradin et al. [6.18], (a), is approximated by a polynomial near the Fermi surface as in (b).

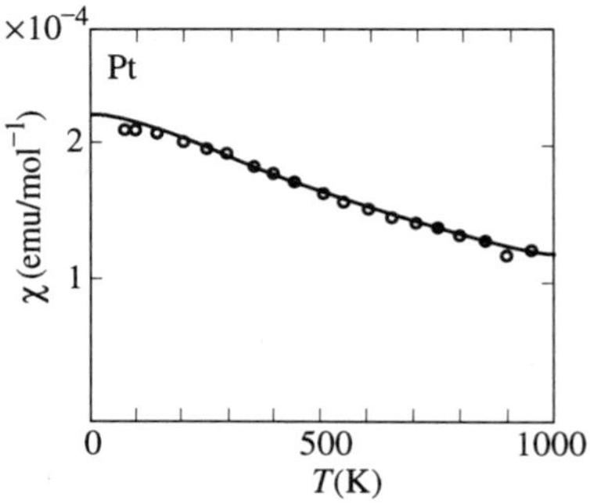

Fig.6.12 Observed magnetic susceptibility of Pt by Budworth et al. [6.22] and a its reproduction with the EPI effect.

There are a number of earlier analyses with the Stoner susceptibility. The analysis of Fradin et al. based on the same electronic density of states of Fig.6.11(a), for instance, reproduces the observation quite well in the lower temperature region; at higher temperatures, however, the agreement appears to be less satisfactory. With the Stoner susceptibility, it seems to be difficult to explain the temperature dependence of χ of Pt in a wide temperature range, as it was with the other cases discussed above.

6.5 SPIN WAVE CONTRIBUTION TO FREE ENERGY: F_{sw}

The effect of phonons on magnetization turns out to be very similar to that of spin waves. In preparing for the discussion, in this section we show that the contribution of spin waves to the free energy takes the same form as that of phonons.

In the ferromagnetic state, the transverse and the longitudinal components of spin fluctuations behave differently: $\chi_{zz} \neq \chi_{xx}, \chi_{yy}$. As we already discussed in 4.4.2, the spin wave excitations are related to the transverse spin fluctuations, or, $\chi_{+-}(\boldsymbol{q}, \omega)$ and $\chi_{-+}(\boldsymbol{q}, \omega)$. We concentrate on the spin wave contribution, leaving out that of the longitudinal spin fluctuation. As we will discuss in Chapter 10, the transverse spin fluctuation contribution ($\Delta\Omega_{\text{el, lad1}}$) to the free energy is given as (see (10.1.26")),

$$F_{sw}(T, M) \equiv \frac{\hbar}{2\pi} \sum_{\kappa} \int_{-\infty}^{\infty} d\omega\, n(\hbar\omega)\, \mathrm{Im}\Big\{\ln\,[1 - \widetilde{V}F_{-+}(\kappa, \omega)] + \ln\,[1 - \widetilde{V}F_{+-}(\kappa, \omega)]\Big\}, \tag{6.5.1}$$

where $n(\hbar\omega)$ is the Bose distribution and the transverse Lindhard function, $F_{\sigma,-\sigma}$, is defined in (4.4.11); we assume $\widetilde{V}(\kappa) = \widetilde{V}$, independent of κ. In discussing spin waves, for later convenience, we assume $n_- > n_+$ as in Fig.3.5(b), that is, $M > 0$ for the ferromagnetic state, differing from 4.4.2.

Now, concerning Bose distribution $n(\hbar\omega)$ appearing in (6.5.1), we note

$$\frac{1}{e^x - 1} = \frac{d}{dx}\ln(1 - e^{-x}) \tag{6.5.2}$$

Then, (6.5.1) can be rewritten as

$$F_{sw}(T, M) = \frac{\hbar}{2\pi}(\beta\hbar)^{-1}\sum_{\kappa}\left[\int_0^{\infty} d\omega \ln(1 - e^{-\beta\hbar\omega}) - \int_{-\infty}^{0} d\omega \ln(e^{-\beta\hbar\omega} - 1)\right]$$

$$\times \frac{d}{d\omega}\mathrm{Im}\Big\{\ln\big(1 - \widetilde{V}F_{-+}(\kappa, \omega + i0^+)\big) + \ln\big(1 - \widetilde{V}F_{+-}(\kappa, \omega + i0^+)\big)\Big\}$$

$$= \frac{\beta^{-1}}{2\pi}\sum_{\kappa}\left[\int_0^{\infty} d\omega \ln(1 - e^{-\beta\hbar\omega}) - \int_{-\infty}^{0} d\omega \ln(e^{-\beta\hbar\omega} - 1)\right]$$

$$\times \mathrm{Im}\left[\frac{-\widetilde{V}\frac{d}{d\omega}F_{-+}(\kappa,\omega+i0^+)}{1-\widetilde{V}F_{-+}(\kappa,\omega+i0^+)} + \frac{-\widetilde{V}\frac{d}{d\omega}F_{+-}(\kappa,\omega+i0^+)}{1-\widetilde{V}F_{+-}(\kappa,\omega+i0^+)} \right]. \tag{6.5.3}$$

In the above integral, we note that for ω in the vicinity of $\omega_{\mathrm{sw}}(\kappa)$,

$$1-\widetilde{V}F_{-+}(\kappa,\omega+i0^+) \cong -\widetilde{V}\frac{\partial}{\partial\omega}F_{-+}(\kappa,\omega)\Big|_{\omega=\omega_{\mathrm{sw}}(\kappa)}[\omega-\omega_{\mathrm{sw}}(\kappa)]$$

$$-\,i\widetilde{V}I_{-+}(\kappa,\omega), \tag{6.5.4}$$

where $\omega_{\mathrm{sw}}(\kappa)$ is the spin wave frequency which was discussed in 4.4.2 (contrary to 4.4.2 here we assumed $n_- > n_+$ for the ferromagnetic state and therefore the pole of $\chi_{-+}(q,\omega)$, in place of $\chi_{-+}(q,\omega)$, gives us the spin wave spectrum) and $I_{-+}(\kappa,\omega)$ is the imaginary part of the transverse Lindhard function $F_{-+}(\kappa,\omega+i0^+)$. If we note the relations

$$F_{+-}(\kappa,\omega+i0^+) = F_{-+}(\kappa,-(\omega+i0^+)), \tag{6.5.5}$$

$$\chi_{+-}(\kappa,\omega+i0^+) = \chi_{-+}(\kappa,-(\omega+i0^+)), \tag{6.5.6}$$

which can be deduced from the definition of transverse Lindhard function, (4.4.11), we also have

$$1-\widetilde{V}F_{+-}(\kappa,\omega+i0^+) \cong \widetilde{V}\frac{\partial}{\partial\omega}F_{+-}(\kappa,\omega)\Big|_{\omega=-\omega_{\mathrm{sw}}(\kappa)}[\omega+\omega_{\mathrm{sw}}(\kappa)]$$

$$-\,i\widetilde{V}I_{+-}(\kappa,\omega), \tag{6.5.7}$$

Putting (6.5.4) and (6.5.7) into (6.5.3) we obtain

$$F_{\mathrm{sw}}(T,M) = \frac{1}{2\pi\beta}\sum_{\kappa}\int_0^{\infty} d\omega \ln(1-e^{-\beta\hbar\omega})\frac{\gamma_{\mathrm{sw}}(\kappa)}{[\omega-\omega_{\mathrm{sw}}(\kappa)]^2+\gamma_{\mathrm{sw}}(\kappa)^2}$$

$$+\frac{1}{2\pi\beta}\sum_{\kappa}\int_{-\infty}^{0} d\omega \ln(e^{-\beta\hbar\omega}-1)\frac{\gamma_{\mathrm{sw}}(\kappa)}{[\omega+\omega_{\mathrm{sw}}(\kappa)]^2+\gamma_{\mathrm{sw}}(\kappa)^2}, \tag{6.5.8}$$

where we put

$$\gamma_{\rm sw}(\kappa) = \left(I_{-+}(\kappa, \omega) \Big/ \frac{\partial}{\partial \omega} R_{-+}(\kappa, \omega) \right) \Bigg|_{\omega = \omega_{\rm sw}(\kappa)} , \tag{6.5.9}$$

where $R_{-+}(\kappa, \omega)$ is the real part of $F_{-+}(\kappa, \omega)$. $\gamma_{\rm sw}(\kappa)$ represents the damping of spin wave propagation similarly to the case of phonon which is discussed in 5.4. In the present case of spin wave, from (4.4.11) we have

$$I_{-+}(\kappa, \omega) = -\pi \sum_{k} \left(f(\varepsilon_{k+\kappa,+}) - f(\varepsilon_{k-}) \right) \delta\left(\hbar\omega - (\varepsilon_{k+\kappa,+} - \varepsilon_{k-}) \right) \tag{6.5.10}$$

and we find

$$I_{-+}(\kappa, \omega_{\rm sw}(\kappa)) = 0 \tag{6.5.11}$$

for small κ, since $\hbar\omega_{\rm sw}(\kappa)$ can not be equal to $(\varepsilon_{k+\kappa,+} - \varepsilon_{k-}) \cong \widetilde{V}(n_- - n_+)$; see Fig.4.12. A spin wave with a small κ can not decay into Stoner excitations. Thus, if we note that

$$\lim_{\gamma \to 0^+} \frac{\gamma}{(\omega - \omega_{\rm sw})^2 + \gamma^2} = \pi\, \delta\left(\omega - \omega_{\rm sw} \right) , \tag{6.5.12}$$

(6.5.8) reduces to

$$F_{\rm sw}(T, M) = \frac{\beta^{-1}}{2\pi} \sum_{\kappa} \int_0^\infty d\omega \ln\left(1 - e^{-\beta\hbar\omega}\right) \frac{\gamma_{\rm sw}(\kappa)}{\left[\omega - \omega_{\rm sw}(\kappa)\right]^2 + \gamma_{\rm sw}(\kappa)^2}$$

$$= \frac{1}{2} \sum_{\kappa} \hbar\omega_{\rm sw}(\kappa) + \beta^{-1} \sum_{\kappa} \ln\left(1 - e^{-\beta\hbar\omega_{\rm sw}(\kappa)}\right) . \tag{6.5.13}$$

This result is exactly of the same form as that for phonons given in (6.2.4). We can now treat phonons and spin waves parallelly.

6.6 THE EFFECT OF PHONONS ON MAGNETIZATION: THE SIMILARITY BETWEEN PHONON AND SPIN WAVE

It is one of the basic premises of the current standard theory of itinerant electron ferromagnetism that the dominant mechanism of the temperature dependence of magnetization at low temperatures is the spin wave excitations. However, Ishikawa et al. [6.23] showed that this premise is not strictly valid.

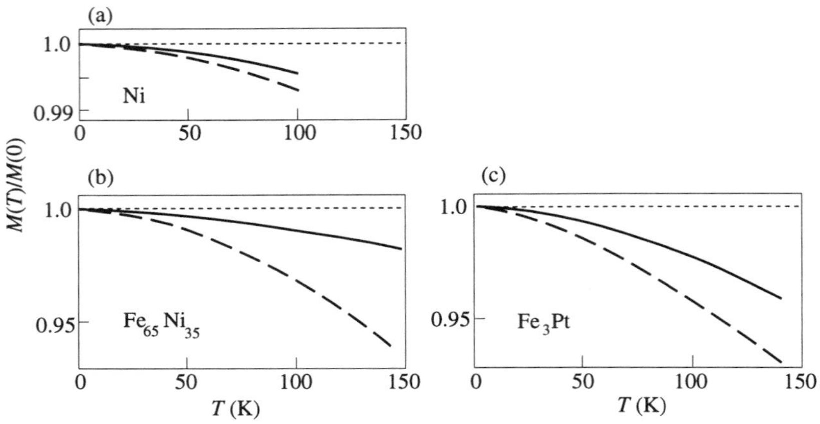

Fig.6.13 The broken lines represent the observed temperature dependence of magnetization of various metallic ferromagnets; the solid lines are what are expected from the spin wave spectra observed by neutron diffraction.

First they determined the spin wave frequency spectrum of a system by neutron diffraction. With that result the number of spin waves to be thermally excited in the system could be straightforwardly calculated by using Bose distribution. Then, by associating a magnetization of $-2\mu_B$ to each of the excited spin waves, which is another fundamental premise, they could predict how magnetization would decrease with increasing temperature. The results they thus obtained, which are shown by the solid lines in Fig.6.13, however, were quite contrary to their anticipation; the observed temperature variation of magnetization in various systems, which are shown by the broken lines, were more than twice as fast as they expected. Discrepancies were particularly large in Invar systems, FeNi and FePt, as shown in (b) and (c), but, even in Ni, shown in (a), which is a non-Invar, discrepancy is significant. Thus, they concluded that there must be some *hidden excitations* to be responsible for the additional decrease of magnetization. Such an experimental situation is being reconfirmed by a number of independent groups [6.24,6.25].

In this section we show that phonon can be such a hidden excitation; we show that each phonon in the ferromagnetic state of a metal carries a magnetization of $\sim 1\ \mu_B$ with negative sign [6.26, 6.27].

In order to make such a conclusion convincing it is essential to discuss simultaneously both of them on the same footing. We do that in this section. We first determine equilibrium magnetization at a given temperature without considering either spin wave or phonon effects, and then look into how that magnetization would be modified by the effects of spin waves and phonons [6.28].

6.6.1 Perturbative Treatment of the Spin Wave and Phonon Effects on Magnetization

If we consider both spin wave and phonon effects, the equilibrium magnetization of a metal is determined from the following condition

$$\frac{\partial}{\partial M}\left\{F_{\mathrm{el,\,m}}(M) + F_{\mathrm{sw}}(M) + F_{\mathrm{ph}}(M)\right\} = 0, \tag{6.6.1}$$

where $F_{\mathrm{el,m}}$ is the mean field (Stoner) approximation part of the electronic free energy. Let us denote by M_{S} the magnetization to be obtained by neglecting the effects of both spin wave and EPI,

$$\left.\frac{\partial F_{\mathrm{el,\,m}}(M)}{\partial M}\right|_{M = M_{\mathrm{S}}} = 0. \tag{6.6.2}$$

The true magnetization M to be obtained from (6.6.1) may, then, be written as

$$M = M_{\mathrm{S}} + \Delta M_{\mathrm{sw}} + \Delta M_{\mathrm{ph}}\,. \tag{6.6.3}$$

Let us assume

$$\left|\Delta M_{\mathrm{sw}}\right|,\ \left|\Delta M_{\mathrm{ph}}\right| << M_{\mathrm{S}}\,. \tag{6.6.4}$$

Under this assumption we can expand $F_{\mathrm{el,\,m}}(M)$ as

$$F_{\mathrm{el,\,m}}(M) = F_{\mathrm{el,\,m}}(M_{\mathrm{S}}) + \frac{1}{2}\left.\frac{\partial^2 F_{\mathrm{el,\,m}}(M)}{\partial M^2}\right|_{M = M_{\mathrm{S}}} (M - M_{\mathrm{S}})^2 + \cdots \tag{6.6.5}$$

By putting (6.6.5) into (6.6.1) we obtain

$$\begin{aligned}\Delta M_\alpha &= -\left.\frac{\partial F_\alpha(M)}{\partial M} \Big/ \frac{\partial^2 F_{\mathrm{el,\,m}}(M)}{\partial M^2}\right|_{M = M_{\mathrm{S}}} \\ &\cong -\left.\frac{\partial F_\alpha(M)}{\partial M}\right|_{M = M_{\mathrm{S}}} \Big/ \left.\frac{\partial^2 F_{\mathrm{el,\,m}}(M)}{\partial M^2}\right|_{M = M_{\mathrm{S}}}\end{aligned} \tag{6.6.6}$$

for α = sw and ph.

Note that the denominator of the last expression gives the inverse of the (Stoner) high field magnetic susceptibility, $\chi_{hf}{}^{S}(M_S)$, given in (3.4.24).

From the result of the preceding subsection, we can now give the contributions of spin waves and phonons to the free energy in the following common form,

$$F_\alpha = \frac{1}{2}\sum_\kappa \hbar\omega_\alpha(\kappa) + k_B T \sum_\kappa \ln\left[1 - \exp\left(-\hbar\omega_\alpha(\kappa)/k_B T\right)\right] \quad (6.6.7)$$

with $\alpha =$ sw or ph. If we set

$$H_\alpha(M_S) = -\left.\frac{\partial F_\alpha(M)}{\partial M}\right|_{M=M_S}$$

$$= -\hbar\sum_q \left[\frac{1}{2} + n(\hbar\omega_\alpha(q,M))\right]\left.\frac{\partial\omega_\alpha(q,M)}{\partial M}\right|_{M=M_S(T)}, \quad (6.6.8)$$

(6.6.6) is rewritten as

$$\Delta M_\alpha = \chi_{hf}^{S}(M_S)\, H_\alpha(M_S). \quad (6.6.9)$$

It is quite natural to call H_{sw} and H_{ph} the *effective magnetic fields due to the effect of, respectively, spin wave and phonon.* Note that here we are taking the ferromagnetic state as given in Fig.3.5(b), namely, as with $M > 0$. (Note that the same notation H_{ph} is used for the bare phonon Hamiltonian; see (5.2.22) and (5.2.25)).

The result of (6.6.6) can be rewritten also as

$$\Delta M_\alpha = \sum_q m_\alpha(q)\left[\frac{1}{2} + n(\hbar\omega_\alpha(q,M))\right], \quad (6.6.10)$$

with

$$m_\alpha(q) = -\chi_{hf}^{S}(M(T))\,\hbar\left.\frac{\partial\omega_\alpha(q)}{\partial M}\right|_{M=M_S(T)}. \quad (6.6.11)$$

$m_\alpha(q)$ is understood as magnetization carried by a spin wave ($\alpha =$ sw) or a phonon ($\alpha =$ ph). With the above formulation we can directly compare the roles of spin wave and phonon on the same footing.

We realize that the reason why spin wave excitation can affect magnetization is because $\partial\omega_{sw}(q)/\partial M \neq 0$, and, then, that phonon excitation also can affect magnetization since $\partial\omega_{ph}(q)/\partial M \neq 0$. Let us discuss the contributions of spin waves and phonons based on (6.6.10)–(6.6.11) separately in the following two subsections.

6.6.2 The Effect of Spin Waves on Magnetization

According to (6.6.10), the magnetization carried by a spin wave is given by

$$m_{\rm sw}(\boldsymbol{q}) = -\chi^{\rm S}_{\rm hf}(M(T))\,\hbar\frac{\partial\omega_{\rm sw}(\boldsymbol{q})}{\partial M}\bigg|_{M=M_{\rm S}(T)} . \qquad (6.6.12)$$

We anticipate to obtain $m_{\rm sw}(\boldsymbol{q}) = -2\mu_{\rm B}$ to reproduce the well-known Bloch result. In order to obtain $\partial\omega_{\rm sw}(\boldsymbol{q})/\partial M$ in (6.6.12), we have to know how $\omega_{\rm sw}(\boldsymbol{q})$ changes with magnetization. Such a spin wave frequency, $\omega_{\rm sw}(\boldsymbol{q}, \Delta M)$, is obtained by solving

$$1 - \widetilde{V}F_{-+}(\boldsymbol{q}, \omega; \Delta M) = 0 , \qquad (6.6.13)$$

where ΔM is the variational magnetization added to the equilibrium one, $M_{\rm S}(T)$. $F_{-+}(\boldsymbol{q}, \omega\,; \Delta M)$ is obtained from $F_{-+}(\boldsymbol{q}, \omega)$ by making the following change, $\varepsilon_{\boldsymbol{k}\sigma} \rightarrow \varepsilon_{\boldsymbol{k}\sigma}(\Delta M) = \varepsilon_{\boldsymbol{k}} + \sigma\eta' = \varepsilon_{\boldsymbol{k}} - Vn_\sigma + \sigma\eta''$, with

$$\eta'' = \frac{F_+(0) + F_-(0)}{4F_+(0)\,F_-(0)}\,\Delta M/\mu_{\rm B} \qquad (6.6.14)$$

to the lowest order in ΔM (*Problem* 6.1).

(6.6.13) is explicitly rewritten as

$$1 = \widetilde{V}\sum_{\boldsymbol{k}}\frac{f(\varepsilon_{\boldsymbol{k}-}; \Delta M) - f(\varepsilon_{\boldsymbol{k}+\boldsymbol{q},+}; \Delta M)}{\varepsilon_{\boldsymbol{k}+\boldsymbol{q}} - \varepsilon_{\boldsymbol{k}} + 2\eta' - \hbar\omega} , \qquad (6.6.13')$$

where

$$n_\sigma(\Delta M) = n_\sigma(\eta') = \sum_{\boldsymbol{k}} f(\varepsilon_{\boldsymbol{k}\sigma}(\Delta M)) = \sum_{\boldsymbol{k}} f(\varepsilon_{\boldsymbol{k}\sigma}(\eta')) . \qquad (6.6.15)$$

Then, corresponding to (4.4.20) we have

$$\hbar\omega - \left[\frac{2\left[F_+(0) + F_-(0)\right]}{4F_+(0)F_-(0)} - \widetilde{V}\right]\frac{\Delta M}{\mu_{\rm B}}$$

$$= \frac{\sum_k \left\{f(\varepsilon_{k-};\Delta M) + f(\varepsilon_{k+};\Delta M)\right\} (\boldsymbol{q}\nabla_k)^2 \varepsilon_k}{2\left(n_-(\Delta M) - n_+(\Delta M)\right)}$$

$$- \frac{1}{\widetilde{V}\left(n_-(\Delta M) - n_+(\Delta M)\right)^2} \sum_k \left\{f(\varepsilon_{k-};\Delta M) - f(\varepsilon_{k+};\Delta M)\right\} (\boldsymbol{q}\nabla_k \varepsilon_k)^2, \quad (6.6.16)$$

retaining terms up to the first order in ΔM. From this we obtain for the spin wave frequency under a variational magnetization ΔM,

$$\hbar\omega_{\rm sw}(\boldsymbol{q};\, \Delta M) \;=\; D(\Delta M) q^2 \;+\; 2\frac{1}{\chi_{\rm hf}^{\rm S}}\,\mu_{\rm B}\Delta M \qquad (6.6.17)$$

with

$$D(\Delta M) \;=\; \frac{n}{n_-(\Delta M) - n_+(\Delta M)}\,\frac{\hbar^2}{2m}$$

$$- \frac{4}{5}\frac{n}{\left(n_-(\Delta M) - n_+(\Delta M)\right)^2}\,\frac{1}{\widetilde{V}}\,\frac{\hbar^2}{2m}\left[\, n_-(\Delta M)\varepsilon_{\rm F_-}(\Delta M) - n_+(\Delta M)\varepsilon_{\rm F_+}(\Delta M)\right],$$

$$(6.6.18)$$

where we assumed that $\varepsilon_k = \hbar^2 k^2/2m$.

Thus we have

$$\frac{\partial}{\partial \Delta M}\hbar\omega_{\rm sw}(\boldsymbol{q};\, \Delta M)\bigg|_{\Delta M = 0} \;=\; \frac{2\mu_{\rm B}}{\chi_{\rm hf}^{\rm S}} + \frac{\partial}{\partial \Delta M} D(\Delta M)\bigg|_{\Delta M = 0} q^2$$

$$\equiv\; \frac{2\mu_{\rm B}}{\chi_{\rm hf}^{\rm S}} + \delta q^2, \qquad (6.6.19)$$

with

$$\delta \;=\; -\; \frac{n}{\left(n_- - n_+\right)^2}\,\frac{1}{\mu_{\rm B}}\,\frac{\hbar^2}{2m} \;+\; \frac{8}{5}\,\frac{\left(n_-\varepsilon_{\rm F_-} - n_+\varepsilon_{\rm F_+}\right)}{\left(n_- - n_+\right)^3}\,\frac{n}{\widetilde{V}}\,\frac{1}{\mu_{\rm B}}\,\frac{\hbar^2}{2m}$$

$$+\; \frac{2}{3}\,\frac{\left(n_-^{2/3} + n_+^{2/3}\right)}{\left(n_- - n_+\right)^2}\left(\frac{3\pi^2}{V}\right)^{\frac{2}{3}}\frac{n}{\widetilde{V}}\,\frac{1}{\mu_{\rm B}}\left(\frac{\hbar^2}{2m}\right)^2, \qquad (6.6.20)$$

and, then

$$m_{sw}(\boldsymbol{q}) = -2\mu_B - \chi_{hf}^{S}(M(T))\delta q^2$$

$$\equiv m_{sw}(\boldsymbol{q})_1 + m_{sw}(\boldsymbol{q})_2. \qquad (6.6.21)$$

$m_{sw}(\boldsymbol{q})_1$, and $m_{sw}(\boldsymbol{q})_2$ correspond, respectively, to the first and second terms on the right hand side of (6.6.19). Whereas $m_{sw}(\boldsymbol{q})_1$ gives us the classic Bloch result, $m_{sw}(\boldsymbol{q})_2$ is a new contribution. As we will see in our numerical example in 6.6.4, $m_{sw}(\boldsymbol{q})_2$, which is of positive sign, can be as large as $\sim 1\mu_B$ for $q \sim k_F$. This is a significant correction.

6.6.3 The Effect of Phonon on Magnetization: Phonon carries Magnetization like Spin Wave does!

According to (6.6.11), the magnetization carried by a phonon is given by [6.26]

$$m_{ph}(\boldsymbol{q}) = -\chi_{hf}^{S}(M(T)) \left. \frac{\hbar \partial \omega_{ph}(\boldsymbol{q})}{\partial M} \right|_{M = M_S(T)} . \qquad (6.6.22)$$

The value of $m_{ph}(\boldsymbol{q})$ for $T < T_C$ of a ferromagnetic metal depends on temperature or magnetization of the metal; both χ_{hf} and $\partial\omega_{ph}(\boldsymbol{q})/\partial M$ depend on the temperature in (6.6.22).

Let us first see that the magnitude of $m_{ph}(\boldsymbol{q})$ can be $\sim 1\mu_B$. If we use the Debye approximation, (6.6.22) is rewritten by using the quantity A which, as defined in (5.5.37), gives the magnetic field dependence of sound velocity in the ferromagnetic state, as

$$m_{ph}(\boldsymbol{q}) = -A \frac{q}{q_D} \frac{\hbar \omega_D(M_S(T))}{W} \mu_B , \qquad (6.6.23)$$

where from (5.5.33), (5.5.36)–(5.5.37) and (6.6.14) we noted that

$$\frac{1}{\omega_D(M)} \frac{\partial \omega_D(M)}{\partial M} = \frac{1}{s} \frac{\partial s(M)}{\partial M} = \frac{f_1}{W} \frac{F_+(0) + F_-(0)}{4F_+(0)F_-(0)} = \frac{\mu_B}{\chi_{hf}} \frac{A}{W} . \qquad (6.6.24)$$

If $|A| \cong O(1)$, we have $|m_{ph}(\boldsymbol{q})| = O[(\hbar\omega_D/\varepsilon_F)(q/q_D)] \cong 10^{-2}(q/q_D)\mu_B$; $|m_{ph}(\boldsymbol{q})|$ can not be larger than $\sim 10^{-2}\mu_B$. As we saw in 5.5.3, however, we can have $|A| = O(10^2)$. With $|A| = O(10^2)$, we have $|m_{ph}(\boldsymbol{q})| = O(q/q_D)\mu_B$; for $q \cong q_D$, we have $|m_{ph}(\boldsymbol{q})| = O(1)\mu_B$.

6.6.4 A Quantitative Comparison of the Phonon and Spin Wave Effects on Magnetization

Now we present numerical examples on our result of the preceding two subsections [6.27,6.29–6.31]. We start from an equilibrium magnetization, $M_S(T)$ that is determined from $F_{el,m}$ alone, and, pursue how that magnetization decreases by the effects of spin waves and phonons at each temperature. In order to compare the effect of phonons to that of spin waves it is essential to use a model which enables us to discuss them on the same footing. It is desirable to introduce as few parameters as possible. At present the only model with such a qualification seems to be a jellium-like model with the electron energy dispersion $\varepsilon_{\boldsymbol{k}} = \hbar^2 \boldsymbol{k}^2/2m$. As for parameters in the model, we assume m/m_o = 1, 5, and 10 [3.15] for the effective mass of electrons, m, with m_o the free electron mass, $k_F = 1.4\times10^8$/cm, $\overline{V} = \widetilde{V}N(0) = 1.1$, $\theta_D(T=0) = 400$ K and $\xi = 1.0$.

First, in Fig.6.14(a) we show $m_{sw}(\boldsymbol{q})_2$ which represents how the magnetization carried by a spin wave deviates from $-2\mu_B$. Here T_C^S is the Curie temperature in the (Stoner) mean field approximation; by the effects of spin waves and phonons the true Curie temperature is drastically reduced from T_C^S as can be seen from Figs.6.16–6.17 below. Although according to Fig. 6.16, due to the effects of spin waves and phonons the magnetization vanishes already for $T/T_C^S < 0.1$, we showed the behavior of $m_{sw}(\boldsymbol{q})_2$ (and $m_{ph}(\boldsymbol{q})$ in Fig.6.15) up to $T/T_C^S < 0.6$, since $m_{sw}(\boldsymbol{q})_2$ (and $m_{ph}(\boldsymbol{q})$) depends more on the spin splitting of the electron energy bands than on temperature. In a non-perturbative treatment, $m_{sw}(\boldsymbol{q})_2$ (and $m_{ph}(\boldsymbol{q})$) would be able to take values corresponding to the spin splitting of the bands up to $T/T_C^S = 1$ in Figs.6.14 and 15.

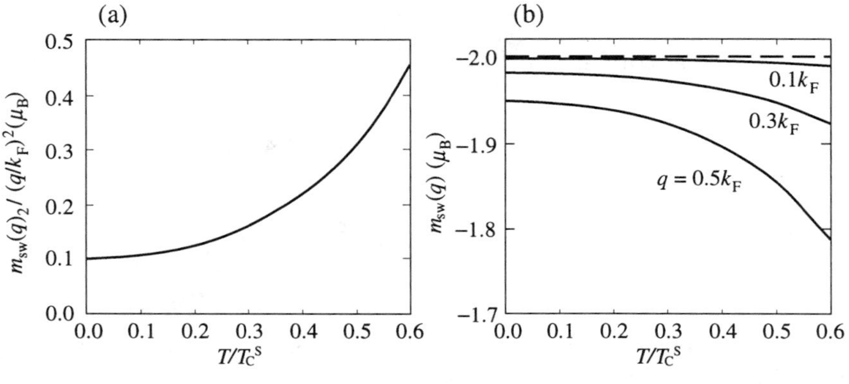

Fig.6.14 Deviations from $-2\mu_B$ of the magnetization carried by a spin wave, (a), and behaviors of $m_{sw}(q) = m_{sw}(q)_1 + m_{sw}(q)_2$, (b).

In Fig.6.14(b) we show the behavior of $m_{sw}(\boldsymbol{q}) = m_{sw}(\boldsymbol{q})_1 + m_{sw}(\boldsymbol{q})_2$ for different values of q.

Next, in Fig.6.15 we show the behavior of $m_{ph}(q)$. The magnetization carried by a phonon can be of the order of $-1\mu_B$ for $q \cong q_D$. In contrast to $m_{sw}(\boldsymbol{q})_2$, the behavior of $m_{ph}(\boldsymbol{q})$ sensitively depends on m/m_o. It is because ε_F, which plays the role of W in (6.6.22), is inversely proportional to m/m_o.

Note, however, that as can be seen from (5.5.37), the quantity A/ε_F or A/W that determines $m_{ph}(\boldsymbol{q})$ depends on the detailed features of the electronic structure. It is impossible to account for the difference in the behaviors of $m_{ph}(\boldsymbol{q})$ between Ni and an FeNi Invar alloy, for instance, simply by choosing different values for ε_F or m/m_o within a simple parabolic electronic density of states. There is a limitation in representing different materials by different values of the parameters within a free electron gas model.

In Fig.6.16 we show how the behavior of magnetization is modified by such spin wave and phonon effects. Here, in calculating spin wave contribution to the free energy we sum over the wave vectors only up to a cut off wave number, q_c; spin waves with wave vectors larger than q_c merge into the Stoner excitations and decay. Since generally $q_c \cong k_F$, we put $q_c / k_F = 1/2$, in our cal-culation The effects are present even at $T = 0$ and quite substantial. Such large effect of the zero point oscillation of spin wave and phonon have been overlooked. The size of the phonon contribution depends on the value of the parameter ξ, and that of spin wave or the cut off wave number q_c, but within the reasonable range of the values of the parameters, phonon effect can be generally as large as that of spin waves at all temperatures.

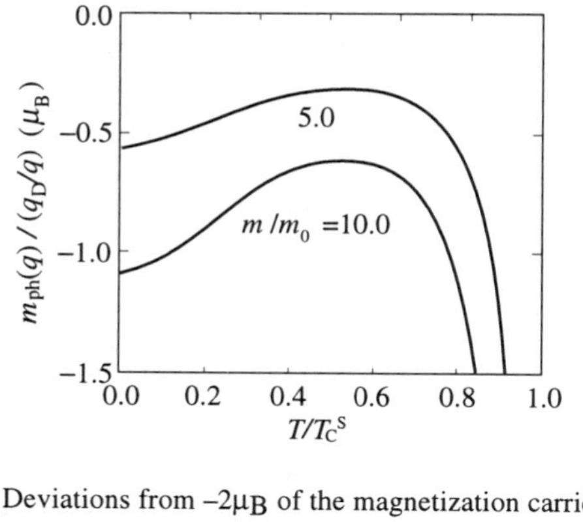

Fig.6.15 Deviations from $-2\mu_B$ of the magnetization carried by a phonon

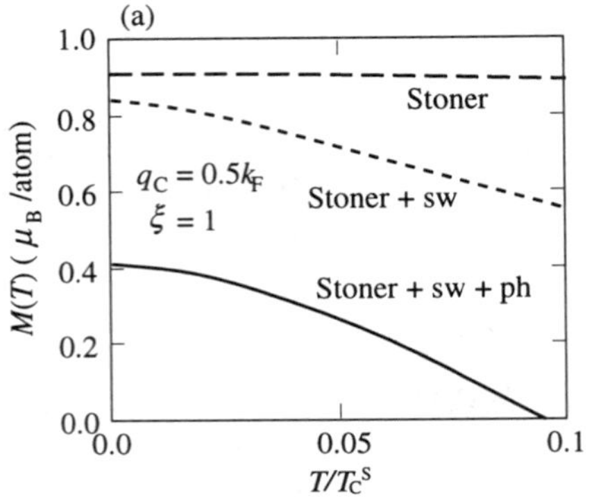

Fig.6.16 The effects of various excitations on the temperature dependence of magnetization at low temperatures.

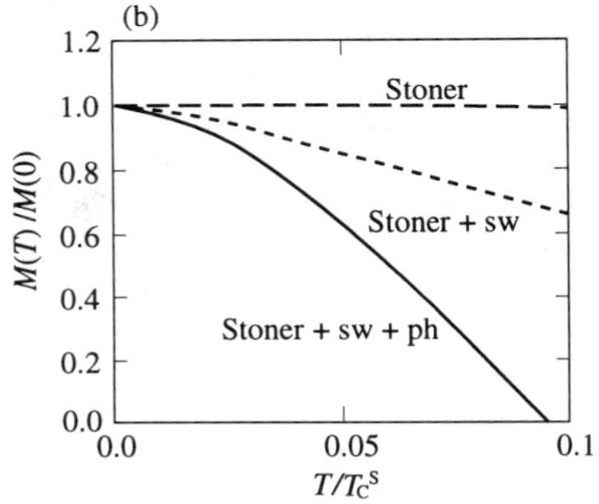

Fig.6.17 Reproduction of Fig.6.16 in a way appropriate for comparison with experiments such as Fig.6.13. $\varepsilon_F/W = 0.5$, $\theta_D(T = 0) = 400$ K, with $\bar{V} = 1.1$, and $\xi = 1$.

In Fig.6.17 we replotted the result of Fig.6.16 in a way convenient to compare with Fig.6.13 [6.28,6.30]. The decrease of magnetization with increasing temperature can be made nearly twice faster by the effect of phonons compared with the case where only spin waves are considered. Such result is exactly we require to understand experimental results of Fig.6.13.

6.7 A NON-PERTURBATIVE ESTIMATION OF PHONON EFFECT ON MAGNETIZATION

The perturbational treatment of the phonon and spin wave effects given in the preceding section was carried out under the assumption of (6.6.4). However, according to Figs.6.14–17, such an assumption is not valid both for the spin wave and phonon effects. A non-perturbative approach is necessary. In this section we carry out a non-perturbative estimation of ΔM_{ph}. Although it is required to consider simultaneously the effect of spin waves, here we leave it out. Concentrating on the phonon effect, we want to reconfirm that it can be as large as $\sim 1\mu_B$/atom even if it is treated non-perturbatively.

We pursue numerically how F_{el} and F_{ph} would change with magnetization. Here, for $F_{el}(M)$ we use the mean field approximation result, $F_{el,m}$, namely, (3.4.10) with (3.4.5), and for F_{ph} the Debye approximation result of (6.2.9). We use the model electronic density of states of Fig.5.5 with $W = 1$ eV. First, we show the result for the case $\varepsilon_F/W = 0.5$, $\theta_D(T = 0) = 400$ K, with $\bar{V} = 1.1$, and $\xi = 1$, in Fig.6.18. We denoted the magnetization dependent part of the free energy by ΔF. By the effect of EPI the minimum of the free energy shifts quite drastically, as we already saw in the perturbational result. By carrying out these calculations for different temperatures we obtain the result of Fig.6.19. The broken and the solid lines show the temperature dependence of the magnetization obtained, respectively, without and with the effect of EPI; note the drastic change in T_C due to the effect of EPI.

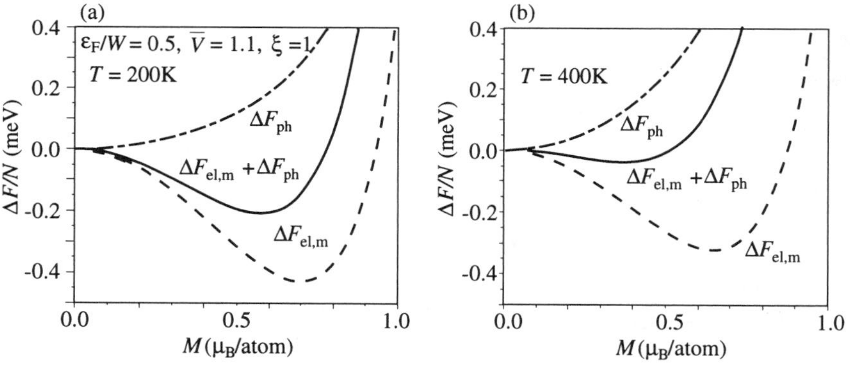

Fig.6.18 The magnetization dependent parts per atom (electron) of $F_{el,m}$ (broken lines), F_{ph} (dot-dashed lines), and $F_{el,m} + F_{ph}$ (solid lines) for $\varepsilon_F/W = 0.5$ in the electronic density of states of Fig.5.5, for (a) $T = 200$K, and (b) $T = 400$K. We put $W = 1$ eV, $\bar{V} = 1.1$, $\theta_D(T = 0) = 400$ K, and $\xi = 1$.

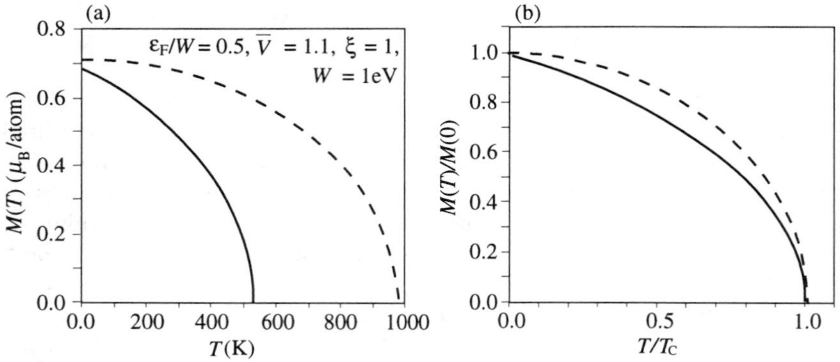

Fig.6.19 The temperature dependence of magnetization corresponding to Fig.6.18(a). The broken line is for the magnetization determined from $F_{\mathrm{el,m}}$ alone, and the solid line is for that determined from $F_{\mathrm{el,m}} + F_{\mathrm{ph}}$. The results of (a) are replotted for T/T_{C} in (b); the notations are the same as in (a).

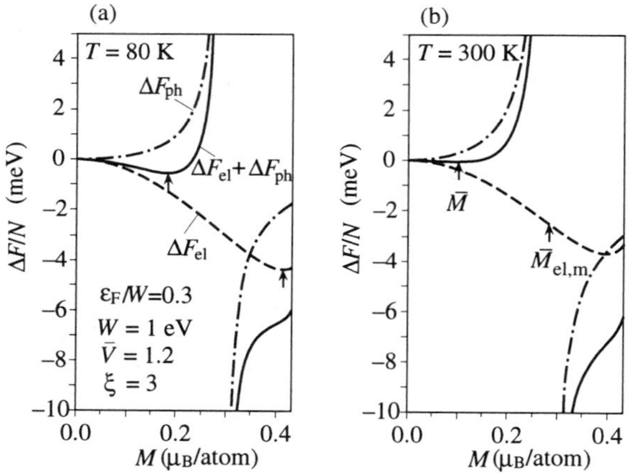

Fig.6.20 The magnetization dependent parts per atom (electron) of F_{el} (broken lines), F_{ph} (dot-dashed lines), and $F_{\mathrm{el,m}} + F_{\mathrm{ph}}$ (solid lines) for $\varepsilon_{\mathrm{F}}/W = 0.3$ in the electronic density of states of Fig.5.5, for (a) $T = 80\mathrm{K}$, and (b) $T = 300\mathrm{K}$. We put $W = 1$ eV, $\bar{V} = 1.2$, $\theta_{\mathrm{D}}(T=0) = 400$ K, and $\xi = 3$. The minima of $F_{\mathrm{el,m}} + F_{\mathrm{ph}}$ are indicated by arrows.

Next, we present the result for the case of $\varepsilon_{\mathrm{F}}/W = 0.3$ in Figs.6.20–6.21. We set $W = 1$ eV, $\bar{V} = 1.2$, $\theta_{\mathrm{D}}(T=0) = 400$ K, and $\xi = 3$. Unlike in Fig.6.18, in Fig.6.20 we observe that $F_{\mathrm{ph}}(M)$ becomes singular at a certain value of M. This singularity originates from the denominator of the second term becoming

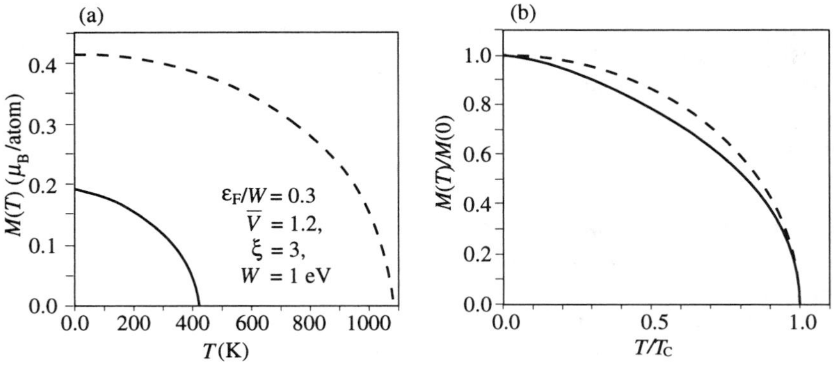

Fig.6.21 The temperature dependence of magnetization corresponding to Fig.6.20(a). The broken line is for the magnetization determined from $F_{el,m}$ alone, and the solid line is for that determined from $F_{el,m} + F_{ph}$. The results of (a) are replotted for T/T_C in (b).

zero in the expression for sound velocity (5.4.27); at this value of M, sound velocity becomes singular. We may view that at the value of M where the singularity occurs the system becomes structurally unstable, and therefore, our discussion becomes invalid. However, before $F_{ph}(M)$ becomes singular the total free energy attains a minimum in each case. The minima of the total free energy are identified as indicated by the arrows in Fig.6.20. Then we obtain the results of Fig.6.21.

Note that the phonon frequencies to be used in the phonon free energy should have been those which are calculated without the adiabatic approximation. As can be seen from (5.1.1) or (5.3.22), the phonon frequencies without the adiabatic approximation never diverge since $\bar{\chi}_{ee}(\boldsymbol{q}, \omega_{\boldsymbol{q}}) \to 0$ for $\omega_{\boldsymbol{q}} \to \infty$. Thus, if the adiabatic approximation is abandoned the behavior of $F_{ph}(M)$ would be drastically modified from those of Fig.6.20 in the region where it is singular.

In conclusion, even if non-perturbatively estimated, the size of the effect of EPI on magnetization becomes as large $\sim 1\mu_B$/atom.

6.8 THE EFFECT OF PHONON AND SPIN WAVE ON THE SPIN SPLITTING OF ELECTRON ENERGY BANDS

Recently a large discrepancy between band calculation and photoemission experiment results is noted on the values of the exchange splitting of the electron energy bands, Δ, for ferromagnetic transition metals [6.32, 6.33]; in the case of Ni, the values for Δ at $T = 0$ are, respectively, ~0.7 eV and ~0.3 eV. This fact may be viewed to imply that it is difficult to understand observed values of Δ by considering only electrons and, within the Stoner picture. In the preceding sections we saw that both spin waves and phonons will have significant effects on Δ even at $T = 0$. In this section we show that such result implies that spin splitting of electron energy bands can not be determined without considering the effects of the spin waves and phonons.

6.8.1 Phonon and Spin Wave Effects on Δ: A New Picture of the Ferromagnetic State

We start from (6.6.1), the equilibrium condition to determine magnetization of a system. It is rewritten as

$$H_S(T, M) + H_{sw}(T, M) + H_{ph}(T, M) = 0, \tag{6.8.1}$$

where H_S, H_{sw}, and H_{ph} are the internal magnetic fields due to, respectively, the one particle part of electrons, spin waves and phonons defined similarly to (6.6.8):

$$H_S(T, M) = -\frac{\partial F_{el,m}(T, M)}{\partial M}. \tag{6.8.2a}$$

$$H_{sw}(T, M) = -\frac{\partial F_{sw}(T, M)}{\partial M}. \tag{6.8.2b}$$

$$H_{ph}(T, M) = -\frac{\partial F_{ph}(T, M)}{\partial M}. \tag{6.8.2c}$$

By using (3.4.14) for H_S given in (6.8.1), we obtain the spin splitting of the bands, $\Delta(T)$, as

$$\begin{aligned} \mu_-(T, M(T)) - \mu_+(T, M(T)) &\equiv \Delta(T, M(T)) \\ &= \widetilde{V} M(T)/\mu_B + 2\mu_B H_{sw}(T, M(T)) + 2\mu_B H_{ph}(T, M(T)) \\ &\equiv \Delta_S(T, M(T)) + \Delta_{sw}(T, M(T)) + \Delta_{ph}(T, M(T)), \end{aligned} \tag{6.8.3}$$

where $M(T)$ is the equilibrium magnetization that is to be self-consistently determined from this equation. Note that here we are assuming that $n_- > n_+$ to make $M(T) > 0$ for the ferromagnetic state as in Fig.6.22.

Our result shows that both spin waves and phonons are directly involved in determining the spin splitting of the electron energy bands of an itinerant electron ferromagnet. With such spin wave and phonon effects on Δ, we have Fig.6.22. Compare it with Fig.3.5(b), the currently prevailing picture of the ferromagnetic state; our result is requiring a fundamental change in such a pictu-

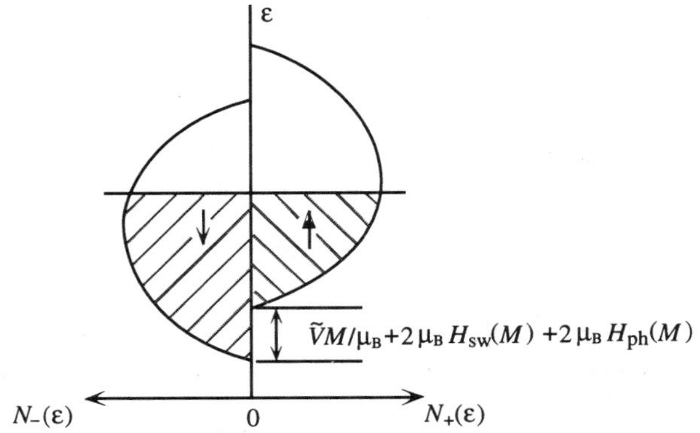

Fig.6.22 A new picture for the ferromagnetic state of metal; Compare with Fig.3.5.

re. We noticed this fact first concerning phonons [6.29]. It is then quite straightforward to notice that spin waves also should contribute to Δ.

Let us estimate how large these effects of spin waves and phonons on Δ can be, first, perturbatively in 6.8.2. Then in 6.8.4 we make a non-perturbative estimation of Δ_{ph}.

6.8.2 Perturbative Estimation of Δ_{sw} and Δ_{ph}

Let us assume that

$$|\Delta_{sw}|, \quad |\Delta_{ph}| << \Delta_S. \tag{6.8.4}$$

This condition corresponds to that of (6.6.4). In this case we may put

$$\Delta_\alpha(M(T)) \cong \Delta_\alpha(M_S(T)) \tag{6.8.5}$$

for α = sw or ph. Recall that $M_S(T)$ is the equilibrium magnetization determined from $F_{el,\,m}$ alone, whereas $M(T)$ is that which is determined from $F_{el,\,m} + F_{sw} + F_{ph}$. Then, from (6.6.9) we have

$$\Delta_\alpha(M_S(T)) = 2\mu_B\, H_\alpha(M_S(T))$$

$$= 2\mu_B\, \Delta M_\alpha / \chi^S_{hf}(M_S). \tag{6.8.6}$$

In 6.6 we found that $\Delta M_\alpha/N \cong -1\ \mu_B$/atom both for α = sw and ph (note that M_α is for a system of N atoms). We have then,

$$\Delta_\alpha \cong -2N\mu_B^2/\chi_{hf}^S(M_S). \tag{6.8.7}$$

The negative sign implies that both of the spin wave and phonon effects reduce the exchange splitting. If we note that $\chi_{hf} = -O(N\mu_B^2\varepsilon_F)$, we have

$$\Delta_\alpha \cong -O(\varepsilon_F). \tag{6.8.8}$$

Actually, however, χ_{hf} is exchange enhanced from the above magnitude and, therefore, Δ_α should be reduced from the above estimation. However, we anticipate the sizes of Δ_α for both spin waves and phonons remain to be a good fraction of ε_F. To see this, we carry out a model numerical calculation in the next subsection.

6.8.3 A Model Calculation of $\Delta_{sw}(M_S(T))$ and $\Delta_{ph}(M_S(T))$

In order to calculate Δ_{sw} and Δ_{ph} on the same footing we use the same jellium-like model with the electron energy dispersion $\varepsilon_k = \hbar^2 k^2/2m$ as in 6.6.4. Then, the required numerical calculation can be immediately carried out by using the result of 6.6.4; we choose the same values for various parameters as there. Thus first we obtain the result of Fig.6.23 for the spin wave and phonon effects on Δ [6.34]. Although the size of the effect depends on the values of the cut off wave number of the spin wave, it is generally of significant size, as we anticipated. Note that the result of Fig.6.23 corresponds to those of Figs.6.14–6.17.

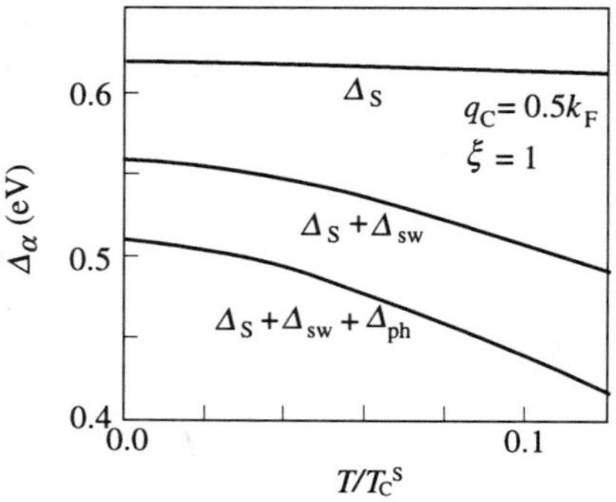

Fig.6.23 The temperature dependence of the exchange splitting in the Stoner approximation, $\Delta_S(T)$, that including the spin wave(sw) effect, $\Delta_S(T) + \Delta_{sw}(T)$, and that including both (spin wave and phonon(ph)) effects, $\Delta_S(T) + \Delta_{sw}(T) + \Delta_{ph}(T)$.

6.8.4 A Non-Perturbative Estimation of Δ_{ph}

In the preceding subsection we saw that the effects of phonons, as well as spin waves, on the exchange splitting is quite significant. That conclusion was derived by a perturbative procedure starting with the assumption of (6.8.4). The obtained result, then, contradicts such an assumption; the effects are very large. In this section we carry out a non-perturbative estimation of the phonon effect on Δ, leaving out, however, that of spin wave. We will find that the conclusion of the preceding subsection remains to be valid even in a non-perturbative estimation.

Because here we concentrate on the phonon effect, neglecting that of spin wave, the equation to be solved is

$$\begin{aligned}\Delta(T) &= \mu_-(T, M(T)) - \mu_+(T, M(T)) \\ &= \widetilde{V} M(T)/\mu_{\mathrm{B}} + 2\mu_{\mathrm{B}} H_{\mathrm{ph}}(T, M(T)),\end{aligned} \tag{6.8.9}$$

in place of (6.8.3). The effective magnetic field due to EPI to be used in (6.8.9) is given from (6.8.2c); the only difference is that here for M in $H_{\mathrm{ph}}(T, M)$, we put $M(T)$ that is to be self-consistently determined from (6.8.9), in place of $M_{\mathrm{S}}(T)$.

With the Debye approximation, we have

$$\begin{aligned}\mu_{\mathrm{B}} H_{\mathrm{ph}}(T) &= -\mu_{\mathrm{B}} \frac{\partial F_{\mathrm{ph}}(M)}{\partial \omega_{\mathrm{D}}(M)} \frac{\partial \omega_{\mathrm{D}}(M)}{\partial M} \\ &= -A(T, M) \frac{k_{\mathrm{B}} \theta_{\mathrm{D}}(T, M)}{W} P\left(\frac{T}{\theta_{\mathrm{D}}(T, M)}\right)\Bigg|_{M = M(T)},\end{aligned} \tag{6.8.10}$$

by noting (6.2.16) and (6.6.24), where A/W is given in (5.5.37).

We carry out a numerical calculation on (6.8.10) with the model electronic density of states of Fig.5.5, exactly in the same way as we obtained the results of Figs.6.18–6.21. Thus we obtain the result given in Fig.6.24. The size of the EPI effect on Δ very sensitively depends upon the electronic structure near the Fermi surface. In the case of Fig.6.24, the effect is drastic and reduces the spin splitting at $T = 0$ nearly by half. Even in the case of $\varepsilon_{\mathrm{F}}/W = 0.5$, the effect is quite significant [6.28]. Thus we confirm that the effect of phonons on the exchange splitting of the electron energy bands in the ferromagnetic state of a metal can be quite significant.

An important notice here is that the size of such phonon effect can be experimentally estimated by measuring the magnetic field dependence of sound velocity as indicated in (6.8.10).

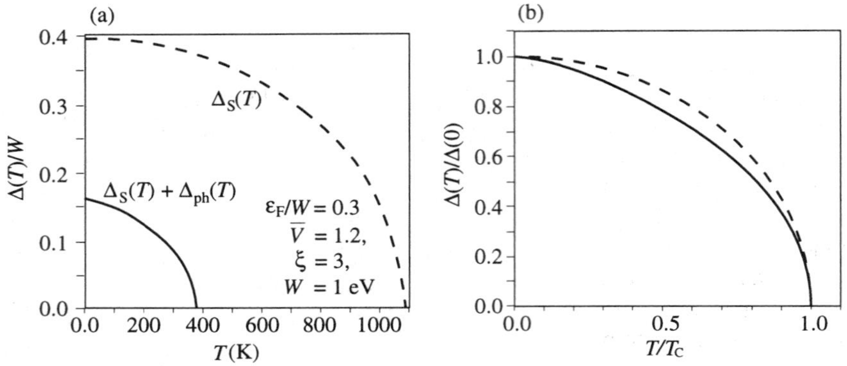

Fig.6.24 The temperature dependence of the exchange splitting in the Stoner approximation, $\Delta_S(T)$, and the phonon effect, $\Delta_S(T) + \Delta_{ph}(T)$,(a). In (b) shown $\Delta_S(T)/\Delta_S(0)$, the broken line, and $(\Delta_S(T) + \Delta_{ph}(T))/(\Delta_S(0) + \Delta_{ph}(0))$, the solid line.

6.9 THE TEMPERATURE DEPENDENCE OF MAGNETIZATION: *dM(T)/dT*

In the preceding sections of this chapter we saw that phonon can play a role as important as that of spin wave in determining spontaneous magnetization of an itinerant ferromagnet. In 6.6.3–6.6.4 we saw that $m_{ph}(\boldsymbol{q})$, the magnetization carried by a phonon with wave vector $\boldsymbol{q}$, can have a magnitude of $-1\ \mu_B$ for $q \cong q_D$. Also we found that $m_{sw}(\boldsymbol{q})$, the magnetization carried by a spin wave, is not exactly $-2\mu_B$; deviations from $-2\mu_B$ can be of the order of 1 μ_B. Those $m_{sw}(\boldsymbol{q})$ and $m_{ph}(\boldsymbol{q})$ appearing there were introduced in the course of a perturbative treatment of the effects of spin waves and phonons on magnetization at a given temperature. In this section we present a general, non-perturbative formulation on the temperature dependence of magnetization, $dM(T)/dT$, and see how naturally $m_{ph}(\boldsymbol{q})$ and $m_{sw}(\boldsymbol{q})$ again emerge. $m_{ph}(\boldsymbol{q})$ and $m_{sw}(\boldsymbol{q})$ to be introduced in this section, however, are different from those of 6.6.3–6.6.4 in that they are to be defined non-perturbatively.

6.9.1 A General Thermodynamical Formulation on the Temperature Dependence of Magnetization

Suppose that at a temperature T, the equilibrium magnetization of a system takes a value $M(T)$. It implies that we have

$$\left.\frac{\partial F(T, M)}{\partial M}\right|_{M = M(T)} = 0 \qquad (6.9.1)$$

for the free energy of the system. Now we raise the temperature of the system to $T + \Delta T$. The change in the equilibrium magnetization, ΔM, is, then, determined by the condition,

$$\frac{\partial}{\partial(\Delta M)} F(T+\Delta T, M(T)+\Delta M) = 0. \tag{6.9.2}$$

We can make the following expansion by assuming that ΔT and, therefore, ΔM are small:

$$F(T+\Delta T,\ M(T)+\Delta M) = F(T, M(T)) + F_T(T, M(T))\,\Delta T$$

$$+ \frac{1}{2} F_{MM}(T, M(T))\,(\Delta M)^2 + F_{MT}(T, M(T))\,\Delta M \Delta T + \cdots, \tag{6.9.3}$$

where we noted (6.9.1) and put

$$F_{MM}(T, M(T)) = \frac{\partial^2}{\partial M^2} F(T, M)\bigg|_{M = M(T)} \quad \text{etc.} \tag{6.9.4}$$

If we put (6.9.3) into (6.9.2), we obtain

$$F_{MT}(T, M(T))\Delta T + F_{MM}(T, M(T))\Delta M = 0$$

or

$$\frac{dM(T)}{dT} = -\frac{F_{MT}(T, M(T))}{F_{MM}(T, M(T))}$$

$$= \chi_{\text{hf}}(M(T)) \frac{\partial}{\partial T} H(T, M)\bigg|_{M = M(T)}, \tag{6.9.5}$$

where $\chi_{\text{hf}} = (\partial^2 F/\partial M^2)^{-1}$ is the high field magnetic susceptibility in the ferromagnetic state of the system, and

$$H(T, M) = -\frac{\partial}{\partial M} F(T, M) \tag{6.9.6}$$

is the effective internal magnetic field of the system. Note that from (6.9.1) although $H(T, M(T)) = 0$, $\partial H(T, M)/\partial T\,|_{M = M(T)} = -F_{TM}(T, M(T)) \neq 0$.

Up until here our discussion was independent of the model. Now we consider the jellium-like model. Then for the free energy we have

$$F(T, M) = F_{el}(T, M) + F_{ph}(T, M)$$

$$= F_{el,m}(T, M) + F_{sw}(T, M) + F_{ph}(T, M), \qquad (6.9.7)$$

as we noted in (6.6.1); for $T < T_C$, we took the spin wave contribution F_{sw} for $F_{el,c}$ of (6.2.2). Corresponding to such division of the free energy, (6.9.5)–(6.9.6) are rewritten as

$$\frac{dM(T)}{dT} = \chi_{hf}(M(T))\sum_{\alpha} \left.\frac{\partial H_\alpha(T, M)}{\partial T}\right|_{M = M(T)} \equiv \sum_{\alpha} \left.\frac{dM(T)}{dT}\right|_{\alpha}, \qquad (6.9.8)$$

$$H_\alpha(T, M) = -\frac{\partial F_\alpha(T, M)}{\partial M}, \quad \alpha = \text{S, sw, or ph}, \qquad (6.9.9)$$

where S stands for the Stoner (mean field) approximation contribution; $F_S = F_{el,m}$ etc. H_S, H_{sw}, and H_{ph} are, respectively, the Stoner excitation, spin wave and phonon contributions to the effective magnetic field. Here it is important to note that the high field susceptibility, $\chi_{hf}(M(T))$, appearing there is that which is actually to be observed, containing the entire effects of electrons and phonons; it is different from that appearing in the perturbational discussions given in 6.6.1 and 6.8.2–6.8.3.

We are most interested in spin wave and phonon contributions. Before going into that subject, we would like to confirm the validity and usefulness of our general formulation of this subsection by obtaining the Stoner contribution in the form as is generally anticipated.

6.9.2 The Stoner Excitation Contribution to $dM(T)/dT$

The Stoner excitation contribution to the temperature dependence of magnetization is given as

$$\left.\frac{dM(T)}{dT}\right|_{S} = \chi_{hf}(M(T)) \left.\frac{\partial H_S(T, M)}{\partial T}\right|_{M = M(T)}. \qquad (6.9.10)$$

From (3.4.14) we note that

$$H_S(T, M) = -\frac{\partial F_{el,m}(T, M)}{\partial M}$$

$$= \frac{1}{2\mu_B}\sum_{\sigma} \sigma\,(\mu_\sigma - \widetilde{V}n_\sigma). \qquad (6.9.11)$$

Then, we have

$$\left.\frac{\partial H_S(T, M)}{\partial T}\right|_{M = M(T)} = \frac{1}{2\mu_B}\sum_\sigma \sigma \left.\frac{\partial \mu_\sigma(T, M)}{\partial T}\right|_{M = M(T)} \tag{6.9.12}$$

The temperature dependence of the chemical potential μ under constant magnetization $M(T)$ is obtained from

$$\left.\frac{\partial}{\partial T} n_\sigma(T, M)\right|_{M = M(T)} = 0. \tag{6.9.13}$$

By putting

$$n_\sigma(T, M) = \sum_k f_\sigma(\varepsilon_k) = \sum_k \frac{1}{e^{\beta(\varepsilon_k - \mu_\sigma(T, M))} + 1} \tag{6.9.14}$$

into (6.9.13) we obtain

$$\begin{aligned}\left.\frac{\partial \mu_\sigma(T, M)}{\partial T}\right|_{M = M(T)} &= -\frac{1}{T}\sum_k \left(-\frac{\partial f_\sigma(\varepsilon_k)}{\partial \varepsilon_k}\right)(\varepsilon_k - \mu_\sigma) \Big/ \sum_k -\frac{\partial f_\sigma(\varepsilon_k)}{\partial \varepsilon_k} \\ &\cong -\frac{\pi^2}{3}\frac{N'(\mu_\sigma)}{N(\mu_\sigma)} k_B^2 T,\end{aligned} \tag{6.9.15}$$

where we used the relation

$$\sum_k \left(-\frac{\partial f_\sigma(\varepsilon_k)}{\partial \varepsilon_k}\right)(\varepsilon_k - \mu_\sigma)^2 \cong \frac{\pi^2}{3} N(\mu_\sigma)(k_B T)^2 + \cdots . \tag{6.9.16}$$

The final result is then obtained as

$$\left.\frac{dM(T)}{dT}\right|_S = \mu_B \frac{\pi^2}{6}\left[\frac{N'(\mu_-)}{N(\mu_-)} - \frac{N'(\mu_+)}{N(\mu_+)}\right]\bar{\chi}_{hf}(T)\, k_B^2 T + \cdots , \tag{6.9.17}$$

where $\bar{\chi}_{hf} = \chi_{hf} / \mu_B^2$. The result agrees with what we anticipated; the decrease of magnetization by Stoner excitation is proportional to $(T/\varepsilon_F)^2$. There is, however, an additional source of temperature dependence through χ_{hf} in the result. Note that χ_{hf} in (6.9.17) is that which includes the effects of both spin waves and phonons.

6.9.3 Spin Wave and Phonon Contribution to $dM(T)/dT$ and Non-perturbative $m_{\text{sw}}(T)$ and $m_{\text{ph}}(T)$

With (6.6.7), (6.6.8)–(6.6.9) for α = sw or ph are rewritten as

$$\left.\frac{dM(T)}{dT}\right|_{\alpha} = \sum_{q} m_{\alpha}(\boldsymbol{q}) \frac{\partial}{\partial T}\left[\frac{1}{2} + n(\hbar\omega_{\alpha}(\boldsymbol{q}))\right], \tag{6.9.18}$$

$$m_{\alpha}(\kappa) = -\chi_{\text{hf}}(M(T))\,\hbar \left.\frac{\partial \omega_{\alpha}(\boldsymbol{q})}{\partial M}\right|_{M=M(T)}, \tag{6.9.19}$$

with $n(\varepsilon)$ the Bose distribution; we neglect the temperature dependence in $\omega_{\alpha}(\boldsymbol{q}) = \omega_{\alpha}(\boldsymbol{q};\, T, M)$, which is relatively of the order of $(k_{\text{B}}T/\varepsilon_{\text{F}})^2$; note that in differentiating with respect to T in (6.9.18), we have to treat T and M as independent variables. $m_{\alpha}(\boldsymbol{q})$ is understood as magnetization carried by a spin wave (α = sw) or a phonon (α = ph), as we already saw in 6.6.1. Note, however, that $m_{\alpha}(\boldsymbol{q})$ of (6.9.19) is a non-perturbative quantity, whereas that of (6.6.11) is a perturbative one. χ_{hf} in (6.9.19) is that which contains the effects of both spin waves and phonons differing from the Stoner high field susceptibility in (6.6.11). Also $M(T)$ in (6.9.19) is the equilibrium magnetization determined by including the effects of both spin waves and phonons, unlike the Stoner magnetization in (6.6.11).

Thus, for spin wave contribution we have

$$\left.\frac{dM(T)}{dT}\right|_{\text{sw}} = \sum_{q} m_{\text{sw}}(\boldsymbol{q}) \frac{\partial}{\partial T} n(\hbar\omega_{\text{sw}}(\boldsymbol{q})), \tag{6.9.20}$$

$$\begin{aligned} m_{\text{sw}}(\boldsymbol{q}) &= -\chi_{\text{hf}}(M(T))\,\hbar \left.\frac{\partial \omega_{\text{sw}}(\boldsymbol{q})}{\partial M}\right|_{M=M(T)} \\ &= -\chi_{\text{hf}}(M(T)) \frac{2\mu_{\text{B}}}{\chi_{\text{hf}}^{\text{S}}(M(T))} - \chi_{\text{hf}}(M(T))\,\delta\, q^2 \\ &= m_{\text{sw}}(\boldsymbol{q})_1 + m_{\text{sw}}(\boldsymbol{q})_2, \end{aligned} \tag{6.9.21}$$

where δ is given by the same expression as (6.6.20) with n_{σ} and $\varepsilon_{\text{F}\sigma}$ corresponding to $M(T)$ in place of $M_{\text{S}}(T)$. If $m_{\text{sw}}(\boldsymbol{q}) = -2\mu_{\text{B}}$, (6.9.20) gives us the well-known Bloch result. As we already noted in 6.6.2 and 6.6.4, however, $m_{\text{sw}}(\boldsymbol{q}) \neq -2\mu_{\text{B}}$.

Our new, non-perturbative result of (6.9.21) is different from (6.6.21). First, on $m_{\text{sw}}(\boldsymbol{q})_1$, we have $m_{\text{sw}}(\boldsymbol{q})_1 = -2\mu_{\text{B}}$ only when we replace χ_{hf} by $\chi_{\text{hf}}^{\text{S}}$

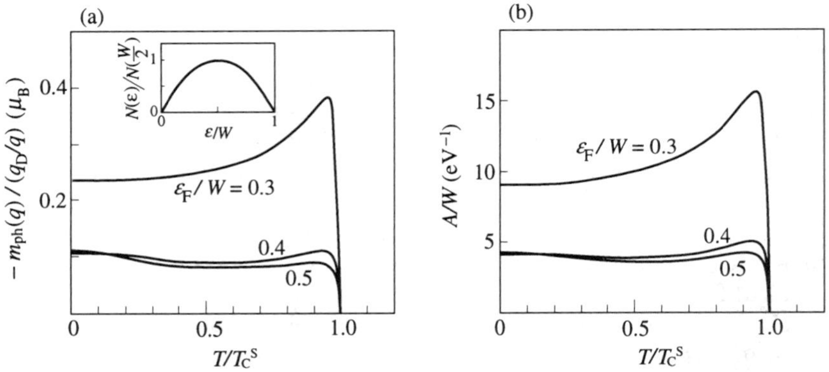

Fig.6.25 In (a), the value of phonon magnetization, $m_{ph}(q)$ and its temperature dependence are calculated from (6.9.23) for different locations of the Fermi energy in the spin unsplit state in the model electronic density of states given in the inset. We put W = 1 eV, $\tilde{V}N(\varepsilon_F) = \tilde{V}N(0) = 1.2$, $\theta_D(T{=}0) = 3.3\times10^2$ K, and $\xi = 1.2$. The corresponding behavior of the quantity A/W is shown in (b).

(or, vice versa) in (6.9.21). Here we note that χ_{hf}^S in (6.9.21) originates from our using of (6.6.13) in determining spin wave frequency. This finding suggests that, in a self-consistent treatment, spin wave frequency should be determined by including the effects of both spin waves and phonons; in such a treatment χ_{hf} would replace χ_{hf}^S in $m_{sw}(\boldsymbol{q})_1$, leading to $m_{sw}(\boldsymbol{q})_1 = -2\mu_B$. $m_{sw}(\boldsymbol{q})_2$ also becomes different from the perturbative one of (6.6.21). However, we anticipate the size of $m_{sw}(\boldsymbol{q})_2$ remains to be of the same order as there.

$m_{ph}(\boldsymbol{q})$ also is similarly modified from the perturbative result of (6.6.23)–(6.6.24) as

$$m_{ph}(\boldsymbol{q}) = -\chi_{hf}(M(T))\,\hbar\frac{\partial\omega_{ph}(\boldsymbol{q})}{\partial M}\bigg|_{M=M(T)} \tag{6.9.22}$$

$$= -A\,\frac{q}{q_D}\,\frac{\hbar\omega_D(M(T))}{W}\bigg|_{M=M(T)}\mu_B, \tag{6.9.23}$$

where the last result is for the Debye approximation. However, we again anticipate the size of $m_{ph}(\boldsymbol{q})$ remains to be the same order as we estimated in 6.6.3–6.6.4.

To confirm such conclusion we present here an example of non-perturbative numerical calculation of $m_{ph}(\boldsymbol{q})$. In this example we treat phonon effect in a self-consistent way as required, however, we entirely neglect the effect of spin waves. We use the model electronic density of states of Fig.5.5 with $W = 1$ eV. We set the values of various parameters as $\tilde{V} = 1.2$, $\theta_D(T = 0) = 300$ K, and $\xi = 1.2$. Then, with the Debye approximation expression of (6.9.23) we obtain the result of Fig.6.25 (a). We give also A/W, which can be measured independently, in Fig.6.25 (b). We find that $m_{ph}(\boldsymbol{q})$ is negative and its size can be a good fraction of 1 μ_B. Such a behavior of $m_{ph}(\boldsymbol{q})$ is similar to that of Fig.6.14(b) for the case of electrons with energy dispersion of $\varepsilon_{\boldsymbol{k}} = \hbar^2\boldsymbol{k}^2/2m$. More closely looking, we note that m_{ph} of Fig.6.24(a) is smaller than that of Fig.6.14(b). This reflects that A/W of Fig.6.25(b) is rather small; compare Fig.6.25(b) with Fig.5.18; although in both figures the same model electronic density of states is used, the values of various parameters are taken differently. Recall that A/W can be measured and that it can be as large as $\sim 10^2$ eV^{-1}.

6.10 EPI EFFECT ON THE HIGH FIELD MAGNETIC SUSCEPTIBILITY

Let us discuss the effect of EPI on the magnetic susceptibility in the ferromagnetic state, χ_{hf}, of a metal. Quite similarly as we derived (3.4.22) and (6.2.7), we have

$$\frac{1}{\chi_{hf}} = \frac{1}{4\mu_B^2}\left[\sum_\sigma \frac{1}{F_\sigma(0)} - 2\tilde{V} - 2J_{ph}(T)\right], \tag{6.10.1}$$

$$J_{ph}(T) = -2\left.\frac{\partial^2 F_{ph}(T, M)}{\partial \overline{M}^2}\right|_{M = M(T)}, \tag{6.10.2}$$

where $M(T)$ is the equilibrium magnetization at temperature T.

Here, if we use the Debye approximation for phonon frequencies as in 6.2 (see (6.2.9)), we have (see (6.2.13))

$$\frac{\partial^2 F_{ph}(T, M)}{\partial \overline{M}^2} = \frac{\partial F_{ph}}{\partial \omega_D}\frac{\partial^2 \omega_D(T, M)}{\partial \overline{M}^2} + \frac{\partial^2 F_{ph}}{\partial \omega_D^2}\left(\frac{\partial \omega_D(T, M)}{\partial \overline{M}}\right)^2$$

$$= N\hbar\, P\left(\frac{T}{\theta_D}\right)\frac{\partial^2 \omega_D}{\partial \overline{M}^2} - \frac{N\hbar}{\omega_D} Q\left(\frac{T}{\theta_D}\right)\left(\frac{\partial \omega_D}{\partial \overline{M}}\right)^2, \tag{6.10.3}$$

where $P(x)$ is defined in (6.2.16) and we set

$$Q\left(\frac{T}{\theta_D}\right) = -\frac{\omega_D}{N\hbar}\frac{\partial^2 F_{ph}(T, M)}{\partial \omega_D^2}$$

$$= 4\frac{T}{\theta_D}D\left(\frac{\theta_D}{T}\right) - \frac{3}{\exp(\theta_D/T) - 1} \tag{6.10.4}$$

$$\cong \begin{cases} \frac{4\pi^4}{5}\left(\frac{T}{\theta_D}\right)^4 & \text{for} \quad T << \theta_D, \\ \frac{T}{\theta_D} & \text{for} \quad T \geq \theta_D/3. \end{cases} \tag{6.10.5}$$

We illustrate the behavior of $Q(T/\theta_D)$ in Fig.6.26.

By setting $M = M(T)+\Delta M$, $s(T, M(T)+\Delta M) = s(T, M(T))+\Delta s(T, \Delta M)$ etc., we note that

$$\left.\frac{\partial \omega_D(T, M)}{\partial \overline{M}}\right|_{M = M(T)} = q_D s \frac{d}{d\Delta \overline{M}}\frac{\Delta s(\Delta \overline{M})}{s} = \omega_D \frac{d}{d\Delta \overline{M}}\frac{\Delta s(\Delta \overline{M})}{s},$$

$$\left.\frac{\partial^2 \omega_D(T, M)}{\partial \overline{M}^2}\right|_{M = M(T)} = \omega_D \frac{d^2}{d\left(\Delta \overline{M}\right)^2}\frac{\Delta s(\Delta \overline{M})}{s}, \tag{6.10.6}$$

where we abbreviated as $s(T, M(T)) = s$, $\omega_D(T, M(T)) = \omega_D$ etc. Here from (5.5.13)–(5.5.19) and Problem 6.1 we have

$$\frac{\Delta s(\Delta \overline{M})}{s} = -\frac{f_1}{W}J_1\Delta \overline{M} + \left(\frac{f_1}{W}J_2 + \frac{f_2}{W^2}J_1^2\right)\left(\Delta \overline{M}\right)^2 + \cdots, \tag{6.10.7}$$

where various notations are given in 5.5 and Problem 6.1. Thus, in (6.10.3) we have

$$\left.\frac{\partial \omega_D}{\partial \overline{M}}\right|_{M = M(T)} = -\omega_D \frac{f_1}{W}J_1\bigg|_{M = M(T)} = \omega_D Y, \tag{6.10.8}$$

$$\left.\frac{\partial^2 \omega_D}{\partial \overline{M}^2}\right|_{M = M(T)} = 2\omega_D\left[\frac{f_1}{W}J_1 + \frac{f_2}{W^2}J_1^2\right]_{M = M(T)}, \tag{6.10.9}$$

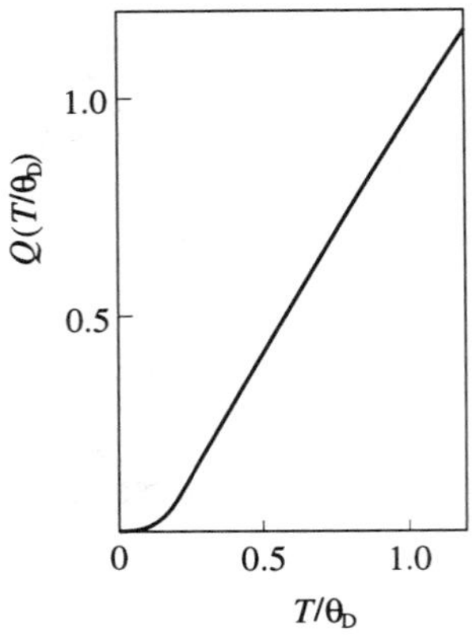

Fig.6.26 Temperature dependence of $Q(T/\theta_D)$ defined by (6.10.4).

where Y is given in (5.5.34). All the quantities appearing in (6.10.8)–(6.10.9) depend on T and M, as ω_D does.

Finally, from these we obtain the effective exchange interaction due to EPI in the ferromagnetic state as

$$J_{\rm ph}(T) = \left[-4N\hbar\omega_D \left(\frac{f_1}{W} J_2 + \frac{f_2}{W^2} J_1^2 \right) P(T/\theta_D) + 2N\hbar\omega_D \left(\frac{f_1}{W} J_1 \right)^2 Q(T/\theta_D) \right]_{M=M(T)} . \qquad (6.10.10)$$

With this in (6.10.1) we have the high field magnetic susceptibility including the effect of EPI. For the paramagnetic state, where $M(T) = 0$ and, therefore, $f_1 = 0$, (6.10.10) reduces to the previous result, (6.2.13)–(6.2.15).

We expect the effect of EPI to play an important role in determining the high field magnetic susceptibility. At higher temperatures, for instance, it will drive the temperature dependence of $\chi_{\rm hf}$ to the Curie-Weiss form through the functions $P(T/\theta_D)$ and $Q(T/\theta_D)$ in (6.10.10), similarly to the case of the paramagnetic susceptibility.

6.11 ISOTOPE EFFECT

If phonons are involved in magnetism of metals, there can be isotope effects in their magnetic properties. In this section we discuss where we may possibly observe such an isotope effect [3.23]. Rather contrary to the earlier and widespread view, we point out that the Curie temperature is not most revealing observable of the isotope effect; instead the magnetization at $T = 0$ of an itinerant electron ferromagnet, and the magnetic susceptibility at $T = 0$ of a strongly exchange enhanced paramagnetic metal may be better quantities for this purpose. We expect to observe an isotope effect also in the volume behavior of a ferromagnetic metal; we discuss it in the next chapter.

6.11.1 Isotope Effect in the Curie Temperature

Recall that as shown (6.2.20)–(6.2.22), the spin susceptibility of a metal containing the effect of EPI can have three characteristically different temperature dependence. Let us discuss the expected isotope effect in T_C for each of these three temperature regions.

A $T_C \geq \theta_D/3$; *the temperature region where (6.2.22) is valid*

Quite interestingly, although phonons may play a crucial role in determining T_C of a ferromagnetic metal, as given by (6.2.28), T_C does not depend on the isotopic mass of ions; both $(s_0/s)^2$ and L are independent of ionic mass M_I. Note that (6.2.28) is derived from (6.2.22) which is valid for $T \geq \theta_D/3$. Thus, we conclude

$$\frac{dT_C}{dM_I} = 0 \qquad \text{for} \quad T_C \geq \theta_D/3. \tag{6.11.1}$$

B $\theta_D \gg T_C > \theta_D(\theta_D/T_F)^{1/2}$; the temperature region where (6.2.21) is valid

If the magnetic susceptibility diverges within this temperature range, it is at

$$T_C = \theta_D\left(\frac{1}{|a|}\frac{W}{k_B\theta_D}\right)^{1/4}\left(\overline{V} - 1 - |b|\frac{k_B\theta_D}{W}\right)^{1/4}, \tag{6.11.2}$$

with

$$a = \frac{\pi^4}{5}\left(\frac{s_0}{s}\right)^2 L, \quad b = \frac{3}{8}\left(\frac{s_0}{s}\right)^2 L, \tag{6.11.3}$$

where we assumed $\overline{V} > 1 + |b|\, k_B\theta_D / W$ and $L < 0$.

Unlike the case of (6.2.28), T_C of (6.11.2) can have an isotope effect. If we assume

$$\theta_D \propto M_I^{-1/2}, \tag{6.11.4}$$

we have

$$\frac{M_I}{T_C}\frac{dT_C}{dM_I} = -\frac{3}{8} + \frac{15}{64\pi^4}\left(\frac{\theta_D}{T_C}\right)^4. \tag{6.11.5}$$

Letting $\theta_D \cong 200\text{K}$, $T_C \cong 30\text{K}$, corresponding to the case of $ZrZn_2$, we have $(M_I/T_C)(dT_C/dM_I) \cong 4$. According to the experiment of Knapp et al. [6.35], however, the size of the possible isotope effect in $ZrZn_2$ appears to be smaller than such a value by one order of magnitude.

Zverev and Silin also discussed the isotope effect of T_C in this temperature region [6.36,6.37]. However, as we will summarize at the end of this chapter, since they postulate that the phonon effect is absent at $T = 0$, their result corresponds to putting $b = 0$ in (6.11.2) and leads to $(M_I/T_C)\, dT_C/dM_I = -3/8$ in place of (6.11.5). Thus, in the case of $ZrZn_2$, their value for the isotope effect is different from ours in both sign and magnitude.

The Curie temperature of a ferromagnet is often determined from an extrapolation to lower temperatures of the inverse of the Curie-Weiss susceptibility. A Curie temperature determined in such a way will not show the isotope effect of (6.11.5) even if $T_C \leq \theta_D/3$. This might be the case with the observation on $ZrZn_2$. Note that in the temperature region where (6.2.21) and, therefore, (6.11.2) and (6.11.5) are valid, the temperature dependence of magnetic susceptibility is not Curie-Weiss like.

C $T_C \ll \theta_D(\theta_D/T_F)^{1/2}$; the temperature region where (6.2.20) is valid

If the Curie temperature is in this temperature range, it will be given by

$$T_C \cong \left[\left\{\widetilde{V} + J_{ph}(T=0)\right\} N(0) - 1\right]^{1/2} T_F. \tag{6.11.6}$$

It is this form of T_C that was used first by Hopfield [6.2], and later, by Fay and Appel [6.38] in discussing the isotope effect in T_C. From (6.11.6) we obtain the following result for the isotope effect

$$\frac{M_I}{T_C}\frac{dT_C}{dM_I} = \pm O\left[\left(\frac{\theta_D}{T_C}\right)\left(\frac{T_F}{T_C}\right)\right]. \tag{6.11.7}$$

The sign of the isotope effect is opposite to that of J_{ph}.

If we use (6.11.7) for $ZrZn_2$, with $\varepsilon_F \cong 1$ eV, $T_C \cong 30$K and $\theta_D \cong 200$K, we obtain, $|(M_I/T_C)\, dT_C/dM_I| \cong 10^3$; even if we assume $\varepsilon_F \cong 0.1$eV, we still have a large magnitude of $\sim 10^2$ for the isotope effect on T_C.

Note, however, that it is quite inappropriate to use (6.11.7) for the case of $ZrZn_2$; we already pointed out that even the result of (6.11.5) may not be used to analyze the case of $ZrZn_2$. For such a large isotope effect of (6.11.7) to be observed in an itinerant ferromagnet, it is required to have a much lower T_C than $ZrZn_2$. A possible candidate may be Sc_3In with $T_C \cong 6$K [6.36].

The experimental result of Knapp et al. [6.32] on the isotope effect of T_C in $ZrZn_2$ has long been viewed as convincingly denying the role of phonons in magnetism of metals. As we have shown in this chapter, such an earlier belief is entirely misleading. The role of phonons in magnetism of metals is quite different from, and much more important than, that previously thought.

6.11.2 Isotope Effect on χ of Non-ferromagnetic Metals at Low Temperatures

For a non-ferromagnetic metal, as $T \to 0$ the magnetic susceptibility of (6.2.19) or (6.2.20) reduces to

$$\chi = \frac{2\mu_B^2 N(0)}{(1-\bar{V}) - 3/8\,(s_0/s)^2\, L\hbar\omega_D / W}. \tag{6.11.8}$$

This can have an isotope effect through the quantity ω_D/W in the denominator:

$$\frac{M_I}{\chi}\frac{d\chi}{dM_I} = -\frac{\chi}{2\mu_B^2 N(0)}\left(\frac{s_0}{s}\right)^2 \frac{3}{16} L\frac{\hbar\omega_D}{W}. \tag{6.11.9}$$

The first factor on the right hand side, which is the exchange enhancement factor including the EPI effect, can have a magnitude of ~ 10. The magnitude of L also can be ~ 10, as we can see in the model calculation of Fig.6.6. Thus, with $\hbar\omega_D/W \cong 10^{-2}$, the magnitude of the isotope effect of (6.11.9) can be ~ 1.

Generally $|L|$ becomes large for $\bar{V} \to 1$. Thus, a good candidate to observe such an isotope effect is a paramagnetic metal with a strong exchange interaction ($\bar{V} \to 1$, but $\bar{V} < 1$).

6.11.3 Isotope Effect on Spontaneous Magnetization at Low Temperatures

We saw in 6.6–6.9 that the spontaneous magnetization of a ferromagnetic metal can be strongly affected by the EPI. Such an effect of the EPI can have an isotope effect at low temperatures, $T \ll \theta_D$.

The EPI effect on magnetization, $\Delta\overline{M}_{\rm ph}$ is obtained by (6.6.9) and (6.8.10),

$$\Delta\overline{M}_{\rm ph} = -A\,\frac{k_B\theta_D}{W}\,P\!\left(\frac{T}{\theta_D}\right). \tag{6.11.10}$$

For $T\to 0$ we have

$$\lim_{T\to 0}\Delta\overline{M}_{\rm ph} = -\frac{3}{8}A\,\frac{\hbar\omega_D}{W}\,. \tag{6.11.11}$$

Clearly this $\Delta M_{\rm ph}$ can have the following isotope effect:

$$\frac{M_I}{\Delta M_{\rm ph}}\frac{d\,\Delta M_{\rm ph}}{dM_I} = -\frac{1}{2}\,. \tag{6.11.12}$$

Note that this isotope effect is expected to disappear for higher temperatures of $T \geq \theta_D/3$.

6.12 REVIEW AND SUMMARY

It seems to be Herring [3.10] and Hopfield [6.2] who first discussed the effect of phonons on the magnetic susceptibility of a metal based on (6.2.1). However, they could not go beyond an order of magnitude estimation; even how the sign of the phonon effect is determined was not known; as to temperature dependence of the phonon effect, they did not notice its importance at all.

This subject concerning magnetic susceptibility, however, still continues to be very controversial. We will summarize and discuss those diverse current views later in 9.5.3. Here, however, we comment, in particular, on the recent work of Zverev and Silin [6.36,6.37] since their approach is very similar to that of ours presented in this chapter. While they agree with our assessment of the importance of the role of phonons, there are some differences between their approach and conclusions and our own. The most outstanding differences may be summarized as follows:

(1) Zverev and Silin postulate a phenomenological form of phonon free energy whose contribution vanishes at $T = 0$. Thus, according to them the EPI can not affect the magnetization at $T = 0$ of a ferromagnetic metal, contrary to our result. According to their view, at $T = 0$ we should have $H_{\rm ph} = 0$ in Fig.6.22. Similarly, according to them, phonons can not play any role in

determining the volume of a metal, irrespective of ferromagnetic or not, at $T = 0$, again contrary to our result to be presented in Chapter 7. In this respect, note that our treatment of the role of phonons is based on a free energy of an interacting electron-phonon system which is microscopically derived as presented in Chapter 10.

(2) The phonon contribution to the free energy of a metal is given in terms of phonon frequencies, and the phonon frequencies are determined by the screening constant of the ion-ion interaction, as we have been seeing. Then, the screening constant for the spin split state of electrons which Zverev and Silin use, either equation (26) or (27) of Ref. [6.37], is fundamentally different from that of ours. They attribute the origin of this difference to the difference in that which is kept constant, the magnetization or the magnetic induction, in calculating the electron response; they claim we erroneously kept the magnetic inductance constant, in place of magnetization. Their claim, however, is unfounded as can be easily checked as follows. In our discussion, the magnetization dependence of the phonon frequency originates from the spin splitting of the electron energy bands. We, then, treated the spin splitting of the bands in a way that would correctly reproduce the Stoner model result. We, thus, think the source of the difference between our and their results lies elsewhere. It might be instead due to the difference between the methods used to calculate phonon frequency. We will present some discussions relevant to this point in Chapter 7.

As we have seen in this chapter, the effect of phonons on the magnetic properties of a metal does not come from either their effect on temperature or volume, contrary to common expectation. Our discussion of this chapter on the role of phonons was carried out for given, fixed temperature and volume.

A very plausible argument against the effect of phonons may be, " Phonon does not have spin (or, magnetization). How can such a phonon have any magnetic effect ? " [6.39]. As we saw in 6.6–6.7, a phonon indeed does have magnetization! And, what is more fundamental than having magnetization is whether the excitation energy of phonon depends on the magnetization of a system. As we saw in this chapter, the origin of a spin wave's having effect on magnetization also lies there.

PROBLEMS FOR CHAPTER 6

6.1 Confirm the following derivation of the relation between an induced spin splitting, 2η, of the electron energy bands of a metal, either in the paramagnetic or ferromagnetic state, and the corresponding change in magnetization, ΔM or $\Delta\overline{M}$. If we note (5.5.6) and (5.5.7), we obtain, in the same way as we derived (5.5.11),

$$\Delta M/\mu_B = -\Delta\overline{M}$$

$$= \int d\varepsilon N(\varepsilon)\left[f_-(\varepsilon;\eta) - f_+(\varepsilon;\eta)\right] - \int d\varepsilon N(\varepsilon)\left[f_-(\varepsilon) - f_+(\varepsilon)\right]$$

$$\equiv I_1\eta + I_2\eta^2 + \cdots, \quad (6P.1)$$

$$I_1 = 4F_+(0)F_-(0)/P(0), \quad (6P.2)$$

$$I_2 = \frac{4}{P(0)^3}\left[F'_+(0)F_-(0)^3 - F'_-(0)F_+(0)^3\right]. \quad (6P.3)$$

Next, to derive the inverse relation between η and ΔM, or $\Delta\overline{M}$. If we set

$$\eta = J_1(\Delta M/\mu_B) + J_2(\Delta M/\mu_B)^2 + \cdots = -J_1\Delta\overline{M} + J_2\Delta\overline{M}^2 + \cdots, \quad (6P.4)$$

and put it into (6P.1) we obtain

$$J_1 = 1/I_1, \quad (6P.5)$$

$$J_2 = -J_1 I_2/I_1^2 = -I_2/I_1^3. \quad (6P.6)$$

In the paramagnetic state we may write $\overline{M}$ for $\Delta\overline{M}$, as in (6.1.3).

6.2 Relate the quantity $(\widetilde{V}N(0) - 1)$ in (6.1.8) to the Stoner Curie temperature given in (3.3.18). Then, what conclusion will we have in place of 6.1.13)?

6.3 Derive the phonon free energy in the Debye approximation of (6.2.9).

6.4 Confirm the Debye approximation result for the effective exchange interaction due to EPI given in (6.2.14)–(6.2.16).

6.5 Confirm the property of (6.2.18)–(6.2.19) on the function $P(T/\theta_D)$ defined in (6.2.16).

Chapter 7

THE VOLUME DEGREE OF FREEDOM IN ITINERANT ELECTRON FERROMAGNETISM

In Chapter 5 we discussed the elastic property of a metal. There we noted that the result (5.4.27) succeeds in qualitatively explaining the various aspects of the elastic behavior of ferromagnetic metals. If we look into the details of experimental results, however, we find that our result in Chapter 5 is not satisfactory. See, for instance, Figs.5.2 (a) and 5.2(b): as temperature lowers toward T_C from above, the elastic constants first increase before decreasing. Such a behavior can not be accounted for by (5.4.27). How can we, then, improve such a result of Chapter 5? In this chapter we show that the goal is achieved by properly taking into account the contributions of thermal phonons to elastic constants; the situation is exactly the same as in Chapter 6 of going beyond the Stoner model in studying itinerant electron magnetism.

The most outstanding problem concerning the elastic properties of an itinerant ferromagnet may be the origin of the Invar behavior. Although various mechanisms have been proposed most of them are based on some ad hoc phenomenological hypotheses. The fundamental mechanism of the Invar behavior is not yet understood. In this chapter we try to understand the mechanism of Invar behavior based only on first principles of physics. We will find the role of thermal phonons to be very important in producing a negative thermal expansion coefficient.

In order to discuss the relation between the elastic and magnetic properties, we have to treat the volume and magnetization variables on the same footing. From such a procedure we also find that the consideration of the interaction between magnetization and volume, the magneto-volume interaction (MVI), is vitally important in understanding the magnetic properties themselves. The high field magnetic susceptibility is enhanced generally by the effect of MVI by $O(1)$, for instance.

7.1 THE INTERACTION BETWEEN MAGNETIZATION AND VOLUME (MVI) IN THE ELASTIC AND MAGNETIC BEHAVIOR OF A FERROMAGNET

Up to this chapter, the volume of a system was always a given, fixed constant. In discussing the elasticity and volume properties, however, it is inevitable to treat volume as a variable of the system. In this section we develop a general theory of various magnetic and elastic responses of a ferromagnet by treating the volume and magnetization variables on the same footing. We systematically derive thermodynamic expressions for those responses, such as the thermal expansion coefficient of a ferromagnet in terms of the various derivatives with respect to magnetization and volume of its free energy. Those thermodynamic expressions are not entirely new, but, perhaps due to lack of a transparent derivation, they are not well understood. Very often, inadequate expressions are used for the thermal expansion coefficient in the ferromagnetic state, for instance.

7.1.1 The Volume Degree of Freedom in Itinerant Electron Magnetism

Suppose that a system is in an equilibrium state at a temperature T, without any external magnetic field ($H = 0$), with volume, V_0, and magnetization, M_0. This implies that the free energy of the system, $F(T, V, M)$, takes its minimum at $V = V_0$ and $M = M_0$. Then we require

$$F_V = 0, \quad F_M = 0, \tag{7.1.1}$$

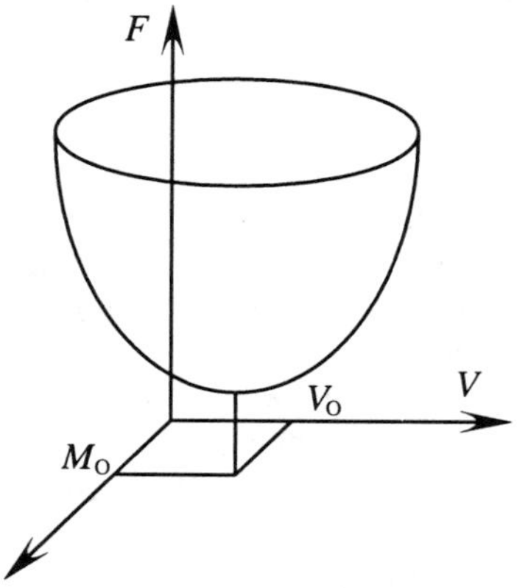

Fig.7.1 Schematic view of the free energy and the equilibrium values of volume V_0 and magnetization M_0.

together with

$$F_{VV} > 0, \quad F_{MM} > 0, \quad F_{VV} F_{MM} - F_{VM}^2 > 0\,, \tag{7.1.2}$$

where we put

$$\left.\frac{\partial F(T, V, M)}{\partial V}\right|_{V = V_o, M = M_o} = F_V \quad \text{etc.} \tag{7.1.3}$$

The above is the only principle on which we base our entire discussion in this section. This is an extension of our earlier Landau procedure, (3.4.1), so as to include the volume degree of freedom.

7.1.2 Response to a Magnetic Field

We now apply a magnetic field H in the direction of z-axis on this system. Then the equilibrium volume and magnetization would change to, say, $V_o + \Delta V_o$ and $M_o + \Delta M_o$. The condition to determine ΔM_o and ΔV_o is given by

$$\left.\frac{\partial F(T, V_o + \Delta V, M_o + \Delta M; H)}{\partial \Delta V}\right|_{\Delta V = \Delta V_o, \Delta M = \Delta M_o} = 0, \tag{7.1.4}$$

$$\left.\frac{\partial F(T, V_o + \Delta V, M_o + \Delta M; H)}{\partial \Delta M}\right|_{\Delta V = \Delta V_o, \Delta M = \Delta M_o} = 0, \tag{7.1.5}$$

where

$$F(T, V, M; H) = F(T, V, M) - HM \tag{7.1.6}$$

is the free energy of the system under the magnetic field. We also understand the additional condition corresponding to (7.1.2). Our task is to obtain ΔV_o and ΔM_o from (7.1.4) and (7.1.5).

In dealing with (7.1.4) and (7.1.5), by assuming that H and, therefore, both ΔV and ΔM are small, we make the following expansion,

$$\begin{aligned} F(T, V_o + \Delta V, M_o + \Delta M; H) &= F(T, V_o + \Delta V, M_o + \Delta M) - (M_o + \Delta M)H \\ &= F(T, V_o, M_o) + \frac{1}{2!} F_{VV} (\Delta V)^2 + F_{VM}\, \Delta V \Delta M \\ &\quad + \frac{1}{2!} F_{MM} (\Delta M)^2 - (M_o + \Delta M)H, \end{aligned} \tag{7.1.7}$$

where we noted (7.1.1). By putting (7.1.7) into (7.1.4) and (7.1.5) we obtain

$$F_{VV}\,\Delta V_o + F_{MV}\,\Delta M_o = 0, \tag{7.1.8}$$

$$F_{MV}\,\Delta V_o + F_{MM}\,\Delta M_o - H = 0, \tag{7.1.9}$$

and then,

$$\Delta M_o = \frac{1}{F_{MM} - F_{MV}^2 / F_{VV}} H, \tag{7.1.10}$$

$$\Delta V_o = -\frac{F_{MV} / F_{VV}}{F_{MM} - F_{MV}^2 / F_{VV}} H. \tag{7.1.11}$$

A. *Magnetic Susceptibility*

From (7.1.10) we obtain the magnetic susceptibility as,

$$\chi = \frac{1}{F_{MM} - F_{MV}^2 / F_{VV}}. \tag{7.1.12}$$

If we do not consider the volume degree of freedom, (7.1.12) reduces to the familiar expression,

$$\chi = \frac{1}{F_{MM}}. \tag{3.4.4}$$

We find, however, that

$$F_{MV} = 0 \quad \text{for} \quad T > T_C. \tag{7.1.13}$$

This comes from the premise that for $T > T_C$ since $F(T, V, M)$ must be symmetric with respect to M it can not have a term linear in M. Then, even if we consider the volume degree of freedom, for $T > T_C$ the magnetic susceptibility is given by (3.4.4). Thus the susceptibility given in (7.1.12) with $F_{MV}^2/F_{VV} \neq 0$ is exclusively for $T < T_C$.

Since $F_{VV} > 0$ (see (7.1.2)) the volume related effect always enhances χ. In a metallic ferromagnet, since $O(F_{MM}) = O(\varepsilon_F / n\mu_B^2)$, $O(F_{VV}) = O(n\varepsilon_F / V_o^2)$, and $O(|F_{MV}|) = O(\varepsilon_F / \mu_B V_o)$, ε_F and n being, respectively, the electron Fermi energy and the total number of electrons, F_{MM} and F_{MV}^2/F_{VV} are generally of the same order of magnitude. Thus this enhancement effect can be quite important. Note that up until here we used (3.4.4) even for the ferromagnetic state.

B. *Change in Volume Caused by a Magnetic Field: Magnetic Expansivity*

We may rewrite (7.1.11) as

$$\frac{\Delta V_{\text{o}}}{V_{\text{o}}} = -\frac{\chi}{V_{\text{o}}}\frac{F_{MV}}{F_{VV}}H = \kappa_{\text{m}}H, \qquad (7.1.14)$$

by noting (7.1.12). Let us call κ_{m} the *magnetic expansivity*. $\kappa_{\text{m}} = 0$ for $T > T_{\text{C}}$, because of (7.1.13).

Depending upon the sign of F_{MV}, volume of a ferromagnet either expands or shrinks by the effect of a magnetic field. Such effect has long been known, as the *forced magnetic striction,* but it was seldom endeavored to calculate microscopically the sign and magnitude of the effect. In the case of an itinerant ferromagnet, (7.1.14) leads to the following order of magnitude estimation: $O(|\Delta V_{\text{o}}/V_{\text{o}}|) = O(\mu_{\text{B}}H/\varepsilon_{\text{F}})$.

7.1.3 Response to Pressure

We again start from the state which satisfies the equilibrium condition, (7.1.1)–(7.1.2). Without an external pressure, the system has volume, V_{o}, and magnetization, M_{o}. When we apply an external pressure, p, on the system, the volume and magnetization would change, to $V_{\text{o}}+\Delta V_{\text{o}}$ and $M_{\text{o}}+\Delta M_{\text{o}}$, respectively. The procedure to obtain ΔV_{o} and ΔM_{o} is quite similar to the previous case of a magnetic field.

The free energy under the pressure is given by

$$F(T, V_{\text{o}}+\Delta V, M_{\text{o}}+\Delta M; p) = F(T, V_{\text{o}}+\Delta V, M_{\text{o}}+\Delta M) - p(V_{\text{o}}+\Delta V), \quad (7.1.15)$$

corresponding to (7.1.6) for the magnetic field case; we took the positive sign of the pressure in the outward direction from the system. Then, from the equilibrium condition similar to (7.1.4) and (7.1.5), we obtain

$$F_{VV}\,\Delta V_{\text{o}} + F_{MV}\,\Delta M_{\text{o}} - p = 0,$$

$$F_{MV}\,\Delta V_{\text{o}} + F_{MM}\,\Delta M_{\text{o}} = 0. \qquad (7.1.16)$$

These equations give

$$\frac{\Delta M_{\text{o}}}{\Delta V_{\text{o}}} = -\frac{F_{MV}}{F_{MM}}, \qquad (7.1.17)$$

$$\Delta V_o = \frac{p}{F_{VV} - F_{MV}^2 / F_{MM}} . \qquad (7.1.18)$$

C. *Compressibility and Bulk Modulus*

The compressibility, κ, and bulk modulus, B, are defined as

$$\frac{\Delta V_o}{V_o} = \kappa p = \frac{1}{B} p. \qquad (7.1.19)$$

Then from (7.1.18) we have

$$B = \frac{1}{\kappa} = V_o \left(F_{VV} - \frac{F_{MV}^2}{F_{MM}} \right) . \qquad (7.1.20)$$

Again, because of (7.1.13), we have

$$B = V_o F_{VV} \quad \text{for} \quad T > T_C . \qquad (7.1.21)$$

The expression of (7.1.21) is often used even in the ferromagnetic state. However, the sizes of the two terms in (7.1.20) are estimated to be of the same order for a metallic ferromagnet. Note also that the effect of this large correction term is always to reduce the bulk modulus.

D. *Change in Magnetization Caused by External Pressure: Magnetic Susceptibility to Pressure*

The result of (7.1.17) gives us the change in magnetization caused by external pressure. We may rewrite the result as

$$\frac{dM}{dV} = \frac{B}{V} \frac{dM}{dp} = - \frac{F_{MV}}{F_{MM}} . \qquad (7.1.17')$$

Let us define the *magnetic susceptibility to pressure*,

$$\chi_p = \frac{dM}{dp} = - V\kappa \frac{F_{MV}}{F_{MM}} . \qquad (7.1.22)$$

Note that the signs of χ_p and κ_m are the same.

7.1.4 Response to Temperature Variation

We start again from the state satisfying the equilibrium condition of (7.1.1)–(7.1.2). We now raise the temperature from T to $T+\Delta T$. The volume and magnetization, then, would change, respectively, to $V_o+\Delta V_o$ and $M_o+\Delta M_o$. The condition to determine ΔV_o and ΔM_o is given by

$$\frac{\partial}{\partial \Delta V} F(T+\Delta T, V_o+\Delta V, M_o+\Delta M)\bigg|_{\Delta V=\Delta V_o,\ \Delta M=\Delta M_o} = 0, \quad (7.1.23)$$

$$\frac{\partial}{\partial \Delta M} F(T+\Delta T, V_o+\Delta V, M_o+\Delta M)\bigg|_{\Delta V=\Delta V_o,\ \Delta M=\Delta M_o} = 0. \quad (7.1.24)$$

Here, by assuming that ΔT and, therefore, ΔV and ΔM are small, we expand the free energy as follows,

$$\begin{aligned} F(T+\Delta T, V_o+\Delta V, M_o+\Delta M) &= F(T, V_o, M_o) \\ &+\frac{1}{2!} F_{VV}(\Delta V)^2 + \frac{1}{2!} F_{MM}(\Delta M)^2 + F_{VM}\,\Delta V\Delta M \\ &+ F_T\,\Delta T + F_{VT}\,\Delta V\Delta T + F_{MT}\,\Delta M\Delta T + \cdots, \end{aligned} \quad (7.1.25)$$

where various differential coefficients of the free energy are defined similarly to (7.1.3) and we noted (7.1.1). Then, by putting (7.1.25) to (7.1.23) and (7.1.24) we obtain

$$F_{VV}\frac{\Delta V_o}{\Delta T} + F_{VM}\frac{\Delta M_o}{\Delta T} + F_{VT} = 0, \quad (7.1.26)$$

$$F_{VM}\frac{\Delta V_o}{\Delta T} + F_{MM}\frac{\Delta M_o}{\Delta T} + F_{MT} = 0. \quad (7.1.27)$$

These equations give

$$\frac{\Delta V_o}{\Delta T} = -\frac{F_{VT}F_{MM} - F_{MT}F_{VM}}{F_{VV}F_{MM} - F_{VM}^2}, \quad (7.1.28)$$

$$\frac{\Delta M_o}{\Delta T} = -\frac{F_{VV}F_{MT} - F_{VT}F_{VM}}{F_{VV}F_{MM} - F_{VM}^2}. \quad (7.1.29)$$

E. *Thermal Expansion Coefficient*

The thermal expansion coefficient, β, is found from (7.1.28) as

$$\beta = \frac{1}{V_o}\frac{\Delta V_o}{\Delta T} = -\frac{1}{V_o}\frac{F_{VT}F_{MM} - F_{MT}F_{VM}}{F_{VV}F_{MM} - F_{VM}^2}. \qquad (7.1.30)$$

If we note (7.1.12), (7.1.14), and (7.1.20), this result can be rewritten as

$$\beta = \kappa\frac{\partial p_{\mathrm{I}}}{\partial T} + \kappa_{\mathrm{m}}\frac{\partial H_{\mathrm{I}}}{\partial T}, \qquad (7.1.31)$$

where we put

$$-F_{VT} = \frac{\partial}{\partial T}(-F_V) \equiv \frac{\partial p_{\mathrm{I}}}{\partial T}; \qquad -F_{MT} = \frac{\partial}{\partial T}(-F_M) \equiv \frac{\partial H_{\mathrm{I}}}{\partial T}. \qquad (7.1.32)$$

p_{I} and H_{I} are understood, respectively, as the *internal pressure* and *magnetic field* (See(6.6.8)); here we put the subscript I to distinguish them from external pressure and magnetic field. Note that at the equilibrium we have $p_{\mathrm{I}} = H_{\mathrm{I}} = 0$, but $\partial p_{\mathrm{I}}/\partial T \neq 0$, and $\partial H_{\mathrm{I}}/\partial T \neq 0$.

In the paramagnetic state, due to (7.1.13), we have the well-known textbook result,

$$\beta = \kappa\frac{\partial p_{\mathrm{I}}}{\partial T}. \qquad (7.1.33)$$

In the ferromagnetic state, the thermal expansion coefficient becomes quite different as given in (7.1.30) or (7.1.31).

The result of (7.1.31) is physically of a very transparent structure; the first and second terms on the right hand side represent, respectively, the effects of the temperature dependence of the internal pressure and the magnetic field.

A very important point to note in (7.1.31) is that the magnetic effect on the thermal expansion is not represented by the second term only; the magnetic effect also enters the first term through both κ and $\partial p_{\mathrm{I}}/\partial T$, as can be seen from (7.1.20) and (7.1.32); since F depends on M, so does F_{VT}.

F. *Temperature Dependence of Magnetization*

The result of (7.1.29) can be rewritten as

$$\frac{dM}{dT} = \chi \frac{\partial H_{\rm I}}{\partial T} + \chi_{\rm p} \frac{\partial p_{\rm I}}{\partial T}. \qquad (7.1.34)$$

This result again is of a very plausible physical structure. The second term on the right hand side represents the effect of volume change on magnetization; note, however, that out of the two kinds of contributions to volume change given in (7.1.31), it is only the first one that enters here. Unless having based our discussion only on the very first principles, we could not have confidently excluded the other contribution.

Note that with only the first term on the right hand side in (7.1.34), we reproduce the result of (6.9.5). Recall how rich consequences even such (6.9.5) produced. With (7.1.34), the role of spin waves, as well as phonons, would be further changed from the textbook result.

In summary, in this section we showed how to calculate various magnetic and elastic responses of a ferromagnet via its free energy by correctly including the interaction between the magnetic and elastic properties. We systematically derived the expressions for the various responses in terms of differential coefficients of the free energy. Prepared with these results, we discuss the effects of spin waves and thermal phonons on those various responses.

7.2 THE RELATION BETWEEN THE DYNAMICAL AND TOTAL ENERGY CALCULATIONS OF ELASTIC CONSTANTS

The bulk modulus, B, which is to be calculated from (7.1.20) is related to the velocity, s, of the longitudinal acoustic sound as

$$s^2 = B / (NM_{\rm I}/V), \qquad (7.2.1)$$

where N and $M_{\rm I}$ are, respectively, the total number of ions and the mass of an ion. In Chapter 5, however, we did not use the procedure of (7.1.20), which we call the *total energy approach*, in calculating the elastic constant. There we calculated the phonon frequency $\omega_{\boldsymbol{q}}$ (= $\omega_{\rm ph}(\boldsymbol{q})$) by pursuing the dynamics of an interacting electron-ion system, and then we took the limit of (5.4.16) to obtain the sound velocity s; we call this procedure of calculating the sound velocity or the elastic constant the *dynamical approach*.

Some quantities, such as thermal expansion coefficient, can be calculated only by the total energy approach. However, bulk modulus and magnetic susceptibility can be calculated by both approaches. Then, a question arises. Would two approaches give the same result?

As regards the magnetic susceptibility within the mean field approximation these two approaches give the same result; see 3.4.2 and 4.3.2. The equivalence of the two approaches, however, does not always hold. We anticipate that they will give the same result if calculations are done exactly. An actual calculation, however, always involves approximations, and, accordingly, two approaches often lead to quite different results even if the approximations used are of similar nature; see, for instance, Refs. [7.1–7.3]. This seems to be the case with the calculation of elastic constants. Recently, many elaborate calculations of elastic constants have been done on ferromagnetic transition metals with the total energy approach, such as that of Moruzzi et al. [7.4]. The results of such calculations, however, are thought not to be directly related to our general result of (5.4.27) or (5.4.32) which is derived by the dynamical approach.

Quite pleasingly, however, we recently found that the results of (5.4.27) and (5.4.32) can be also derived by the total energy approach [7.5], although under some restrictions. With this finding, it becomes clear how to improve the result on the elastic constants of Chapter 5: Include in the free energy to be used in (7.1.20) the contribution from phonons, $F_{\rm ph}$, and spin waves, $F_{\rm sw}$. We carry out such a procedure in later sections.

In this section we demonstrate that the total energy approach can indeed reproduce the result of (5.4.27). In the course of the demonstration we also learn how to deal with the volume variable.

According to the result of the preceding section, the bulk modulus is calculated from a free energy as

$$B = V\left(F_{VV} - \frac{F_{MV}^2}{F_{MM}}\right). \tag{7.1.20}$$

Here, for the free energy we use

$$\begin{aligned} F_{\rm m}(T, M, V) &= F_{\rm el,\,m}(T, M, V) + E_\xi \\ &= \Omega_0 - \frac{1}{2}\widetilde{V}\sum_\sigma n_\sigma^2 + \sum_\sigma n_\sigma \mu_\sigma + E_\xi\,, \end{aligned} \tag{7.2.2}$$

the same one as we used in 3.4.2 except the last term, E_ξ. Ω_0 is the thermodynamic potential of free electrons given in (3.4.5). E_ξ, which may be written as

$$E_\xi = \frac{1}{2}\sum_{i \neq j} W_\xi(\boldsymbol{R}_i - \boldsymbol{R}_j), \tag{7.2.3}$$

and represents the interaction energy between ions outside that of jellium. It is the origin of the ξ defined in (5.3.9). Recall that in the pure jellium model with the uniform distribution of both the negative electron charge and the positive ion charge, the three kinds of direct Coulomb interaction energies, between electron and electron, ion and ion, and electron and ion, cancel each other and, therefore, do not appear in (7.2.2).

Of the three differential coefficients required in (7.1.20), F_{MM}, or, $F_{\mathrm{m},MM}$ is already given in (3.4.22); it is the inverse of the Stoner high field magnetic susceptibility. Note that E_ξ is independent of magnetization.

In obtaining $F_{\mathrm{m},MV}$, we use (3.4.14) for $F_{\mathrm{m},M}$. Then we have

$$F_{\mathrm{m},MV} = \sum_\sigma \frac{\sigma}{2}\left(\frac{\partial \mu_\sigma}{\partial V} - \frac{\partial \widetilde{V}}{\partial V} n_\sigma\right). \tag{7.2.4}$$

Here, for the volume dependence of the effective exchange interaction, we put

$$\frac{\partial}{\partial V}\widetilde{V} = -\frac{1}{V}\widetilde{V}, \tag{7.2.5}$$

by the following reason. We understand $\widetilde{V}(\boldsymbol{q})$ to be the Fourier transform of a distance dependent effective exchange interaction, $\widetilde{V}(\boldsymbol{r})$:

$$\widetilde{V}(\boldsymbol{q}) = \frac{1}{V}\int_V \widetilde{V}(\boldsymbol{r})\, e^{-i\boldsymbol{q}\boldsymbol{r}}\, d\boldsymbol{r}. \tag{7.2.6}$$

Suppose, for simplicity, that $\widetilde{V}(\boldsymbol{r})$ is short-ranged and its property is independent of volume, and, therefore, the integral does not depend on V. Then, we have $\widetilde{V}(\boldsymbol{q}) \propto 1/V$, and

$$\frac{d}{dV}\widetilde{V}(\boldsymbol{q}) = -\frac{1}{V}\widetilde{V}(\boldsymbol{q}); \qquad \frac{d^2}{dV^2}\widetilde{V}(\boldsymbol{q}) = \frac{2}{V^2}\widetilde{V}(\boldsymbol{q}). \tag{7.2.7}$$

Next, the volume dependence of chemical potential is obtained by requiring

$$\frac{\partial n_\sigma}{\partial V} = \frac{\partial}{\partial V}\int d\varepsilon\, N(\varepsilon) f_\sigma(\varepsilon) = 0; \tag{7.2.8}$$

(see (3.4.12)–(3.4.13)). Then if we note that $N(\varepsilon) \propto V$ for the free electron energy dispersion (see (1.3.5)), we have

$$\frac{n_\sigma}{V} + F_\sigma(0)\frac{\partial \mu_\sigma}{\partial V} = 0, \tag{7.2.9}$$

or

$$\frac{\partial \mu_\sigma}{\partial V} = -\frac{n_\sigma}{VF_\sigma(0)}, \tag{7.2.9'}$$

where $F_\sigma(0)$ is defined in (4.3.18). With (7.2.5) and (7.2.9) in (7.2.4) we have

$$F_{\mathrm{m},MV} = -\frac{1}{2V}\sum_\sigma \sigma\, n_\sigma \frac{1 - \widetilde{V}F_\sigma(0)}{F_\sigma(0)}. \tag{7.2.10}$$

Let us proceed to obtain $F_{\mathrm{m},VV}$. First we have

$$F_{\mathrm{m},V} = -k_{\mathrm{B}}T\sum_\sigma \int d\varepsilon \frac{\partial N(\varepsilon)}{\partial V}\ln\left[1 + e^{-\beta(\varepsilon-\mu_\sigma)}\right]$$

$$-k_{\mathrm{B}}T\sum_\sigma \int d\varepsilon\, N(\varepsilon)\frac{e^{-\beta(\varepsilon-\mu_\sigma)}}{1 + e^{-\beta(\varepsilon-\mu_\sigma)}}\left(\beta\frac{\partial \mu_\sigma}{\partial V}\right)$$

$$+\sum_\sigma n_\sigma \frac{\partial \mu_\sigma}{\partial V} - \frac{1}{2}\frac{\partial \widetilde{V}}{\partial V}(n_+^2 + n_-^2) + \frac{\partial E_\xi}{\partial V}. \tag{7.2.11}$$

If we note that the second and third terms on the right hand side cancel each other, and use (7.2.5) we have

$$F_{\mathrm{m},V} = -\frac{1}{V}k_{\mathrm{B}}T\sum_\sigma \int d\varepsilon\, N(\varepsilon)\ln\left[1 + e^{-\beta(\varepsilon-\mu_\sigma)}\right]$$

$$+\frac{\widetilde{V}}{2V}(n_+^2 + n_-^2) + \frac{\partial E_\xi}{\partial V}. \tag{7.2.12}$$

By differentiating again with respect to V we have

$$F_{\mathrm{m},VV} = -\frac{1}{V}\sum_\sigma\left(-\frac{n_\sigma}{VF_\sigma(0)}\right)n_\sigma - \frac{\widetilde{V}}{V^2}\sum_\sigma n_\sigma^2 + \frac{\partial^2 E_\xi}{\partial V^2}$$

$$= \frac{1}{V^2}\sum_\sigma \frac{n_\sigma^2}{\widetilde{F}_\sigma(0)} + \frac{\partial^2 E_\xi}{\partial V^2}. \tag{7.2.13}$$

Thus, with (7.2.10), (7.2.13) and (3.4.22) in (7.1.20) we obtain

$$B = \frac{1}{V}\frac{n^2}{\widetilde{F}_+(0)\,\widetilde{F}_-(0)} \Big/ \left[\frac{1}{\widetilde{F}_+(0)} + \frac{1}{\widetilde{F}_-(0)}\right] + V\frac{\partial^2 E_\xi}{\partial V^2}$$

$$= \frac{1}{V}\frac{n^2}{\sum_\sigma \widetilde{F}_\sigma(0)} + V\frac{\partial^2 E_\xi}{\partial V^2}, \tag{7.2.14}$$

where $\widetilde{F}(0)$ is defined in (5.4.29). If we note (7.2.1) and (5.4.20), (7.2.14) is rewritten as

$$\left(\frac{s}{s_0}\right)^2 = B\left(\frac{V}{NM_\mathrm{I}}\right)\left(\frac{3M_\mathrm{I}}{mZ\mathrm{v}_\mathrm{F}^2}\right) = B\frac{3V}{2n\varepsilon_\mathrm{F}}$$

$$= \frac{2N(0)}{\sum_\sigma \widetilde{F}_\sigma(0)} + \frac{3V^2}{2n\varepsilon_\mathrm{F}}\frac{\partial^2 E_\xi}{\partial V^2}. \tag{7.2.14'}$$

Then, if we can put

$$\frac{3V^2}{2n\varepsilon_\mathrm{F}}\frac{\partial^2 E_\xi}{\partial V^2} = \xi, \tag{7.2.15}$$

(7.2.14'), which is the result of the total energy approach with the free energy of (7.2.2), coincides with (5.4.27), the dynamical approach result. Thus, it remains only to show the validity of (7.2.15).

Let us review how ξ comes about in the dynamical approach. In the general discussion on lattice vibration given in 5.2.1 we may divide the inter-ionic interaction potential as

$$W(\boldsymbol{R}_i - \boldsymbol{R}_j) = W_\mathrm{C}(\boldsymbol{R}_i - \boldsymbol{R}_j) + W_\xi(\boldsymbol{R}_i - \boldsymbol{R}_j), \tag{7.2.16}$$

where the first term $W_\mathrm{C}(\boldsymbol{R}_i - \boldsymbol{R}_j)$ is the screened Coulombic ion-ion interaction and the second term $W_\xi(\boldsymbol{R}_i - \boldsymbol{R}_j)$ is the interaction between ions outside such Coulombic one, such as the direct repulsion between ion cores. Then, for the present case of jellium ions, for which lattice structure is neglected, the longitudinal acoustic phonon frequency is given from (5.2.29) as

$$\omega_{\rm ph}(\boldsymbol{q})^2 = (N/M_{\rm I})\left[W_{\rm C}(\boldsymbol{q}) + W_\xi(\boldsymbol{q})\right] q^2, \tag{7.2.17}$$

where $W_{\rm C}(\boldsymbol{q})$ and $W_\xi(\boldsymbol{q})$ are the Fourier transformations of $W_{\rm C}(\boldsymbol{R}_i - \boldsymbol{R}_j)$ and $W_\xi(\boldsymbol{R}_i - \boldsymbol{R}_j)$ is defined similarly to (5.2.14). Unlike in Chapter 5, we use $\omega_{\rm ph}(\boldsymbol{q})$ for phonon frequency. We can rewrite (7.2.17) as

$$\omega_{\rm ph}(\boldsymbol{q})^2 = \underline{s}^2 q^2 + (N/M_{\rm I})\, W_\xi(\boldsymbol{q})\, q^2, \tag{7.2.17'}$$

with the screened sound velocity $\underline{s}$ of the pure jellium model; in the pure jellium model we have $W_\xi = 0$. If we put $\omega_{\rm ph}(\boldsymbol{q}) = sq$ in (7.2.17') for small q, we have

$$s^2 = \underline{s}^2 + (N/M_{\rm I})\, W_\xi(\boldsymbol{q}). \tag{7.2.18}$$

Since $W_\xi(\boldsymbol{R}_i - \boldsymbol{R}_j)$ is considered to be short-ranged, we may assume that $W_\xi(\boldsymbol{q})$ does not sensitively depend on $\boldsymbol{q}$: $W_\xi(\boldsymbol{q}) \cong W_\xi$. Then, if we recall that the Bohm-Staver sound velocity $s_{\rm o}$ is given by (5.4.20), (7.2.18) is rewritten as

$$\left(\frac{s}{s_{\rm o}}\right)^2 = \left(\frac{\underline{s}}{s_{\rm o}}\right)^2 + \frac{3M_{\rm I}}{mZ}\frac{1}{v_{\rm F}^2}\left(\frac{N}{M_{\rm I}}\right) W_\xi$$

$$= \left(\frac{\underline{s}}{s_{\rm o}}\right)^2 + \frac{3}{2}\frac{1}{n\varepsilon_{\rm F}} N^2 W_\xi . \tag{7.2.18'}$$

From our discussion in 5.1–5.3, it is obvious that $(\underline{s}/s_{\rm o})^2 = 2N(0)/\Sigma\tilde{F}_\sigma(0)$. Then, ξ in the dynamical approach is given as

$$\xi = \frac{3}{2} N^2 W_\xi / n\varepsilon_{\rm F} . \tag{7.2.19}$$

For this ξ to be equivalent to that of (7.2.15), we require

$$\frac{\partial^2 E_\xi}{\partial V^2} = \frac{N^2}{V^2} W_\xi . \tag{7.2.20}$$

Let us see how this comes about.

If we introduce the density of ions, $N(\boldsymbol{R})$, E_ξ of (7.2.2) can be rewritten as

$$E_\xi = \frac{1}{2}\int W_\xi(\boldsymbol{R}-\boldsymbol{R}')\,N(\boldsymbol{R})\,N(\boldsymbol{R}')\,d\boldsymbol{R}\,d\boldsymbol{R}'$$

$$= \frac{1}{2}\sum_{\boldsymbol{q}} W_\xi(\boldsymbol{q})\,N_{\boldsymbol{q}}N_{-\boldsymbol{q}}\,, \tag{7.2.21}$$

where

$$N_{\boldsymbol{q}} = \int N(\boldsymbol{R})\,e^{-i\boldsymbol{q}\boldsymbol{R}}\,d\boldsymbol{R}, \tag{7.2.22}$$

and $W_\xi(\boldsymbol{q})$ is the Fourier transform of $W_\xi(\boldsymbol{r})$ as we already defined similarly to (7.2.6); for $W_\xi(\boldsymbol{q})$ we assume the same V dependence as that in (7.2.7). If the ionic distribution is characterized by a wave vector $\boldsymbol{Q}$, we have $N_{\boldsymbol{q}} = N\delta_{\boldsymbol{q},\boldsymbol{Q}}$ and, accordingly

$$E_\xi = \frac{1}{2}N^2 W_\xi(\boldsymbol{Q}), \tag{7.2.23}$$

and, then, (7.2.20). Note that in the jellium, we have $\boldsymbol{Q} = 0$. Note also that from (7.2.23), (7.2.19) and (5.4.20), we have

$$E_\xi = \frac{1}{2}\xi\,NM_{\mathrm{I}}\,s_0^2. \tag{7.2.24}$$

The above demonstration of the equivalence of the dynamical and the total energy approaches is by no means general, however. In order to arrive at the equivalence we had to assume some specific forms of volume dependence for a few physical quantities. It is quite puzzling that in the dynamical approach we nowhere needed to know the volume dependence of physical quantities such as $\widetilde{V}(\boldsymbol{q})$, while in the total energy approach the result depends crucially on the detailed volume dependence of such physical quantities. We do not yet fully understand relation between these two approaches.

However, the discussion of this section convinces us that the dynamical approach for the phonon and elasticity properties of Chapter 5 is essentially equivalent to considering only F_{m} of (7.2.2) in carrying out the procedure of (7.1.20) or (7.1.28). Hereafter we assume this equivalence without any further general proof. Then it becomes clear how to go beyond Chapter 5: Include the contribution of F_{ph} in carrying out the procedure of (7.1.20) for elastic constants, and that of (7.1.1) for volume.

Thus, F_{ph} (and F_{sw}) must be considered in determining $\omega_{\mathrm{ph}}(\boldsymbol{q})$. On the other hand, F_{ph} is determined by $\omega_{\mathrm{ph}}(\boldsymbol{q})$. Therefore a self-consistent procedure is required to obtain F_{ph} and $\omega_{\mathrm{ph}}(\boldsymbol{q})$. Here, however, for the phonon spectrum required in $F_{\rm ph}$ we use the result of Chapter 5, namely, that derived from $F_{\rm m}$ alone, as in Chapter 6.

Note that phonons are involved also in determining $F_{\rm sw}$, since phonons are involved in determining magnetic properties of a metal as we already saw in Chapter 6. We neglect such phonon effect too.

Thus, our handling in this chapter of the effects of $F_{\rm ph}$ and $F_{\rm sw}$ on the elastic and magnetic properties is essentially a perturbational one.

7.3 MAGNETO-VOLUME EFFECTS IN THE MEAN FIELD THEORY OF ITINERANT ELECTRON MAGNETISM

Our ultimate goal is to discuss the magnetic and elastic properties of a metallic ferromagnet by including all of the effects of spin waves and phonons, together with consideration of the volume degree of freedom. All of these effects need to be treated simultaneously and self-consistently (non-perturbatively) as noted at the end of the preceding section. At this time, however, such a goal is not yet achieved. Thus, in the remainder of this chapter we present only the beginning steps of such a grand endeavor. First in this section we explore within the Stoner model the role of the volume degree of freedom in determining the magnetic and elastic properties of a metal.

7.3.1 Jellium-like Model with Volume Degree of Freedom

In the numerical examples to be given in this section we use the jellium-like model with the free electron energy dispersion, $\varepsilon_{\boldsymbol{k}} = \hbar^2 \boldsymbol{k}^2/2m^*$. Our numerical study starts from determining equilibrium volume and magnetization by the conditions (7.1.1)–(7.1.2) at each temperature. Thus we find that to realize an equilibrium volume within a reasonable range we need to elaborate the model.

As we noted repeatedly, in the pure jellium model where both the electron and the ion charges are assumed to be uniformly distributed, the Coulomb interaction energies among electrons, among ions, and between electrons and ions cancel out. However, if we assume that each point charge ion is located on a lattice point such a cancellation does not occur. The simplest way to estimate the Coulomb interaction energy of such an electron-ion system may be that of Wigner and Seitz [7.6]. In that model, each ion is assumed to be surrounded by uniformly distributed electrons so as to screen out the ionic charge Ze within a Wigner-Seitz cell. Then the size of the attractive interaction energy between a positive point charge and surrounding electrons is larger than that of the repulsive interaction energy between electrons within a Wigner-Seitz cell. Based on this observation we phenomenologically introduce a potential energy for an electron by

$$U_{\mathrm{WS}} = -f\frac{e^2}{2a_{\mathrm{B}}}\left(\frac{V}{V_{\mathrm{r}}}\right)^{-d}, \tag{7.3.1}$$

where V_{r} is a reference volume, which is chosen to be close to the equilibrium volume, and d and f are parameters which are considered to be positive and $O(1)$. Since in a real metal, especially, in a transition metal, the assumption of uniform distribution of electronic charge is not valid, the corresponding original Wigner-Seitz value of $d = 1/3$ may not be a good approximation. The same is the case with f. Thus, we treat both d and f in (7.3.1) as parameters.

The effect of U_{WS} is to shift downward the origin of the one-particle energy of an electron. The size of the shift increases with decreasing inter-atomic distance. This is a well-known, important mechanism of metallic cohesion. Let us call U_{WS} the Wigner-Seitz potential.

Besides (7.3.1), we need to assume the following forms of volume dependence for the key quantities in the model;

$$W = W_{\mathrm{r}}\left(\frac{V}{V_{\mathrm{r}}}\right)^{-a}, \tag{7.3.2a}$$

$$\widetilde{V} = \widetilde{V}_{\mathrm{r}}\left(\frac{V}{V_{\mathrm{r}}}\right)^{-b}, \tag{7.3.2b}$$

$$\xi s_0^2 = \left(\xi s_0^2\right)_{\mathrm{r}}\left(\frac{V}{V_{\mathrm{r}}}\right)^{-c}, \tag{7.3.2c}$$

similarly to (7.3.1), with positive constants a, b and c of $O(1)$, where W is the width of the electron energy band. Although W does not appear in this section, we here present (7.3.2a) for later convenience.

In 7.2 we assumed $b = c = 1$. However, in this section we vary their values. Concerning c, for instance, if we consider the effect of screening, it is no wonder that its value can be different from the often assumed value of 1/3. Note also that (7.3.1)–(7.3.2) do not represent the intrinsic volume dependence of the physical quantities; they represent only the volume dependence *phenomenologically* within some narrow region around the reference volume, V_{r}. For different materials, the values of such parameters can be different. Thus, we take the position that a set of different values of the parameters represents a different material. We will see in the numerical examples given in this and the succeeding two sections, 7.4–7.5, based on the present jellium-like model, that a small difference in the values of those parameters results in a qualitative difference in the behavior of all the physical quantities. (In 7.4 where we use the model electronic density of states of Fig.5.5, however, we take a different position. There, a different material is to be represented by a different location of the Fermi energy in the model electronic density of states; the values of those parameters, b, c etc. are kept common.)

In the numerical calculations in this section we choose the values of those parameters in the following two different ways.

Case 1,

r_s (at $V = V_r$) = 3.0 (k_F =1.2×10^8/cm), $m/m_o = 4$, $(\widetilde{V}N(0))_r = 1.02$,
$(b, c, d, f, \xi_r) = (1/3, 1, 1/3, 1.03/3, 2.4)$, $T_C{}^S = 2550$ K
($T_C{}^S$ is the Stoner Curie temperature).

Case 2,

r_s (at $V = V_r$) = 3.0 (k_F =1.2×10^8/cm), $m/m_o= 4$, $(\widetilde{V}N(0))_r = 1.02$,
$(b, c, d, f, \xi_r) = (2/3, 1, 1, 0.3, 2.5)$, $T_C{}^S = 2220$ K

Here, r_s is the mean distance between electrons expressed in distance between in units of Bohr radius, $a_B = \hbar^2/ me^2$,

$$r_s = \frac{r_0}{a_B} = \left(\frac{9\pi}{4}\right)^{1/3}\frac{1}{k_F a_B}; \quad \frac{n}{V} = \left(\frac{4}{3}\pi r_0^3\right)^{-1}. \tag{7.3.3}$$

m and m_o are, respectively, the effective and free masses of an electron.

We now carry out numerical calculations on various quantities for the above two sets of parameters for the jellium-like model with the parabolic electron energy dispersion within the mean field approximations.

7.3.2 Magnetic Susceptibility, χ, with and without MVI

In Fig.7.2 we show the results of the numerical calculations of the magnetic susceptibility χ according to the formula (7.1.12)

$$\chi = \frac{1}{F_{MM} - F_{MV}^2 / F_{VV}}. \tag{7.1.12}$$

Fig.7.2(a) and (b) correspond to Case 1 and Case 2, respectively. Dotted and solid lines represent the results obtained without and with consideration of the effect of MVI. We see that the effect of MVI on the susceptibility sensitively depends on the values of the parameters and is large for the Case 1.

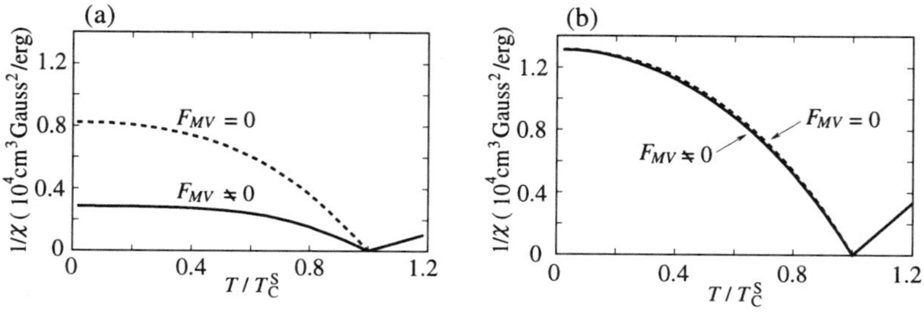

Fig.7.2 The temperature dependence of the magnetic susceptibility. Solid and dotted lines represent the results with and without the MVI effect. Figs.(a) shows the results for Case 1 and (b) for Case 2

7.3.3 Bulk Modulus, B, or Compressibility, κ, with and without MVI: The Origin of the ΔE Effects

In Fig.7.3 we show the bulk modulus B, or the inverse of the compressibility obtained by the numerical calculations.

$$B = \frac{1}{\kappa} = V_0\left(F_{VV} - \frac{F_{MV}^2}{F_{MM}}\right), \tag{7.1.20}$$

$$= B_1 + B_2 .$$

In Fig.7.3(a) B is much reduced from B_1 since the magnitude of B_2 is nearly equal to that of B_1. On the other hand, in Fig.7.3(b) the effect of MVI is very small, but two values of M with and without MVI are never equal.

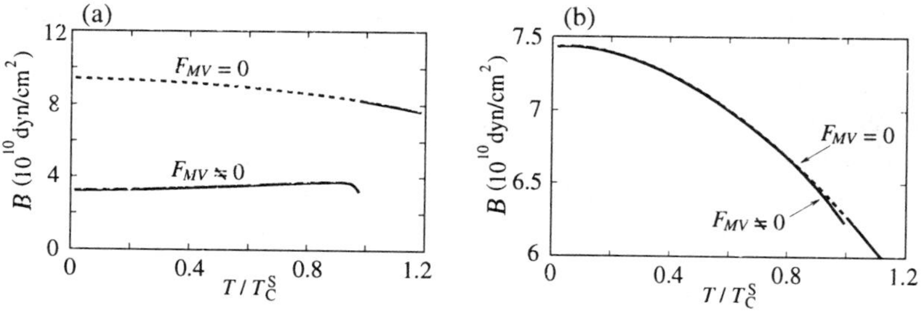

Fig.7.3 Temperature dependence of the bulk modulus for two Cases 1 (a) and 2 (b). Solid and dotted lines represent the results with and without MVI, respectively.

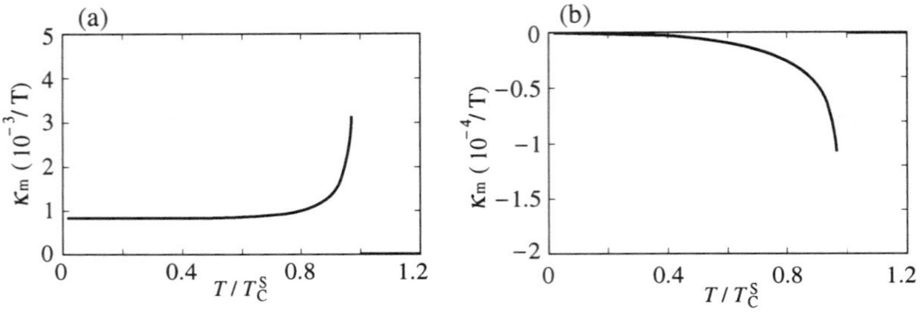

Fig.7.4 The temperature dependence of the magnetic expansivity. When a ferromagnetic metal approaches to the Curie temperature, the volume increases rapidly in the Case 1, (a), and decreases in the other Case 2, (b). Remark that the scales of ordinates are different for Figs.(a) and (b).

Another interesting point in our results is that B drops abruptly as we decrease temperature across Tc. Such behavior of B is often actually observed [7.7] and called ΔE effect. ΔE effect has long been believed to be caused by magnetic domains [5.19], but now it seems to be realized as an intrinsic property that is not caused by domains [5.20].

7.3.4 Magnetic Expansivity, κ_m

In Fig.7.4 we show the temperature dependence of the magnetic expansivity, κ_m.

$$\frac{\Delta V_o}{V_o} = -\frac{\chi}{V_o}\frac{F_{MV}}{F_{VV}}H = \kappa_m H, \qquad (7.1.14)$$

The results show that ferromagnetic metal can expand or shrink when it approaches to the magnetic phase transition temperature. The behavior is particularly remarkable near T_C. The result of Case 1 agrees with actual observations that expansivity with increasing temperature exhibits a peak at T_C in Invar alloys. We should remark that although the expansivity κ_m in Case 2 is negative in contrast with Case 1, the absolute value is very small except temperature near T_C. As we shall see in the succeeding subsections, this feature of κ_m originates from an anomaly of the thermal expansion.

7.3.5 Magnetic Response to Pressure, χ_p

In Fig.7.5 we show the temperature dependence of the derivative *dM/dp,* obtained by numerical calculations. Ferromagnetic metals can respond to appli-

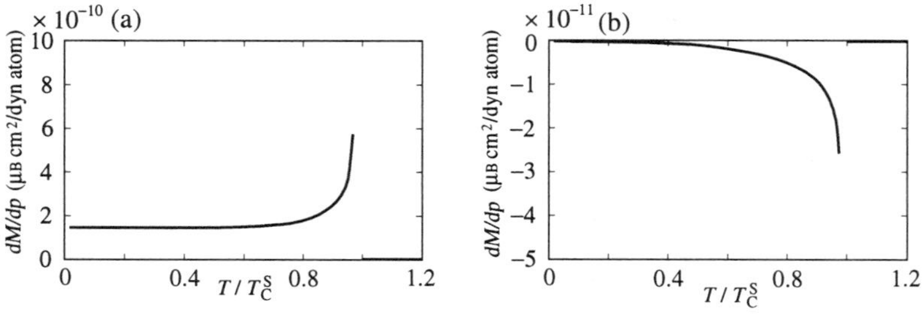

Fig.7.5 The temperature dependence of the magnetic response to applied pressure. The signs of the magnetic response and the temperature derivative of it are different for two cases, (a) for Case 1 and (b) for Case 2.

ed pressure very anomalously. An interesting point to note is that the behavior of *dM/dp* and that of κ_m are the same.

$$\chi_p = \frac{dM}{dp} = -V\kappa\frac{F_{MV}}{F_{MM}}. \tag{7.1.22}$$

7.3.6 Thermal Expansion Coefficient, β

We saw above qualitatively different behavior of several quantities such as bulk modulus B, magnetic expansivity κ_m, and magnetic response to pressure χ_p. We now calculate the thermal expansion coefficient β, and see how the combined effect gives to the resultant β.

The results of the numerical calculations of the thermal expansion coefficient β are shown in Fig.7.6. Two kinds of origin for the effect, internal pressure and internal magnetic field, show quite different temperature dependence.

$$\beta = \kappa\frac{\partial p_I}{\partial T} + \kappa_m\frac{\partial H_I}{\partial T}, \tag{7.1.31}$$

$$= \beta_1 + \beta_2.$$

We see from Fig.7.6 that behavior of β is very different for Case 1 and for Case 2. Further, in the Case 1, we see that the component β_2, which arises through the internal magnetic field, has negative value and seems to contribute to make the value β smaller. This means the Case 1 is more likely to be an Invar. Unfortunately, however, the resultant β is positive for both Cases.

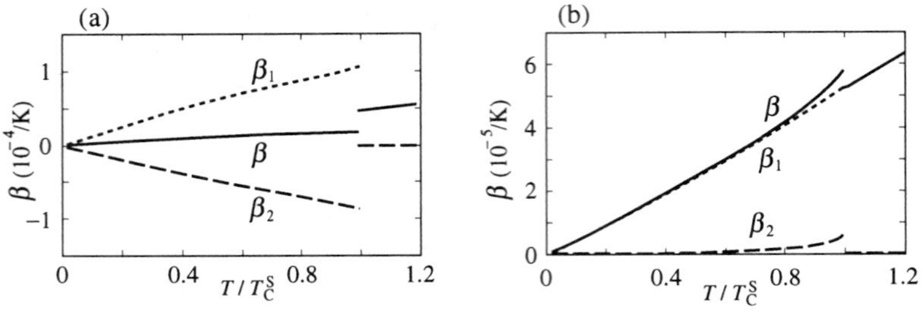

Fig.7.6 (a) shows the results of the numerical calculations for the Case 1 for the thermal expansion coefficient and (b) shows those for the Case 2.

We have tried several other numerical calculations choosing physically acceptable values of various parameters, but we have not yet succeeded to obtain the negative value for β which agrees with experiments shown in Fig.5.1, as far as we stay in this simple model.

7.3.7 Temperature Dependence of Magnetization, d*M*/d*T*

The temperature derivative of magnetization

$$\frac{dM}{dT} = \chi \frac{\partial H_I}{\partial T} + \chi_p \frac{\partial p_I}{\partial T}$$

$$= \left.\frac{dM}{dT}\right|_1 + \left.\frac{dM}{dT}\right|_2 . \tag{7.1.34}$$

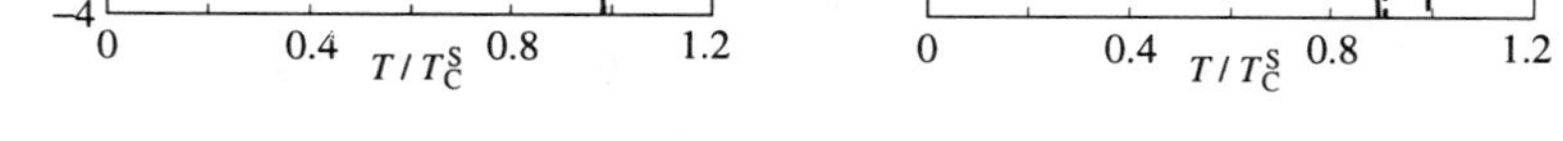

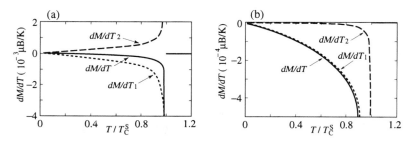

Fig.7.7 The temperature dependence of the various effects on the temperature derivative of the magnetization d*M*/d*T*, (a) for Case 1 and (b) for Case 2.

can also behave anomalously with temperature as shown in Fig.7.7. The result occurs from a combination of effects which occur through the internal pressure and the internal magnetic field. Those two effects behave quite differently in the real metals. We should remark that while dM/dT_2 has positive value for Case 1, it has negative value for Case 2.

7.4 THE ROLE OF THERMAL PHONONS IN THE ELASTIC AND MAGNETIC PROPERTIES: THE PHONON MECHANISM OF THE INVAR EFFECT

As shown in Fig.5.1, the behavior of thermal expansion coefficient, β, of a metallic ferromagnet is often anomalous; β becomes either very small or negative below T_C. It is without doubt that the cause of such anomalous volume behavior, which is called the Invar effect, is " magnetic" one. However, in spite of continuing efforts we have not yet succeeded in pinpointing the specific mechanism.

For the mechanism of the Invar effect, the two-spin-state model, originally proposed by Weiss [7.8], may be the most well-known. It postulates the presence of two (atomic) states separated by a small energy, say, ~10 meV; the lower energy state, which is called the high-spin state, has a larger spin and larger volume, and the higher energy state, which is called the low-spin state, has a smaller spin and smaller volume. Then, due to the thermal excitation of electrons (atoms) from the high-spin state to the low-spin state, the volume of a system will shrink with increasing temperature. Very interestingly, recent band calculation results [7.9] seem to give a support to such hypothesis; however, there still lies a long way before we ascertain whether such results at zero temperature would actually lead to negative β [7.10].

Another well-known recent proposal, which we may call the local moment hypothesis, is to postulate that the magnetic effect on volume is proportional to the thermal average of the squared atomic magnetic moment, $\langle m^2 \rangle$ [7.11.7.12]. The temperature dependence of volume can then be related to that of magnetization, or $\langle m^2 \rangle$, which can be independently measured. However, according to recent analyses [7.13], it seems difficult to understand the anomalous volume behavior within this hypothesis.

Thus we have no single valid theory for the Invar behavior. As mentioned in the above, it is not yet shown how the band calculation results, which are considered to give a support to the two-spin-state hypothesis, would actually lead to a negative β. As for the localized moment hypothesis, whereas it does not work any way, it is only a phenomenological hypothesis.

Also there is a fundamental question in the traditional manner to analyze experimental data on the thermal expansion. Without exceptions, it has been assumed that the contribution of phonons to β of a system is independent of its magnetic properties. Thus from the total β, which is anomalous, they subtract the phonon part by assuming it to be normal. The remaining anomalous part is attributed entirely to electrons. The phonon part can not be measured separately, and, therefore, such an assumption on the phonon contribution has never been verified, however. Thus, even if some theory, which considers, of course, only electrons, could account for such an anomalous part of β, which is what all the past theories aspired to achieve, we can not accept it as a real success. As we will demonstrate it in this section, the phonon contribution to β is not always normal; it can be negative and anomalous.

As for such possible role of phonons, Zverev and Silin [7.14] previously pointed out the possibility of the phonon contribution via their effect on magnetization. However, experimentally, as noted above, volume and magnetization are not simply related. One of us [7.15,7.16], on the other hand, showed that γ_D for $T < T_C$ can become negative. However, there, discussion was confined to $T = 0$ K and it was not able to calculate β properly.

The central theme of this section is to show that phonons play a crucial role in the Invar mechanism. In the course of arriving at such conclusion, we also find that it is essential to consider the roles the phonons and the MVI in understanding not only the elastic properties but also the magnetic properties of a metal. We already discussed the role of phonons in the magnetic behavior in Chapter 6. There, however, we assumed the volume of a system to be constant. Now we reconsider the same subject with consideration to the volume change of freedom.

7.4.1 Stoner Model with Phonons

The model we use in this section is essentially the same as that of 6.2. We consider the free energy of a system as the sum of the contributions from electrons, F_{el}, and phonons, F_{ph}; see (6.1.1). We treat F_{el} with the mean field approximation (the Stoner model), as in (7.2.2) and F_{ph} with the Debye approximation, again similarly to 6.2. The only and important difference from Chapter 6 is that here we treat the volume degree of freedom on the same footing as that of magnetization [7.17].

For convenience let us write down the free energy we use in this section;

$$F(V, M) = F_{el}(V, M) + F_{ph}(V, M), \tag{7.4.1}$$

$$\begin{aligned} F_{el}(V, M) &= F_{el,\,m}(V, M) + E_\xi = F_m(V, M) \\ &= -k_B T \sum_\sigma \int d\varepsilon\ N(\varepsilon) \ln\left(1 + e^{-(\varepsilon - \mu_\sigma)/k_B T}\right) \end{aligned}$$

$$-\frac{1}{2}\widetilde{V}\sum_{\sigma} n_{\sigma}^2 + \sum_{\sigma} \mu_{\sigma} n_{\sigma} + \frac{1}{2}\xi M_{\mathrm{I}} s_{\delta}^2, \tag{7.4.2}$$

$$F_{\mathrm{ph}}(V, M) = \frac{3}{8} N \hbar s(M)\, q_{\mathrm{D}} + \frac{3Nk_{\mathrm{B}}T}{q_{\mathrm{D}}^3} \int_0^{q_{\mathrm{D}}} dq\, q^2 \ln\left(1 - e^{-\hbar s(V,M)q/k_{\mathrm{B}}T}\right). \tag{7.4.3}$$

where the sound velocity $s(V,M)$ is given by (5.4.27) and other notations are the same as in Chapters 5 and 6. For our numerical calculations in this section we use the following model electronic density of states,

$$N(\varepsilon) = (30N/W)\left[1/4 - (\varepsilon/W)^2\right], \tag{7.4.4}$$

which is illustrated in Fig.7.9. This band can accommodate up to five electrons per spin per atom.

There is an important difference between the above model density of states and that of (5.4.31). The bottom of this new $N(\varepsilon)$ is set at $\varepsilon = -W/2$ independent of the volume, unlike in that of (5.4.31). This setting embodies the effect of the Wigner-Seitz energy. With decreasing inter-atomic distance, the band width W increases, and, accordingly, the bottom of the electron energy band is lowered. Because of this cohesion mechanism built in (7.4.4), unlike in the preceding section, we can realize equilibrium volume of realistic size without introducing the Wigner-Seitz potential [7.18].

Similarly to the preceding section, we assume the volume dependence of key physical quantities as in (7.3.2).

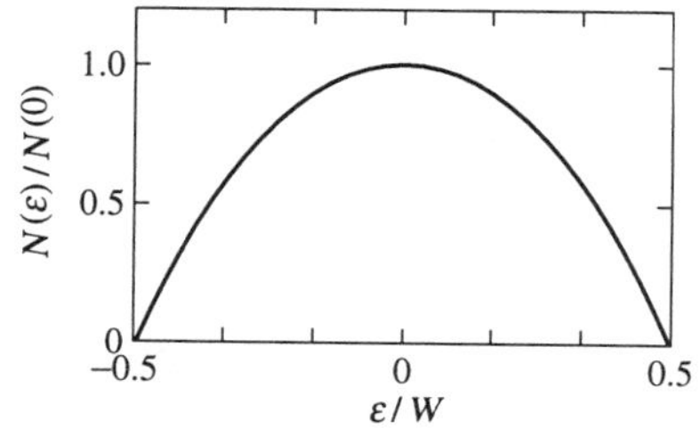

Fig.7.8 Model electronic density of states used in the numerical calculations in this section.

In the numerical examples to be given below we use the following values for various parameters:

$a = 1$, $b = 1/2$, and $c = 3$. $N/V_r = 5.96684 \times 10^{22}/\text{cm}^3$ ($r_s = 3$). $M_I = 10^5 m_o$, with M_I and m_o, respectively, the masses of ion and free electron. $\xi_r = 1.2$. These values are fixed. What we vary is the value of ε_F/W_r, the location of the Fermi energy, ε_F, within the model electronic density of states. We choose the following two sets of values.

Case 3,

$\varepsilon_F/W_r = 0.62$ ($n/V_r = 4.50937 \times 10^{22}/\text{cm}^3$), with $\widetilde{V}_r N(0)_r = 0.92$

($\widetilde{V}_o N(0)_o = 1.05828$; with $N(0)_o$ etc., $N(0)$ etc. at the equilibrium state);

Case 4,

$\varepsilon_F/W_r = 0.73$ ($n/V_r = 5.1391 \times 10^{22}/\text{cm}^3$), with $\widetilde{V}_r N(0)_r = 1.03$

($\widetilde{V}_o N(0)_o = 1.1102$).

The values of $\widetilde{V}_r N(0)_r$ are chosen so as to realize at $T = 0$ the equilibrium volume within appropriate range.

The numerical calculations are carried out in the following way: First we determine the equilibrium volume, $V(T)$ and magnetization, $M(T)$ at a given temperature T, by the condition of (7.1.1)–(7.1.2) for a given value of ε_F / W_0. Then, we calculate various magnetic and elastic responses from relevant differential coefficients of the free energy at the equilibrium volume and magnetization.

Now we proceed quite in parallel to the preceding section. In the following subsections we carry out numerical calculations on χ, κ_m, B, dM/dP, β, and dM/dT for the above two Cases. What differs from 7.3 is that here we are interested in knowing the contributions of the thermal phonons in addition to the effect of MVI.

7.4.2 Magnetic Susceptibility, χ.

In Fig.7.10, we show the temperature dependence of the inverse susceptibility obtained by the numerical calculations for two choices of ε_F/W_r shown by the insets in Figs.(a1) and (b1), which correspond to Case 3 and Case 4, respectively.

Figs.(a1) and (b1) show the inverse of susceptibility,

$$\chi = \frac{1}{F_{MM} - F_{MV}^2 / F_{VV}}, \tag{7.1.12}$$

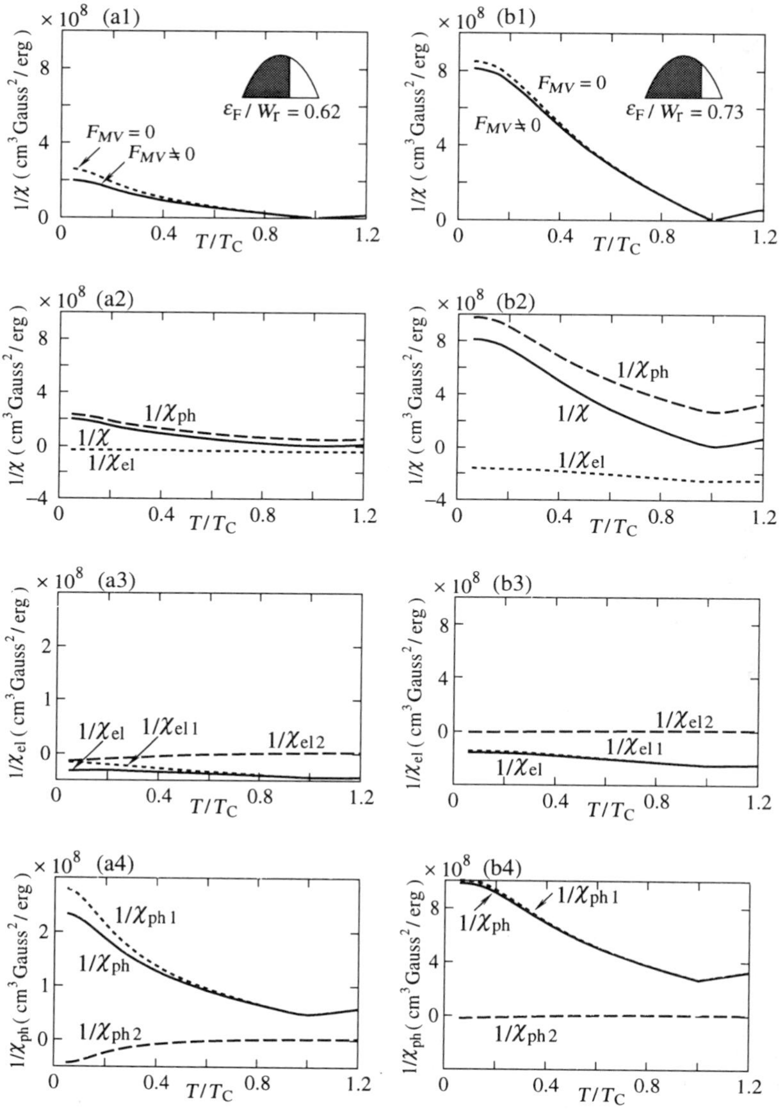

Fig.7.9 The results of numerical calculations of the inverse of the magnetic susceptibility. Figs.(a1) and (b1) the values for $1/\chi$ for Cases 3 and 4, respectively. Figs.(a2)(b2) show the decomposition of them into contributions due to phonon and electron and Figs.(a3)(b3)–(a4)(b4) show the composition to the parts without MVI, $1/\chi_{\alpha 1}$, and due to MVI, $1/\chi_{\alpha 2}$.

with and without MVI by solid and dotted lines. Although the effect of MVI reaches up to 20% for Case 3, it is very small, less than 5%, for Case 4.

In Figs.(a2) and (b2), we decompose $1/\chi$ into two parts due to electrons and phonons. The figures suggest that phonons are more effective to $1/\chi$.

$$\frac{1}{\chi} = \left.\frac{1}{\chi}\right|_{\text{el}} + \left.\frac{1}{\chi}\right|_{\text{ph}}. \tag{7.4.5}$$

Figs.(a3) (b3) (a4) and (b4) show the decomposition of each component to further sub-components without MVI and the effect of MVI.

$$\left.\frac{1}{\chi}\right|_{\alpha} = F_{\alpha,\,MM} - \frac{F_{MV}}{F_{VV}} F_{\alpha,\,MV}\,, \tag{7.4.6}$$

$$= \left.\frac{1}{\chi}\right|_{\alpha 1} + \left.\frac{1}{\chi}\right|_{\alpha 2}.$$

These Figs. show that the MVI effect mediated by phonon gives an important contribution to $1/\chi$ in Case 3, in particular.

7.4.3 Bulk Modulus, B, or Compressibility, κ.

In Fig.7.11, we show the results of numerical calculations of the temperature dependence of the bulk modulus,

$$B = \frac{1}{\kappa} = V_{\text{o}}\left(F_{VV} - \frac{F_{MV}^2}{F_{MM}}\right). \tag{7.1.21}$$

(a) and (b) for Cases 3 and 4, respectively. Figs.(a1) and (b1) show the bulk modulus with and without MVI by solid and dotted lines, respectively.

Figs.(a2)–(b3) show how the resultant bulk modulus are decomposed to the effects due to electrons and phonons and in Figs.(a4) and (b4) we show separate contributions to the bulk modulus from pure internal pressure effect, $B_{\alpha 1}$, and from the MVI effect, $B_{\alpha 2}$..

$$B = B_{\text{el}} + B_{\text{ph}} \tag{7.4.7}$$

$$B_{\alpha} = -V_{\text{o}}\left(\frac{\partial p_{\alpha}}{\partial V} + \frac{\partial p_{\alpha}}{\partial M}\frac{dM}{dV}\right). \tag{7.4.8}$$

$$= B_{\alpha 1} + B_{\alpha 2}$$

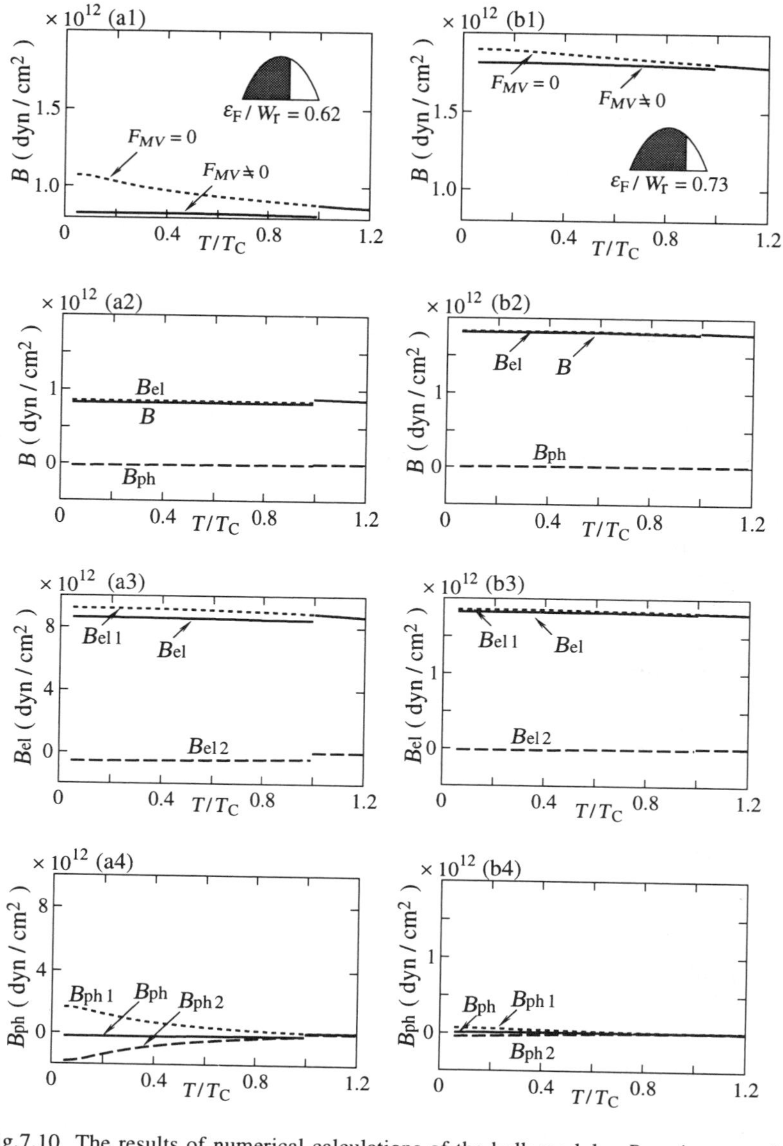

Fig.7.10 The results of numerical calculations of the bulk modulus B, or inverse of the compressibility κ. Figs.(a1) and (b1) show values for B for Cases 3 and 4, respectively. Figs.(a2)(b2)–(a3)(b3) show the decomposition of them into two parts due to phonon and electron and Figs.(a4)(b4) show origin of the bulk modulus due to the pure pressure effect, $B_{\alpha 1}$, and due to the MVI effect, $B_{\alpha 2}$.

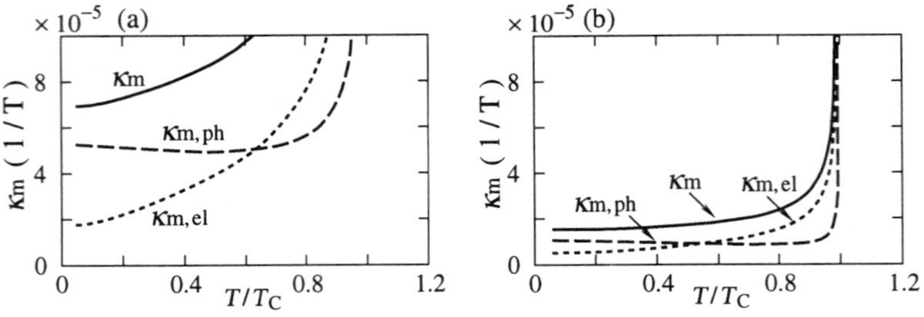

Fig.7.11 The temperature dependence of κ_m and their decomposition into electron and phonon contributions for the Case 3 (a) and Case 4 (b).

In Fig.(a1) we see MVI strongly reduces the bulk modulus more than 20%. However, in the Case 4 of Fig.(b1), although the reduction exists we see only a very small reduction. In this respect, it is interesting to note that for magnetic transition metal element, Moruzzy and Marcus's recently calculated values for B [7.19] are larger than the observed ones by 20%. Since B_{ph} is not included in their calculation, the origin of the difference might be the negative contribution of the second term of B_{ph}.

7.4.4 Magnetic Expansivity, κ_m

In Fig.7.12 we show the temperature dependence of magnetic expansivity, κ_m and their decomposition into parts due to electrons and phonons,

$$\kappa_m = \kappa_{m,el} + \kappa_{m,ph} , \tag{7.4.9}$$

$$\kappa_{m,\alpha} = -\frac{\chi}{V_o F_{VV}} F_{\alpha,MV} , \tag{7.4.10}$$

by noting that $\kappa_{m,ph} H \cong \kappa\, \partial p_{ph}/\partial M\, \chi H$, for instance, $\kappa_{m,ph}$ is understood to represent volume change caused by the effect of an external magnetic field on the phonon pressure.

We emphasize here that, in the low temperature region in Case 3, the contribution due to phonon is larger than that due to electron. We shall see in the next subsection that this feature originates from the large negative value of the thermal expansion coefficient β_2, the part which arises from the effect of phonon connected with enhancement of the internal magnetic field.

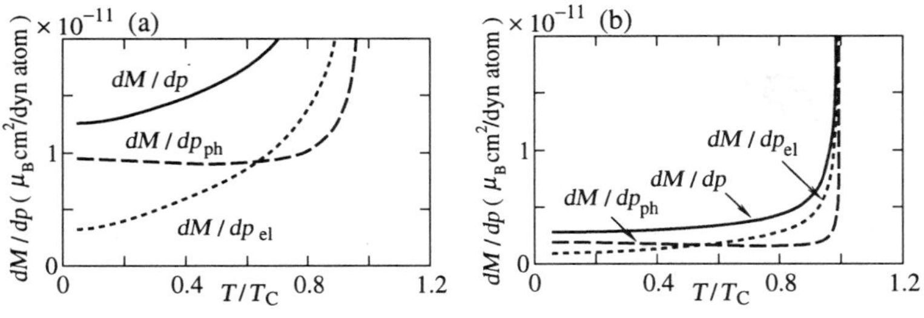

Fig.7.12 The temperature dependence of *dM/dP* and their decomposition into electron and phonon contributions for the Case 3 (a) and Case 4 (b).

7.4.5 Magnetic Response to Pressure

From (7.1.22) we note that magnetic response to pressure *dM/dp* can also be decomposed into the electron and phonon contributions.

$$\frac{dM}{dp} = \left.\frac{dM}{dp}\right|_{el} + \left.\frac{dM}{dp}\right|_{ph} , \qquad (7.4.11)$$

$$\left.\frac{dM}{dp}\right|_{\alpha} = -\frac{V\kappa}{F_{MM}} F_{\alpha,MV} . \qquad (7.4.12)$$

We see in Figs.7.13 (a) and (b) a marked difference between Cases 3 and 4; the size of *dM/dp* is significantly larger in the Case 3. This result agrees with the general observation that the pressure effect on magnetization is larger in an Invar than in a non-Invar. We find from Figs.(a) and (b) that phonon contribution is as important as that of electrons in producing such difference in the behavior of *dM/dp* between an Invar and a non-Invar.

7.4.6 Thermal Expansion Coefficient, β

Now, we discuss the results of our numerical calculations on the thermal expansion coefficient β. In our viewpoint in this section the thermal expansion of a ferromagnet arises through the change of internal pressure and internal magnetic field, as given in (7.1.31) and further they are composed of the effect of electrons β_{el} and of phonons β_{ph} [7.20,7.21].

$$\beta = \beta_{el} + \beta_{ph} .$$

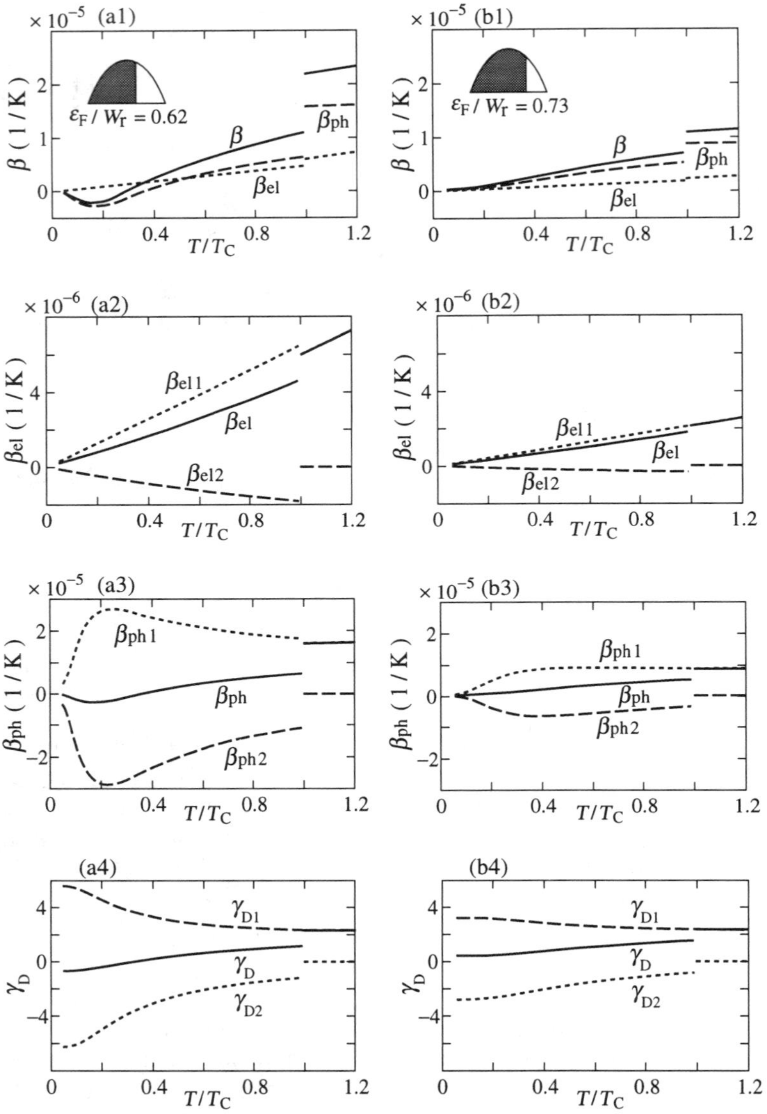

Fig.7.14 The results of the numerical calculations [7.24,7.25] of the thermal expansion β for the two Cases 3 (a) and 4 (b), respectively. Figures that follow are the decomposition according to the channels through electrons β_{el} and phonons β_{ph} and to the origins through internal pressure β_1 and internal magnetic field β_2. In Figs.(a4)(b4) we show the results of the phonon Grüneisen constant γ_{D} for the both Cases. γ_{D2} includes the MVI effect.

$$\beta_{\rm ph} = \kappa \frac{\partial p_{\rm ph}}{\partial T} + \kappa_{\rm m} \frac{\partial H_{\rm ph}}{\partial T}$$

$$\equiv \beta_{\rm ph1} + \beta_{\rm ph2} \, . \tag{7.4.13}$$

where

$$p_{\rm ph} = - \left. \frac{\partial F_{\rm ph}(T, V, M)}{\partial V} \right|_{V = V(T),\, M = M(T)} ;$$

$$H_{\rm ph} = - \left. \frac{\partial F_{\rm ph}(T, V, M)}{\partial M} \right|_{V = V(T),\, M = M(T)} ;$$

are, respectively, the internal pressure and the magnetic field due to phonons. Similarly we can write the electron contribution as $\beta_{\rm el} = \beta_{\rm el1} + \beta_{\rm el2}$.

Although we should also take into account the effect of spin fluctuations, we had not yet succeeded to carry out the calculation until now, and we present here the result which does not include the effect of spin fluctuations.

If we use the Debye approximation for $F_{\rm ph}$, (7.4.13) is rewritten as [7.22].

$$\beta_{\rm ph} = \frac{N k_{\rm B} \theta_{\rm D} \kappa \gamma_{\rm D}}{V} \frac{\partial P(T/\theta_{\rm D})}{\partial T} \, . \tag{7.4.14}$$

Here

$$\gamma_{\rm D} = - \frac{V}{\omega_{\rm D}} \left[\frac{\partial \omega_{\rm D}}{\partial V} + \frac{\partial \omega_{\rm D}}{\partial M} \frac{dM}{dV} \right]$$

$$\equiv \gamma_{\rm D1} + \gamma_{\rm D2} \, , \tag{7.4.15}$$

is the phonon Grüneisen constant for $T < T_{\rm C}$. Note that this division of $\gamma_{\rm D}$ corresponds to that of β given in (7.1.31); $\gamma_{\rm D2} = 0$ for $T > T_{\rm C}$. $P(T/\theta_{\rm D})$ is the function defined in (6.2.16).

In Fig.7.14 we show the results of our numerical calculations for both cases. The model electronic density of states used in this calculation, with the occupied region shaded, is shown in Figs.(a1) and (b1) for each case. In Figs.(a1) and (b1) we note that whereas $\beta_{\rm ph}$ dominates over $\beta_{\rm el}$ at not too low temperatures, the behavior of $\beta_{\rm ph}$ for $T < T_{\rm C}$ is not normal at all, and very sensitively depends on the location of $\varepsilon_{\rm F}$ in $N(\varepsilon)$. Thus, for $\varepsilon_{\rm F}/W_{\rm r} = 0.62$, case (a), β becomes negative by the effect of phonons, but $\beta > 0$ for $\varepsilon_{\rm F}/W_{\rm r} = 0.73$, case (b).

The behavior of β_{ph} is analyzed in Figs.(a3) and (b3) of Fig.7.14. We find that it is β_{ph2} that is responsible for making β_{ph} and, then, β negative. In Figs.(a2) and (b2) we present the behavior of β_{el}. In both cases, although β_{el2} certainly becomes negative as usually anticipated, it is overcompensated by the positive β_{el1}.

As can be seen from (7.4.14), the sign of β_{ph} is determined by that of γ_D. Thus, if β_{ph} dominates over β_{el}, which is generally the case except at very low temperatures, we would have $\gamma_D < 0$ when $\beta < 0$. We think that what Manosa et al. [7.23] observed is exactly this situation.

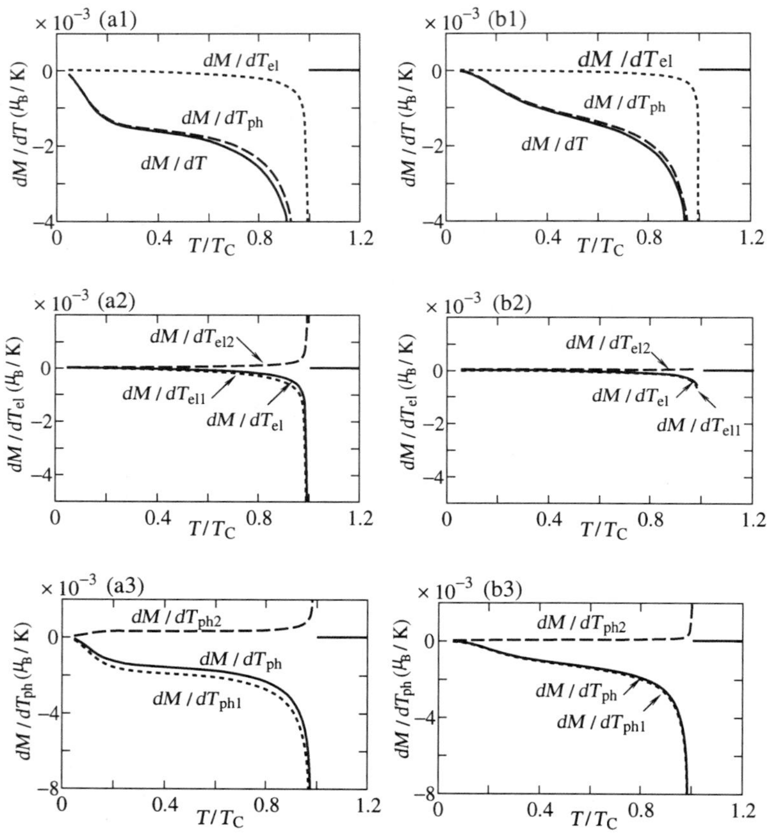

Fig.7.15 Temperature dependence of *dM/dT* and its analysis into electron and phonon parts and further to internal pressure and magnetic field effects.

What, then, is responsible for such a difference in the behavior of γ_D and β_{ph}? Our analysis shows it is the difference in κ_m. To make γ_{D2} and β_{ph2} large with negative signs, κ_m must be large with positive sign. According to (7.1.14), for κ_m to be large, we require a large χ and/or F_{MV} (that is, MVI). Such conclusion seems to be satisfied in actual observations.

7.4.7 Temperature Dependence of Magnetization, d*M*/d*T*

By noting that $H = H_{el} + H_{ph}$ and $p = p_{el} + p_{ph}$, the temperature dependence of magnetization given in (7.1.34) can also be decomposed as

$$\frac{dM}{dT} = \left.\frac{dM}{dT}\right|_{el} + \left.\frac{dM}{dT}\right|_{ph} . \tag{7.4.16}$$

$$\left.\frac{dM}{dT}\right|_{ph} = \chi \frac{\partial H_{ph}}{\partial T} + \chi_p \frac{\partial p_{ph}}{\partial T}$$

$$\equiv \left.\frac{dM}{dT}\right|_{ph1} + \left.\frac{dM}{dT}\right|_{ph2} . \tag{7.4.17}$$

In the Figs.7.15 (a1) and (b1), we see that the magnitude of dM/dT is significantly larger in the Case 3 than in the Case 4. This result agrees with the general observation that the decrease of magnetization with increasing temperature is more rapid in an Invar than in a non-Invar.

In Figs.(a2)–(b3) we show that what is responsible for such a large difference is the phonon contribution, and further show the importance of considering dM/dT_2 both for electrons and phonons; note that both of them have positive signs.

7.5 PHONON VERSUS SPIN WAVE EFFECTS ON VOLUME AND MAGNETIZATION AT LOW TEMPERATURES

The detailed analyses in the preceding section 7.4 show that in order to understand the Invar character shown in Fig.5.1 it is necessary to take into account the other elementary excitations, such as spin waves and phonons. We discuss their effects in this section within the same model with MVI using the parabolic band.

As to the phonon and spin wave effects on magnetization we already discussed it in 6.8 and 6.9. There, however, volume of a system was assumed to be constant. In this chapter we treat the volume and magnetization variables on the same footing, and, therefore, the MVI comes into the problem; we will see how important the MVI effects through phonons and spin waves are.

7.5.1 General Formulation of the Phonon and Spin Wave Effects

The free energy of a system we consider here is the same as (6.6.7) of Chapter 6, except E_ξ and E_{ws}:

$$F(T, V, M) = F_m(T, V, M) + F_{sw}(T, V, M) + F_{ph}(T, V, M) . \quad (7.5.1)$$

There is, however, a difference from that of Chapter 6 in that here the free energy is volume dependent. What we then do is similar to that we did in 6.6. The only difference here is that we pursue how the volume of the system, in addition to the magnetization, is affected by phonons and spin waves.

If we neglect the effects of spin waves and phonons, the volume and magnetization of a metal at a given temperature T are determined by

$$\frac{\partial F_m(M,V)}{\partial V} = 0 ; \qquad \frac{\partial F_m(M,V)}{\partial M} = 0 . \quad (7.5.2)$$

Let us call such volume and magnetization Stoner volume and Stoner magnetization and denote, respectively, as V_S and M_S. If we consider the spin wave or phonon effects, the equilibrium volume and magnetization would be changed, respectively, to $V_S+\Delta V_\alpha$ and $M_S+\Delta M_\alpha$ with α = sw, or ph. ΔV_α and ΔM_α are determined by the following conditions

$$\left.\frac{\partial F(M_S + \Delta M,\, V_S + \Delta V)}{\partial \Delta V}\right|_{\substack{\Delta M = \Delta M_\alpha \\ \Delta V = \Delta V_\alpha}} = 0 ;$$

$$\left.\frac{\partial F(M_S + \Delta M,\, V_S + \Delta V)}{\partial \Delta M}\right|_{\substack{\Delta M = \Delta M_\alpha \\ \Delta V = \Delta V_\alpha}} = 0 . \quad (7.5.3)$$

If we assume

$$|\Delta V / V_S| << 1, \quad |\Delta M / M_S| << 1, \quad (7.5.4)$$

we can expand F_m in the following way up to the second order:

$$F_m(M_S + \Delta M,\, V_S + \Delta V) = F_m(M_S, V_S) + \frac{1}{2} F_{m,\, VV} (\Delta V)^2$$

$$+ F_{m,\, MV}\, \Delta M \Delta V + \frac{1}{2} F_{m,\, MM} (\Delta M)^2 . \quad (7.5.5)$$

F_α can be expanded similarly; note, however, that F_α can have non-vanishing first order derivatives. By putting these into (7.5.3) we obtain

$$F_{VV}\,\Delta V + F_{MV}\,\Delta M + \sum_\alpha F_{\alpha,V} = 0, \qquad (7.5.6)$$

$$F_{VM}\,\Delta V + F_{MM}\,\Delta M + \sum_\alpha F_{\alpha,M} = 0, \qquad (7.5.7)$$

and then, for $\Delta V = \Sigma_\alpha \Delta V_\alpha$, $\Delta M = \Sigma_\alpha \Delta M_\alpha$,

$$\Delta V_\alpha = -\frac{F_{MM}\,F_{\alpha,V} - F_{VM}\,F_{\alpha,M}}{F_{MM}\,F_{VV} - F_{MV}^2}, \qquad (7.5.8)$$

$$\Delta M_\alpha = -\frac{F_{VV}\,F_{\alpha,M} - F_{MV}\,F_{\alpha,V}}{F_{MM}\,F_{VV} - F_{MV}^2}. \qquad (7.5.9)$$

Note that all the derivatives are to be taken at $V = V_S$ and $M = M_S$.

Here we recall our discussion in 7.1. We immediately notice that (7.5.8)–(7.5.9) can be rewritten as

$$\frac{\Delta V_\alpha(T)}{V} = \kappa p_\alpha + \kappa_m H_\alpha, \qquad (7.5.10)$$

$$\Delta M_\alpha(T) = \chi H_\alpha + \chi_p p_\alpha. \qquad (7.5.11)$$

Here, various responses include the effects of EPI and spin waves, and they are to be evaluated at $M = M_S$ and $V = V_S$.

We already introduced the effective magnetic fields due to spin waves, H_{sw}, and phonons, H_{ph}, in Chapter 6 (see (6.6.8)). Similarly, if we note that the contributions of spin waves and phonons to the free energy of a system are given by (6.6.7), the effective pressure due to the spin waves and that due to phonons are obtained in the following form

$$p_\alpha = -\left.\frac{\partial F_\alpha(T, V, M)}{\partial V}\right|_{V = V_S,\, M = M_S}$$

$$= -\hbar \sum_{\boldsymbol{q}} \left[\frac{1}{2} + n(\hbar\omega_\alpha(\boldsymbol{q}, M))\right] \left.\frac{\partial \omega_\alpha(\boldsymbol{q})}{\partial V}\right|_{V = V_S,\, M = M_S}, \qquad (7.5.12)$$

where we used the abbreviations $\omega_\alpha(q; T, M, V) = \omega_\alpha(q)$ etc.

With (6.6.8) and (7.5.12), we can rewrite (7.5.11) as

$$\Delta M_\alpha = \sum_{\boldsymbol{q}} m_\alpha(\boldsymbol{q}) \left[\frac{1}{2} + n(\hbar\omega_\alpha(\boldsymbol{q}, M)) \right], \tag{6.6.10}$$

with

$$m_\alpha(\boldsymbol{q}) = -\chi\hbar \left.\frac{\partial\omega_\alpha(\boldsymbol{q})}{\partial M}\right|_{\substack{M = M_S \\ V = V_S}} - \chi_p \hbar \left.\frac{\partial\omega_\alpha(\boldsymbol{q})}{\partial V}\right|_{\substack{M = M_S \\ V = V_S}}. \tag{7.5.13}$$

This result is an extension of (6.6.10)–(6.6.11); if we keep the volume of a system constant, (7.5.13) reduces to (6.6.11). Even with only the first term on the right hand side of (7.5.13), that is, with (6.6.11), we already derived a number of interesting consequences in 6.2. Now with (7.5.13), our understanding of the roles of spin waves and phonons will be further deepened.

We want to make a quantitative estimation of ΔM_α and ΔV_α for α = sw and ph and to compare them. For such a purpose we have to treat phonon and spin wave on the same footing. Thus we use the same jellium-like model as in 6.6.

In order to obtain the spin wave and phonon effects on magnetization and volume from (7.5.10)–(7.5.11), we need to know H_{sw}, p_{sw}, H_{ph} and p_{ph}. H_{sw} and H_{ph} were already discussed in Chapter 6. We formulate how to obtain p_{sw} and p_{ph} from the free energy, respectively, in 7.5.2 and 7.5.3.

7.5.2 The Pressure due to spin waves, p_{sw}

The pressure due to spin waves, p_{sw}, is given from the spin wave free energy of (6.5.13) as

$$p_{sw} = -\left.\frac{\partial F_{sw}(T, V, M)}{\partial V}\right|_{V = V_S,\, M = M_S}$$

$$= -\hbar \sum_q \left[\frac{1}{2} + n(\hbar\omega_{sw}(q)) \right] \left.\frac{\partial\omega_{sw}}{\partial V}\right|_{V = V_S,\, M = M_S} \tag{7.5.14}$$

We can calculate $\partial\omega_{sw}(q)/\partial V$ by a procedure similar to that which we used in 6.6.2 to calculate $\partial\omega_{sw}(q)/\partial M$. First, corresponding to (6.6.13′) now we have,

$$1 = \widetilde{V}(\Delta V) \sum_{\boldsymbol{k}} \frac{f(\varepsilon_{\boldsymbol{k}-}; \Delta V) - f(\varepsilon_{\boldsymbol{k}+\boldsymbol{q},+}; \Delta V)}{\varepsilon_{\boldsymbol{k}+\boldsymbol{q}}(\Delta V) - \varepsilon_{\boldsymbol{k}}(\Delta V) + \widetilde{V}(\Delta V)(n_- - n_+) - \hbar\omega}, \tag{7.5.15}$$

where $\varepsilon_{\boldsymbol{k}}(\Delta V)$ etc. are $\varepsilon_{\boldsymbol{k}}$ etc. under a variational volume change ΔV. Then corresponding to (4.4.20) or (6.6.16) we have

$$\hbar\omega_{sw}(\boldsymbol{q};\Delta V) = \frac{1}{2(n_- - n_+)}\sum_{\boldsymbol{k}} [\, f(\varepsilon_{\boldsymbol{k}-};\, \Delta V) + f(\varepsilon_{\boldsymbol{k}+};\, \Delta V)\,]\,(\,(\boldsymbol{q}\nabla_{\boldsymbol{k}})^2 \varepsilon_{\boldsymbol{k}}(\Delta V)\,)$$

$$- \frac{1}{\widetilde{V}(\Delta V)(n_- - n_+)^2}\sum_{\boldsymbol{k}} [\, f(\varepsilon_{\boldsymbol{k}-};\, \Delta V) - f(\varepsilon_{\boldsymbol{k}+};\, \Delta V)\,](\boldsymbol{q}\nabla_{\boldsymbol{k}}\varepsilon_{\boldsymbol{k}}(\Delta V))^2. \quad (7.5.16)$$

By changing the sums over $\boldsymbol{k}$ in (7.5.16) to integrals over the electron energy $\varepsilon_k = \hbar^2 k^2/2m = \varepsilon$ we have

$$\hbar\omega_{sw}(q;\, \Delta V) = \frac{1}{2(n_- - n_+)}\frac{\hbar^2}{m}\, q^2 \sum_{\sigma} \int d\varepsilon\, N(\varepsilon;\Delta V)\, f_\sigma(\varepsilon;\Delta V)$$

$$- \frac{1}{\widetilde{V}(\Delta V)(n_- - n_+)^2}\frac{2\hbar^2}{3\,m}\, q^2 \sum_{\sigma} \sigma \int d\varepsilon\, N(\varepsilon;\, \Delta V)\, \varepsilon f_\sigma(\varepsilon;\Delta V). \quad (7.5.17)$$

Next, it is required to calculate $(\partial\omega_{sw}(q;\Delta V)/\partial V)_{\Delta V=C}$. As we noted below (7.2.8), $(\partial N(\varepsilon;\Delta V)/\partial V)_{\Delta V=0} = N(\varepsilon)/V$. As for $f_\sigma(\varepsilon;\Delta V) = f(\varepsilon_{k\sigma};\Delta V)$, unlike in 7.2, we take it in the following form.

$$f_\sigma(\varepsilon;\, \Delta V) = \frac{1}{1 + \exp\{\beta\,[\varepsilon - U_{ws}(\Delta V) - \widetilde{V}(\Delta V)n_\sigma - \mu_\sigma(\Delta V)]\}}. \quad (7.5.18)$$

Here we are starting from an equilibrium state at a given temperature T, with the equilibrium values $M = M_S$ and $V = V_S$. Thus the meaning of μ_σ in (7.5.18) is different from that of 7.2; here, for $\Delta V = 0$, we have $\mu_+ = \mu_-$. Corresponding to (7.2.9) for (7.5.18) we have

$$\frac{n_\sigma}{V} + F_\sigma(0)\left[\frac{\partial\widetilde{V}}{\partial\Delta V}n_\sigma - \frac{\partial}{\partial\Delta V}U_{WS} + \frac{\partial\mu_\sigma}{\partial\Delta V}\right]\Bigg|_{\Delta V=0} = 0, \quad (7.5.19)$$

or

$$\frac{\partial\mu_\sigma}{\partial\Delta V} = -\frac{n_\sigma}{VF_\sigma(0)} + \frac{\partial}{\partial\Delta V}U_{WS} - \frac{\partial\widetilde{V}}{\partial\Delta V}n_\sigma\,. \quad (7.5.19')$$

Then we have

$$\frac{\partial f_\sigma(\varepsilon;\, \Delta V)}{\partial\Delta V}\Bigg|_{\Delta V=0} = -\frac{n_\sigma}{VF_\sigma(0)}\frac{\partial f_\sigma(\varepsilon)}{\partial\varepsilon} \equiv \frac{\partial f_\sigma(\varepsilon)}{\partial V}. \quad (7.5.20)$$

By using (7.5.20), we can calculate $\partial\omega_{sw}(q;\, \Delta V)/\partial\Delta V$ from (7.5.16).

7.5.3 The Pressure due to Phonons, p_{ph}

From the phonon free energy of (6.2.4) or (6.6.7) for α = ph, we have

$$p_{ph}(T, M_S, V_S) = -\left.\frac{\partial F_{ph}(T, M, V)}{\partial V}\right|_{\substack{M=M_S\\ V=V_S}}$$

$$= -\hbar \sum_{q} \left(\frac{1}{2} + n(\hbar\omega_{ph}(\boldsymbol{q}))\right) \left.\frac{\partial \omega_{ph}(\boldsymbol{q}, M_S, V)}{\partial V}\right|_{V=V_S} . \quad (7.5.21)$$

In the Debye approximation, with which the numerical calculation will be done, we have

$$p_{ph}(T, M_S, V_S) = -\frac{\partial F_{ph}}{\partial \omega_D}\frac{\partial \omega_D}{\partial V}$$

$$= -N\hbar\omega_D(T, M_S, V_S)\, P\left(\frac{T}{\theta_D(T, M_S, V_S)}\right) \left.\frac{\partial}{\partial V}\omega_D(T, M, V)\right|_{\substack{M=M_S\\ V=V_S}} \quad (7.5.22)$$

where we noted (6.2.16).

For our jellium-like model, the Debye frequency is given as $\omega_D = sq_D$ with the sound velocity of (5.4.27). Thus we have

$$\frac{\partial}{\partial V}\omega_D = \left(\frac{\partial}{\partial V}s\right) q_D + s\frac{\partial}{\partial V}q_D$$

$$= \frac{1}{2}\left(\frac{s_0}{s}\right)\frac{\partial}{\partial V}\left(\frac{s}{s_0}\right)^2 s_0 q_D - \frac{1}{3V}\omega_D , \quad (7.5.23)$$

where we noted (6.2.11). It is, thus, required to differentiate the right hand side of (5.4.27):

$$\frac{\partial}{\partial V}\left(\frac{s}{s_0}\right)^2 = \frac{\partial \xi}{\partial V} + 2\frac{\partial}{\partial V}N(0) \Big/ \sum_{\sigma} \widetilde{F}_\sigma(0)$$

$$-\left[2N(0) \Big/ \left(\sum_{\sigma} \widetilde{F}_\sigma(0)\right)^2\right] \frac{\partial}{\partial V}\sum_{\sigma} \widetilde{F}_\sigma(0), \quad (7.5.24)$$

by keeping n_σ constant as in 7.5.2. First, concerning $\partial\xi/\partial V$ we note (7.3.2c). Next, we note that for the present jellium-like model, since $N(0) = Vm^*k_F/2\pi^2\hbar^2$,

$$\frac{\partial N(0)}{\partial V} = \frac{2}{3V} N(0). \tag{7.5.25}$$

Finally,

$$\begin{aligned}\frac{\partial}{\partial V} \widetilde{F}_\sigma(0) &= \frac{\partial}{\partial V} \frac{F_\sigma(0)}{1 - \widetilde{V} F_\sigma(0)} \\ &= \frac{\partial}{\partial V} F_\sigma(0) \frac{1}{1 - \widetilde{V} F_\sigma(0)} \\ &+ \frac{F_\sigma(0)}{\left(1 - \widetilde{V} F_\sigma(0)\right)^2} \left[\frac{\partial \widetilde{V}}{\partial V} F_\sigma(0) + \widetilde{V} \frac{\partial}{\partial V} F_\sigma(0) \right],\end{aligned} \tag{7.5.26}$$

where $\partial\widetilde{V}/\partial V$ is put as (7.3.2b). Thus, the only new quantity is $\partial F_\sigma(0)/\partial V$.

If we note (4.2.18) we have

$$\frac{\partial F_\sigma(0)}{\partial V} = -\int d\varepsilon \frac{\partial N(\varepsilon)}{\partial V} \frac{\partial f_\sigma(\varepsilon)}{\partial \varepsilon} - \int d\varepsilon\, N(\varepsilon) \frac{\partial^2 f_\sigma(\varepsilon)}{\partial \varepsilon\, \partial V}. \tag{7.5.27}$$

The second term on the right hand side is reduced as

$$-\int d\varepsilon\, N(\varepsilon) \frac{\partial^2 f_\sigma(\varepsilon)}{\partial \varepsilon\, \partial V} = \int d\varepsilon \frac{\partial N(\varepsilon)}{\partial \varepsilon} \frac{\partial f_\sigma(\varepsilon)}{\partial V},$$

with $\partial f_\sigma(\varepsilon)/\partial V$ given by (7.5.20). Thus we have

$$\frac{\partial F_\sigma(0)}{\partial V} = -\int d\varepsilon \frac{\partial N(\varepsilon)}{\partial V} \frac{\partial f_\sigma(\varepsilon)}{\partial \varepsilon} + \frac{F'_\sigma(0)}{F_\sigma(0)} \int d\varepsilon \frac{\partial N(\varepsilon)}{\partial V} f_\sigma(\varepsilon) \tag{7.5.28}$$

$$= \frac{1}{V} F_\sigma + \frac{n_\sigma}{V} \frac{F'_\sigma(0)}{F_\sigma(0)}, \tag{7.5.29}$$

where $F'(0) \cong -N'(0)$ is defined in (5.5.10).

By putting (7.5.25), (7.5.26) and (7.5.29) into (7.5.24), we can calculate $\partial\omega_D/\partial V$ of (7.5.23) and, then, p_{ph} of (7.5.22). With the above p_{ph}, and H_{ph} of (6.6.8) or (6.8.10), we can now calculate the effect of phonons on volume and magnetization from (7.5.10)–(7.5.11).

7.5.5 Numerical Calculations of Spin Wave and Phonon Effects on Volume at Low Temperature

In carrying out a model numerical analysis on (7.5.10)–(7.5.11) and (7.5.13), we simplify our task by introducing a linear approximation; we look for to obtain ΔV_{sw} and ΔV_{ph} only to the first order in the spin wave and phonon effects. That is, we replace all of κ, κ_m, χ and χ_p by those without the effect of either spin wave or phonon effect; to avoid confusion let us write them κ^S etc.

In the following numerical calculations we use the same model with the same sets of values for various parameters as given in 7.3.1.

A. ΔV_{sw} and ΔV_{ph} at low temperatures

We evaluate changes of the equilibrium volume using the formula (7.5.8), (7.5.10) and (7.5.12). In this perturbational treatment, the fractional change $\Delta V/V$ can be represented as linear combination of contributions from spin waves and phonons, and further they can be decomposed to the effects through internal pressure, $\Delta V_\alpha(T)_1/V$, and internal magnetic field, $\Delta V_\alpha(T)_2/V$.

$$\frac{\Delta V_\alpha(T)}{V} = \kappa^S p_\alpha + \kappa_m^S H_\alpha , \qquad (7.5.10a)$$

$$= \frac{\Delta V_\alpha(T)_1}{V} + \frac{\Delta V_\alpha(T)_2}{V} \qquad (7.5.10b)$$

First in Fig.7.16(a) we show the behavior of the phonon effect. Besides the ordinary positive pressure effect (ΔV_{ph1}), we have a negative magnetic field effect (ΔV_{ph2}) of comparable size; this ΔV_{ph2} has been hitherto completely neglected. Resultantly, ΔV_{ph} is much reduced from ΔV_{ph1}. Such result suggests the possibility of even having $\Delta V_{ph} < 0$, if we can generalize the calculation for the model of 7.5 to the non-perturbational treatment.

In Fig.7.16(b) we show the behavior of spin wave effects. Here the magnitude of negative ΔV_{sw2} is larger than that of positive ΔV_{sw1} and, therefore, $\Delta V_{sw} < 0$. Note, however that there is no a priori reason for ΔV_{sw} to be always negative. The sum of the phonon and spin effects is shown in Fig.7.16(c). Since here $|\Delta V_{\rm ph}| > |\Delta V_{\rm sw}|$, the resultant total thermal expansion is positive, but much reduced from, say, $\Delta V_{\rm ph1}$. In making that reduction, $\Delta V_{\rm ph2}$ plays a role more important than that of $\Delta V_{\rm sw}$; whereas both $\Delta V_{\rm ph2}$ and $\Delta V_{\rm sw}$ are negative, we have $|\Delta V_{\rm ph2}| > |\Delta V_{\rm sw}|$. As regards the temperature dependence of the Stoner volume, we found that $|V_{\rm S}(T) - V_{\rm S}(0)|$ is smaller than $|\Delta V_{\rm ph} + \Delta V_{\rm sw}|$ by one order of magnitude.

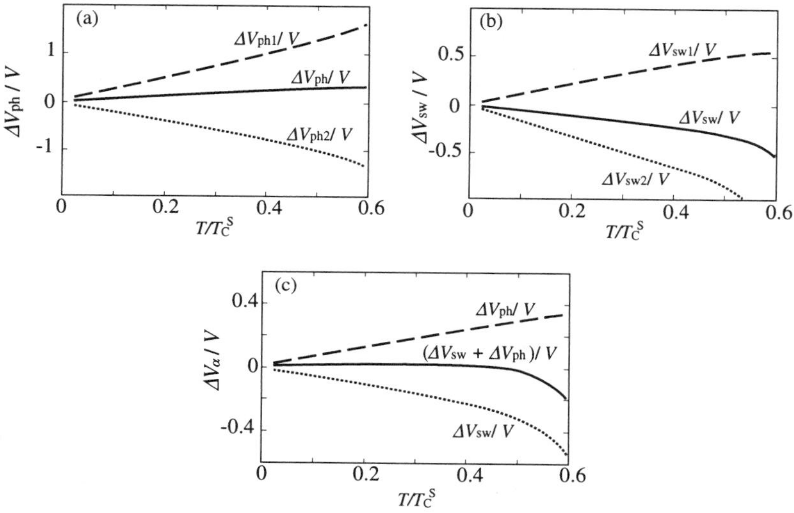

Fig.7.16 In (a), the temperature dependence of the effect of phonons on volume $\Delta V_{\rm ph}$ (solid line), and its decomposition into two components, $\Delta V_{\rm ph1}$ (broken line) and $\Delta V_{\rm ph2}$ (dotted line), according to (7.5.10a). In (b), the temperature dependence of the effect of spin waves on volume $\Delta V_{\rm sw}$ (solid line), and its decomposition into two components, $\Delta V_{\rm sw1}$ (broken line) and $\Delta V_{\rm sw2}$ (dotted line), according to (7.5.10a). In (c), the temperature dependence of phonon (broken line) and spin wave (dotted line) effects on volume and their sum (solid line) ΔV.

In conclusion, we find that we could not obtain negative β in the perturbational treatment even if we take into account the effects of spin waves, although the same result suggests effects of phonons coupled with magnetic expansivity $\kappa_{\rm m}$ play the important role in manifesting the Invar phenomena in ferromagnetic metals. We try to improve the discussion of this section by a generalization of perturbational correction of volume ΔV and magnetization ΔM.

7.6 Concluding Remarks

In this chapter we tried to demonstrate that it is essential to consider the role of phonons to understand the anomalous thermal expansion behavior of itinerant electron ferromagnets. We started from a new derivation of an exact expression for thermal expansion coefficient in the ferromagnetic state in terms of various differential coefficients of the free energy of a system. In the cource of such procedure, we obtained exact expressions which are also useful for other key physical quantities such as magnetic susceptibilities, bulk modulus etc.

Based on those results and by using a simple model electronic density of states we carried out numerical calculation of thermal expansion coefficient and other quantities by correctly treating temperature effect. We tried not to introduce any unjustified assumption in calculating the free energy. Thus we showed that the phonons can be the principal cause of negative thermal expansion which is often observed below Tc of an itinerant electron ferromagnet.

We found that the contribution of the spin waves to β also can be negative at low temperatures. Within our numerical examples of previous section, however, the size of β_{sw} is generally smaller than that of β_{ph} and, therefore, a negative β_{sw} does not bring about a negative β. However, there might be situations where the spin waves can be the principal cause of a negative β.

Recently it has been increasingly noticed that both volume and elasticity behaviors become anomalous around T_C. β takes either a peak or dip[17]. Bulk modulus also often softens as temperature is lowered from above[1]. Although we do not present in this paper, we recently found that such observations also can be understood from the same first principles approach as we used in this paper by considering the effects of spin fluctuations in addition to phonons. Note that in the results of Fig.1, although our treatment there was not quite satisfactory for that temperature region, β discontinuously changes at T_C.

Throughout this chapter we found that the behaviours of various physical quantities depend very sensitively on the location of the Fermi energy in a given electronic density of states. One origin of such sensitivity may be that, particularly in the ferromagnetic states, a difference in the electronic structures are much amplified in the phonon properties. The volume and elasticity properties can be very sensitive probe of the electronic structure and electron interaction.

PROBLEM FOR CHAPTER 7

7.1 Estimate the size of the pressure of a free electron gas confined in a box. Assume the electron density of $10^{22}/\text{cm}^3$ and give the answer in units of *bar*.

7.2 In a metal, what is mainly responsible to counter the pressure due to electron kinetic energy ? Discuss, in particular, the role of the exchange interaction between electrons.

7.3 Derive the thermodynamic expressions for the specific heat at constant pressure when the entoropy depending on temperature, magnetization, and volume is given as $S(T, M, V)$

7.4 Derive the magnetic susceptibility which incudes the volume degree of freedom using (7.2.2). Remember how we derived the bulk modulus in 7.2.

Chapter 8

GREEN'S FUNCTIONS AND FEYNMAN DIAGRAMS

In the preceding chapters we studied how the various linear responses of a metal are modified by the effect of the interaction between electrons, and then, how the electron-phonon interaction together with the electron-electron interaction modify the phonon properties and the electronic properties, particularly, the magnetic ones in a metal. In those discussions we used the equation of motion approach with the mean field approximation in dealing with the interaction effects. In this chapter we introduce the method of Green's functions and Feynman diagrams which is indispensable in further pursuing those problems, as will be illustrated in the succeeding chapters. As for the method of Green's functions and Feynman diagrams there are a number of excellent references. See, for instance, Refs. [1.2–1.6, 7.9, 8.1–8.7]. The purpose of this chapter is to present a simple introduction so as to enable those readers who are not familiar with this method to proceed to the succeeding chapters.

8.1 TWO-TIME GREEN'S FUNCTIONS

8.1.1 Retarded Two-Time Green's Function and Linear Response

There are two kinds of Green's functions to be used for a finite temperature system; one is the two-time Green's function, and the other is the temperature Green's function. As we will see shortly, what is directly related to physical quantities such as one-particle excitation energy and various linear responses is the two-time Green's function. It is, however, the temperature Green's function for which the convenient method of Feynman diagrams can be used. There is a very simple relation between these two kinds of Green's functions. Thus, usually we first obtain a temperature Green's function by the Feynman diagram method and then transform it to the corresponding two-time Green's function.

Let us begin by defining the two-time Green's function. For convenience, from now on we put $\hbar = 1$ in most places. Let A and B be operators such as the particle creation and annihilation operators and magnetization or spin density operator. A retarded two-time Green's function is defined as

$$G^{\mathrm{r}}_{AB}(t) = \mp i\theta(t)\left\langle \{A(t), B\}_{\pm} \right\rangle, \tag{8.1.1}$$

with

$$\{A, B\}_{\pm} = AB \pm BA, \tag{8.1.2}$$

that is, $\{A, B\}_{+} = \{A, B\}$ and $\{A, B\}_{-} = [A, B]$. When A and B are the creation or annihilation operators of a particle, it is convenient to use that with the anticommutator for fermions and that with the commutator for bosons. Thus they are often called, respectively, fermion and boson type Green's functions. The superscript r stands for *retarded*, and $A(t)$'s are the Heisenberg representation of the operator A defined with $K = H - \mu\tilde{n}$,

$$A(t) = e^{iKt} A \, e^{-iKt}, \tag{8.1.3}$$

where H is the total Hamiltonian of the system, μ is the chemical potential, and $\tilde{n}$ is the operator representing the total number of electrons; we often write $\tilde{n}$ simply as n. The grand canonical thermal average with the density matrix of (1.4.13) is represented by $\langle \cdots \rangle$, and $\theta(t)$ is the step function,

$$\theta(t) = \begin{cases} 1 & \text{for } t > 0\,, \\ 0 & \text{for } t < 0\,. \end{cases} \tag{8.1.4}$$

The reason we call (8.1.1) a retarded Green's function is because it represents the response at time $t > 0$ of a system to a perturbation applied at time $t = 0$. We will see this more below.

Note that in most literature the retarded Green's function for bosons is defined with the same negative sign as for fermions in front of the right hand side (8.1.1). However, we find it more convenient to define as in (8.1.1). The Green's function of (8.1.1) is Fourier transformed with respect to time as

$$G^{\rm r}_{AB}(\omega) = \int_{-\infty}^{\infty} G^{\rm r}_{AB}(t)\, e^{i(\omega+i0^+)t}\, dt = \mp i \int_{-\infty}^{\infty} \theta(t) \left\langle \{A(t), B\}_{\pm} \right\rangle e^{i(\omega+i0^+)t}\, dt$$

$$= \mp i \int_{0}^{\infty} \left\langle \{A(t), B\}_{\pm} \right\rangle e^{i(\omega+i0^+)t}\, dt$$

$$\equiv \langle\!\langle A; B \rangle\!\rangle^{\rm r}_{\omega}, \qquad (8.1.5)$$

where the infinitesimal positive imaginary part $i0^+$ is required to assure the convergence of the integral. The inverse transformation is given as

$$G^{\rm r}_{AB}(t) = \frac{1}{2\pi} \int_{-\infty}^{\infty} G^{\rm r}_{AB}(\omega)\, e^{-i\omega t}\, d\omega. \qquad (8.1.5')$$

We then notice that in order to make $G^{\rm r}_{AB}(\omega) = 0$ for $t < 0$, $G^{\rm r}_{AB}(\omega)$ is required to be analytic in the upper half, above the real axis, of the complex ω-plane; the poles of $G^{\rm r}_{AB}(\omega)$ should be in the lower half plane. (*Problem* 8.1)

We already encountered expressions like (8.1.5) in the Kubo theory of linear response; see (4.1.23). Note that what we called the interaction representation in Kubo theory is identical to the Heisenberg representation, (8.1.3). Also $_0\langle\cdots\rangle$ appearing in Kubo theory is identical to $\langle\cdots\rangle$ in the present chapter. Thus, the magnetic susceptibility of (4.1.33), for instance, turns out to be given by the boson type retarded two-time Green's function of the following form,

$$\chi_{\mu\nu}(\boldsymbol{q},\omega) = \langle\!\langle M_{\mu}(\boldsymbol{q})\, ;\, M_{\nu}(-\boldsymbol{q}) \rangle\!\rangle^{\rm r}_{\omega}. \qquad (8.1.6)$$

Although we usually derive a two-time Green's function via the temperature Green's function, note that in principle the two-time Green's function can be obtained directly by solving its equation of motion. The equation of motion is obtained by differentiating both sides of (8.1.1) with respect to t as

$$i \frac{\partial}{\partial t} G^{\rm r}_{AB}(t) = \mp\, i\theta(t) \left\langle \{ [A(t), K]\, , B\}_{\pm} \right\rangle \pm\, \delta(t) \left\langle \{A,\, B\}_{\pm} \right\rangle, \qquad (8.1.7)$$

where we used the relations

$$i\frac{\partial}{\partial t}A(t) = [A(t), K], \qquad (8.1.8)$$

$$\frac{\partial}{\partial t}\theta(t) = \delta(t). \qquad (8.1.9)$$

On the right hand side of (8.1.7), the second term is generally of a simple nature, but the first term comprises a new Green's function with a structure generally more complicated than the original one of the left hand side. We are required to set up a new equation of motion for this new Green's function and thus the procedure continues. If we introduce some approximation, however, we can truncate the chain of the equations of motion and obtain the desired Green's function in an approximate form.

8.1.2 One-Particle Green's Function and the Energy Spectrum of Electrons

In the preceding subsection we saw that a retarded two-time Green's function can represent a linear response. A Green's function such as (8.1.6) is called a *two-particle Green's function* since the quantity $M_\mu(\boldsymbol{q})$ involves simultaneous excitation of two particles (an electron and a hole). Now in this subsection we will see that the pole of a one-particle (electron) retarded two-time Green's function gives us the energy spectrum of the particle (electron).

Let us consider the following Green's function

$$G^{\mathrm{r}}_{\boldsymbol{k}\sigma}(t) = -i\theta(t)\langle\{a_{\boldsymbol{k}\sigma}(t), a^{+}_{\boldsymbol{k}\sigma}\}_{+}\rangle, \qquad (8.1.10)$$

where $a^{+}_{\boldsymbol{k}\sigma}$ and $a_{\boldsymbol{k}\sigma}$ are respectively the creation and annihilation operators of an electron $(\boldsymbol{k},\sigma)$. For simplicity, let us abbreviate $(\boldsymbol{k},\sigma)$ by k; $a_{\boldsymbol{k}\sigma} = a_k$, etc. If we assume that the Hamiltonian of the electrons is given by the one-particle energy H_o of (1.5.14) alone, the equation of motion for the Green's function is given by

$$i\frac{\partial}{\partial t}G^{\mathrm{ro}}_{k}(t) = (\varepsilon_k - \mu)\, G^{\mathrm{ro}}_{k}(t) + \delta(t). \qquad (8.1.11)$$

Thus, for the Fourier transform of the Green's function for a free electron we obtain

$$G^{\mathrm{ro}}_{k}(\omega) = \frac{1}{\omega - \xi_k + i0^{+}}, \qquad (8.1.12)$$

where we put

$$\xi_k = \varepsilon_k - \mu. \tag{8.1.13}$$

Obviously, the pole, $\omega = \xi_k$, represents the energy of an electron in the state $(\boldsymbol{k},\sigma)$ as measured from the chemical potential. Let us see how such result comes out.

For simplicity, let us consider $T = 0$, the absolute zero of temperature, and suppose we put an electron with $k > k_F$ outside the Fermi sea of an electron system. Then the state would evolve with time as

$$e^{-iKt} a_k^+ \mid \mathrm{Fs} \rangle = e^{-i(E + \xi_k)t} a_k^+ \mid \mathrm{Fs} \rangle,$$

where $\mid \mathrm{Fs} \rangle$ represents the Fermi sea of the system of electrons with energy E (the eigenvalue of K). The additional time dependence of this state due to the added electron, $e^{-i\,\xi_k t}$, can be extracted if we project it onto the state

$$a_k^+ e^{-iKt} \mid \mathrm{Fs} \rangle = e^{-iEt} a_k^+ \mid \mathrm{Fs} \rangle$$

which has evolved without the effect of the added electron up until the time t. The Green's function represents such projection. Note that out of the two terms coming from the anticommutator of (8.1.10) it is the term $a_k(t)\, a_k^+$ that is relevant, here.

Similarly, for $k < k_F$, the Green's function of (8.1.10) pursues the effect of creating a hole in the Fermi sea of electrons. In this case the relevant term in the Green's function is the term $a_k^+ a_k(t)$, and we can extract the time dependence of $\exp(-i\xi_k t) = \exp(-i|\xi_k|(-t))$ with $\xi_k < 0$; a hole is viewed to propagate backward in time with a positive energy.

Of course it is unnecessary to use the above procedure to obtain the energy spectrum of free electrons. It is in complicated situations in which the electron-electron and the electron-phonon interactions are involved when the Green's function method becomes particularly useful. In such a situation the one-particle Green's function takes the following form

$$G_k^{\mathrm{r}}(\omega) = \frac{1}{\omega - \xi_k - \Sigma(k, \omega + i0^+) + i0^+}, \tag{8.1.14}$$

where the effect of various interactions is contained in $\Sigma(k, \omega + i0^+)$ which is called the *self-energy*. The energy spectrum of an electron with the wave number $\boldsymbol{k}$ and spin σ is determined from

$$\omega - \xi_k - \Sigma(k, \omega + i0^+) = 0. \tag{8.1.15}$$

Generally $\Sigma(k, \omega + i0^+)$ has a finite imaginary part with negative sign. We already in 5.4 discussed how to treat an equation of the form of (8.1.15) for the case of phonons. As for the case of electrons, examples will be given later.

8.1.3 Relation between $\langle BA \rangle$ and G^r_{AB}

In 4.1.3 we derived the relation of (4.1.52) between a correlation function $\langle A(t)\, B \rangle$ and a linear response function $\chi_{AB}(t)$ and called it the fluctuation-dissipation theorem. Such a relation can be translated to relation between a correlation function and a retarded two-time Green's function as follows. It is useful to introduce a two-time *advanced* Green's function by

$$G^a_{AB}(t) = \pm\, i\theta(-t) \left\langle \{A(t),\, B\}_\pm \right\rangle, \tag{8.1.16}$$

corresponding to the retarded Green's function of (8.1.1). The Fourier transform is defined similarly to (8.1.5) as

$$G^a_{AB}(\omega) = \pm\, i \int_{-\infty}^{\infty} \theta(-t) \left\langle \{A(t),\, B\}_\pm \right\rangle e^{i(\omega - i0^+)t}\, dt$$

$$= \pm\, i \int_{-\infty}^{0} \left\langle \{A(t),\, B\}_\pm \right\rangle e^{i(\omega - i0^+)t}\, dt. \tag{8.1.17}$$

If we note the relation (4.1.48) and (4.1.51), and

$$\theta(t) = -\frac{i}{2\pi} \int_{-\infty}^{\infty} \frac{e^{ixt}}{x - i0^+}\, dx, \tag{8.1.18}$$

$$\theta(-t) = -\frac{i}{2\pi} \int_{-\infty}^{\infty} \frac{e^{-ixt}}{x - i0^+}\, dx, \tag{8.1.19}$$

the Fourier transforms of the Green's functions are given as

$$G^{\mathrm{r}}_{AB}(\omega) = \pm\frac{1}{2\pi}\int_{-\infty}^{\infty} \frac{e^{\beta\omega'}\pm 1}{\omega-\omega'+i0^+} J'(\omega')\,d\omega' , \qquad (8.1.20a)$$

$$G^{\mathrm{a}}_{AB}(\omega) = \pm\frac{1}{2\pi}\int_{-\infty}^{\infty} \frac{e^{\beta\omega'}\pm 1}{\omega-\omega'-i0^+} J'(\omega')\,d\omega' . \qquad (8.1.20b)$$

Then, if we use the relation (5.4.12) we have

$$G^{\mathrm{r}}_{AB}(\omega) - G^{\mathrm{a}}_{AB}(\omega) = \mp i\,(e^{\beta\omega}\pm 1)\,J'(\omega) \qquad (8.1.21)$$

and, then,

$$\begin{aligned} \langle BA \rangle &= \frac{1}{2\pi}\int_{-\infty}^{\infty} J'(\omega)\;d\omega \\ &= \mp\frac{1}{2\pi i}\int_{-\infty}^{\infty}\left[G^{\mathrm{r}}_{AB}(\omega) - G^{\mathrm{a}}_{AB}(\omega)\right]\frac{d\omega}{e^{\beta\omega}\pm 1} . \end{aligned} \qquad (8.1.22)$$

This is a very useful relation.

Now, corresponding to (8.1.20a) and (8.1.20b) let us define a function

$$G_{AB}(z) = \pm\frac{1}{2\pi}\int_{-\infty}^{\infty} \frac{e^{\beta\omega'}\pm 1}{z-\omega'} J'(\omega')\,d\omega'. \qquad (8.1.23)$$

This function of z is analytic except on the real axis and, in the upper-half of the complex z-plane, it can be viewed as the analytic continuation of $G^{\mathrm{r}}(\omega)$ and, in the lower-half of the z plane, as that of $G^{\mathrm{a}}(\omega)$. Thus we note

$$G^{\mathrm{r}}(\omega) = G^{\mathrm{r}}(\omega+i0^+) \qquad (8.1.24a)$$

$$G^{\mathrm{a}}(\omega) = G^{\mathrm{r}}(\omega-i0^+) \qquad (8.1.24b)$$

for a real ω. This relation is evident if we compare (8.1.20a) and (8.1.20b). Thus, it is not necessary separately to obtain $G^{\mathrm{a}}_{AB}(\omega)$ if $G^{\mathrm{r}}_{AB}(\omega)$ is known.

If we apply the procedure of (8.1.22) to the free-electron Green's function of (8.1.12), we obtain

$$\langle a_k^+ a_k \rangle = -\frac{1}{2\pi i}\int_{-\infty}^{\infty} d\omega \left[\frac{1}{\omega - \xi_k + i0^+} - \frac{1}{\omega - \xi_k - i0^+}\right]\frac{1}{e^{\beta\omega} + 1}$$

$$= \int_{-\infty}^{\infty} d\omega\, \delta(\omega - \xi_k)\frac{1}{e^{\beta\omega} + 1} = f(\xi_k), \tag{8.1.25}$$

where we used the relation of (5.4.12) ; $f(\xi_k)$ is the Fermi distribution function.

8.1.4 One-Particle Green's Function of Interacting Electrons

If we apply (8.1.22) on the retarded two-time Green's function of interacting electrons, (8.1.14), we obtain

$$\langle a_k^+ a_k \rangle = \int_{-\infty}^{\infty} A(k, \omega) f(\omega)\, d\omega, \tag{8.1.26}$$

$$A(k, \omega) = -\frac{1}{\pi}\operatorname{Im} G_k^{\mathrm{r}}(\omega) = -\frac{1}{\pi}\frac{\operatorname{Im}\Sigma}{(\omega - \xi_k - \operatorname{Re}\Sigma)^2 + (\operatorname{Im}\Sigma)^2}, \tag{8.1.27}$$

where we put $\Sigma = \Sigma(k, \omega + i0^+)$. We call $A(k, \omega)$ the spectral density.

Let us explore the basic properties of the spectral density and the Green's function of interacting electrons. Since $G_k^{\mathrm{r}}(\omega)$ is analytic in the upper-half of the complex ω-plane as noted in conjunction with (8.1.5') (*Problem* 8.1), we have the *Kramers-Kronig* relation between its real and imaginary parts,

$$\operatorname{Re} G_k^{\mathrm{r}}(\omega) = \frac{1}{\pi}\int_{-\infty}^{\infty} \mathrm{P}\,\frac{\operatorname{Im} G_k^{\mathrm{r}}(\omega')}{\omega' - \omega}\, d\omega'. \tag{8.1.28}$$

Then, if we use (5.4.12), (8.1.28) is rewritten as

$$G_k^{\mathrm{r}}(\omega) = \int_{-\infty}^{\infty} \frac{A(k, \omega')}{\omega - \omega' + i0^+}\, d\omega'. \tag{8.1.29}$$

The spectral density is proved to satisfy the following conditions (*Problem* 8.5),

$$A(k,\,\omega)\;\geq\;0\,; \qquad \int_{-\infty}^{\infty} d\omega\, A(k,\,\omega) \;=\; 1. \tag{8.1.30}$$

Here if we recall (8.1.22) and (8.1.20), we obtain an alternative expression for $A(k, \omega)$.

The meaning of the spectral density may be clear from either (8.1.26) or (8.1.29) together with (8.1.30). When there is no interaction between electrons, the spectral density is given by a single delta function with a peak at the free electron energy ξ_k as

$$A^{\mathrm{o}}(k,\,\omega) \;=\; -\frac{1}{\pi}\,\mathrm{Im}\,G_k^{\mathrm{ro}}(\omega) \;=\; \delta(\omega - \xi_k). \tag{8.1.31}$$

This case corresponds to assuming

$$\mathrm{Re}\,\Sigma \;=\; 0\;, \qquad \mathrm{Im}\,\Sigma \;=\; -\,0^+, \tag{8.1.32}$$

in (8.1.27). (To be exact, we put $\Sigma = 0$ and recover the $i0^+$ in (8.1.12)).

When there is interaction between electrons, both $\mathrm{Re}\,\Sigma$ and $\mathrm{Im}\,\Sigma$ generally do not vanish in (8.1.27). In such a situation the energy of an electron with wave vector $\boldsymbol{k}$ and spin σ is not fixed but spreads over a non-zero range of ω. Often $A(k, \omega)$ consists of a delta function-like sharp peak and a broad background; the peak corresponds to a one-particle-like excitation with an energy modified by $\mathrm{Re}\,\Sigma(k, \omega)$:

$$A(k,\,\omega) \;=\; \text{a sharp peak + a broad background.} \tag{8.1.33}$$

Concerning the sharp peak, its location, $\bar{\xi}_k$, is determined from

$$\omega - \xi_k - \mathrm{Re}\;\Sigma(k,\,\omega + i\,0^+) \;=\; 0, \tag{8.1.34}$$

where we assumed $\mathrm{Im}\;\Sigma(k,\,\omega + i0^+) \to 0$ for simplicity. Then for the spectral density corresponding to this peak we have

$$\begin{aligned} &\delta(\omega - \xi_k - \mathrm{Re}\;\Sigma(k,\,\omega + i\,0^+)) \\ &\quad = \; Z\,\delta(\omega - \bar{\xi}_k), \end{aligned} \tag{8.1.35}$$

where

$$Z = \left|1 - \frac{\partial}{\partial \omega} \mathrm{Re}\ \Sigma(k, \omega + i\,0^+)\right|^{-1}_{\omega = \bar{\xi}_k}$$

is called the *renormalization factor*.

If we note (8.1.30) and (8.1.33), we require $Z < 1$. This implies that the new one-particle state with the energy $\bar{\xi}_k$ can not accommodate one full electron. Corresponding to (8.1.33) the one-particle Green's function of interacting electrons may be written as

$$G^{\mathrm{r}}_k(\omega) = \frac{Z}{\omega - \bar{\xi}_k + i\,0^+} + \text{background}. \tag{8.1.36}$$

As to $\bar{\xi}_k$ to be obtained from (8.1.34) we may assume in the following form

$$\bar{\xi}_k = \frac{\boldsymbol{k}^2}{2m^*} - \mu, \tag{8.1.37}$$

with an effective mass of an electron m^*, and the chemical potential μ. Recall that $\xi_k = \boldsymbol{k}^2/2m - \mu$. Then from (8.1.34) we obtain

$$\frac{\partial \bar{\xi}_k}{\partial \xi_k} - 1 - \left[\frac{\partial\, \mathrm{Re}\ \Sigma(k, \omega + i\,0^+)}{\partial \xi_k} + \frac{\partial\, \mathrm{Re}\ \Sigma(k, \omega + i\,0^+)}{\partial \omega} \frac{\partial \bar{\xi}_k}{\partial \xi_k}\right]_{\omega = \bar{\xi}_k} = 0$$

or

$$\frac{m^*}{m} = \left[\frac{1 - \partial\, \mathrm{Re}\, \Sigma(k, \omega)/\partial \omega}{1 + \partial\, \mathrm{Re}\ \Sigma(k, \omega)/\partial \xi_k}\right]_{\omega = \bar{\xi}_k}, \tag{8.1.38}$$

where we noted $\partial \bar{\xi}_k/\partial \xi_k = m/m^*$.

There may be a case in which we can assume the solution of (8.1.15) in the form

$$\omega = \xi_k + \mathrm{Re}\ \Sigma + i\ \mathrm{Im}\ \Sigma, \tag{8.1.39}$$

or, put

$$G^{\mathrm{r}}_k(\omega) = \frac{1}{\omega - \xi_k - \mathrm{Re}\, \Sigma - i\, \mathrm{Im}\, \Sigma}, \tag{8.1.40}$$

by suppressing the ω dependence in $\mathrm{Re}\,\Sigma$ and $\mathrm{Im}\,\Sigma$. Then by Fourier Inverse transforming (8.1.40) we obtain

$$G_k^r(t) = \frac{1}{2\pi}\int_{-\infty}^{\infty} d\omega \frac{e^{-i\omega t}}{\omega - \xi_k - \mathrm{Re}\,\Sigma - i\,\mathrm{Im}\,\Sigma}$$

$$= -i\,e^{-i(\xi_k + \mathrm{Re}\,\Sigma)t}\,e^{-|\mathrm{Im}\,\Sigma|\,t}, \qquad (8.1.41)$$

where we assumed $\mathrm{Im}\,\Sigma < 0$, and closed the contour of the integration by adding a semicircle in the lower-half of the complex ω-plane. The effect of the self-energy is, first, to shift the energy of an electron by $\mathrm{Re}\,\Sigma$ then to make the life-time of the state finite. We already encountered a situation such as above with phonon in 5.4.1; see also 9.2.

8.2 TEMPERATURE GREEN'S FUNCTIONS

8.2.1 The Definition of the Temperature Green's Functions

The one-particle temperature Green's function of electron is defined as follows,

$$G_\sigma(\boldsymbol{k}, \tau, \tau') = G(k, \tau, \tau') = -\left\langle T_\tau\, a_k(\tau)\, a_k^+(\tau') \right\rangle, \qquad (8.2.1)$$

where we put

$$a_k(\tau) = e^{\tau K}\, a_k\, e^{-\tau K}. \qquad (8.2.2)$$

If we compare (8.2.2) with (8.1.3) we note that $a_k(\tau)$ corresponds to the Heisenberg representation for the imaginary time $t = -i\tau\hbar$ with $\hbar = 1$. T_τ is the *chronological operator* which dictates to order the operators according to the "time" τ. In the case of fermion operators we have to attach (−1) for each permutation of an operator with another one; in the case of boson operators, such a change of sign does not occur. Thus, for electrons we have

$$T_\tau(\,a_k(\tau)\,a_k^+(\tau')\,) = \begin{cases} a_k(\tau)\,a_k^+(\tau') & \text{for} \quad \tau > \tau', \\ \\ -\,a_k^+(\tau')\,a_k(\tau) & \text{for} \quad \tau < \tau'. \end{cases} \qquad (8.2.3)$$

The case of $\tau = \tau'$ in (8.2.1) is to be understood as

$$G(k, \tau, \tau) = G(k, \tau, \tau + 0^+) = \langle a_k^+(\tau)\, a_k(\tau) \rangle. \tag{8.2.4}$$

For bosons such as phonons,

$$T_\tau(\, b_k(\tau)\, b_k^+(\tau')\,) = \begin{cases} b_k(\tau) b_k^+(\tau') & \text{for } \tau > \tau', \\ \\ b_k^+(\tau') b_k(\tau) & \text{for } \tau < \tau'. \end{cases} \tag{8.2.5}$$

$\langle \ \rangle$ stands for the thermal average with the grand canonical distribution.

The Green's function of (8.2.1) is explicitly written as

$$G(k, \tau, \tau') = -\,\mathrm{tr}\left[e^{-\beta(K-\Omega)}\, T_\tau\!\left(e^{\tau K}\, a_k\, e^{-(\tau-\tau')K}\, a_k^+\, e^{-\tau' K}\right)\right], \tag{8.2.6}$$

where Ω is the thermodynamic potential defined in (1.4.18). With the cyclic property of (3.4.8) for the trace, we find that the Green's function depends only on the time difference,

$$G(k, \tau, \tau') = G(k, \tau - \tau'). \tag{8.2.7}$$

Thus it suffices to pursue the Green's function of the following form

$$G(k, \tau) = -\langle T_\tau\, a_k(\tau)\, a_k^+ \rangle. \tag{8.2.8}$$

8.2.2 Fourier Transform of Temperature Green's Function

We find it convenient to define $G(k, \tau)$ for

$$-\beta \le \tau \le \beta\,. \tag{8.2.9}$$

The Fourier transform of $G(k, \tau)$ is, then, defined as

$$G(k, i\omega_n) = \frac{1}{2}\int_{-\beta}^{\beta} d\tau\, G(k, \tau)\, e^{i\omega_n \tau}, \tag{8.2.10}$$

$$G(k, \tau) = \frac{1}{\beta}\sum_n G(k, i\omega_n)\, e^{-i\omega_n \tau}\,. \tag{8.2.11}$$

The frequency ω_n appearing in the Fourier transform is an integral multiple of $2\pi/2\beta = \pi/\beta$. For the one-particle Green's function of fermions the following relation holds for $-\beta < \tau < 0$,

$$G(k, \tau + \beta) = -G(k, \tau). \tag{8.2.12}$$

This relation can be proved by using the cyclic property of the trace. Similarly for $\beta > \tau > 0$, we can prove the following relation,

$$G(k, \tau - \beta) = -G(k, \tau). \tag{8.2.13}$$

We find then that the Fourier transform vanishes for even multiples of π/β. Thus, the components $G_\sigma(\boldsymbol{k}, i\omega_n)$ which do not vanish are those with the following form of frequency,

$$\omega_n = (2n+1)\pi/\beta \tag{8.2.14}$$

which is called the *fermion Matsubara frequency.*

8.2.3 Relation between Temperature Green's Function and Retarded Two-Time Green's Function

Although we demonstrated in 8.1 the physical meaning of the retarded two-time Green's function, we do not yet know the physical meaning of the temperature Green's function. The answer to this question is given by the following relation,

$$G(k, i\omega_n)\,|_{\,i\omega_n \to \omega + i0^+} = G_k^{\mathrm{r}}(\omega). \tag{8.2.15}$$

Once we know a temperature Green's function, the corresponding retarded two-time Green's function is straightforwardly obtained by the procedure of (8.2.15).

The above relation is proved as follows. Let us represent the eigenvalues and the eigenfunctions of K respectively by E_α and $|\,\alpha\rangle$, so that

$$K|\,\alpha\rangle = E_\alpha|\,\alpha\rangle. \tag{8.2.16}$$

Then, for $\tau > 0$ the temperature Green's function of (8.2.8) takes the following explicit form

$$G(k, \tau) = -\,\mathrm{tr}\left[e^{\beta(\Omega - K)}\, e^{\tau K}\, a_k\, e^{-\tau K}\, a_k^+\right]$$

$$= - \sum_{\alpha',\alpha''} e^{\beta(\Omega - E_{\alpha'})}$$

$$\times \left\langle \alpha' \middle| a_k \middle| \alpha'' \right\rangle \left\langle \alpha'' \middle| a_k^+ \middle| \alpha' \right\rangle e^{\tau(E_{\alpha'} - E_{\alpha''})}. \tag{8.2.17}$$

If we Fourier transform this result by the prescription of (8.2.10), we obtain

$$G(k, i\omega_n) = \sum_{\alpha',\alpha''} e^{\beta(\Omega - E_{\alpha'})} \frac{e^{\beta(E_{\alpha'} - E_{\alpha''})} + 1}{i\omega_n + E_{\alpha'} - E_{\alpha''}}$$

$$\times \langle \alpha' | a_k | \alpha'' \rangle \langle \alpha'' | a_k^+ | \alpha' \rangle. \tag{8.2.18}$$

On the other hand, if we rewrite the corresponding retarded two-time Green's function (8.1.10) in a form similar to (8.2.17) by using (8.2.16), and then Fourier transform the result by the prescription (8.1.5), we obtain

$$G_k^{\mathrm{r}}(\omega) = \sum_{\alpha',\alpha''} e^{\beta(\Omega - E_{\alpha'})} \frac{e^{\beta(E_{\alpha'} - E_{\alpha''})} + 1}{E_{\alpha'} - E_{\alpha''} + \omega + i0^+}$$

$$\times \langle \alpha' | a_k | \alpha'' \rangle \langle \alpha'' | a_k^+ | \alpha' \rangle. \tag{8.2.19}$$

It is straightforward to confirm (8.2.15) by comparing (8.2.18) and (8.2.19). We can similarly confirm (8.2.15) also by assuming $\tau < 0$ in (8.2.8).

From the above demonstration it is obvious that a relation of the form of (8.2.15) holds also for Green's function of bosons (or, two-particle Green's function such as that of magnetization $M_\mu(\boldsymbol{q})$ etc.). Actually the signs of various Green's functions are defined so as to be consistent with (8.2.15).

As we already noted, the Feynman diagram method can be used for a temperature Green's function. Then, the corresponding retarded two-time Green's function, which bears direct physical meaning, can be obtained from the relation (8.2.15). As a simple example, the temperature Green's function of a free electron corresponding to the retarded two-time Green's function of (8.1.12) is obtained as

$$G_k^{\mathrm{o}}(i\omega_n) = \frac{1}{i\omega_n - \xi_k}. \tag{8.2.20}$$

8.3 TEMPERATURE GREEN'S FUNCTION AND THE THERMODYNAMIC POTENTIAL

We have understood that we can obtain the energy spectrum of interacting electrons if we know the one-particle temperature Green's function. In this section we show how the temperature Green's function of interacting electrons can be obtained by the method of perturbation expansion. In the course of the discussion we obtain also the prescription for the perturbational expansion of the thermodynamic potential of interacting electrons.

Let us divide the total Hamiltonian of interacting electrons into the free electron part and the interaction part,

$$K = H - \mu n = H_o + H_i - \mu n = K_o + H_i \, , \tag{8.3.1}$$

$$K_o = H_o - \mu n. \tag{8.3.2}$$

H_o is the sum of one-particle energy and H_i is either the Coulomb interaction between electrons, H_C, or the electron-phonon interaction, $H_{el\text{-}ph}$, or the sum of them, $H_C + H_{el\text{-}ph.}$ We introduce an operator $U(\tau)$ as follows,

$$e^{-\tau K} = e^{-\tau K_o} U(\tau); \quad e^{\tau K} = U^{-1}(\tau)\, e^{\tau K_o}. \tag{8.3.3}$$

This is an analogue of the time evolution operator in quantum mechanics; it may be viewed as the time evolution operator for imaginary time τ. Note that generally K_o and H_i, and, therefore, $e^{-\tau K_o}$ and $U(\tau)$ do not commute with each other. We can easily prove the relations

$$U(\tau)\, U^{-1}(\tau) = U^{-1}(\tau)\, U(\tau) = 1. \tag{8.3.4}$$

Let us define the interaction representation of an operator A as

$$\widehat{A}(\tau) = e^{\tau K_o} A\, e^{-\tau K_o}. \tag{8.3.5}$$

Then (8.2.8) is rewritten as

$$G(k, \tau) = -\frac{\mathrm{tr}\left[T_\tau\, e^{-\beta K_o}\, U(\beta)\, U^{-1}(\tau)\, \widehat{a}_k(\tau)\, U(\tau)\, a_k^+\right]}{\mathrm{tr}\left[e^{-\beta K_o}\, U(\beta)\right]} \tag{8.3.6}$$

For $U(\tau)$, we have the following equation of motion,

$$\frac{\partial}{\partial \tau} U(\tau) = -\widehat{H}_{\rm i}(\tau)\, U(\tau). \tag{8.3.7}$$

If we note that $U(0) = 1$, the above differential equation can be solved by iteration in the following way,

$$U(\tau) = 1 - \int_0^{\tau} d\tau_1\, \widehat{H}_{\rm i}(\tau_1)\, U(\tau_1)$$

$$= 1 - \int_0^{\tau} d\tau_1\, \widehat{H}_{\rm i}(\tau_1) + (-1)^2 \int_0^{\tau} d\tau_1 \int_0^{\tau_1} d\tau_2\, \widehat{H}_{\rm i}(\tau_1)\, \widehat{H}_{\rm i}(\tau_2) + \cdots$$

$$= \sum_{n=0}^{\infty} (-1)^n \int_0^{\tau} d\tau_1 \cdots \int_0^{\tau_{n-2}} d\tau_{n-1} \int_0^{\tau_{n-1}} d\tau_n\, \widehat{H}_{\rm i}(\tau_1) \cdots \widehat{H}_{\rm i}(\tau_n). \tag{8.3.8}$$

In the above integral, the imaginary times are ordered as

$$\tau \geq \tau_1 \geq \tau_2 \cdots \geq \tau_n. \tag{8.3.9}$$

If we use the chronological operator T_τ defined in (8.2.3), we can rewrite (8.3.8) in the following form,

$$U(\tau) = \sum_{n=0}^{\infty} (-1)^n \frac{1}{n!} \int_0^{\tau} d\tau_1 \cdots \int_0^{\tau} d\tau_n \left[T_\tau \widehat{H}_{\rm i}(\tau_1) \cdots \widehat{H}_{\rm i}(\tau_n) \right]$$

$$\equiv T_\tau \exp\left[-\int_0^{\tau} d\tau'\, \widehat{H}_{\rm i}(\tau') \right]. \tag{8.3.10}$$

The origin of the factor $1/n!$ is that the integral of (8.3.10) contains $n!$ equivalent contributions corresponding to $n!$ different labelling of times in (8.3.9). Since $H_{\rm i}$ contains even number of fermion operators, a reordering of $\widehat{H}_{\rm i}(\tau_j)$'s does not require a change in the sign.

Now note that such $U(\tau)$ can be commuted with $\hat{a}_k(\tau)$ in (8.3.6); both in $H_{\rm C}$ and $H_{\text{el-ph}}$ the electron creation and annihilation operators appear pair-wise and therefore under T_τ operation $\hat{a}_k(\tau)$ and $H_{\rm i}(\tau_j)$ can be commuted without any change in sign. Then, by using (8.3.4) and noting

$$\hat{a}_k^+(0) \equiv \hat{a}_k^+ = a_k^+, \tag{8.3.11}$$

(8.3.6) is reduced to

$$G(k, \tau) = -\frac{\langle T_\tau U(\beta)\, \hat{a}_k(\tau)\, \hat{a}_k^+ \rangle_0}{\langle U(\beta) \rangle_0}, \tag{8.3.12}$$

where we introduced the thermal average without the effect of interaction as

$$\langle A \rangle_0 = \frac{\mathrm{tr}\left(e^{-\beta K_0} A\right)}{\mathrm{tr}\left(e^{-\beta K_0}\right)}. \tag{8.3.13}$$

$U(\beta)$ appearing in (8.3.12) can be expanded as in (8.3.10). Then, the Green's function is obtained as a perturbation series with respect to the interaction H_i. As we will see shortly, however, it is quite complicated to actually evaluate the series to higher orders. From now on we are going to present a standard prescription for carrying out such a calculation systematically and efficiently.

Note that the denominator of (8.3.12) can be written as

$$\langle U(\beta) \rangle_0 = \frac{\mathrm{tr}\, e^{-\beta K}}{\mathrm{tr}\, e^{-\beta K_0}} = \exp\left(-\beta\left(\Omega - \Omega_0\right)\right), \tag{8.3.14}$$

where Ω and Ω_0 are the thermodynamic potentials, respectively, with and without the effect of the interaction. Thus, the evaluation of the denominator of (8.3.12) comprises a calculation of the thermodynamic potential of interacting electrons. In the next section we first show the usefulness of the method of temperature Green's functions and Feynman diagrams in calculating the thermodynamic potential from (8.3.14). Then in 8.5 we proceed to see how to obtain Green's functions of interacting electrons by using Feynman diagrams.

8.4 THE METHOD OF FEYNMAN DIAGRAMS FOR THE THERMODYNAMIC POTENTIAL

In this section, for the interaction H_i we consider only the Coulomb interaction between electrons H_C. The contribution of $H_{el\text{-}ph}$ to Ω will be fully discussed in Chapter 10. Then, from (8.3.10), (8.3.14) is expanded in the following form of series

$$\left\langle U(\beta)\right\rangle_o = e^{-\beta(\Omega-\Omega_o)} \equiv e^{-\beta\Delta\Omega}$$

$$= 1 - \int_0^\beta d\tau \left\langle \widehat{H}_C(\tau)\right\rangle_o + \frac{1}{2!}\int_0^\beta d\tau_1 \int_0^\beta d\tau_2 \left\langle T_\tau \widehat{H}_C(\tau_1)\,\widehat{H}_C(\tau_2)\right\rangle_o + \cdots$$

$$= 1 - \frac{1}{2}\sum_{l,l',\kappa}{}' \, v(\kappa)\int_0^\beta d\tau \left\langle T_\tau \widehat{a}_l^{+}(\tau)\,\widehat{a}_{l'}^{+}(\tau)\,\widehat{a}_{l'+\kappa}(\tau)\,\widehat{a}_{l-\kappa}(\tau)\right\rangle_o$$

$$+ \frac{1}{2!}\left(\frac{1}{2}\right)^2 \sum_{\substack{\kappa_1,l_1,l_1' \\ \kappa_2,l_2,l_2'}}{}' v(\kappa_1)v(\kappa_2)\int_0^\beta d\tau_1 \int_0^\beta d\tau_2 \Big\langle T_\tau \widehat{a}_{l_1}^{+}(\tau_1)\,\widehat{a}_{l'_1}^{+}(\tau_1)\,\widehat{a}_{l'_1+\kappa_1}(\tau_1)\,\widehat{a}_{l_1-\kappa_1}(\tau_1)$$

$$\times\, \widehat{a}_{l_2}^{+}(\tau_2)\,\widehat{a}_{l_2'}^{+}(\tau_2)\,\widehat{a}_{l'_2+\kappa_2}(\tau_2)\,\widehat{a}_{l_2-\kappa_2}(\tau_2)\Big\rangle_o + \cdots . \tag{8.4.1}$$

We wrote down the perturbation series only up to the second order terms, but it is already quite complicated. The problem we face now is how to evaluate the thermal average of the form of $\langle T_\tau \hat{a}_{l1}{}^+(\tau_1)\, \hat{a}_{l'1}{}^+(\tau_1) \cdots \hat{a}_{l1-\kappa1}(\tau_2)\rangle_o$ appearing in the integrand. Wick's theorem tells us how such a thermal average can be decomposed into the sum of products of quantities $\langle T_\tau \hat{a}_l{}^+(\tau)\, \hat{a}_{l'}(\tau')\rangle_o$. Then, each term of the decomposition can be conveniently represented by a Feynman diagram, and vice versa. In this way it becomes possible to sum systematically a perturbation series such as (8.4.1) to higher orders. Let us begin with Wick's theorem.

8.4.1 Wick's Theorem

Let A_i represent either $\hat{a}_{k_i}(\tau_i)$ or $\hat{a}_{k_i}{}^+(\tau_i)$ and consider a quantity of the form of

$$\left\langle T_\tau A_1\, A_2 \cdots A_{2n}\right\rangle_o \equiv W_n. \tag{8.4.2}$$

Wick's theorem tells us that W_n can be decomposed as

$$W_n = \sum_{P_i} \delta(P_i)\left\langle T_\tau A_{i_1} A_{i_2}\right\rangle_o \left\langle T_\tau A_{i_3} A_{i_4}\right\rangle_o \cdots \left\langle T_\tau A_{i_{2n-1}} A_{i_{2n}}\right\rangle_o, \tag{8.4.3}$$

where P_i stands for a permutation $(1, 2, \cdots, 2n) \to (i_1, i_2, \cdots, i_{2n})$ and $\delta(P_i) = 1$ or -1 according whether the permutation P_i is even or odd. Σ_{P_i} indicates to sum over all the possible permutations. We call $\langle T_\tau A_{i1}\, A_{i2}\rangle_o$ the *contraction* of A_{i1} and A_{i2}. Note that the following two products,

$$\langle T_\tau A_{i_1} A_{i_2}\rangle_o \langle T_\tau A_{i_3} A_{i_4}\rangle_o \cdots \langle T_\tau A_{i_{2n-1}} A_{i_{2n}}\rangle_o,$$

$$\langle T_\tau A_{i_3} A_{i_4}\rangle_o \langle T_\tau A_{i_1} A_{i_2}\rangle_o \cdots \langle T_\tau A_{i_{2n-1}} A_{i_{2n}}\rangle_o,$$

are to be treated as equivalent and correspond to a single P_i; they should not be doubly counted. Also, note that

$$\langle T_\tau A_{i_2} A_{i_1}\rangle_o \langle T_\tau A_{i_3} A_{i_4}\rangle_o \cdots \langle T_\tau A_{i_{2n-1}} A_{i_{2n}}\rangle_o$$

is to be treated as equivalent to the above ones and, therefore, should not be counted separately (the difference in signs is taken care by $\delta(\boldsymbol{P}_i)$). In the next subsection we show how to avoid such possible double counting by relating each product of contractions to a Feynman diagram.

A proof of Wick's theorem is given in Appendix B. Here we concentrate on how to use the theorem. As an example, let us first discuss the first order term of (8.4.1). According to Wick's theorem we have

$$\langle T_\tau \hat{a}_l^+ \hat{a}_{l'}^+ \hat{a}_{l'+\kappa} \hat{a}_{l-\kappa}\rangle_o$$

$$= \langle T_\tau \hat{a}_l^+ \hat{a}_{l-\kappa}\rangle_o \langle T_\tau \hat{a}_{l'}^+ \hat{a}_{l'+\kappa}\rangle_o - \langle T_\tau \hat{a}_l^+ \hat{a}_{l'+\kappa}\rangle_o \langle T_\tau \hat{a}_{l'}^+ \hat{a}_{l-\kappa}\rangle_o . \quad (8.4.4)$$

There appears also a term $\langle T_\tau \hat{a}_l^+ \hat{a}_{l'}^+\rangle_o \langle T_\tau \hat{a}_{l-\kappa} \hat{a}_{l'+\kappa}\rangle_o$, but obviously it vanishes. Also, since in the present case the time τ is common we wrote $\hat{a}_l^+(\tau)$ etc. simply as $\hat{a}_l^+$ etc. It is important to note that the decomposition of (8.4.4) is an exact result.

The contractions at equal time appearing in (8.4.4) can be easily calculated as

$$\langle T_\tau \hat{a}_l^+(\tau)\, \hat{a}_{l-\kappa}(\tau)\rangle_o = \langle a_l^+ a_{l-\kappa}\rangle_o = f(\varepsilon_l)\delta_{\kappa,0}\,, \quad (8.4.5)$$

where $f(\varepsilon_l)$ is the Fermi distribution function.

By putting (8.4.4) together with (8.4.5) into the first order term in (8.4.1), we obtain the first order contribution of H_C to the thermodynamic potential, as

$$e^{-\beta\Delta\Omega} = 1 + \frac{\beta}{2} \sum_{l,l',\sigma} \mathrm{v}(\boldsymbol{l}-\boldsymbol{l}')\, f(\varepsilon_{l\sigma})\, f(\varepsilon_{l'\sigma}) + \cdots \quad (8.4.6)$$

The contribution coming from the first term on the right hand side of (8.4.4), $(\beta\, v(0)/2)\ \Sigma_{l\sigma,l'\sigma'} f(\varepsilon_{l\sigma}) f(\varepsilon_{l'\sigma'})$, does not exist since there should not appear v(0) as we discussed in 3.2. It is obvious that (8.4.6) represents the effect of the exchange interaction. Take logarithm of the both sides of (8.4.6) and then expand the right hand side in terms of $v(\boldsymbol{l}-\boldsymbol{l}')$; the first order term gives us the result of (3.4.6).

If we proceed to the second order term in (8.4.1), there appear a much greater number of terms with more complicated structures. Also there appear contractions between different times in addition to those between equal times. We now need the method of Feynman diagrams to deal with such complicated situations.

8.4.2 Wick's Theorem and Feynman Diagrams

In the higher order contributions in (8.4.1), there appear terms of the form of

$$\left\langle T_\tau \hat{a}^+(1)\,\hat{a}^+(1')\,\hat{a}(1')\,\hat{a}(1)\,\hat{a}^+(2)\,\hat{a}^+(2')\,\hat{a}(2')\,\hat{a}(2)\cdots\hat{a}^+(n)\,\hat{a}^+(n')\,\hat{a}(n')\,\hat{a}(n)\right\rangle_0$$

$$\equiv X_n\,, \qquad (8.4.7)$$

where we put $\hat{a}^+(i) = \hat{a}_{l_i}{}^+(\tau_i)$, $\hat{a}(i) = \hat{a}_{l_i-\kappa_i}(\tau_i)$ etc. Wick's theorem tells us how to decompose X_n into products of contractions with proper signs in all possible ways. Here note that since $\langle\,\hat{a}(i)\,\hat{a}(j)\,\rangle_0 = \langle\,\hat{a}^+(i)\,\hat{a}^+(j)\,\rangle_0 = 0$, there can not appear terms which contain such contractions. Also in making a contraction we always put the creation operator to the left of the annihilation operator. In this way we avoid double counting the contraction. In the following we show how we can carry out this complicated procedure correctly and efficiently.

Suppose first that $\hat{a}^+(1)$ is to be contracted with $\hat{a}(i)$. Then we rewrite (8.4.7) as

$$X_n = \left\langle T_\tau \hat{a}(1)\,\hat{a}^+(1)\,\hat{a}(1')\,\hat{a}^+(1')\cdots\hat{a}(i)\,\hat{a}^+(i)\cdots\hat{a}(n')\,\hat{a}^+(n')\right\rangle_0$$

$$= \left\langle T_\tau \hat{a}(1)\,\hat{a}^+(1)\ \hat{a}(i)\,\hat{a}^+(i)\ \hat{a}(1')\,\hat{a}^+(1')\cdots\hat{a}(n')\,\hat{a}^+(n')\right\rangle_0 . \qquad (8.4.8)$$

Since involved permutations are all even ones, a change in sign is not required. Now we contract $\hat{a}^+(1)$ with $\hat{a}(i)$:

$$\left\langle T_\tau \hat{a}(1)\ \underbrace{\hat{a}^+(1)\,\hat{a}(i)}\ \hat{a}^+(i)\,\hat{a}(1')\,\hat{a}^+(1')\cdots\hat{a}(n')\,\hat{a}^+(n')\right\rangle_0 , \qquad (8.4.9)$$

where we connected the pair of operators to be contracted by $\sqcup$.

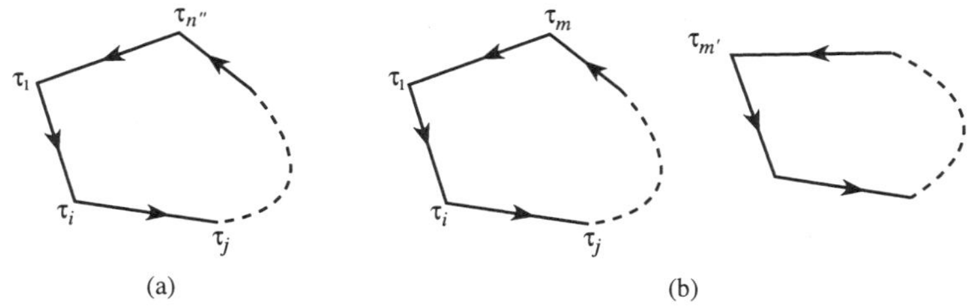

Fig.8.1 Contractions corresponding to (8.4.11) and (8.4.14) resulting in a single closed electron loop, (a), or more than one closed loops, (b).

Next, suppose $\hat{a}^+(i)$ is then to be contracted with $\hat{a}(j)$. We bring $\hat{a}(j)$ $\hat{a}^+(j)$ to the immediate right of $\hat{a}^+(i)$ and make a contraction as follows,

$$\left\langle T_\tau \hat{a}(1)\ \underbrace{\hat{a}^+(1)\,\hat{a}(i)}\ \underbrace{\hat{a}^+(i)\,\hat{a}(j)}\ \hat{a}^+(j)\,\hat{a}(1')\,\hat{a}^+(1')\cdots \right\rangle_0 .$$

Since the involved permutation is again an even one, we don't need a change in sign.

In this way finally we arrive at

$$\left\langle T_\tau \hat{a}(1)\ \underbrace{\hat{a}^+(1)\,\hat{a}(i)}\ \underbrace{\hat{a}^+(i)\,\hat{a}(j)}\ \underbrace{\hat{a}^+(j)\ \cdots}\ \hat{a}(n'')\,\hat{a}^+(n'') \right\rangle_0 . \qquad (8.4.10)$$

We make the final contraction between $\hat{a}^+(n'')$ and $\hat{a}(1)$ as

$$-\left[\ \underbrace{\hat{a}^+(1)\,\hat{a}(i)}\ \underbrace{\hat{a}^+(i)\,\hat{a}(j)}\ \cdots\ \underbrace{\hat{a}^+(n'')\,\hat{a}(1)}\ \right] . \qquad (8.4.11)$$

Here the sign changes because the permutation required to move $\hat{a}(1)$ from the position in (8.4.10) to that in (8.4.11) is an odd one.

The above procedure of making contractions can be represented as in Fig.8.1(a), a closed loop starting at 1 and ending at the starting point. A diagram such as Fig.8.1(a) is called a *Feynman diagram.*

There can be situations with more than 2 closed loops. In (8.4.10), if $\hat{a}(1)$ is contracted with some $\hat{a}^+(m)$ appearing to the left of $\hat{a}^+(n'')$, there one loop is completed, as shown in Fig.8.1(b). Then we begin to construct another loop with remaining members of operators. Thus, for a Feynman diagram with n_l closed loops, we need the sign factor of $(-1)^{n_l}$.

In this way we can exhaust all the possible ways of decomposing X_n of (8.4.7) into products of contractions. Finally, we note

$$\underbrace{\hat{a}^+(i)\,\hat{a}(j)} = \left\langle T_\tau\, \hat{a}^+(i)\,\hat{a}(j) \right\rangle_0 = -\left\langle T_\tau\, \hat{a}_{l_j}(\tau_j)\,\hat{a}^+_{l_i}(\tau_i) \right\rangle_0$$

$$= G^{o}(l_i, \tau_j - \tau_i)\delta_{l_j, l_i}, \tag{8.4.12}$$

where we assumed the system (electron gas) is spatially homogeneous. When it is not homogeneous because of an applied external field, for instance, (8.4.12) may not always vanish for $l_i \neq l_j$.

Let us note that (8.4.12) implies

$$\underbrace{\hat{a}^{+}(i)\,\hat{a}(j)} = \begin{cases} -\langle \hat{a}(j)\,\hat{a}^{+}(i)\rangle_{o} & \text{if} \quad \tau_i < \tau_j, \quad \text{(a)} \\ \\ \langle \hat{a}^{+}(i)\,\hat{a}(j)\rangle_{o} & \text{if} \quad \tau_i \geq \tau_j. \quad \text{(b)} \end{cases} \tag{8.4.13}$$

In the case (a), an electron l_i is created at time τ_i and later it is annihilated at time τ_j. On the other hand, in the case (b), first at time τ_j an electron is annihilated (a hole is created) and later at time τ_i the electron is created (the hole is annihilated). If we neglect the order of times τ_i and τ_j, however, both cases can be viewed commonly as the process of creating an electron at time τ_i and annihilating it at time τ_j, and are expressed commonly by the Green's function $G^{o}(l_i, \tau_j, \tau_i) = G^{o}(l_i, \tau_j - \tau_i)$.

Thus, to the diagram of Fig.8.1(a) corresponds the following expression,

$$- G^{o}(i, 1)\, G^{o}(j, i) \cdots G^{o}(1, n''), \tag{8.4.14}$$

where we put

$$G^{o}(j, i) = G^{o}(l_i, \tau_j - \tau_i)\, \delta_{l_i, l_j}. \tag{8.4.15}$$

Let us summarize what we have learned up to here. The problem is how to handle the thermal expectation of the form of X_n (8.4.7). We have found that X_n can be decomposed into the sum of the products of Green's functions of the form of (8.4.14), and that each of such products can be represented by a Feynman diagram such as Figs.8.1(a) or (b) and vice versa. This is the essence of the method of Green's functions and Feynman diagrams.

8.4.3 Linked Cluster Expansion of Thermodynamic Potential

Based on (8.4.1), let us proceed actually to calculate the thermodynamic potential of interacting electrons. Firstly, the first order contributions can be represented by Figs.8.2 (a) and (b). The broken line which represents the Coulomb interaction is illustrated in (c).

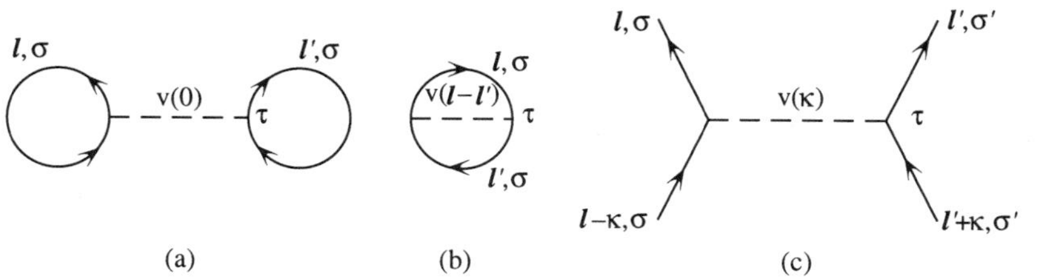

Fig.8.2 The Coulomb interaction between electrons, (c), and its first order contributions in the expansion of (8.4.1) for the thermodynamic potential of an electron gas, (a) and (b).

As we saw in the preceding subsection, electron lines must form closed loops. Thus, in the case of the first order contribution, there can be only these two diagrams given in Figs.8.2(a) and (b). It is obvious that the first and second terms on the right hand side of (8.4.4) correspond, respectively, to Figs.8.2(a) and 8.2(b).

Those diagrams in which time appears we call Feynman diagrams in the *time representation*. We will shortly encounter Feynman diagrams in the frequency representation.

As for the contribution of the first order terms, we already derived the explicit result of (8.4.6). Thus, let us proceed to the second order contributions. In the second order, there appear many kinds of Feynman diagrams as shown in Fig.8.3. The rule to obtain those second order Feynman diagrams is as follows: (1) Draw two horizontal Coulomb broken lines (two vertices). (2) At each end of the vertex, one electron line terminates and another electron line emerges. (3) By connecting electron lines, construct closed loops; the number of closed loops can be more than one.

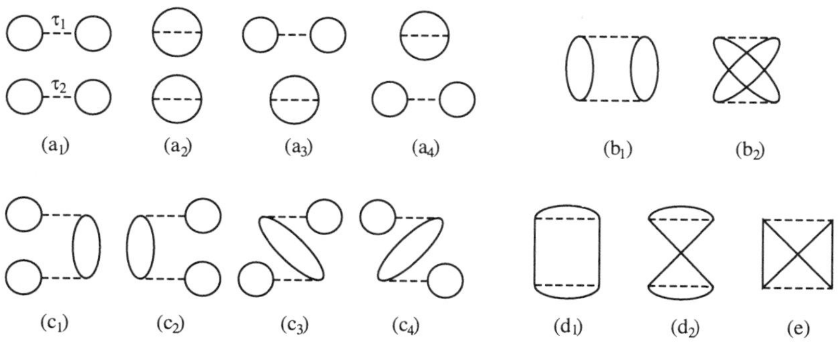

Fig.8.3 2nd order diagrams in the expansion of (8.4.1).

The Feynman diagrams of Fig.8.3 are drawn following the above rules. Note that the diagrams (a_1)–(a_4) are qualitatively different from the remaining diagrams (b_1)–(e); the diagrams such as (a_1)–(a_4) are called *disconnected diagrams*, while the diagrams such as (b_1)–(e) are called *connected diagrams*.

In this way, with increasing order, the right hand side of (8.4.1) gives rise to rapidly increasing number of diagrams, either connected or disconnected. Here comes up the *linked cluster expansion theorem*. The theorem states that the right hand side of (8.4.1) can be expressed as

$$\exp\left[-\beta(\Omega-\Omega_0)\right]$$

$$= \exp\left[\sum_{n=1}^{\infty}\frac{1}{n!}(-1)^n\int_0^\beta d\tau_1\cdots\int_0^\beta d\tau_n\left\langle T_\tau \widehat{H}_C(\tau_1)\cdots\widehat{H}_C(\tau_n)\right\rangle_{oc}\right], \quad (8.4.16)$$

where $\langle\ \rangle_{oc}$ indicates to take only the connected diagrams; "linked" means "connected". This theorem greatly simplifies our job.

The linked cluster expansion theorem is proved as follows. Introducing a parameter g, we rewrite (8.4.1) as

$$\exp\left[-\beta(\Omega-\Omega_0)\right]$$

$$= \sum_{n=0}^{\infty}\frac{g^n}{n!}(-1)^n\int_0^\beta d\tau_1\cdots\int_0^\beta d\tau_n\left\langle T_\tau \widehat{H}_C(\tau_1)\cdots\widehat{H}_C(\tau_n)\right\rangle_0\Big|_{g=1}$$

$$\equiv \sum_{n=0}^{\infty} g^n V_n\big|_{g=1}\,. \quad (8.4.17)$$

The n-th order contribution consists of various products of connected diagrams; if m_j-th order connected diagrams appear p_j times in one of such n-th order terms, we require $\Sigma_j\, p_j\, m_j = n$. Let us put the contribution of such an m_j-th order connected diagrams as U_{m_j}, namely,

$$U_{m_j} = \frac{(-1)^{m_j}}{m_j!}\int_0^\beta d\tau_1\cdots\int_0^\beta d\tau_{m_j}\left\langle T_\tau \widehat{H}_C(\tau_1)\cdots\widehat{H}_C(\tau_{m_j})\right\rangle_{oc}. \quad (8.4.18)$$

Then (8.4.17) is rewritten as

$$\sum_{n=0}^{\infty} g^n V_n = \sum_{m_j, p_j} \frac{(g^{m_1} U_{m_1})^{p_1}}{p_1!} \frac{(g^{m_2} U_{m_2})^{p_2}}{p_2!} \cdots$$

$$= \prod_{m_j} \sum_{p_j} \frac{(g^{m_j} U_{m_j})^{p_j}}{p_j!} = \exp\left(\sum_{m=1}^{\infty} g^m U_m \right), \qquad (8.4.17')$$

where we noted that the number of different ways to put n items (τ_i) into p_j bundles, $j = 1, 2, \cdots$, each containing m_j items, is $n! \,/\, \prod_j p_j!\, (m_j!)^{p_j}$. By putting $g = 1$ we obtain (8.4.16), namely,

$$-\beta(\Omega - \Omega_0) \equiv -\beta \Delta\Omega = -\beta \sum_n \Delta\Omega_n$$

$$= \ln \langle U(\beta) \rangle_0 = \sum_{m=1}^{\infty} U_m . \qquad (8.4.19)$$

Of the many different diagrams of the second order given in Fig.8.3, the diagrams (a_1)–(a_4) do not contribute to $\Delta\Omega$, since they are "not connected". The diagrams (c_1)–(c_4) also do not contribute since they involve v(0) which is excluded from the Hamiltonian H_C. The diagrams such as (b_1) and (b_2) are called the *ring diagrams* and their contributions are most important as we will show shortly. The diagrams (d_1), (d_2) and (e) are connected but do not belong to ring diagrams.

Now, note that for each of the m_j-th order connected ring diagrams such as Figs.8.3 (b_1)–(b_2), there are $(m_j - 1)!\; 2^{m_j - 1}$ entirely equivalent diagrams. The first factor $(m_j - 1)!$ comes from the fact that the integral of (8.4.18) is independent of time ordering of $(\tau_1, \tau_2, ..., \tau_{m_j})$ and that in a connected diagram a cyclic permutation of $(\tau_1, \tau_2, ..., \tau_{m_j})$ does not produce any new diagrams. The second factor $2^{m_j - 1}$ comes from the fact that rotating the Coulomb broken line by 180° around an axis vertical to the page, or, changing the polarity of the Coulomb vertex, leaves the integral invariant; why it is not 2^{m_j} but $2^{m_j - 1}$ is because if we change the polarity of all of m_j vertices in a ring diagram simultaneously, it reduces to the original one.

In the second order connected ring diagrams, $(m_j - 1)! = 1$ and this factor does not matter. However, $2^{m_j - 1} = 2$, and this factor stands for the fact that the contributions of the two diagrams Figs.8.3 (b_1) and (b_2), for instance, are identical.

From the above we find it convenient to introduce the *unlabelled diagrams* in which all such equivalent diagrams are represented by a single diagram. Then, for the case of ring diagrams, (8.4.18), which is to be put in (8.4.19), is rewritten as

$$U_{m,\,\text{ring}} = \frac{(-1)^m\, 2^{m-1}}{m} \int_0^\beta d\tau_1 \cdots \int_0^\beta d\tau_m \left\langle T_\tau \widehat{H}_C(\tau_1) \cdots \widehat{H}_C(\tau_m) \right\rangle_{oL} . \quad (8.4.20)$$

To distinguish them from the time and polarity labelled diagrams we used the subscript "L" for the unlabelled connected diagrams.

8.4.4 The Sum of Ring Diagram Contributions to Infinite Order

Let us calculate the contribution to $\Delta\Omega$ of the diagram (b_1) of Fig.8.3 which is redrawn as diagram (a) of Fig.8.4. If we call the contribution of this diagram $\Delta\Omega_b$, from (8.4.18)–(8.4.19) and what we learned in 8.4.2, we have

$$-\beta\,\Delta\Omega_b = \frac{1}{2!}(-1)^2\left(\frac{1}{2}\right)^2 \sum_{\kappa,l,l'} v(\kappa)\, v(-\kappa) \int_0^\beta d\tau_1 \int_0^\beta d\tau_2$$

$$\times \left\{-G^o(l,\, \tau_2-\tau_1)\, G^o(l+\kappa,\, \tau_1-\tau_2)\right\}$$

$$\times \left\{-G^o(l',\, \tau_2-\tau_1)\, G^o(l'-\kappa,\, \tau_1-\tau_2)\right\}. \quad (8.4.21)$$

We noted that in order to conserve the momenta of electrons, the momentum to be exchanged at the vertices should be related as given in Fig.8.4(a). Each of two curly brackets represents the contribution of a closed loop.

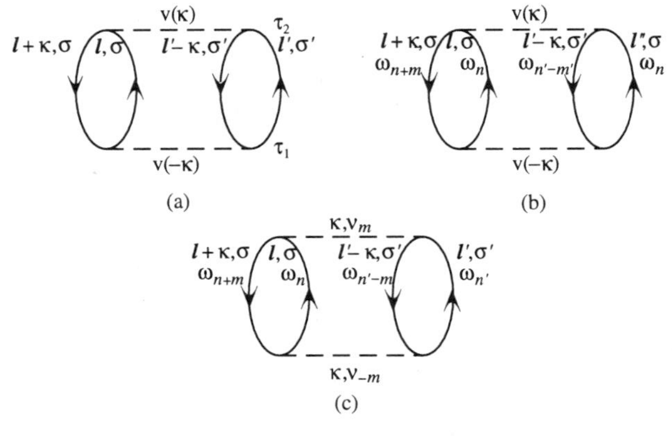

Fig.8.4 Momentum and energy conservation in the second order ring diagram.

Let us rewrite (8.4.21) as

$$-\beta\,\Delta\Omega_b = \frac{1}{8}\sum_{\kappa} \mathrm{v}(\kappa)\,\mathrm{v}(-\kappa)\,\pi_2(\kappa), \tag{8.4.22}$$

$$\pi_2(\kappa) = \sum_{\sigma,\sigma'}\int_0^\beta d\tau_1 \int_0^\beta d\tau_2\, F_\sigma(\kappa,\tau_2,\tau_1)\, F_{\sigma'}(-\kappa,\tau_2,\tau_1), \tag{8.4.23}$$

$$F_\sigma(\kappa,\tau_2,\tau_1) = -\sum_{l} G^0_\sigma(\boldsymbol{l},\tau_2-\tau_1)\, G^0_\sigma(\boldsymbol{l}+\kappa,\tau_1-\tau_2), \tag{8.4.24}$$

where we made the spin of electrons explicit; we have been using the convention $l = (\boldsymbol{l},\sigma)$ (see (8.2.1)).

Evidently $F_\sigma(\kappa,\tau_2,\tau_1)$ corresponds to the left hand bubble or ring diagram in Fig.8.4(a). If we note (8.2.11) and (8.2.20) on the Fourier transform of temperature Green's function, (8.4.24) is rewritten as

$$F_\sigma(\kappa,\tau_2,\tau_1) = -\sum_{l,n,m}\frac{1}{\beta^2}\,\frac{1}{i\omega_n-\xi_l}\,\frac{1}{i\omega_{n+m}-\xi_{l+\kappa}}$$

$$\times\, e^{-i\omega_n(\tau_2-\tau_1)}\, e^{-i\omega_{n+m}(\tau_1-\tau_2)}, \tag{8.4.25}$$

where we chose various frequencies as in Fig.8.4(b). If we choose various frequencies of Fourier transform as in Fig.8.4(b) also for the right hand bubble, the integral in (8.4.23) becomes as follows,

$$\int_0^\beta d\tau_1\int_0^\beta d\tau_2 \exp\Big[-i\big(\omega_n-\omega_{n+m}\big)\tau_2 - i\big(\omega_{n+m}-\omega_n\big)\tau_1$$

$$-\, i\big(\omega_{n'}-\omega_{n'-m'}\big)\tau_2 - i\big(\omega_{n'-m'}-\omega_{n'}\big)\tau_1\Big] = \beta^2\delta_{m,m'}\,. \tag{8.4.26}$$

Thus, in Fig.8.4(b) the only non-vanishing contributions are those for $m = m'$. This result is understood as the requirement of *energy conservation* at the Coulomb vertices at times τ_1 and τ_2; at time τ_2 for instance, we require.

$$\omega_n + \omega_{n'} = \omega_{n+m} + \omega_{n'-m'}\,. \tag{8.4.27}$$

As for momentum conservation, since it is built into the Hamiltonian there cannot appear any diagrams which violate it.

In this way we arrive at Fig.8.4(c); this is the final Feynman diagram which satisfies the requirement of conservation of energy and momentum. The upper broken line representing the Coulomb interaction may be viewed to carry the momentum κ and energy

$$\nu_m = 2m\pi/\beta = \omega_{n+m} - \omega_n . \tag{8.4.28}$$

Later we will find ν_m to be the *boson Matsubara frequency* .

In Fig.8.4(c) there does not appear the time τ. Instead, there appear frequencies ω_n and ν_m and the requirement of their conservation. We call such a diagram as Feynman diagram in the *frequency representation*. Note that while an electron line has its direction as shown by an arrow, the Coulomb broken line does not have a direction. It is because a Coulomb broken line of either direction can be represented by the same $v(\kappa) = v(-\kappa)$. The same situation holds for phonons as we will see later. (In Fig.8.4(c), however, it might be convenient to assume the broken lines are directed left-ward.)

Thus, (8.4.23) and (8.4.24) are reduced to

$$\pi_2(\kappa) = \sum_{m,\sigma,\sigma'} F_\sigma(\kappa, i\nu_m) F_{\sigma'}(-\kappa, -i\nu_m) , \tag{8.4.29}$$

$$F_\sigma(\kappa, i\nu_m) = -\frac{1}{\beta}\sum_{n,l} G^0_\sigma(l, i\omega_n) G^0_\sigma(l+\kappa, i\omega_{n+m})$$

$$= -\frac{1}{\beta}\sum_{n,l} \frac{1}{i\omega_n - \xi_{l\sigma}} \frac{1}{i\omega_{n+m} - \xi_{l+\kappa,\sigma}} \tag{8.4.30}$$

$F_\sigma(\kappa, i\nu_m)$ represents a bubble or ring of electron line with spin σ and it is often called the polarization function. Obviously it satisfies the relation

$$F_\sigma(\kappa, i\nu_m) = F_\sigma(-\kappa, -i\nu_m). \tag{8.4.31}$$

It is convenient to put

$$P(\kappa, i\nu_m) = \sum_\sigma F_\sigma(\kappa, i\nu_m). \tag{8.4.32}$$

Note that the contributions of diagrams (b_1) and (b_2) of Fig.8.3 are identical. Then the sum of the contributions to the thermodynamic potential of these two second order ring diagrams is obtained from (8.4.22) as

$$-2\beta\,\Delta\Omega_b = \frac{(-1)^2}{2\times 2}\sum_{\kappa,m} v(\kappa)^2\,P(\kappa,\,i\nu_m)^2$$

$$\equiv -\beta\,\Delta\Omega_{2,\,\mathrm{ring}}\,, \qquad (8.4.33)$$

where we used (8.4.31) and $v(\kappa) = v(-\kappa)$. This result confirms the general rule of (8.4.20) with unlabelled ring diagrams. From now on, let us use (8.4.20) in obtaining the contributions from the higher order ring diagrams.

The unlabelled 3rd order ring diagram is shown by the second diagram in Fig.8.5. In the same way as we obtained (8.4.22) and (8.4.33), for this contribution we have

$$-\beta\,\Delta\Omega_{3,\,\mathrm{ring}} = \frac{(-1)^3}{3}\,2^{3-1}\sum_{\kappa,m}\left(\frac{1}{2}\,v(\kappa)\right)^3 P(\kappa,\,i\nu_m)^3$$

$$= \frac{(-1)^3}{2\times 3}\sum_{\kappa,m} v(\kappa)^3 P(\kappa,\,i\nu_m)^3. \qquad (8.4.34)$$

Similarly, the contribution of the n-th order ring diagram in Fig.8.5 is readily obtained as

$$-\beta\,\Delta\Omega_{n,\,\mathrm{ring}} = \frac{(-1)^n}{2n}\sum_{\kappa,m} v(\kappa)^n P(\kappa,\,i\nu_m)^n. \qquad (8.4.35)$$

We, then, note that the contributions of the ring diagrams to infinite order can be summed as

$$\Delta\Omega'_{\mathrm{ring}} \equiv \Delta\Omega_{2,\,\mathrm{ring}} + \Delta\Omega_{3,\,\mathrm{ring}} + \cdots$$

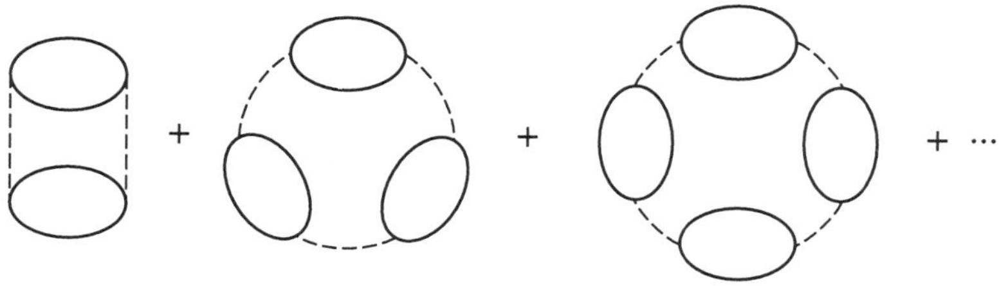

Fig.8.5 The sum of unlabelled ring diagrams.

$$= -\frac{1}{2}\frac{1}{\beta}\sum_{\kappa,m}\left[\frac{1}{2}\{\mathrm{v}(\kappa)\,P(\kappa,i\nu_m)\}^2 - \frac{1}{3}\{\mathrm{v}(\kappa)\,P(\kappa,i\nu_m)\}^3 + \cdots\right]$$

$$= \frac{1}{2}\frac{1}{\beta}\sum_{\kappa,m}\left[\ln\{\,1+\mathrm{v}(\kappa)\,P(\kappa,i\nu_m)\,\} - \mathrm{v}(\kappa)\,P(\kappa,i\nu_m)\right]. \qquad (8.4.36)$$

The contribution from diagrams other than rings, such as diagrams (d_1), (d_2), and (e) in Fig.8.3, for instance, are not included in the above sum. The reason why the ring diagrams are most important is as follows. As can be seen from (8.4.36), the contribution of an n-th order ring diagram gives rise to a factor $1/\kappa^{2n}$ which diverges for $\kappa \to 0$; any n-th order contribution other than that of ring diagrams has lower degree of singularity with respect to small κ.

The result of (8.4.36) given in terms of $P(\kappa, i\nu_m)$ is quite formal. In the next subsection we show how in (8.4.30) the summation over ω_n can be carried out to make the polarization function $P(\kappa, i\nu_m)$ more tractable.

8.4.5 Summation over Frequencies $i\omega_n$ and $i\nu_m$

If we note that the function $1/(e^{\beta z}+1)$ has poles at the fermion Matsubara frequencies, $z = i\omega_n = i(2n+1)\pi/\beta$, with the common residues, $-1/\beta$, the summation over ω_n of $\mathrm{F}(i\omega_n)$ can be carried out by using Cauchy's theorem:

$$\sum_{n=-\infty}^{\infty} \mathrm{F}(i\omega_n) = -\frac{\beta}{2\pi i}\int_C \frac{\mathrm{F}(z)}{e^{\beta z}+1}\,dz\,, \qquad (8.4.37)$$

where we take the contour of the integral C as in Fig.8.6(a); the fermion and boson Matsubara frequencies are noted, respectively, by crosses and dots.

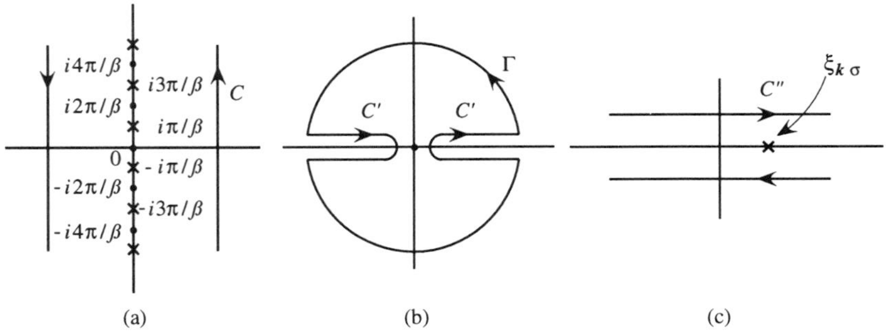

Fig.8.6 The contours of integration to be used in (8.4.37) (8.4.37') etc.

Similarly, for the summation over the boson frequencies, $i\nu_m = i2\pi m/\beta$ of $B(i\nu_m)$, if we note that the function $1/(e^{\beta z} - 1)$ has poles at $z = i\nu_m$ with residues $1/\beta$, we have

$$\sum_{m=-\infty}^{\infty} B(i\nu_m) = \frac{\beta}{2\pi i} \int_C \frac{B(z)}{e^{\beta z} - 1} dz . \tag{8.4.38}$$

If the contribution from Γ can be neglected we can transform the contour C in (8.4.37) to C'' via C', as shown in Fig.8.6. Then we have

$$\sum_{n=-\infty}^{\infty} F(i\omega_n) = -\frac{\beta}{2\pi i} \int_{-\infty}^{\infty} d\omega f(\omega) \left[F(\omega + i0^+) - F(\omega - i0^+)\right], \tag{8.4.37'}$$

where $f(z) = 1/(e^{\beta z} + 1)$ is the Fermi distribution function.

If $B(i\nu_m)$ is not singular at $i\nu_m = 0$, a similar relation also holds for summation over boson frequencies:

$$\sum_{m=-\infty}^{\infty} B(i\nu_m) = \frac{\beta}{2\pi i} \int_{-\infty}^{\infty} d\omega\, n(\omega) \left[B(\omega + i0^+) - B(\omega - i0^+)\right] + B(0), \tag{8.4.38'}$$

where $n(z) = 1/(e^{\beta z} - 1)$ is the Bose distribution function.

As an illustration, let us derive the following fundamental result for electrons using (8.4.37) (c.f.(8.2.4) and (8.2.11)):

$$\langle a^+_{k\sigma} a_{k\sigma} \rangle_0 = G^0_\sigma(\boldsymbol{k}, \tau = -0^+) = \frac{1}{\beta} \sum_{n=-\infty}^{\infty} G^0_\sigma(\boldsymbol{k}, i\omega_n)\, e^{i\omega_n 0^+}$$

$$= \frac{1}{\beta} \sum_{n=-\infty}^{\infty} \frac{e^{i\omega_n 0^+}}{i\omega_n - \xi_k} = -\frac{1}{2\pi i} \int_{C''} \frac{1}{z - \xi_k} \frac{1}{e^{\beta z} + 1} dz$$

$$= \frac{1}{e^{\beta \xi_k} + 1} = f(\xi_{k\sigma}). \tag{8.4.39}$$

Note that the contour C'' runs clockwise and that now within the contour C'' the integrand has a pole at $z = \xi_k$. Also note the role of the convergence factor $\exp(0^+ z)$; it assures that the integral over Γ vanishes even in the region Re $z < 0$.

Another important example which requires the summation over $i\omega_n$ is the polarization function given in (8.4.30). In transforming the sum to an integral by the procedure of (8.4.37), since the integrand is proportional to $1/z^2$, the contribution from Γ of Fig.8.6(b) can be safely neglected. Then rewriting (8.4.30) as

$$F_\sigma(\kappa, i\nu_m) = -\frac{1}{\beta}\sum_l \frac{1}{i\nu_m - \left(\xi_{l+\kappa,\sigma} - \xi_{l\sigma}\right)} \sum_n \left[\frac{1}{i\omega_n - \xi_{l\sigma}} - \frac{1}{i\omega_{n+m} - \xi_{l+\kappa,\sigma}}\right]$$

and using the result of (8.4.39) we obtain

$$F_\sigma(\kappa, i\nu_m) = \sum_l \frac{f(\xi_{l+\kappa,\sigma}) - f(\xi_{l\sigma})}{i\nu_m - \left(\xi_{l+\kappa,\sigma} - \xi_{l\sigma}\right)} . \tag{8.4.40}$$

If we put $i\nu_m \to \omega + i0^+$ with real ω as in (8.2.15), we obtain the familiar Lindhard response function, namely,

$$F_\sigma(\kappa, i\nu_m)\big|_{i\nu_m \to \omega + i0^+} = F_\sigma(\kappa, \omega). \tag{8.4.41}$$

We will later see the application of (8.4.38) for the summation over boson frequencies in connection with phonons and spin fluctuations.

8.4.6 General Rules to Obtain the Thermodynamic Potential of an Electron Gas Using Feynman Diagrams

Let us formulate the above procedure in terms of general rules. The diagrams here are labelled ones in the frequency representation.

(1) We represent an electron with spin σ and 4 momentum $(\boldsymbol{k}, i\omega_n)$ by a solid line and a Coulomb interaction by a broken line. With these solid lines and broken lines we construct a connected (or, linked) diagram. The 4 momenta of electrons are conserved at each Coulomb broken line and the spin of an electron does not change at each end of a Coulomb broken line.

(2) We associate an electron line with spin σ and 4 momentum $(\boldsymbol{k}, i\omega_n)$ with a factor

$$G^0_\sigma(\boldsymbol{k}, i\omega_n) = \frac{1}{i\omega_n - \xi_{\boldsymbol{k}\sigma}} .$$

However, when the two end points of an electron line are connected to the same Coulomb broken line either as (a) or (b) of Fig.8.2, where $G^{\mathrm{o}}_{\sigma}(\boldsymbol{k},\ \tau)$ with $\tau = -0^+$ comes up (see (8.2.4)), the electron line is associated with a factor

$$G^{\mathrm{o}}_{\sigma}(\boldsymbol{k},\ i\omega_n)\ e^{i\omega_n 0^+}.$$

(3) A Coulomb broken line by which two electrons are scattered as $\boldsymbol{k} \rightarrow \boldsymbol{k}'$ and $\boldsymbol{l} \rightarrow \boldsymbol{l}'$ is associated with a factor

$$\mathrm{v}(\boldsymbol{k}' - \boldsymbol{k}) = \mathrm{v}(\boldsymbol{l} - \boldsymbol{l}') = \frac{4\pi e^2}{V\left|\boldsymbol{k}' - \boldsymbol{k}\right|^2},$$

where V is the volume of the system.

(4) An n-th order (n Coulomb broken lines) connected diagram with n_l closed loops is associated with a factor

$$\left(-1\right)^{n+1}\left(-1\right)^{n_l}\frac{1}{\beta^{n+1}n!}\left(\frac{1}{2}\right)^n.$$

(5) If we represent those factors which come out of the above rules (2), (3) and (4) by $\mathrm{K_n}(2)$, $\mathrm{K_n}(3)$, and $\mathrm{K_n}(4)$, we sum over the spins and the 4 momenta of all the electron lines in the product $\mathrm{K_n}(2)\ \mathrm{K_n}(3)\ \mathrm{K_n}(4)$:

$$\sum_{\sigma_1 k_1 \omega_1}\ \sum_{\sigma_2 k_2 \omega_2} \cdots [\mathrm{K_n}(2)\ \mathrm{K_n}(3)\ \mathrm{K_n}(4)].$$

We apply the procedure of (1)–(5) to other topologically inequivalent (labelled) diagrams. Then we sum all these contributions over the order n.

Let us see how the factor $1/\beta^{n+1}$ comes out in the above rule (4). According to (8.4.18), in a contribution which is n-th order in H_{C}, there appear $2n$ pairs of creation and annihilation operators of electrons; we have, thus, a product of $2n$ one-electron Green's functions. If we rewrite each Green's function in the product by Fourier transformed one by (8.2.11), there appears the factor $1/\beta^{2n}$. Then we carry out the integral over all the times $\tau_1, \tau_2, \cdots, \tau_n$ for the interval $(0, \beta)$ as required in (8.4.18), we obtain the factor β^n together with the energy conservation requirement such as (8.4.26). Finally, to obtain $\Delta\Omega$ we have to divide both sides of (8.4.19) by β. In this way we end up with the factor $1/\beta^{n+1} : = (1/\beta^{2n})\ \beta^n\ (1/\beta) = 1/\beta^{n+1}$

Particularly, in the case of the ring diagrams, the rules are simplified by introducing unlabelled diagrams as in (8.4.20).

As a simple example let us calculate the first order contribution of Fig.8.2(b) using the above procedure. If we put various Fourier variables as in Fig.8.7, we have

$$\Delta\Omega_1 = (-1)^{1+1}(-1)^1 \frac{1}{\beta^{1+1}2}$$

$$\times \sum_{\substack{k,n,\kappa \\ m,\sigma}} v(\kappa)\, G^0_\sigma(k, i\omega_n)\, e^{i\omega_n 0^+}\, G^0_\sigma(k-\kappa, i\omega_{n-m})\, e^{i\omega_{n-m}0^+}$$

$$= \frac{1}{2\beta} \sum_{\kappa,m,\sigma} v(\kappa)\, F_\sigma(\kappa, i\nu_m) \tag{8.4.42}$$

$$= -\frac{1}{2} \sum_{k,l,\sigma} v(k-l)\, f(\xi_{k\sigma})\, f(\xi_{l\sigma}). \tag{8.4.42'}$$

The results of (8.4.42) and (8.4.42′) are obtained by using, respectively, (8.4.30) and (8.4.39). The result (8.4.42′) reproduces the earlier result of (8.4.6).

Note that if we add this $\Delta\Omega_1$ to $\Delta\Omega'_{\mathrm{ring}}$ of (8.4.36), the last term of $\Delta\Omega'_{\mathrm{ring}}$ is cancelled, and we have

$$\Delta\Omega_1 + \Delta\Omega'_{\mathrm{ring}} = \frac{1}{2\beta}\sum_{\kappa,m} \ln\{\, 1 + v(\kappa)\, P(\kappa, i\nu_m)\,\}$$

$$\equiv \Delta\Omega_{\mathrm{el,\ ring}}\,. \tag{8.4.43}$$

We will discuss the physical implication of the above result in Chapter 10 (see 10.1.1). As will be shown there, $\Delta\Omega_{\mathrm{el,ring}}$ represents the effect of the existence of electronic plasma oscillations, and that of the screening of the Coulomb interaction between electrons of the same spin.

The energy of a system beyond the Hartree-Fock approximation result is called the *correlation energy*. In the present case of an electron gas, the evaluation of the correlation energy at $T = 0$ equivalent to the result of (8.4.43) was first correctly carried out by Gell-Mann and Brueckner [8.7]. The result is often expressed in terms of the mean distance between electrons in units of Bohr radius, $a_B = \hbar^2/me^2$,

$$r_s = \frac{r_0}{a_B} = \left(\frac{9\pi}{4}\right)^{1/3}\frac{1}{k_F\, a_B}\,; \quad \frac{n}{V} = \left(\frac{4}{3}\pi r_0^3\right)^{-1}. \tag{8.4.44}$$

Then the total energy per electron at the ground state in units of Rydberg, $e^2/2a_B = me^4/2\hbar^2$, is given by

$$\varepsilon = \frac{2.21}{r_s^2} - \frac{0.916}{r_s} + 0.062 \ln r_s - 0.096, \tag{8.4.45}$$

where on the right hand side the first term is the kinetic energy, the second term is the exchange energy, $\Delta\Omega_1$, the last two terms represent the correlation energy which is the sum of the contributions of $\Delta\Omega'_{\text{ring}}$ and Fig.8.3(e), the last of which being a constant, -0.096. It is known that the contributions from the diagrams of the type of Figs.8.3(d_1) and (d_2) cancel out when they are summed to infinite order [7.9].

As to the correlation effect, or, energy, we give a more extensive discussion in Chapter 10. The above Gell-Mann and Brueckner's estimation by no means exhausts all the important aspects of correlation effects.

8.5 THE METHOD OF FEYNMAN DIAGRAMS FOR GREEN'S FUNCTION

We have extensively used thermal Green's functions of electrons in obtaining the thermodynamic potential of interacting electrons. Those Green's functions G^o's, however, were those of non-interacting electrons. In this section we study how to obtain Green's functions of electrons, G's, which contain the effect of H_C and/or H_{el-ph}. We discuss also the Green's functions of phonons. From such Green's functions we can obtain the energy spectra of interacting electrons and phonons.

8.5.1 Perturbational Method for the Electron Green's Function

Our starting point to formulate the perturbational method with Feynman diagrams in obtaining the thermal Green's functions of interacting electrons is (8.3.12). As for the denominator of (8.3.12), we already studied in the preceding section how it can be obtained by the linked cluster expansion.

In (8.3.12), the only difference between the numerator and the denominator is that the former contains the operator $\hat{a}_k(\tau)\hat{a}_k^\dagger$ within the bracket. Thus, the electron Green's function of (8.3.12) may be represented by the Feynman diagram of Fig.8.8, where we assumed $\tau > 0$. An electron ($\boldsymbol{k}$, σ) is created at $\tau = 0$ and after interacting with other electrons and phonons in the box, it is annihilated at time τ. We call the electron lines coming in and going out of the box the *external lines*.

When the box in Fig.8.8 is empty, that is, when there is no interaction, the Green's function of an electron is given by G^o.

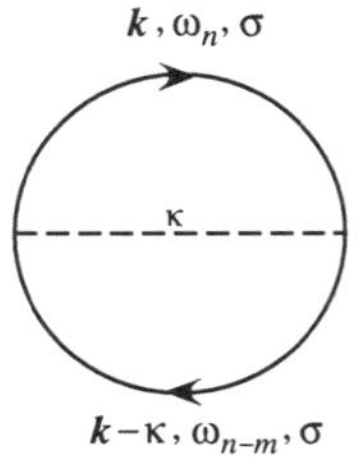

Fig.8.7 Fourier variables in the 1st order ring diagram which contributes to the thermodynamic potential of an electron gas.

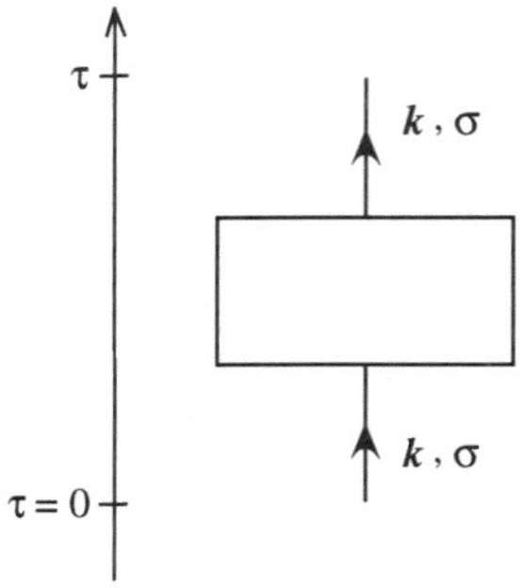

Fig.8.8 The general structure of one electron Green's function.

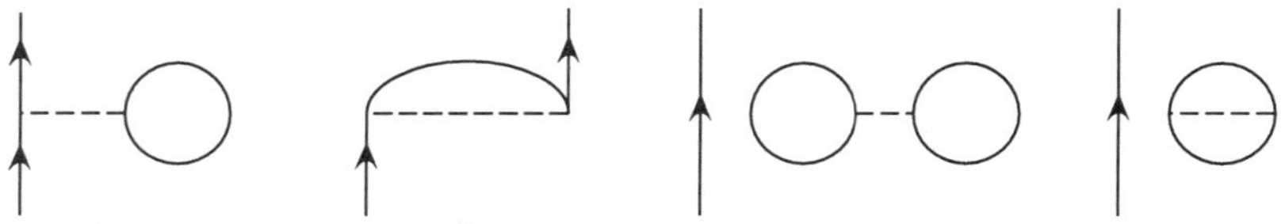

Fig.8.9 The first order diagrams in the expansion of $-\langle T_\tau U(\beta)\, \hat{a}_k(\tau)\, \hat{a}_k^+ \rangle_0$. While (a) and (b) are connected to the external lines, (c) and (d) are not.

Next, the contribution which is first order in H_C in $-\langle T_\tau U(\beta)\, \hat{a}_k(\tau)\, \hat{a}_k^+ \rangle_0$ the numerator of (8.3.12) is represented as in Fig.8.9. This can be confirmed by putting the first order term of the expansion of $U(\beta)$, (8.3.8) or (8.3.10), in it and then decomposing the time ordered expectation by Wick's theorem.

Of the 4 diagrams of Fig.8.9, (a) and (b) are qualitatively different from (c) and (d); while in (a) and (b) all the lines are connected to the external lines, in (c) and (d), parts of diagrams are not connected to the external lines. Diagrams for higher order processes are also expected to be divided into such two kinds. Actually we find

$$
\begin{aligned}
-\langle T_\tau U(\beta)\, \hat{a}_k(\tau)\, \hat{a}_k^+ \rangle_0 &= (\text{ all the diagrams connected to the external lines}) \\
&\quad \times (1 + \text{all the diagrams not connected to the external lines }) \\
&= -\langle T_\tau U(\beta)\, \hat{a}_k(\tau)\, \hat{a}_k^+ \rangle_{0c} \langle U(\beta) \rangle_0 . \qquad (8.5.1)
\end{aligned}
$$

Thus, from (8.3.12) we obtain

$$G(k,\tau) = (\text{ all the diagrams connected to the external lines })$$

$$= -\left\langle T_\tau\, U(\beta)\, \hat{a}_k(\tau)\, \hat{a}_k^+ \right\rangle_{\mathrm{oc}} . \tag{8.5.2}$$

The subscript c indicates that we must include only the connected diagrams. We can immediately check the validity of (8.5.1) or (8.5.2) for the zero-th and first order terms. We can prove its general validity as follows.

For the moment, for simplicity, we consider only electrons, interacting with the Coulomb interaction H_{C}. Then $U(\beta)$ in (8.5.1) is the same as in the preceding section and we have

$$-\left\langle T_\tau\, U(\beta)\, \hat{a}_k(\tau)\, \hat{a}_k^+ \right\rangle_{\mathrm{o}} = \sum_{n=0}^{\infty} \frac{1}{n!}(-1)^{n+1} \int_0^\beta d\tau_1 \cdots \int_0^\beta d\tau_n$$

$$\times \left\langle T_\tau\, \hat{H}_{\mathrm{C}}(\tau_1) \cdots \hat{H}_{\mathrm{C}}(\tau_n)\, \hat{a}_k(\tau)\, \hat{a}_k^+ \right\rangle_{\mathrm{o}} . \tag{8.5.3}$$

By applying Wick's theorem we can decompose the contribution of each order into many different terms. Suppose that one of the n-th order contribution consists of an m-th order diagram connected to the external line and the remaining part of the $(n-m)$-th order. The contribution of such a term is given as

$$(-1)\frac{(-1)^n}{n!} \int_0^\beta d\tau_1 \cdots \int_0^\beta d\tau_m \left\langle T_\tau\, \hat{a}_k(\tau)\, \hat{a}_k^+\, \hat{H}_{\mathrm{C}}(\tau_1) \cdots \hat{H}_{\mathrm{C}}(\tau_m) \right\rangle_{\mathrm{oc}}$$

$$\times \int_0^\beta d\tau_{m+1} \cdots \int_0^\beta d\tau_n \left\langle T_\tau\, \hat{H}_{\mathrm{C}}(\tau_{m+1}) \cdots \hat{H}_{\mathrm{C}}(\tau_n) \right\rangle_{\mathrm{o}} . \tag{8.5.4}$$

There are ${}_nC_m = n!\,/\,(m!\,(n-m)!)$ ways to choose m of $\hat{H}_{\mathrm{C}}(\tau_i)$'s to be put to the first factor in the integrand of (8.5.4) and each of them gives rise to identical contribution. Thus, considering all such contributions in (8.5.3) we have

$$(-1)\frac{(-1)^m}{m!} \int_0^\beta d\tau_1 \cdots \int_0^\beta d\tau_m \left\langle T_\tau\, \hat{a}_k(\tau)\, \hat{a}_k^+\, \hat{H}_{\mathrm{C}}(\tau_1) \cdots \hat{H}_{\mathrm{C}}(\tau_m) \right\rangle_{\mathrm{oc}}$$

$$\times \frac{(-1)^{n-m}}{(n-m)!} \int_0^\beta d\tau_{m+1} \cdots \int_0^\beta d\tau_n \left\langle T_\tau \widehat{H}_C(\tau_{m+1}) \cdots \widehat{H}_C(\tau_n) \right\rangle_0 . \quad (8.5.5)$$

If we note (8.4.1), by summing (8.5.5) over n and m, we arrive at (8.5.1), and finally, (8.5.2).

Starting from (8.5.2) we can formulate general Feynman rules for obtaining Green's function of electrons, as we will present in 8.5.3. Before that, in the next subsection we introduce the Green's functions of phonons.

8.5.2 Green's Functions of Phonons

The temperature Green's function of phonons is defined as

$$D(\boldsymbol{q}, \tau) = \left\langle T_\tau \, \widehat{\phi}_{\boldsymbol{q}}(\tau) \, \widehat{\phi}_{\boldsymbol{q}}^+ \right\rangle , \quad (8.5.6)$$

with

$$\phi_{\boldsymbol{q}} = b_{\boldsymbol{q}} + b_{-\boldsymbol{q}}^+ . \quad (8.5.7)$$

Note that $\phi_{\boldsymbol{q}} \propto Q_{\boldsymbol{q}}$ (see, (5.2.24)). $\widehat{\phi}_{\boldsymbol{q}}(\tau)$ is the Heisenberg representation of $\phi_{\boldsymbol{q}}$. Compare (8.5.6) with the Green's function of electron defined in (8.2.1); note the difference in the signs. (Some authors define the phonon Green's function with a negative sign on the right hand side of (8.5.6).)

Since phonons are bosons, a reordering of $\phi_{\boldsymbol{q}}$'s does not involve a change in sign. Thus, unlike (8.2.12)–(8.2.13), we have

$$\begin{aligned} D(\boldsymbol{q}, \tau+\beta) &= D(\boldsymbol{q}, \tau) \quad \text{for} \quad -\beta \le \tau \le 0, \\ D(\boldsymbol{q}, \tau-\beta) &= D(\boldsymbol{q}, \tau) \quad \text{for} \quad 0 < \tau \le \beta. \end{aligned} \quad (8.5.8)$$

With these properties, if we define the Fourier transform of phonon Green's function similarly to (8.2.10) we have

$$D(\boldsymbol{q}, i\nu_m) = \int_0^\beta d\tau \, D(\boldsymbol{q}, \tau) \, e^{i\nu_m \tau}, \quad (8.5.9a)$$

$$D(\boldsymbol{q}, \tau) = \frac{1}{\beta} \sum_m D(\boldsymbol{q}, i\nu_m) \, e^{-i\nu_m \tau}, \quad (8.5.9b)$$

where $\nu_m = 2\pi m/\beta$ is the boson Matsubara frequency defined in (8.4.28).

In the absence of the electron-phonon interaction, $H_{\text{el-ph}}$, namely, when there is only H_{ph} of (5.2.25), the phonon Green's function is given as

$$D^{\circ}(\boldsymbol{q}, i\nu_m) = \frac{2\Omega_q}{\nu_m^2 + \Omega_q^2}. \tag{8.5.10}$$

If we set $i\nu_m = \omega + i0^+$ in this temperature Green's function, its poles give us the (unscreened) phonon spectrum, $\omega = \pm\, \Omega_q$,.

Let us see how the result of (8.5.10) is derived from the definition of (8.5.6). First we note

$$\widehat{b}_q(\tau) \equiv e^{\tau H_{\text{ph}}}\, b_q\, e^{-\tau H_{\text{ph}}} = e^{-\Omega_q \tau}\, b_q, \tag{8.5.11}$$

$$\widehat{b}_q^{+}(\tau) \equiv e^{\tau H_{\text{ph}}}\, b_q^{+}\, e^{-\tau H_{\text{ph}}} = e^{\Omega_q \tau}\, b_q^{+}. \tag{8.5.12}$$

The result of (8.5.11) can be obtained by solving the following equation of motion

$$\frac{\partial}{\partial \tau}\widehat{b}_q(\tau) = \left[H_{\text{ph}}, \widehat{b}_q(\tau)\right] = -\,\Omega_q\, \widehat{b}_q(\tau), \tag{8.5.13}$$

which directly comes out of the definition of $\hat{b}_q(\tau)$ in accord with the initial condition $\hat{b}_q(0) = b_q$. The result of (8.5.12) also can be obtained similarly. If we put (8.5.11) and (8.5.12) into (8.5.6), for $\tau > 0$ we have

$$\begin{aligned} D^{\circ}(\boldsymbol{q}, \tau) &= \left\langle T_\tau \left(\widehat{b}_q(\tau) + \widehat{b}_{-q}^{+}(\tau) \right)\left(\widehat{b}_q^{+} + \widehat{b}_{-q} \right)\right\rangle_{\!\circ} \\ &= e^{-\Omega_q \tau} \left\langle b_q\, b_q^{+} \right\rangle_{\!\circ} + e^{\Omega_q \tau}\left\langle b_{-q}^{+}\, b_{-q} \right\rangle_{\!\circ} \\ &= e^{-\Omega_q \tau}\left(1 + n(\Omega_q)\right) + e^{\Omega_q \tau}\, n(\Omega_q), \end{aligned} \tag{8.5.14}$$

where $n(\Omega_q)$ is the Bose distribution for phonons (c.f. (2.5.37)) and we noted $\Omega_q = \Omega_{-q}$. If we put this result into (8.5.9a) and carry out the Fourier transform we obtain (8.5.10).

Equation (8.5.14) shows that a phonon Green's function $D^{\circ}(\boldsymbol{q}, \tau)$ represents the effect of creating a phonon with a wave number $\boldsymbol{q}$, or, annihilating a phonon with a wave number $-\,\boldsymbol{q}$ in a system.

In the presence of the electron-phonon interaction, it is clear from our discussion in the preceding subsection that the phonon Green's function is given as

$$D(\boldsymbol{q},\tau) = \left\langle T_\tau\, U(\beta) \left(\widehat{b}_{\boldsymbol{q}}(\tau) + \widehat{b}^{+}_{-\boldsymbol{q}}(\tau) \right) \left(\widehat{b}^{+}_{\boldsymbol{q}} + \widehat{b}_{-\boldsymbol{q}} \right) \right\rangle_{\mathrm{oc}} . \qquad (8.5.15)$$

Note that here the external lines are phonon lines and that in the expression for

$$H_{\mathrm{i}} = H_{\mathrm{C}} + H_{\mathrm{el\text{-}ph}} . \qquad (8.5.16)$$

8.5.3 Feynman Rules for Green's Functions of Electrons and Phonons

Before presenting a set of general rules, let us see how such rules come out. In order to make comparison with the Feynman procedure of obtaining the thermodynamic potential of interacting electrons given in the preceding section, we consider the electron Green's function in the presence of only H_{C}. Then $U(\beta)$ in (8.5.1) is the same as in the preceding section. What we have to calculate now is, corresponding to X_n, (8.4.7), the following form of quantities $U(\tau)$ such as (8.3.7) and (8.3.10) we have to set

$$- \Big\langle T_\tau\, \widehat{a}^{+}(1)\, \widehat{a}^{+}(1')\, \widehat{a}(1')\, \widehat{a}(1)\, \widehat{a}^{+}(2)\, \widehat{a}^{+}(2')\, \widehat{a}(2')\, \widehat{a}(2)$$

$$\times \cdots\ \widehat{a}^{+}(n)\, \widehat{a}^{+}(n')\, \widehat{a}(n')\, \widehat{a}(n)\, \widehat{a}(\tau)\, \widehat{a}^{+}(0) \Big\rangle_{\mathrm{oc}} = Y_n . \qquad (8.5.17)$$

The difference from X_n is the presence of the last two factors on the left hand side. We decompose this Y_n by using Wick's theorem. Suppose that one term of such decomposition is represented by Fig.8.10. Let us see how we can produce the product of contractions so as to correspond to this diagram. We proceed as we did concerning Fig.8.1 in the preceding section.

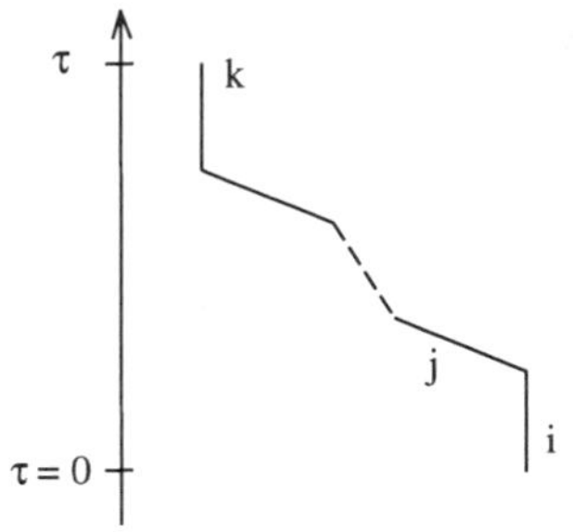

Fig.8.10 The general structure of the Feynman diagrams for the one-electron Green's function corresponding to the constructions of (8.5.20).

First we rearrange (8.5.17) as

$$Y_n = -\langle T_\tau\, \hat{a}(1)\, \hat{a}^+(1)\, \hat{a}(1')\, \hat{a}^+(1') \cdots \hat{a}(\tau)\, \hat{a}^+(0) \rangle_{\text{oc}}. \tag{8.5.18}$$

The sign does not change since the necessary permutation is an even one. Next, in order to contract as in Fig.8.10, we rearrange (8.5.18) as

$$Y_n = -\langle T_\tau\, \hat{a}(i)\, \hat{a}^+(i)\, \hat{a}(j)\, \hat{a}^+(j) \cdots \hat{a}(k)\, \hat{a}^+(k)\, \hat{a}(\tau)\, \hat{a}^+(0) \rangle_{\text{oc}}. \tag{8.5.19}$$

This is again an even permutation. Finally, after moving $\hat{a}^+(0)$ to the left most position, which is an odd permutation this time, we make contractions in the following way:

$$\underbrace{\hat{a}^+(0)\ \hat{a}(i)}\ \underbrace{\hat{a}^+(i)\ \hat{a}(j)} \cdots \underbrace{\hat{a}^+(k)\ \hat{a}(\tau)}. \tag{8.5.20}$$

Here recall (8.4.13).

As can be seen in Fig.8.10, a Green's function is represented by an open line of which both ends are external lines. To each junction of two electron lines, a Coulomb broken line is connected, as in Figs.8.9(a) and (b). The line can have parts containing closed loops.

If we consider the electron-phonon interaction $H_{\text{el-ph}}$ in addition, there appear diagrams like Fig.9.11(a) in Chapter 9, where the wavy line stands for phonon Green's function D^{o}. Note that the structure of Fig.9.11(a) is the same as that of Fig.8.9(b); simply the Coulomb broken line in the latter is replaced by the phonon wavy line in the former.

For the Green's function of interacting phonons, (8.5.15), there appear diagrams as given in Fig.9.8. We present an extensive discussion on the phonon Green's function in the next chapter (cf. 9.2).

Note that the thermal average $\langle \cdots \rangle_{\text{o}}$ of the mixed product of electron and phonon operators is factorized into the separate averages of the electron and the phonon operators. Also note that since phonons are bosons, in applying Wick's theorem as (8.4.3) by putting $A(i) = \phi_{qi}(\tau_i)$, we have always $\delta(P_i) = 1$.

With the above background, we now summarize the general rules for obtaining the Green's functions of electrons and phonons by using Feynman diagrams in the presence of both H_{C} and $H_{\text{el-ph}}$. First for electrons:

(1) While an electron with spin σ and 4 momenta $(\boldsymbol{k}, i\omega_n)$ comes in from below and goes out above as in Fig.8.8, we draw all the possible diagrams representing the interaction with other electrons and phonons within the box. The Coulomb interaction is represented by a broken line and a phonon is represented by a wavy line. We draw only those

diagrams which are connected to the external line. At each vertex, 4 momenta and spins are conserved.

(2) To an electron line with spin σ and 4 momenta $(\boldsymbol{k}, i\omega_n)$ we associate

$$G^{o}_{\sigma}(\boldsymbol{k}, i\omega_n) = G^{o}(k, i\omega_n) = 1/(i\omega_n - \xi_{k\sigma}).$$

However, to an electron line whose two end points are connected to the same Coulomb broken line we attach a factor $\exp(i\omega_n 0^+)$ to the above $G^{o}(k, i\omega_n)$.

(3) A phonon wavy line with the 4 momenta $(\boldsymbol{q}, i\nu_m)$ is associated with a factor

$$D^{o}(\boldsymbol{q}, i\nu_m) = 2\Omega_q/(\nu_m^2 + \Omega_q^2) .$$

(4) A Coulomb broken line at which two electrons are scattered as $\boldsymbol{k} \to \boldsymbol{k}'$, $\boldsymbol{p} \to \boldsymbol{p}'$, is associated with a factor

$$\mathrm{v}(\boldsymbol{p}' - \boldsymbol{p}) = \mathrm{v}(\boldsymbol{k}' - \boldsymbol{k}) = 4\pi e^2/V\,|\boldsymbol{p} - \boldsymbol{p}'|^2.$$

(5) An electron-phonon interaction vertex, at which either a phonon with momentum $\boldsymbol{q}$ is emitted or a phonon with momentum $-\boldsymbol{q}$ is absorbed, is associated with the factor $\alpha(\boldsymbol{q})$ of (5.2.41) and (5.2.38').

(6) To an "unlabelled" diagram which is n-th order in $\mathrm{v}(\kappa)$, $2m$-th order in $\alpha(\boldsymbol{q})$, and with n_l closed electron loops, we attach a factor

$$\beta^{-(n+m)}(-1)^n(-1)^{n_l} .$$

(7) If we represent those factors coming from the above rules (2), ··· ,(6), by F[2], ··· , F[6], we sum over the 4 momenta $(\boldsymbol{k}_i, i\omega_i)$, $(\boldsymbol{q}_j, i\nu_j)$ and spin σ of all the internal lines in the product of these factors:

$$\sum_{\boldsymbol{k}_1,\omega_1,\sigma_1} \cdots \sum_{\boldsymbol{q}_1,\nu_1} \cdots \mathrm{F}[2]\cdots\mathrm{F}[6] .$$

(8) We repeat the above procedure for all the topologically different diagrams of the $(n+2m)$-th order, and sum them up. Then we sum the contribution of all orders.

Although the above rules are for electrons, rules for phonon Green's function are essentially the same. The only difference is that for a phonon Green's function the external line is a wavy line.

Note that unlike in the rule (4) for the thermodynamic potential of 8.4.6, in the rule (6) above for the Green's function there does not appear the factor $1/(n+2m)!$ This is because, here the rule is for "unlabelled" diagrams, and in the case of Green's function we obtain $(n+2m)!$ different Feynman diagrams, not $(n+2m-1)!$ ones, by permuting the order of $(n+2m)$ times τ_i's, as can be seen from the procedure (8.5.18)–(8.5.20) corresponding to Fig.8.10. Further, the factor $1/2^n$ is cancelled by the 2^n different diagrams to be obtained by changing the polarities of n Coulomb vertices.

Also, the reason why in the rule (6) above we have the factor $(1/\beta)^m$ for a diagram which is $2m$-th order in $H_{\text{el-ph}}$ is as follows. For such a diagram, in connection with the electron-phonon interaction there appear $2m$ electron Green's functions and m phonon Green's functions. If we rewrite all of these Green's functions by Fourier transforming them as (8.2.11) and (8.5.9), there appears a factor $1/\beta^{3m}$. Then, if we integrate such an expression over $2m$ times $\tau_1,\cdots,\tau_{2m}$ we get a factor β^{2m} and the energy conservation as in (8.4.26). In this way we end up with the factor $1/\beta^m$ in the rule (6).

We already discussed the origin of the factor $1/\beta^n$ concerning H_C in the rule (4) in 8.4.6.

In the next subsection we present a simple example involving only H_C. Examples of a more extensive application of the above general rules will be given in Chapter 9.

8.5.4 The Effect on the Electron Green's Function of the Coulomb Interaction between Electrons

The contribution to an electron Green's function of the diagram (a) of Fig.8.11, where $n=1$, $m=0$, and, $n_l=0$, is obtained from the rules given in 8.5.3 as

$$\Delta_1 G_\sigma(\boldsymbol{k}, i\omega_n) = G_\sigma^0(\boldsymbol{k}, i\omega_n)\, \Sigma_\sigma^{\text{HF}}(\boldsymbol{k}, i\omega_n)\, G_\sigma^0(\boldsymbol{k}, i\omega_n), \tag{8.5.21}$$

with

$$\Sigma_\sigma^{\text{HF}}(\boldsymbol{k}, i\omega_n) = -\frac{1}{\beta}\sum_{\kappa,m} \text{v}(\kappa)\, G_\sigma^0(\boldsymbol{k}+\kappa, i\omega_{n+m})\, e^{i\omega_{n+m}0^+}$$

$$= -\sum_{\kappa} \text{v}(\kappa)\, f(\xi_{\boldsymbol{k}+\kappa,\sigma}), \tag{8.5.22}$$

where we used (8.4.39). Obviously Σ_σ^{HF} corresponds to the diagram of Fig.8.11(d).

Similarly, for the contribution of the 2nd order diagram given in Fig.8.11(b) we have

$$\Delta_2 G_\sigma(\boldsymbol{k}, i\omega_n) = G_\sigma^0(\boldsymbol{k}, i\omega_n)\, \Sigma_\sigma^{HF}(\boldsymbol{k}, i\omega_n)\, G_\sigma^0(\boldsymbol{k}, i\omega_n)\, \Sigma_\sigma^{HF}(\boldsymbol{k}, i\omega_n) \times G_\sigma^0(\boldsymbol{k}, i\omega_n). \tag{8.5.23}$$

We can sum up the contributions of this type of diagrams to infinite order as follows,

$$\begin{aligned} G_\sigma(\boldsymbol{k}, i\omega_n) &= G_\sigma^0 + G_\sigma^0 \Sigma_\sigma^{HF} G_\sigma^0 + G_\sigma^0 \Sigma_\sigma^{HF} G_\sigma^0 \Sigma_\sigma^{HF} G_\sigma^0 + \cdots \\ &= \frac{G_\sigma^0(\boldsymbol{k}, i\omega_n)}{1 - G_\sigma^0(\boldsymbol{k}, i\omega_n)\, \Sigma_\sigma^{HF}(\boldsymbol{k}, i\omega_n)} = \frac{1}{G_\sigma^0(\boldsymbol{k}, i\omega_n)^{-1} - \Sigma_\sigma^{HF}(\boldsymbol{k}, i\omega_n)} \\ &= \frac{1}{i\omega_n - \xi_k - \Sigma_\sigma^{HF}(\boldsymbol{k}, i\omega_n)}. \end{aligned} \tag{8.5.24}$$

We can rewrite (8.5.24) also in the following form

$$G_\sigma(\boldsymbol{k}, i\omega_n) = G_\sigma^0(\boldsymbol{k}, i\omega_n) + G_\sigma^0(\boldsymbol{k}, i\omega_n)\, \Sigma_\alpha^{HF}(\boldsymbol{k}, i\omega_n)\, G_\sigma(\boldsymbol{k}, i\omega_n), \tag{8.5.25}$$

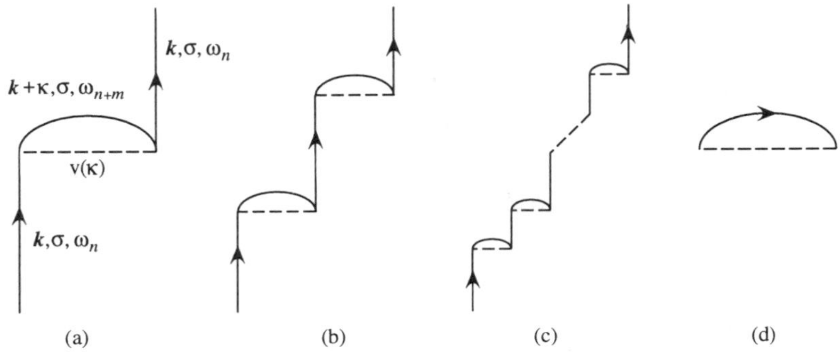

Fig.8.11 The effect of the Coulomb interaction on an electron Green's function of the first order, (a), the second order, (b), and higher order, (c). The sum of these effects can be represented by the Hartree-Fock self-energy of an electron, (d).

which may be represented as in Fig.8.12, where the thick solid line represents $G_\sigma(\boldsymbol{k}, i\omega_n)$, while the thin solid line stands for $G^o_\sigma(\boldsymbol{k}, i\omega_n)$. An equation like (8.5.25) is called *Dyson equation* and a quantity like $\Sigma^{\rm HF}_\sigma$ which appears in a Dyson equation for a one-particle Green's function as in (8.5.25) is called the *irreducible self-energy*.

An irreducible self-energy part of an electron has such a structure that it can not be separated into two parts by cutting an electron line in it. The self-energy part of Fig.8.11(b), which is obtained by eliminating the two external lines connected to it, is, thus, not irreducible, unlike that of Fig.8.11(d).

If we replace $i\omega_n$ by $\omega + i0^+$ in the temperature Green's function of an electron, we obtain the energy spectrum of electron from its pole. Thus from (8.5.24) we have a pole at

$$\omega = \xi_k - \sum_\kappa \mathrm{v}(\kappa) f(\xi_{k+\kappa,\sigma}). \qquad (8.5.26)$$

This reproduces the earlier result such as (3.3.1) on the mean field treatment of the exchange interaction between electrons. The second term on the right hand side, or, (8.5.22) is called the *Hartree-Fock*, or, the *exchange self-energy*.

In the Hartree-Fock self-energy, the effect of $H_{\rm C}$ appears only to the lowest order. There the contributions of diagrams such as given in Fig.8.13 are not included. Those higher order diagrams represent the following physical processes. An external electron ($\boldsymbol{k}$, σ) produces an electron density polarization by exciting electron-hole pairs in the Fermi sea of electrons; the electron density polarization (electron-hole pair) produces another electron density polarization by the Coulomb interaction. This chain process goes on and finally the initial external electron is scattered by the last electron polarization.

In summing up the contributions of the diagrams of the kinds of Fig.8.13, in addition to those given in Fig.8.11, let us first define $\mathrm{v_{sc}}(\kappa, i\nu_m)$ by the thick broken lines as in Fig.8.14. By the rules given in 8.5.3, corresponding Dyson equation is written as

$$\mathrm{v_{sc}}(\kappa, i\nu_m) = \mathrm{v}(\kappa) - \mathrm{v}(\kappa)\, P(\kappa, i\nu_m)\, \mathrm{v_{sc}}(\kappa, i\nu_m), \qquad (8.5.27)$$

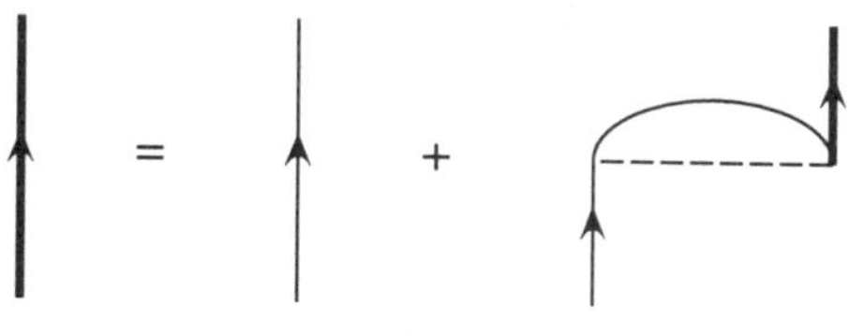

Fig.8.12 Dyson equation for electron Green's function corresponding to (8.5.25). The thick and thin solid lines, respectively, represent the electron Green's functions with, and without the effect of Coulomb interaction.

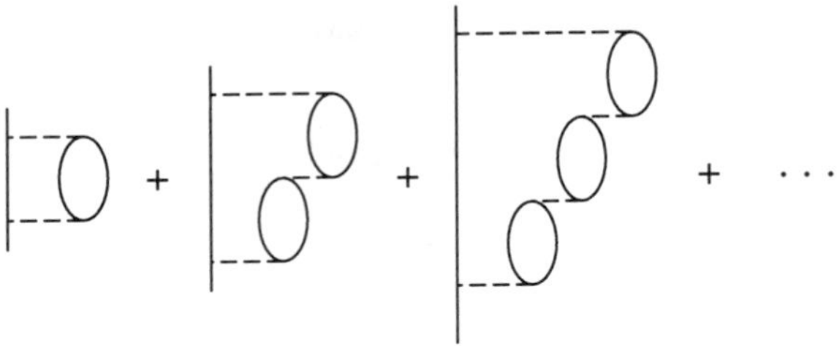

Fig.8.13 Feynman diagrams for the one-electron Green's function which were not included in Figs.8.11 and 8.12.

where $P(\kappa, i\nu_m)$, which represents a ring, is defined in (8.4.32). From (8.5.27) we have

$$\mathrm{v}_{\mathrm{sc}}(\kappa, i\nu_m) \;=\; \frac{\mathrm{v}(\kappa)}{\varepsilon_{\mathrm{o}}(\kappa, i\nu_m)} \,, \tag{8.5.28}$$

where

$$\varepsilon_{\mathrm{o}}(\kappa, i\nu_m) \;=\; 1 + \mathrm{v}(\kappa)\, P(\kappa, i\nu_m) \tag{8.5.29}$$

is the screening constant of (4.3.30). Thus $\mathrm{v}_{\mathrm{sc}}(\kappa, i\nu_m)$ is a screened Coulomb interaction.

Then the sum of the contributions given in Fig.8.11 or 8.12 and Fig.8.13 to infinite order is represented as in Fig.8.15(a) with the new electron self-energy of Fig.8.15(b) involving a thick broken line. The corresponding Dyson equation takes the same structure as (8.5.24) simply with the Hartree-Fock self-energy there replaced by the new self-energy

$$\Sigma_\sigma^{\mathrm{HF}}(\boldsymbol{k}, i\omega_n) \;=\; -\frac{1}{\beta}\sum_{m,\kappa} \mathrm{v}_{\mathrm{sc}}(\kappa, i\nu_m)\, G_\sigma^{\mathrm{o}}(\boldsymbol{k}+\kappa, i\omega_{n+m}). \tag{8.5.30}$$

Note that unlike in (8.5.22) we do not here need the factor $\exp(i\omega_{n+m}0^+)$ on the right hand side, since the two ends of a screened Coulomb interaction have different times. Thus, corresponding to (8.5.24) we obtain

$$G_\sigma(\boldsymbol{k}, i\omega_n) \;=\; \frac{1}{i\omega_n - \xi_{k\sigma} - \Sigma_\sigma^{\mathrm{RPA}}(\boldsymbol{k}, i\omega_n)} \,. \tag{8.5.31}$$

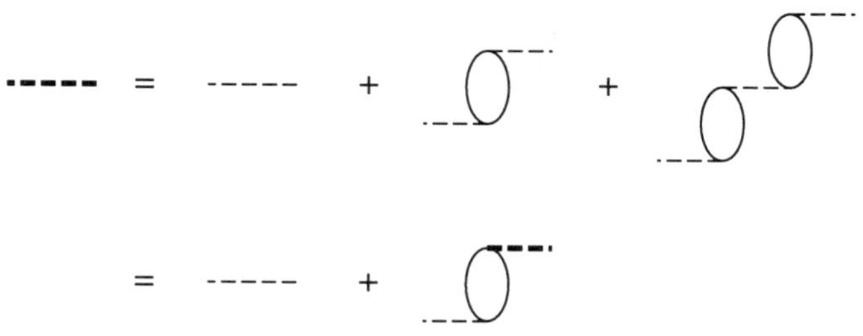

Fig.8.14 The Dyson equation for the screened Coulomb interaction (thick broken line) corresponding to (8.5.32).

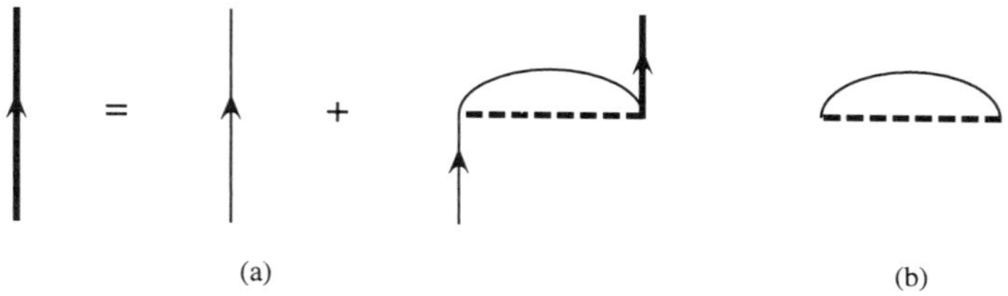

Fig.8.15 Dyson equation for one-electron Green's function in the RPA which includes the contributions of diagrams of Fig.8.13 in addition to that of Fig.8.11 or Fig.8.12, (a), and the corresponding RPA self-energy of an electron, (b). The thick broken lines represents the screened Coulomb interaction defined in Fig.8.14.

All the broken lines in Figs.8.13 and 8.14 carry the same common κ. An electron density polarization (electron-hole pairs) with a wave vector κ is assumed to interact only with other electron density polarization (electron-hole pairs) with the same wave vector κ; the effect of the interaction with other electron density polarizations with $\kappa' \neq \kappa$ is assumed to cancel out. We call such an approximation the *random phase approximation* (RPA).

If we approximate (8.5.28) as

$$\mathrm{v}_{\mathrm{sc}}(\kappa, i\nu_m) \cong \frac{\mathrm{v}(\kappa)}{\varepsilon_{\mathrm{o}}(\kappa, 0)} = \mathrm{v}_{\mathrm{sc}}(\kappa, 0) \equiv \mathrm{v}_{\mathrm{sc}}(\kappa), \tag{8.5.32}$$

the RPA electron self-energy of (8.5.30) is approximated as

$$\Sigma_\sigma^{\mathrm{HF}}(\boldsymbol{k}, i\omega_n) \cong -\sum_{\kappa} \mathrm{v}_{\mathrm{sc}}(\kappa) f(\xi_{\boldsymbol{k}+\kappa,\sigma}). \tag{8.5.33}$$

From this result we realize that an important aspect of the RPA consists in replacing $v(\kappa)$ with $v_{sc}(\kappa)$ in the Hartree-Fock approximation result of (8.5.22). Note, however, that such result is the consequence of simplifying the dynamic screening constant in (8.5.32) by the static one. A very important physics involving the electron plasma oscillation, which is originally contained in (8.5.30)–(8.5.31) is lost by such simplification. For a very vivid discussion on the physical consequence of the RPA on an electron gas the reader is referred to Pines [5.8]. We also discuss this subject in Chapter 10.

8.5.5 Relation between the Thermodynamic Potential and the Green's Function

If we compare Fig.8.5 for $\Delta\Omega'_{\text{ring}}$ with Fig.8.13 for $G_\sigma(\boldsymbol{k}, i\omega_n)$, we anticipate some close relation between the ring approximation for the thermodynamic potential and the RPA for the electron Green's function. An easy way to explore such a relation is to use the following relation

$$\Delta\Omega = \frac{1}{2\beta} \sum_{\sigma,\boldsymbol{k},n} \int_0^1 \frac{dg}{g} \Sigma_\sigma(\boldsymbol{k}, i\omega_n)_g \, G_\sigma(\boldsymbol{k}, i\omega_n)_g \,, \tag{8.5.34}$$

where $\Sigma_\sigma(\boldsymbol{k}, i\omega_n)_g$ and $G_\sigma(\boldsymbol{k}, i\omega_n)_g$ are what we obtain by putting $v(\kappa) \to g v(\kappa)$ in the exact expressions for, respectively, the self-energy and the Green's function of electrons. A proof of this relation is given in Appendix C. With (8.5.34), we can actually establish

$$\Delta\Omega_{\text{el, ring}} = \frac{1}{2\beta} \sum_{\sigma,\boldsymbol{k},n} \int_0^1 \frac{dg}{g} \Sigma_\sigma^{\text{RPA}}(\boldsymbol{k}, i\omega_n)_g \, G_\sigma^0(\boldsymbol{k}, i\omega_n)_g \,. \tag{8.5.35}$$

8.1 Prove that when $G^r(\omega)$ does not have any singularities (poles) in the upper-half of the complex ω-plane, we have

$$G^r(t) = 0 \quad \text{for} \quad t < 0 . \tag{8.P1}$$

(Change the integral of the form of (8.1.5′) to that over a closed contour such as given in Fig.8.P1.)

8.2 Derive (8.1.12) from (8.1.11).

8.3 Prove the results of (8.2.12) and (8.2.13) for the one-particle temperature Green's function of fermions.

8.4 Confirm that the Fourier transform of the one-electron Green's function as defined in (8.2.10) vanishes for a frequency not equal to the fermion Matsubara frequency, (8.2.14).

8.5 Confirm the general properties of (8.1.30) for the spectral density of a system of interacting electrons. The first property is proved by extracting the explicit expression for $A(k,\omega)$ by comparing (8.2.18) and (8.1.29) as

$$A(k, \omega) = \sum_{\alpha',\alpha''} [e^{\beta(\Omega - E_{\alpha''})} + e^{\beta(\Omega - E_{\alpha'})}] \langle \alpha' | a_k | \alpha'' \rangle$$

$$\times \langle \alpha'' | a_k^+ | \alpha' \rangle \, \delta(\omega - E_{\alpha'} + E_{\alpha''}).$$

The second property is then derived by rewriting the left hand side as

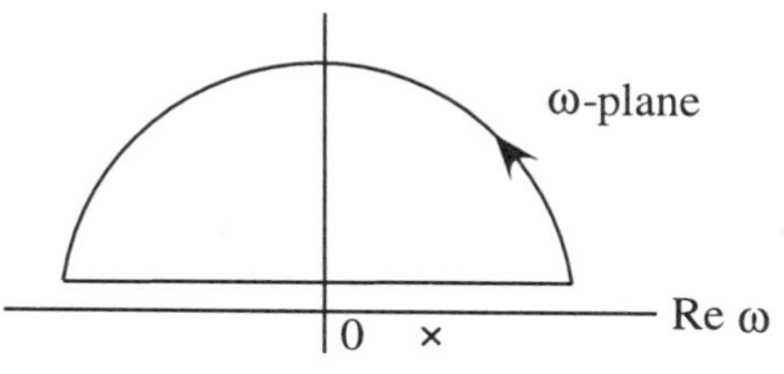

Fig.8.P.1 The contour of integration for a retarded two-time Green's function. The cross represents the possible pole of the Green's function.

$$\frac{1}{2\pi}\int_{-\infty}^{\infty} d\omega\, A(k,\omega) = \sum_{\alpha',\alpha''} [e^{\beta(\Omega - E_{\alpha''})} + e^{\beta(\Omega - E_{\alpha'})}] \langle \alpha' | a_k | \alpha'' \rangle \langle \alpha'' | a_k^+ | \alpha' \rangle$$

$$= \sum_{\alpha} e^{\beta(\Omega - E_{\alpha})} \langle \alpha | a_k a_k^+ + a_k^+ a_k | \alpha \rangle = 1 .$$

8.6 Derive the result (8.4.39) by using (8.4.37′).

8.7 Carry out the Fourier transform of the phonon Green's function given in (8.5.14) to obtain the result of (8.5.10).

8.8 Obtain the one-particle temperature Green's function of a free electron, (8.2.20), by a procedure similar to that used for the phonon Green's function in 8.5.2.

8.9 Confirm the relation (8.5.35).

Chapter 9

FEYNMAN DIAGRAMS AND GREEN'S FUNCTIONS IN ITINERANT ELECTRON MAGNETISM

In this chapter we explore some basic problems of itinerant electron magnetism by applying the method of Green's functions and Feynman diagrams. First we rederive the magnetic susceptibility and the phonon spectrum by using Feynman diagrams. Then we study how magnetism is involved in the electron self-energy due to the electron-phonon interaction (EPI). Such an observation leads to a new insight on the role of magnetism in superconductivity. We discuss also the electron self-energy due to spin waves, and spin fluctuations. Finally, we critically review various current viewpoints on the role of the EPI in magnetism of a metal.

9.1 MAGNETIC SUSCEPTIBILITY $\chi_{zz}(q, \omega)$ AND CHARGE SUSCEPTIBILITY $\chi_{ee}(q, \omega)$: TWO-PARTICLE GREEN'S FUNCTION

As already noted in (8.1.6), the longitudinal magnetic susceptibility is given in terms of a boson type retarded two-time Green's function as

$$\chi_{zz}(\boldsymbol{q}, \omega) = \langle\langle M_z(\boldsymbol{q}); M_z(-\boldsymbol{q}) \rangle\rangle^{\mathrm{r}}_{\omega}. \tag{9.1.1}$$

If we put the explicit expression (4.2.6) for the magnetization density, it is rewritten as

$$\chi_{zz}(\boldsymbol{q}, \omega) = \mu_{\mathrm{B}}{}^2 \sum_{\sigma,\sigma'} \sigma\sigma' \langle\langle n_{\sigma'}(\boldsymbol{q}); n_{\sigma}(-\boldsymbol{q}) \rangle\rangle^{\mathrm{r}}_{\omega}, \tag{9.1.2}$$

where, as already given in (4.2.27),

$$n_\sigma(\boldsymbol{q}) = \sum_{\boldsymbol{k}} a^+_{\boldsymbol{k}\sigma} a_{\boldsymbol{k}+\boldsymbol{q},\sigma} \cdot \tag{9.1.3}$$

Thus, the required two-time Green's function is of the following form,

$$G^{\mathrm{r}}_{\sigma'\sigma}(\boldsymbol{q}, \omega) = \left\langle\!\left\langle n_{\sigma'}(\boldsymbol{q}, \tau); n_\sigma(-\boldsymbol{q}) \right\rangle\!\right\rangle^{\mathrm{r}}_\omega. \tag{9.1.4}$$

The corresponding temperature Green's function is

$$G_{\sigma'\sigma}(\boldsymbol{q}, \tau) = \left\langle T_\tau\, n_{\sigma'}(\boldsymbol{q}, \tau)\, n_\sigma(-\boldsymbol{q}) \right\rangle$$

$$= \sum_{\boldsymbol{k},\boldsymbol{k}'} \left\langle T_\tau\, a^+_{\boldsymbol{k}'\sigma'}(\tau)\, a_{\boldsymbol{k}'+\boldsymbol{q},\sigma'}(\tau)\, a^+_{\boldsymbol{k}\sigma} a_{\boldsymbol{k}-\boldsymbol{q},\sigma} \right\rangle \equiv \sum_{\boldsymbol{k},\boldsymbol{k}'} G_{\boldsymbol{k}'\sigma',\boldsymbol{k}\sigma}(\boldsymbol{q}, \tau). \tag{9.1.5}$$

These Green's functions are two-particle Green's functions. The two-particle Green's function $G_{\boldsymbol{k}'\sigma',\boldsymbol{k}\sigma}(\boldsymbol{q}, \tau)$ can be represented as in Fig.9.1. Here, the external lines consist of a pair of an electron line and a hole line; the hole propagates backward in time τ. Thus, $G_{\boldsymbol{k}'\sigma',\boldsymbol{k}\sigma}(\boldsymbol{q}, \tau)$ represents the propagation of an electron-hole pair.

Let us start with the zero-th order term in the perturbational expansion of $G_{\boldsymbol{k}'\sigma',\boldsymbol{k}\sigma}(\boldsymbol{q}, \tau)$ with respect to H_{C}. If we assume $\boldsymbol{q} \neq 0$ and use Wick's theorem, we have,

$$G^{\mathrm{o}}_{\boldsymbol{k}'\sigma',\boldsymbol{k}\sigma}(\boldsymbol{q}, \tau) = \left\langle T_\tau\, \hat{a}^+_{\boldsymbol{k}'\sigma'}(\tau)\, \hat{a}_{\boldsymbol{k}'+\boldsymbol{q},\sigma'}(\tau)\, \hat{a}^+_{\boldsymbol{k}\sigma}\, \hat{a}_{\boldsymbol{k}-\boldsymbol{q},\sigma} \right\rangle_{\mathrm{o}}$$

$$= \left\langle T_\tau\, \hat{a}_{\boldsymbol{k}'+\boldsymbol{q},\sigma'}(\tau)\, \hat{a}^+_{\boldsymbol{k}\sigma} \right\rangle_{\mathrm{o}} \left\langle T_\tau\, \hat{a}^+_{\boldsymbol{k}'\sigma'}(\tau)\, \hat{a}_{\boldsymbol{k}-\boldsymbol{q},\sigma} \right\rangle_{\mathrm{o}}$$

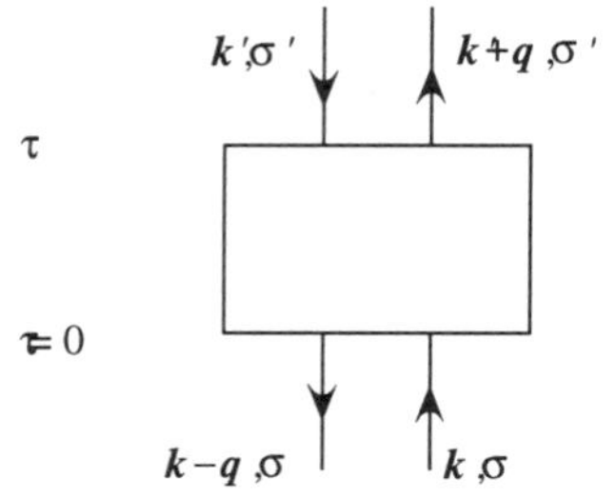

Fig.9.1 General structure for a two-particle (electron-hole) Green's function (9.1.6).

$$= -G^{\text{o}}_{\sigma'}(\boldsymbol{k},\tau)\, G^{\text{o}}_{\sigma}(\boldsymbol{k}-\boldsymbol{q},-\tau)\; \delta_{\sigma,\sigma'}\; \delta_{\boldsymbol{k},\boldsymbol{k}'+\boldsymbol{q}}, \tag{9.1.6}$$

where we noted that $\langle a^{+}_{\boldsymbol{k}\sigma}\, a_{\boldsymbol{k}-\boldsymbol{q},\sigma} \rangle_{\text{o}} = 0$ for $\boldsymbol{q} \neq 0$. If we Fourier transform this expression by noting (8.4.24)–(8.4.30), we find

$$G^{\text{o}}_{\sigma'\sigma}(\boldsymbol{q}, i\nu_m) = \sum_{\boldsymbol{k},\boldsymbol{k}'} G^{\text{o}}_{\boldsymbol{k}'\sigma',\boldsymbol{k}\sigma}(\boldsymbol{q}, i\nu_m) = F_{\sigma}(\boldsymbol{q}, i\nu_m)\, \delta_{\sigma,\sigma'}\,. \tag{9.1.7}$$

The corresponding retarded two-time Green's function is obtained by putting $i\nu_m \to \omega + i0^{+}$, and we have

$$\chi^{\text{o}}_{zz}(\boldsymbol{q}, \omega) = \mu_{\text{B}}{}^{2} \sum_{\sigma} F_{\sigma}(\boldsymbol{q}, \omega) \tag{9.1.8}$$

for the longitudinal magnetic susceptibility of free electrons, reproducing our earlier result of (4.2.16).

The zero-th order two-particle Green's function, $G^{\text{o}}_{\boldsymbol{k}\sigma,\boldsymbol{k}\sigma}(\boldsymbol{q}, i\nu_m)$ or $G^{\text{o}}_{\sigma\sigma}(\boldsymbol{q}, i\nu_m)$, may be represented as either (a) or (b) of Fig.9.2, the box being empty in Fig.9.1; this is identical with what we called a ring or bubble in Chapter 8.

Note that if we connect the upper ends of the pair of external lines in Fig.9.1, it reduces to the Feynman diagram for a one-particle Green's function. Thus, the general rules of Feynman diagram method for two-particle Green's function are essentially similar to that for one-particle Green's function discussed in 8.5. A two-particle Green's function including the effect of interaction between particles can be obtained by summing the contributions of the diagrams *connected* to one or both of the pair of external lines.

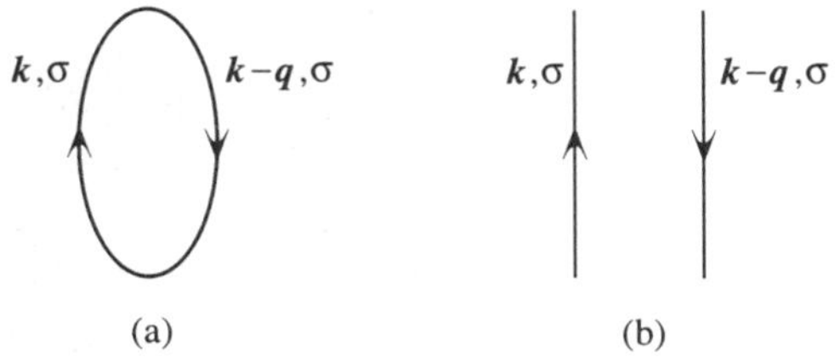

Fig.9.2 The zero-th order two-particle (electron-hole) Green's function, (9.1.6) or (9.1.7), can be represented either as in (a), or (b).

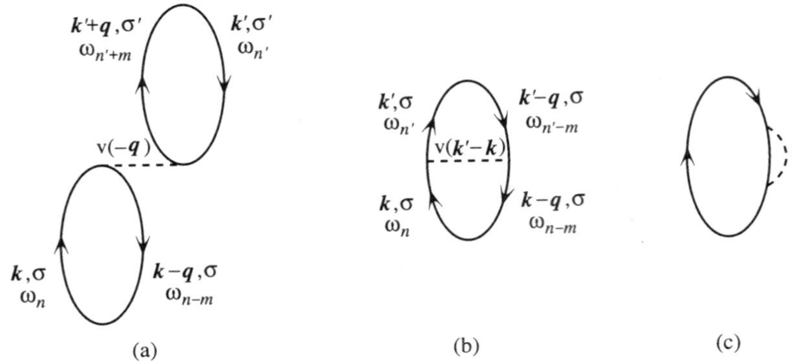

Fig.9.3 Three kinds of the 1st order diagrams appearing in the calculation of the longitudinal spin susceptibility.

In Fig.9.3 we give three kinds of the first order diagrams. We associate to each solid electron line an electron Green's function and to each broken line the Coulomb potential $v(\boldsymbol{q})$, exactly in the same way as in the case of one-particle Green's function. The only change from the case of one-particle Green's function is that in the rule (6) of 8.5.3, we have to modify the numerical factor as

$$\frac{1}{\beta^{n+m}}(-1)^{n+2m}(-1)^{n_l} \rightarrow \frac{1}{\beta^{n+m+1}}(-1)^{n+2m}(-1)^{n_l}. \tag{9.1.9}$$

When we consider only the Coulomb interaction ($m = 0$), the required change is

$$\frac{1}{\beta^{n}}(-1)^{n}(-1)^{n_l} \rightarrow \frac{1}{\beta^{n+1}}(-1)^{n}(-1)^{n_l}. \tag{9.1.9'}$$

Let us see why such a change is necessary for the case of $m = 0$; we want to know how the factor $1/\beta^{n+1}$ comes out. An n-th order connected diagram in the expansion of a two-particle Green's function should contain the product of $(2n + 2)$ one-particle Green's functions $G^0_{ki}(\tau_i)$. If we rewrite each of these one-particle Green's functions by its Fourier transform as (8.2.11), there comes out a factor $1/\beta^{2n+2}$. When we integrate such an expression over $(n + 1)$ "times" $(\tau_1, \tau_2, \cdots, \tau_n, \tau)$ each for the interval $(0, \beta)$, there comes out a factor β^{n+1}. Thus, eventually there results the factor $(1/\beta^{2n+2})\beta^{n+1} = 1/\beta^{n+1}$.

Based on these general rules let us see how each diagram of Fig.9.3 contributes to the two-particle Green's function of (9.1.5). First, the contribution of the diagram (a), which we set $\Delta_a G_{\sigma\sigma'}(\boldsymbol{q}, i\nu_m)$, is calculated as

$$\Delta_a G_{\sigma\sigma'}(\boldsymbol{q}, i\nu_m) = \frac{1}{\beta^{1+1}} (-1)(-1)^2 \sum_{n,n',\boldsymbol{k},\boldsymbol{k}'} \mathrm{v}(\boldsymbol{q})$$

$$\times G_\sigma^0(\boldsymbol{k}, i\omega_n)\, G_\sigma^0(\boldsymbol{k}-\boldsymbol{q}, i\omega_{n-m})\, G_{\sigma'}^0(\boldsymbol{k}', i\omega_{n'})\, G_{\sigma'}^0(\boldsymbol{k}'+\boldsymbol{q}, i\omega_{n'+m})$$

$$= -\,\mathrm{v}(\boldsymbol{q})\, F_\sigma(-\boldsymbol{q}, -i\nu_m)\, F_{\sigma'}(\boldsymbol{q}, i\nu_m). \qquad (9.1.10)$$

Thus, the contribution to the magnetic susceptibility of this diagram is

$$\Delta_a \chi_{zz}(\boldsymbol{q}, i\nu_m) = \mu_B{}^2 \sum_{\sigma,\sigma'} \sigma\sigma'\, \Delta_a G_{\sigma\sigma'}(\boldsymbol{q}, i\nu_m)$$

$$= -\,\mu_B^2\, \mathrm{v}(\boldsymbol{q}) \sum_{\sigma,\sigma'} \sigma\sigma'\, F_\sigma(\boldsymbol{q}, i\nu_m)\, F_{\sigma'}(\boldsymbol{q}, i\nu_m), \qquad (9.1.11)$$

where we noted the relation of (8.4.31).

For the paramagnetic state where

$$F_+(\boldsymbol{q}, i\nu_m) = F_-(\boldsymbol{q}, i\nu_m) = F(\boldsymbol{q}, i\nu_m), \qquad (9.1.12)$$

we have

$$\Delta_a \chi_{zz}(\boldsymbol{q}, i\nu_m) = 0. \qquad (9.1.13)$$

Thus, in the paramagnetic state, the diagram of Fig.9.3(a) does not contribute to the magnetic susceptibility. In the ferromagnetic state, however, (9.1.12) and, accordingly, (9.1.13) do not hold.

Next, let us proceed to Fig.9.3(b). If we choose various Fourier variables as in the figure, we obtain its contribution as

$$\Delta_b G_{\sigma\sigma'}(\boldsymbol{q}, i\nu_m) = 1/\beta^2 (-1)(-1) \sum_{n,n',\boldsymbol{k},\boldsymbol{k}'} \mathrm{v}(\boldsymbol{k}'-\boldsymbol{k})\, G_\sigma^0(\boldsymbol{k}, i\omega_n)\, G_\sigma^0(\boldsymbol{k}', i\omega_{n'})$$

$$\times G_\sigma^0(\boldsymbol{k}'-\boldsymbol{q}, i\omega_{n'-m})\, G_\sigma^0(\boldsymbol{k}-\boldsymbol{q}, i\omega_{n-m})\, \delta_{\sigma,\sigma'}$$

$$= \sum_{n,n',\boldsymbol{k},\boldsymbol{k}'} \mathrm{v}(\boldsymbol{k}'-\boldsymbol{k}) \left\{ -1/\beta\, G_\sigma^0(\boldsymbol{k}, i\omega_n)\, G_\sigma^0(\boldsymbol{k}-\boldsymbol{q}, i\omega_{n-m}) \right\}$$

$$\times \left\{ -1/\beta\, G_\sigma^0(\boldsymbol{k}', i\omega_{n'})\, G_\sigma^0(\boldsymbol{k}'-\boldsymbol{q}, i\omega_{n'-m}) \right\} \delta_{\sigma,\sigma'}. \qquad (9.1.14)$$

It is not feasible to carry out the summation over $\boldsymbol{k}$ and $\boldsymbol{k}'$ analytically in the above result. Thus, as in (4.3.5), we introduce $\widetilde{V}(\boldsymbol{q})$ in the following way.

$$\Delta_{\rm b} G_{\sigma\sigma'}(\boldsymbol{q}, i\nu_m) \cong \widetilde{V}(\boldsymbol{q}) \sum_{n,n',\boldsymbol{k},\boldsymbol{k}'} \left\{ -1/\beta\, G^{\rm o}_{\sigma}(\boldsymbol{k}, i\omega_n)\, G^{\rm o}_{\sigma}(\boldsymbol{k}-\boldsymbol{q}, i\omega_{n-m}) \right\}$$

$$\times \left\{ -1/\beta\, G^{\rm o}_{\sigma}(\boldsymbol{k}', i\omega_{n'})\, G^{\rm o}_{\sigma}(\boldsymbol{k}'-\boldsymbol{q}, i\omega_{n'-m}) \right\} \delta_{\sigma,\sigma'}$$

$$= \widetilde{V}(\boldsymbol{q})\, F_{\sigma}(\boldsymbol{q}, i\nu_m)\, F_{\sigma}(\boldsymbol{q}, i\nu_m)\, \delta_{\sigma,\sigma'} \,. \tag{9.1.15}$$

The corresponding contribution to the magnetic susceptibility is

$$\Delta_{\rm b}\, \chi_{zz}(\boldsymbol{q}, i\nu_m) = \mu_{\rm B}^2\, \widetilde{V}(\boldsymbol{q}) \sum_{\sigma} F_{\sigma}(\boldsymbol{q}, i\nu_m)^2 \,. \tag{9.1.16}$$

Finally, as for the contribution of Fig.9.3(c), if we recall Fig.8.12 or 8.15, we immediately notice that such contribution can be included up to infinite order by putting

$$G^{\rm o}_{\sigma}(\boldsymbol{k}, i\omega_n) \rightarrow G_{\sigma}(\boldsymbol{k}, i\omega_n) \tag{9.1.17}$$

in Fig.9.2(a). $G_{\sigma}(\boldsymbol{k}, i\omega_n)$ is the one-particle Green's function of an electron containing the self-energy effect of $H_{\rm C}$ by the Hartree-Fock approximation or the RPA. As we did in Figs.8.12 and 8.15(a), such effect can be represented by thickening electron lines.

Apart from such self-energy effect of (9.1.17), what are the contributions to χ_{zz} of higher order effects of $H_{\rm C}$? It is much simpler to consider such a problem in terms of diagrams directly.

First we note that Fig.9.4 represents the sum up to infinite order of the contributions of the type of Fig.9.3(b); the zero-th and the first order ones are given, respectively, in (9.1.6) and (9.1.15). The sum of all such *ladder* diagram contributions, which we represent by a *shaded ring*, can be obtained straightforwardly as follows:

$$G^{\rm b}_{\sigma\sigma'}(\boldsymbol{q}, i\nu_m) = \delta_{\sigma,\sigma'} \Big[F_{\sigma}(\boldsymbol{q}, i\nu_m) + F_{\sigma}(\boldsymbol{q}, i\nu_m)\, \widetilde{V}(\boldsymbol{q})\, F_{\sigma}(\boldsymbol{q}, i\nu_m)$$

$$+ F_{\sigma}(\boldsymbol{q}, i\nu_m)\, \widetilde{V}(\boldsymbol{q})\, F_{\sigma}(\boldsymbol{q}, i\nu_m)\, \widetilde{V}(\boldsymbol{q})\, F_{\sigma}(\boldsymbol{q}, i\nu_m) + \cdots \Big]$$

$$= \frac{F_{\sigma}(\boldsymbol{q}, i\nu_m)}{1 - \widetilde{V}(\boldsymbol{q})\, F_{\sigma}(\boldsymbol{q}, i\nu_m)}\, \delta_{\sigma,\sigma'} \equiv \widetilde{F}_{\sigma}(\boldsymbol{q}, i\nu_m)\, \delta_{\sigma,\sigma'} . \tag{9.1.18}$$

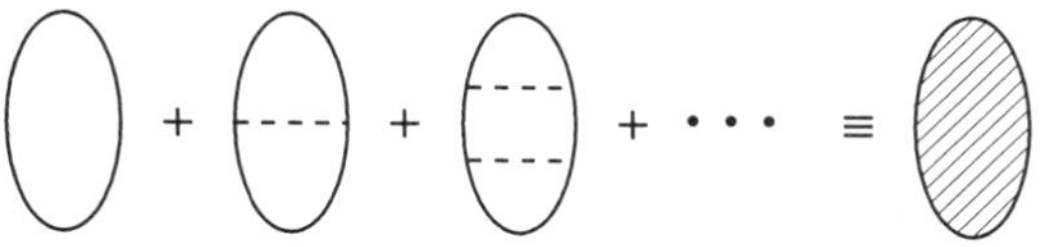

Fig.9.4 A shaded ring is defined as the sum to infinite order of the contributions of the type of Fig.9.3(b), the Coulomb interaction ladder.

In the right hand side of the first equality, each $\widetilde{V}(\boldsymbol{q})$ should have accompanied a minus sign, but the minus sign is absorbed in each $F_\sigma(\boldsymbol{q}, i\nu_m)$ appearing to the right of each $\widetilde{V}(\boldsymbol{q})$. The corresponding contribution to the magnetic susceptibility is

$$\chi_{zz}^{\mathrm{b}}(\boldsymbol{q}, i\nu_m) = \mu_{\mathrm{B}}^2 \widetilde{P}(\boldsymbol{q}, i\nu_m), \tag{9.1.19}$$

where, corresponding to (8.4.32) and (9.1.18) we put

$$\widetilde{P}(\boldsymbol{q}, i\nu_m) = \sum_\sigma \widetilde{F}_\sigma(\boldsymbol{q}, i\nu_m). \tag{9.1.20}$$

In the paramagnetic state it is only those ladder diagrams of Fig.9.4 that contribute to the magnetic susceptibility. In the ferromagnetic state, however, diagrams such as given in Fig.9.5 also contribute; the first diagram of Fig.9.5 is the same as that of Fig.9.3(a).

If we refer to Fig.9.4, we immediately note that the sum to infinite order of the contributions of the form of Fig.9.5 can be represented as in Fig.9.6(a). If we denote the thick broken line by $\overline{\mathrm{v}}_{\mathrm{sc}}$, from Fig.9.6(b) we have

$$\overline{\mathrm{v}}_{\mathrm{sc}} = \mathrm{v}(\boldsymbol{q}) - \mathrm{v}(\boldsymbol{q})\, \widetilde{P}(\boldsymbol{q}, i\nu_m)\, \overline{\mathrm{v}}_{\mathrm{sc}} \,,$$

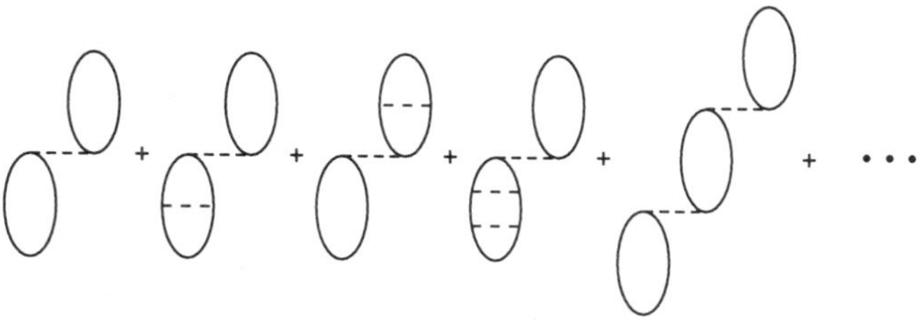

Fig.9.5 The higher order diagrams of the kind of Fig.9.3(a) which contribute to the spin susceptibility in the ferromagnetic state of a metal.

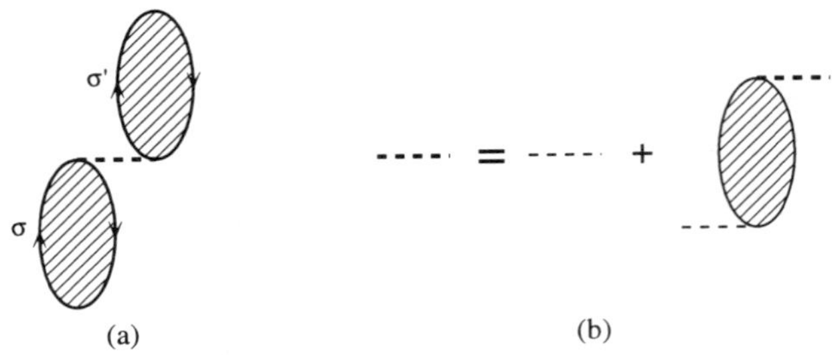

Fig.9.6 The sum to infinite order of the contributions of the kinds of Fig.9.5 is represented in terms of the shaded rings as in (a), and the screened Coulomb interaction is defined in (b).

or

$$\overline{\mathrm{v}}_{\mathrm{sc}}(\boldsymbol{q}, i\nu_m) = \frac{\mathrm{v}(\boldsymbol{q})}{1 + \mathrm{v}(\boldsymbol{q})\,\widetilde{P}(\boldsymbol{q}, i\nu_m)} . \tag{9.1.21}$$

This is a screened Coulomb interaction. Differing from that of (8.5.28), however, here the screening is done with the screening constant $\varepsilon_{\mathrm{e}}(\boldsymbol{q}, \omega)$ of (4.3.53) in place of $\varepsilon_0(\boldsymbol{q}, \omega)$. With this, the contribution of Fig.9.6(a) is readily written down as

$$G^{\mathrm{a}}_{\sigma\sigma'}(\boldsymbol{q}, i\nu_m) = -\widetilde{F}_\sigma(\boldsymbol{q}, i\nu_m)\,\overline{\mathrm{v}}_{\mathrm{sc}}(\boldsymbol{q}, i\nu_m)\,\widetilde{F}_{\sigma'}(\boldsymbol{q}, i\nu_m) . \tag{9.1.22}$$

The corresponding contribution to the magnetic susceptibility is

$$\chi^{\mathrm{a}}_{zz}(\boldsymbol{q}, i\nu_m) = -\mu_{\mathrm{B}}^2 \sum_{\sigma, \sigma'} \sigma\sigma'\,\overline{\mathrm{v}}_{\mathrm{sc}}(\boldsymbol{q}, i\nu_m)\,\widetilde{F}_\sigma(\boldsymbol{q}, i\nu_m)\,\widetilde{F}_{\sigma'}(\boldsymbol{q}, i\nu_m). \tag{9.1.23}$$

Finally, summing (9.1.19) and (9.1.23) we obtain the longitudinal magnetic susceptibility as

$$\begin{aligned}\chi_{zz}(\boldsymbol{q}, i\nu_m) &= \chi^{\mathrm{b}}_{zz}(\boldsymbol{q}, i\nu_m) + \chi^{\mathrm{a}}_{zz}(\boldsymbol{q}, i\nu_m) \\ &= \mu_{\mathrm{B}}^2 \frac{\sum_\sigma \widetilde{F}_\sigma(\boldsymbol{q}, i\nu_m) + 4\mathrm{v}(\boldsymbol{q})\,\widetilde{F}_+(\boldsymbol{q}, i\nu_m)\,\widetilde{F}_-(\boldsymbol{q}, i\nu_m)}{1 + \mathrm{v}(\boldsymbol{q}) \sum_\sigma \widetilde{F}_\sigma(\boldsymbol{q}, i\nu_m)} . \end{aligned} \tag{9.1.24}$$

If we put $i\nu_m \to \omega + i0^+$ in (9.1.24), it reproduces the earlier result of (4.3.16).

We can similarly derive the result of (4.3.25) for electron charge susceptibility by the method of Feynman diagram. The Kubo formula for the electron charge susceptibility is given by

$$\chi_{\text{ee}}(\boldsymbol{q},\,\omega) \;=\; e^2 \langle\!\langle\, n(\boldsymbol{q});\, n(-\boldsymbol{q})\,\rangle\!\rangle^{\text{r}}_{\omega}$$

$$= e^2 \sum_{\sigma,\sigma'} \langle\!\langle\, n_{\sigma'}(\boldsymbol{q});\, n_{\sigma}(-\boldsymbol{q})\,\rangle\!\rangle^{\text{r}}_{\omega}\,. \tag{9.1.25}$$

We note that the diagrams to be considered are the same as that for χ_{zz}. Then it is straightforward to reproduce the earlier result of (4.3.25).

It is another good exercise to reproduce the result of (4.3.21) on the charge and magnetism response functions $\chi_{\text{em}}(\boldsymbol{q},\,\omega) = \chi_{\text{me}}(\boldsymbol{q},\,\omega)$ by means of the Feynman diagram method.

For later convenience and as an exercise, let us rederive by using Feynman diagrams the transverse magnetic susceptibility discussed in 4.4.1. First note that corresponding to (9.1.1), the Kubo formula for the transverse magnetic susceptibility is given by

$$\chi_{+-}(\boldsymbol{q},\,\omega) \;=\; \langle\!\langle\, M_{+}(\boldsymbol{q});\, M_{-}(-\boldsymbol{q})\,\rangle\!\rangle^{\text{r}}_{\omega}$$

$$= \mu_{\text{B}}^2 \langle\!\langle\, \sigma_{+}(\boldsymbol{q});\, \sigma_{-}(-\boldsymbol{q})\,\rangle\!\rangle^{\text{r}}_{\omega}\,, \tag{9.1.26}$$

where $\sigma_{\pm}(\boldsymbol{q})$ is defined in (4.4.3). If we put (4.4.3) into (9.1.26), we have

$$\chi_{+-}(\boldsymbol{q},\,\omega) \;=\; \mu_{\text{B}}^2 \sum_{\boldsymbol{k},\boldsymbol{k}'} \langle\!\langle\, a^{+}_{\boldsymbol{k}+}\, a_{\boldsymbol{k}+\boldsymbol{q},-};\, a^{+}_{\boldsymbol{k}'-}\, a_{\boldsymbol{k}'-\boldsymbol{q},+}\,\rangle\!\rangle^{\text{r}}_{\omega}\,. \tag{9.1.26'}$$

It is obvious that we have only those diagrams given in Fig.9.7, which are similar to Figs.9.2 and 9.3(b), or Fig.9.4, and the self-energy diagrams of the kind of Fig.9.3(c); diagrams similar to Figs.9.3(a), or 9.6 cannot appear. Then it is straightforward to derive the following result

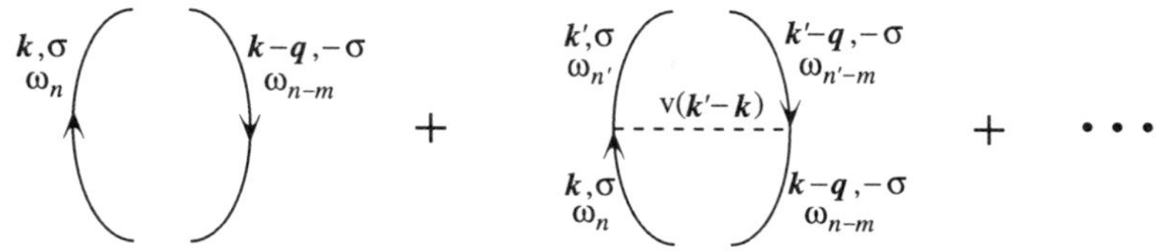

Fig.9.7 Diagrams for the transverse spin susceptibility $\chi_{\sigma,-\sigma}(\boldsymbol{q},\,i\omega_n)$.

$$\chi_{+-}(\boldsymbol{q}, i\nu_m) = \mu_B^2 \frac{F_{+-}(\boldsymbol{q}, i\nu_m)}{1 - \widetilde{V}(\boldsymbol{q})\, F_{+-}(\boldsymbol{q}, i\nu_m)}$$

$$= \mu_B^2 \widetilde{F}_{+-}(\boldsymbol{q}, i\nu_m), \tag{9.1.27}$$

where, corresponding to (8.4.30) and (8.4.40), we defined

$$F_{+-}(\boldsymbol{q}, i\nu_m) = -\frac{1}{\beta} \sum_n \sum_k G_+^{\mathrm{o}}(\boldsymbol{k}, i\omega_n)\, G_-^{\mathrm{o}}(\boldsymbol{k}+\boldsymbol{q}, i\omega_{n+m})$$

$$= \sum_{\boldsymbol{k}} \frac{f(\xi_{\boldsymbol{k}+\boldsymbol{q},-}) - f(\xi_{\boldsymbol{k}+})}{i\nu_m - \left(\xi_{\boldsymbol{k}+\boldsymbol{q},-} - \xi_{\boldsymbol{k}+}\right)}\,. \tag{9.1.28}$$

If we put $i\nu_m \to \omega + i0^+$ in (9.1.27), it reproduces the earlier result of (4.4.12).

Even with the method of Green's functions and Feynman diagrams, the results we obtain for various response functions are exactly the same as those of Chapter 4 which were derived with a more elementary method. Both methods have their own merits. The Feynman diagram method seems to reveal the nature of involved physics and used approximations more transparently. For instance, it is quite natural to think of the extensions, Fig.9.3(b)→Fig.9.4, and Fig.9.3(a)→Fig.9.5→Fig.9.6. Similarly, concerning the thermodynamic potential of an electron gas which was discussed in Chapter 8, we are naturally led to extend Fig.8.7 to Fig.10.1, and we can write down the result immediately (c.f. Chapter 10).

We will see more of the usefulness of the Feynman diagram method in the remaining part of this book.

9.2 PHONON GREEN'S FUNCTION AND PHONON FREQUENCY

We already saw that the Green's function of a free phonon, $D^{\mathrm{o}}(\boldsymbol{q}, i\nu_m)$, is given by (8.5.10) (or, (9.2.1) below). In this section we pursue the phonon Green's function $D(\boldsymbol{q}, i\nu_m)$ which includes the effect of both of the electron-phonon interaction, (5.2.40), and the Coulomb interaction between electrons, (1.5.24). Such an approach to determine the phonon spectrum in a metal was first started by Migdal [9.1]. For further development, see Refs. [1.1–1.3, 2.5, 7.2, 8.2, 9.2–9.5].

The perturbational expansion of a phonon Green's function can be represented as in Fig.9.8. The thick wavy line on the left hand of the equality stands for $D(\boldsymbol{q}, i\nu_m)$, the phonon Green's function containing the effect of interaction. In the perturbation series of the right hand side, the thin wavy line of (a) represents the free phonon Green's function

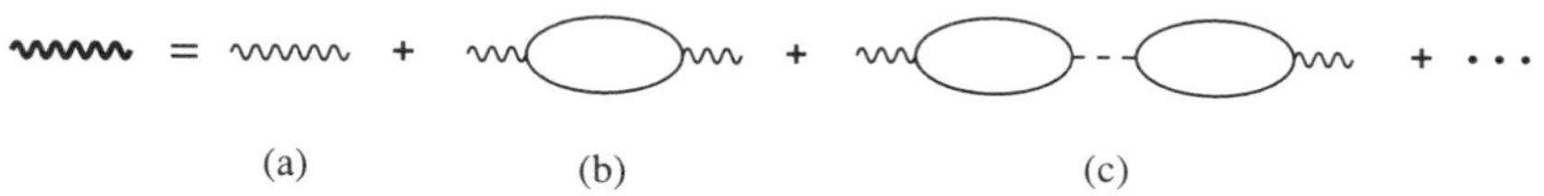

Fig.9.8 The basic lowest order diagrams for a phonon Green's function. A thin wavy line represents a phonon.

$$D^{\circ}(\boldsymbol{q}, i\nu_m) = \frac{2\Omega_q}{\nu_m^2 + \Omega_q^2}, \qquad (9.2.1)$$

which we derived in 8.5.2.

The lowest order diagram with respect to $H_{\text{el-ph}}$ is the second order one given in Fig.9.8(b). The physical processes described by this diagram is clear; a phonon is absorbed to produce an electron polarization; the electron polarization then produces another phonon. From the general rule of 8.5.3, the contribution of this diagram is obtained as

$$\Delta_b D(\boldsymbol{q}, i\nu_m) = \frac{1}{\beta}(-1)^2(-1)|\alpha(\boldsymbol{q})|^2 D^{\circ}(\boldsymbol{q}, i\nu_m)$$

$$\times \sum_{\boldsymbol{k},\sigma,n} G_\sigma^{\circ}(\boldsymbol{k}, i\omega_n)\, G_\sigma^{\circ}(\boldsymbol{k}+\boldsymbol{q}, i\omega_{n+m})\, D^{\circ}(\boldsymbol{q}, i\nu_m)$$

$$= |\alpha(\boldsymbol{q})|^2 D^{\circ}(\boldsymbol{q}, i\nu_m) \sum_\sigma F_\sigma(\boldsymbol{q}, i\nu_m)\, D^{\circ}(\boldsymbol{q}, i\nu_m), \qquad (9.2.2)$$

where we chose various Fourier variables as in Fig.9.9 and noted the relation $\alpha(\boldsymbol{q})\alpha(-\boldsymbol{q}) = |\alpha(\boldsymbol{q})|^2$.

Next, let us turn to Fig.9.8(c). This diagram is first order in H_C and second order in $H_{\text{el-ph}}$. It is clear what kind of physical process is represented by this diagram. The contribution of this diagram can also be readily written down from the general rule of 8.5.3 as

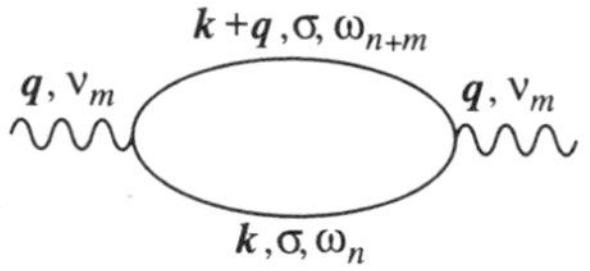

Fig.9.9 Fourier variables used in calculating the contribution of Fig.9.8(b).

$$\Delta_c D(\boldsymbol{q}, i\nu_m) = -\, \mathrm{v}(\boldsymbol{q})\, |\alpha(\boldsymbol{q})|^2 D^{\mathrm{o}}(\boldsymbol{q}, i\nu_m)$$

$$\times \sum_{\sigma,\sigma'} F_\sigma(\boldsymbol{q}, i\nu_m)\, F_{\sigma'}(\boldsymbol{q}, i\nu_m)\, D^{\mathrm{o}}(\boldsymbol{q}, i\nu_m). \qquad (9.2.3)$$

Above two diagrams (b) and (c) of Fig.9.8 represent the fundamental processes to determine the Green's function of a phonon. Then it is straightforward to conceive of extending Fig.9.8 to Fig.9.10, where the shaded bubbles are the same as in Fig.9.4 and the thick broken line stands for $\overline{\mathrm{v}}_{\mathrm{sc}}$ introduced in Fig.9.6(b).

From the results (9.2.1)–(9.2.3) for Fig.9.8, it is evident that the Dyson equation corresponding to Fig.9.10 can be written down as

$$D(\boldsymbol{q}, i\nu_m) = D^{\mathrm{o}}(\boldsymbol{q}, i\nu_m) + D^{\mathrm{o}}(\boldsymbol{q}, i\nu_m)\, |\alpha(\boldsymbol{q})|^2\, \widetilde{P}(\boldsymbol{q}, i\nu_m)\, D(\boldsymbol{q}, i\nu_m)$$

$$-\, \overline{\mathrm{v}}_{\mathrm{sc}}(\boldsymbol{q}, i\nu_m)\, |\alpha(\boldsymbol{q})|^2\, D^{\mathrm{o}}(\boldsymbol{q}, i\nu_m)\, \widetilde{P}(\boldsymbol{q}, i\nu_m)^2\, D(\boldsymbol{q}, i\nu_m), \qquad (9.2.4)$$

where $\widetilde{P}$ and $\overline{\mathrm{v}}_{\mathrm{sc}}$ are defined, respectively, in (9.1.20) and (9.1.21). From (9.2.4) we obtain the phonon Green's function including the effect of the EPI and the Coulomb interaction between electrons as

$$D(\boldsymbol{q}, i\nu_m) = \frac{1}{D^{\mathrm{o}}(\boldsymbol{q}, i\nu_m)^{-1} - |\alpha(\boldsymbol{q})|^2 \widetilde{P}(\boldsymbol{q}, i\nu_m) + |\alpha(\boldsymbol{q})|^2\, \overline{\mathrm{v}}_{\mathrm{sc}}(\boldsymbol{q}, i\nu_m)\, \widetilde{P}(\boldsymbol{q}, i\nu_m)^2}$$

$$= \frac{2\Omega_q}{\nu_m^2 + \Omega_q^2 \left[1 - \dfrac{2}{\Omega_q} |\alpha(\boldsymbol{q})|^2 \dfrac{\widetilde{P}(\boldsymbol{q}, i\nu_m)}{1 + \mathrm{v}(\boldsymbol{q})\, \widetilde{P}(\boldsymbol{q}, i\nu_m)} \right]} \qquad (9.2.5)$$

If we put $i\nu_m \to \omega + i0^+$ in this Green's function, we obtain the corresponding retarded two-time Green's function and, then, from its pole we obtain the phonon spectrum. Thus from (9.2.5) we obtain

Fig.9.10 Dyson equation for a phonon in a metal. The thick wavy and the thick broken lines represent, respectively, the screened phonons and the screened Coulomb interaction.

$$\omega_q^2 = \Omega_q^2 - 2\,\frac{\Omega_q}{\hbar}\,|\,\alpha(q)\,|^2\,\frac{\widetilde{P}(q,\,\omega_q)}{1 + \mathrm{v}(q)\,\widetilde{P}(q,\,\omega_q)}\ , \qquad (9.2.6)$$

where we recovered $\hbar$ in the final expression. If we note the relation of (5.2.41), this result is exactly the same as that of (5.1.1)–(5.1.3). This result should be understood as we discussed in 5.4.1; as shown there, phonon frequency should necessarily be a complex quantity.

Most of the previous discussions on phonons in a metal were confined to the paramagnetic state. Our result given in this section [5.6] is an extension of such a result so as to encompass the ferromagnetic state.

9.3 THE EFFECT OF EPI ON ELECTRONS

In 9.2 we discussed how the phonon spectrum is affected by the effect of the EPI in a metal. Next, in this section we discuss how the energy spectrum of electrons will be affected by the effect of EPI in a metal. The method of Green's functions and Feynman diagrams turns out to be indispensable in this discussion.

9.3.1 Electron Self-Energy due to EPI

If we replace the Coulomb broken line of Fig.8.11(a) by a wavy phonon line we obtain Fig.9.11(a). The meaning of Fig.9.11(a) is clear; an electron emits a phonon and later reabsorbs it, or, inversely, first absorbs a phonon and later reemits it. The contribution of such processes to the electron Green's function can be calculated by the rule of 8.5.3. We now know that the sum of all the higher order contributions such as given in Fig.8.11(c), with all the broken lines replaced by wavy lines, can be represented by the electron self-energy of Fig.9.11(b). From the rule of 8.5.3 we can readily write down the following

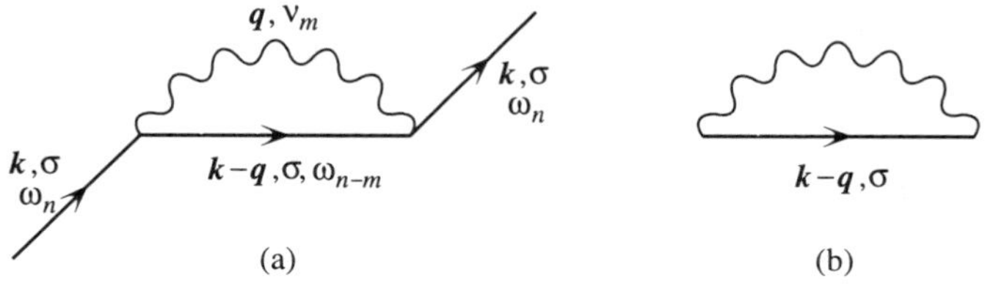

Fig.9.11 The lowest order contribution of EPI to the electron Green's function, (a), and the corresponding electron self-energy due to EPI, (b).

expression for the self-energy

$$\Sigma_\sigma^{\rm ph}(\boldsymbol{k}, i\omega_n) = \frac{1}{\beta}\sum_{\boldsymbol{q},m} |\alpha(\boldsymbol{q})|^2 D^{\rm o}(\boldsymbol{q}, i\nu_m)\, G_\sigma^{\rm o}(\boldsymbol{k}-\boldsymbol{q}, i\omega_{n-m}) . \quad (9.3.1)$$

The electron self-energy due to EPI of Fig.9.11(a) can be extended to that of Fig.9.12(a). Note that in the latter diagram both the electron line and the phonon wavy line are thick. The thick wavy line is defined in Fig.9.10 and is associated with the phonon Green's function $D(\boldsymbol{q}, i\nu_m)$ of (9.2.5). The thick electron line is associated with the electron Green's function $G_\sigma(\boldsymbol{k}-\boldsymbol{q}, i\omega_{n-m})$ which includes the effect of the Coulomb interaction as in (8.5.24) or (8.5.31). The EPI vertex which is screened as in (b) and (c) of Fig.9.12 is represented by a large dot.

The self-energy corresponding to Fig.9.12(a) is obtained by replacing various quantities in (9.3.1) as $G_\sigma^{\rm o} \to G_\sigma$, $D^{\rm o} \to D$, and $\alpha(\boldsymbol{q}) \to$

$$\overline{\alpha}_\sigma(\boldsymbol{q}, i\nu_m) = \frac{\alpha(\boldsymbol{q})}{\varepsilon_\sigma(\boldsymbol{q}, i\nu_m)} , \quad (9.3.2)$$

$\varepsilon_\sigma(\boldsymbol{q}, i\nu_m)$ being the screening constant of (4.3.33), as

$$\Sigma_\sigma^{\rm ph}(\boldsymbol{k}, i\omega_n) = \frac{1}{\beta}\sum_{\boldsymbol{q},m} |\overline{\alpha}_\sigma(\boldsymbol{q}, i\nu_m)|^2 D(\boldsymbol{q}, i\nu_m)\, G_\sigma(\boldsymbol{k}-\boldsymbol{q}, i\omega_{n-m}). \quad (9.3.3)$$

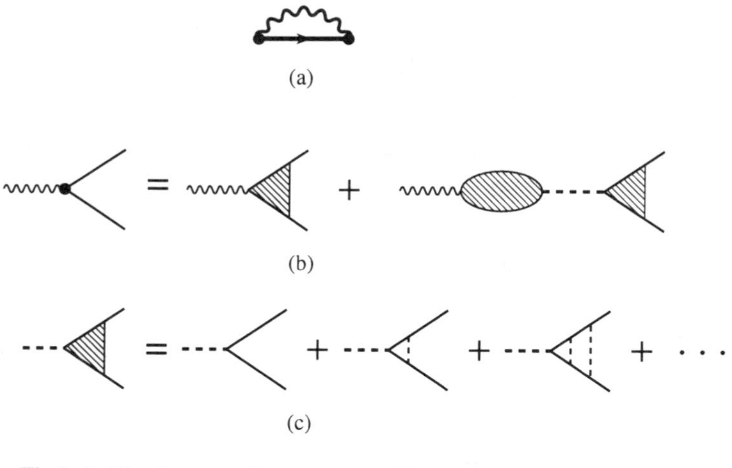

Fig.9.12 The electron self-energy due to EPI, (a), including the effect of the screening of the EPI vertex as shown in (b) and (c), as well as that of phonon frequency. The screened EPI vertex is represented by a large dot.

In carrying out an explicit evaluation of the above electron self-energy due to EPI, here we set

$$G_\sigma(\boldsymbol{k}, i\omega_n) = \frac{1}{i\omega_n - \xi_{\boldsymbol{k}\sigma}}, \tag{9.3.4}$$

$$D(\boldsymbol{q}, i\nu_m) = \frac{2\Omega_{\boldsymbol{q}}}{\nu_m^2 + \omega_{\boldsymbol{q}}^2}, \tag{9.3.5}$$

$$\overline{\alpha}_\sigma(\boldsymbol{q}, i\nu_m) \cong \frac{\alpha(\boldsymbol{q})}{\varepsilon_\sigma(\boldsymbol{q}, 0)} \equiv \overline{\alpha}_\sigma(\boldsymbol{q}), \tag{9.3.6}$$

where $\xi_{\boldsymbol{k}\sigma} = \varepsilon_{\boldsymbol{k}\sigma} - \widetilde{V}n_\sigma - \mu$, and $\omega_{\boldsymbol{q}}^2$ is given in (9.2.6). (Note that it is a simplification to assume the Green's functions of interacting particles in the forms of (9.3.4) and (9.3.5): see 8.1.4. See also problem 9.4.) Then in (9.3.3) we can carry out the summation over m by the method of 8.4.5 as follows,

$$\Sigma_\sigma^{\mathrm{ph}}(\boldsymbol{k}, i\omega_n) = \frac{1}{\beta}\sum_{m,\boldsymbol{q}} |\overline{\alpha}_\sigma(\boldsymbol{q})|^2 \frac{1}{i\omega_{n-m} - \xi_{\boldsymbol{k}-\boldsymbol{q},\sigma}} \left(\frac{1}{\omega_{\boldsymbol{q}} + i\nu_m} + \frac{1}{\omega_{\boldsymbol{q}} - i\nu_m}\right)\frac{\Omega_{\boldsymbol{q}}}{\omega_{\boldsymbol{q}}}$$

$$= \frac{1}{\beta}\sum_{n',\boldsymbol{q}} \frac{\Omega_{\boldsymbol{q}}}{\omega_{\boldsymbol{q}}} |\overline{\alpha}_\sigma(\boldsymbol{q})|^2 \frac{1}{i\omega_{n'} - \xi_{\boldsymbol{k}-\boldsymbol{q},\sigma}} \left(\frac{1}{\omega_{\boldsymbol{q}} + i\omega_n - i\omega_{n'}} + \frac{1}{\omega_{\boldsymbol{q}} - i\omega_n + i\omega_{n'}}\right)$$

$$= \frac{1}{\beta}\sum_{\boldsymbol{q}} \frac{\Omega_{\boldsymbol{q}}}{\omega_{\boldsymbol{q}}} |\overline{\alpha}_\sigma(\boldsymbol{q})|^2 \left(-\frac{\beta}{2\pi i}\right) \int_{C'''} \frac{dz}{e^{\beta z} + 1}$$

$$\times \frac{1}{z - \xi_{\boldsymbol{k}-\boldsymbol{q},\sigma}} \left(\frac{1}{\omega_{\boldsymbol{q}} + i\omega_n - z} + \frac{1}{\omega_{\boldsymbol{q}} - i\omega_n + z}\right). \tag{9.3.7}$$

Here the contour of integration C''' given in Fig.9.13 is obtained by starting from the contour C of Fig.8.6. The contribution from Γ vanishes when we make the radius of Γ infinite. Thus, from the residues at the three poles within the contour, we obtain

$$\Sigma_\sigma^{\mathrm{ph}}(\boldsymbol{k}, i\omega_n) = \sum_{\boldsymbol{q}} |\overline{\alpha}_\sigma(\boldsymbol{q})|^2 \frac{\Omega_{\boldsymbol{q}}}{\omega_{\boldsymbol{q}}} \left[\frac{n(\omega_{\boldsymbol{q}}) + 1 - f(\xi_{\boldsymbol{k}-\boldsymbol{q},\sigma})}{i\omega_n - \xi_{\boldsymbol{k}-\boldsymbol{q},\sigma} - \omega_{\boldsymbol{q}}} + \frac{n(\omega_{\boldsymbol{q}}) + f(\xi_{\boldsymbol{k}-\boldsymbol{q},\sigma})}{i\omega_n - \xi_{\boldsymbol{k}-\boldsymbol{q},\sigma} + \omega_{\boldsymbol{q}}}\right], \tag{9.3.8}$$

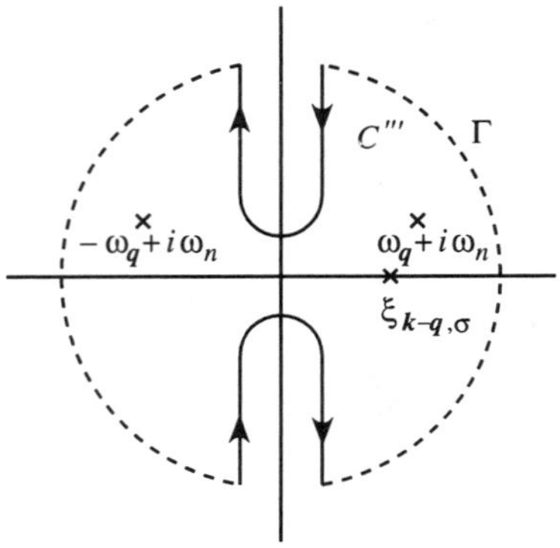

Fig.9.13 The contour of integral used in (9.3.7).

where $n(\omega_q)$ is the Bose distribution function of phonons. The Bose function in (9.3.8) emerged from the following expression which appears in the residues at the poles $z = i\omega_n \pm \omega_q$,

$$\frac{1}{e^{\beta(i\omega_n \pm \omega_q)} + 1} = \frac{1}{-e^{\pm\beta\,\omega_q} + 1} = \begin{cases} -n\,(\omega_q) \\ 1 + n\,(\omega_q). \end{cases} \tag{9.3.9}$$

With the electron self-energy of (9.3.8), the energy spectrum of an electron is obtained from the following equation

$$\omega - \xi_{k,\sigma} - \mathrm{Re}\,\Sigma_\sigma^{\mathrm{ph}}(\boldsymbol{k},\,\omega + i0^+) = 0. \tag{9.3.10}$$

Before solving this equation, let us first discuss the physical nature of the result of (9.3.8). For simplicity we consider in low temperature region such as $T \ll \theta_{\mathrm{D}}$, where we can put $n(\omega_q) = 0$. Then the real part of (9.3.8) becomes

$$\mathrm{Re}\,\Sigma_\sigma^{\mathrm{ph}}(\boldsymbol{k},\,\omega + i0^+) = \sum_{\boldsymbol{q}} |\,\overline{\alpha}_\sigma(\boldsymbol{q})|^2\,\frac{\Omega_q}{\omega_q}\left[\frac{1 - f(\xi_{k-q,\sigma})}{\omega - (\xi_{k-q,\sigma} + \omega_q)} - \frac{f(\xi_{k-q,\sigma})}{\xi_{k-q,\sigma} - (\omega + \omega_q)}\right]. \tag{9.3.11}$$

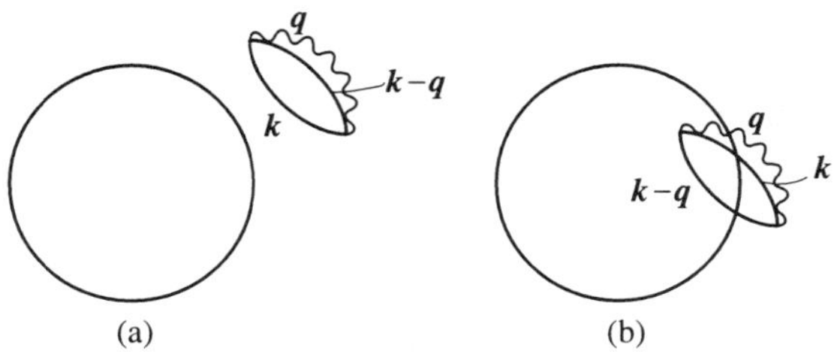

Fig.9.14 The 2nd order process which gives rise to an energy gain of an electron added outside of the Fermi surface, (a), and the same kind of process now suppressed for the electrons inside the Fermi surface because of the added electron, (b).

Suppose that we bring in an electron ($\boldsymbol{k}$, σ) to the outside of the Fermi surface as in Fig.9.14(a). By this addition of one electron how much will the total energy of the system increase? Obviously the first contribution is $\xi_{\boldsymbol{k},\sigma}$. The next contribution is the second order perturbation energy due to the process shown in Fig.9.14(a); the electron ($\boldsymbol{k}$, σ) makes a transition to the state ($\boldsymbol{k}-\boldsymbol{q}$, σ) by emitting a phonon $\omega_{\boldsymbol{q}}$, and then returns to the state ($\boldsymbol{k}$, σ) by reabsorbing the phonon. The first term on the right hand side of (9.3.11) represents such contribution. The factor $(1-f(\xi_{\boldsymbol{k}-\boldsymbol{q},\sigma}))$ appears since for such a process to be possible the electron state ($\boldsymbol{k}-\boldsymbol{q}$, σ) must be unoccupied.

The second process shown in Fig.9.14(b) is of more subtle nature. Now, let us consider an electron ($\boldsymbol{k}-\boldsymbol{q}$, σ) inside the Fermi sphere ($f(\xi_{\boldsymbol{k}-\boldsymbol{q},\sigma}) \neq 0$). The energy to be associated with this electron is originally the sum of $\xi_{\boldsymbol{k}-\boldsymbol{q},\sigma}$ and the second order perturbation energy discussed in the above. If we put an electron ($\boldsymbol{k}$, σ) outside of the Fermi sphere, as we have assumed, however, the transition of the electron ($\boldsymbol{k}-\boldsymbol{q}$, σ) to that particular state ($\boldsymbol{k}$, σ) is now prohibited. Thus, the presence of the electron ($\boldsymbol{k}$, σ) eliminates the possible gain of such second order perturbation energy. The second term on the right hand side of (9.3.11) represents such an effect.

Although the gross nature of the result of (9.3.8) or (9.3.11) is understood as above, it actually reflects quite complicated physical processes. Consider, for instance, the processes involved in screening $\alpha(\boldsymbol{q})$ to $\overline{\alpha}(\boldsymbol{q})$ and $\Omega_{\boldsymbol{q}}$ to $\omega_{\boldsymbol{q}}$. All such processes are properly included in the (9.3.8) or (9.3.11) result. The self-energy $\Sigma_\sigma^{\mathrm{ph}}(\boldsymbol{k}, \omega+i0^+)$ has also an imaginary part and, similarly to the case of phonons which was discussed in 5.4.1, it is related to the life-time of an electron.

In most literature they discuss the electron self-energy due to EPI only for the paramagnetic state, where $\Sigma_+^{\mathrm{ph}}(\boldsymbol{k}, i\omega_n) = \Sigma_-^{\mathrm{ph}}(\boldsymbol{k}, i\omega_n) = \Sigma^{\mathrm{ph}}(\boldsymbol{k}, i\omega_n)$. Here

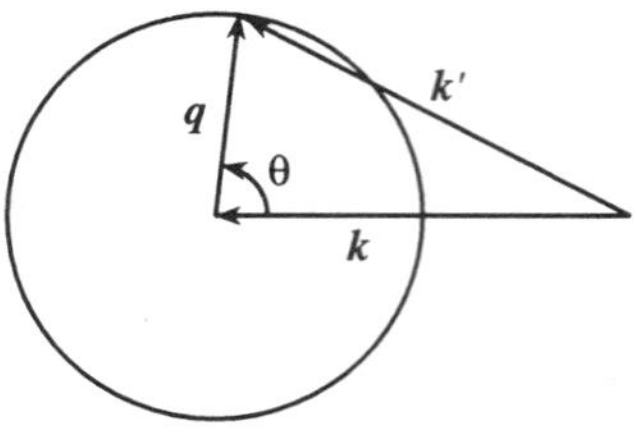

Fig.9.15 Coordinate system used in the integration of (9.3.14).

we extended such an earlier result so as to encompass the ferromagnetic state, by fully including exchange effects [9.6]. In the ferromagnetic state we have

$$\Sigma_+^{\mathrm{ph}}(\boldsymbol{k},\, i\omega_n) \;\neq\; \Sigma_-^{\mathrm{ph}}(\boldsymbol{k},\, i\omega_n). \tag{9.3.12}$$

The electron self-energy due to EPI becomes spin dependent. Also we will find shortly that even in the paramagnetic state the exchange effect plays a very important role in the electron self-energy due to EPI.

9.3.2 Electron-Phonon Coupling Constant $\lambda_{\mathrm{ph}\sigma}$

In the temperature region $T \ll \theta_{\mathrm{D}}$, where $(k_{\mathrm{B}}T/\varepsilon_{\mathrm{F}})^2 \ll 1$ too, (9.3.11) reduces to

$$\mathrm{Re}\,\Sigma_\sigma^{\mathrm{ph}}(\boldsymbol{k},\, \omega) = \sum_{\boldsymbol{q}} |\overline{\alpha}_\sigma(\boldsymbol{q})|^2 \frac{\Omega_{\boldsymbol{q}}}{\omega_{\boldsymbol{q}}} \left[\frac{\theta(\xi_{\boldsymbol{k}-\boldsymbol{q},\sigma})}{\omega - \xi_{\boldsymbol{k}-\boldsymbol{q},\sigma} - \omega_{\boldsymbol{q}}} + \frac{\theta(-\xi_{\boldsymbol{k}-\boldsymbol{q},\sigma})}{\omega - \xi_{\boldsymbol{k}-\boldsymbol{q},\sigma} + \omega_{\boldsymbol{q}}} \right], \tag{9.3.13}$$

where $\theta(x)$ is the step function defined in (8.1.4). Let us estimate the effect of this self-energy on the energy spectrum of electrons in a metal.

In carrying out the summation over $\boldsymbol{q}$ in (9.3.13), it is convenient to use the following procedure [9.1]:

$$\sum_{\boldsymbol{q}} \varphi(\boldsymbol{q},\, |\boldsymbol{k}+\boldsymbol{q}|) = \frac{V}{(2\pi)^3} \int d^3\boldsymbol{q}\; \varphi(\boldsymbol{q},\, |\boldsymbol{k}+\boldsymbol{q}|)$$

$$= \frac{V}{4\pi^2} \frac{1}{k} \int k'\, dk' \int q\, dq\; \varphi(q,\, k'). \tag{9.3.14}$$

This relation is derived as follows. As shown in Fig.9.15, in integrating over $\boldsymbol{q}$, first we integrate over the spherical surface with a fixed radius $|\boldsymbol{q}| = q$, and then integrate over q, the radius of the sphere:

$$\frac{V}{(2\pi)^3}\int d^3\boldsymbol{q}\ \varphi(\boldsymbol{q}, |\boldsymbol{k}+\boldsymbol{q}|) = \frac{V}{(2\pi)^3}\int dq \int_0^{\pi} \left(2\pi\, q \sin\theta\right) q\, d\theta\, \varphi(\boldsymbol{q}, |\boldsymbol{k}+\boldsymbol{q}|)\,.$$

If we introduce $\boldsymbol{k}'$ as in Fig.9.15, we have $|\boldsymbol{k}'|^2 = (\boldsymbol{k}+\boldsymbol{q})^2 = k^2+q^2 - 2kq\cos\theta$. Differenting this relation under constant k and q, we have $2k'dk'=2kq\ sin\theta\ d\theta$, or $\sin\theta\, d\theta = (k'/\,kq)\, dk'$. By putting this last relation into the above integral we obtain (9.3.14).

In applying (9.3.14) on (9.3.13), note that the latter expression is invariant for the replacement $\boldsymbol{q} \rightarrow -\boldsymbol{q}$. Then, by changing the integral over k' to that over $\xi_{k'}$ by the relation $k'dk' = m\ \ d\xi_{k',\sigma}$, with the interval of integration $(-\infty, \infty)$, we obtain

$$\mathrm{Re}\,\Sigma_\sigma^{\mathrm{ph}}(\boldsymbol{k}, \omega) = -\,V m \frac{1}{4\pi^2}\frac{1}{k}\int_0^{2k_{\mathrm{F}\sigma}} q\, dq\, |\bar{\alpha}_\sigma(\boldsymbol{q})|^2 \frac{\Omega_q}{\omega_q} \ln\left|\frac{\omega+\omega_q}{\omega-\omega_q}\right| . \quad (9.3.15)$$

We put the upper bound of the integral over q as $2k_{\mathrm{F}\sigma}$ although if the Debye wave number q_D is smaller than $2k_{\mathrm{F}\sigma}$ we should choose q_D as the upper bound.

In order to see how the EPI affects the energy spectrum of electrons, we are required to put (9.3.15) back into (9.3.10) and solve it for ω. In Fig.9.16 we illustrate the expected result of such a procedure in the paramagnetic state. The energy spectra of electrons are markedly modified in the region near the Fermi level over the width $\sim\omega_\mathrm{D}$.

In the immediate vicinity of the Fermi level such as $|\omega| \ll \omega_\mathrm{D}$, since

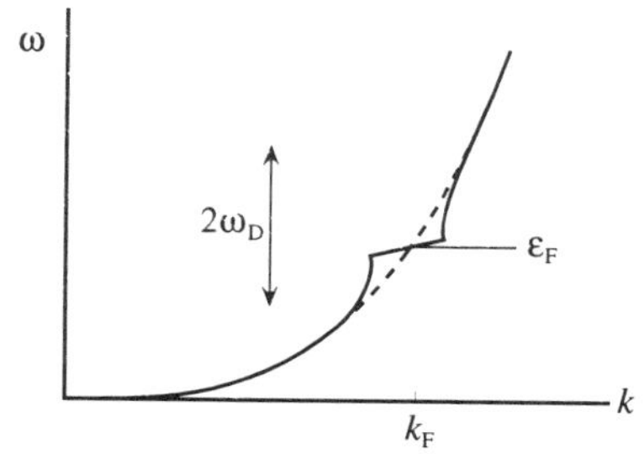

Fig.9.16 A schematic illustration of the effect of EPI on the energy dispersion of electrons in a metal.

$$\ln\left|\frac{\omega+\omega_q}{\omega-\omega_q}\right| \cong 2\frac{\omega}{\omega_q},$$

(9.3.15) can be approximated as

$$\mathrm{Re}\,\Sigma_\sigma^{\mathrm{ph}}(k_{\mathrm{F}\sigma},\omega) = -\lambda_{\mathrm{ph}\sigma}\omega \tag{9.3.16}$$

with

$$\lambda_{\mathrm{ph}\sigma} = -\frac{\partial}{\partial\omega}\mathrm{Re}\,\Sigma_\sigma^{\mathrm{ph}}(k_{\mathrm{F}\sigma},\omega)\Big|_{\omega=0} = \frac{1}{2\pi^2}V\frac{m}{k_{\mathrm{F}\sigma}}\int_0^{2k_{\mathrm{F}\sigma}} q\,dq\,|\overline{\alpha}_\sigma(\boldsymbol{q})|^2\frac{\Omega_q}{\omega_q^2}$$

$$= 2N_\sigma(0)\,\langle\langle|\overline{\alpha}_\sigma(\boldsymbol{q})|^2\frac{\Omega_q}{\omega_q^2}\rangle\rangle_\sigma, \tag{9.3.17}$$

where

$$\langle\langle h(\boldsymbol{q})\rangle\rangle_\sigma = \frac{1}{2k_{\mathrm{F}\sigma}{}^2}\int_0^{2k_{\mathrm{F}\sigma}} q\,dq\,h(\boldsymbol{q}) \tag{9.3.18}$$

is the average of $h(\boldsymbol{q})$ over the change in the wave vector $\boldsymbol{q}$ on the Fermi surface of σ spin electrons, and $N_\sigma(0) = Vm\,k_{\mathrm{F}\sigma}/2\pi^2$ is the density of states at the Fermi surface of σ spin electrons. We call $\lambda_{\mathrm{ph}\sigma}$ the *electron-phonon coupling constant*. As we will see below, $\lambda_{\mathrm{ph}\sigma}$ is a very important quantity in metal physics.

By putting (9.3.16) into (9.3.10), for $k \cong k_{\mathrm{F}\sigma}$ we obtain

$$\omega = \frac{1}{1+\lambda_{\mathrm{ph}\sigma}}\xi_{k\sigma}. \tag{9.3.19}$$

If we introduce the effective mass m^* of an electron near the Fermi surface, (9.3.19) leads to

$$\frac{1}{m_\sigma^*} = \frac{\partial^2}{\partial k^2}\omega = \frac{1}{(1+\lambda_{\mathrm{ph}\sigma})m}, \tag{9.3.20}$$

or,

$$m_\sigma^* = (1+\lambda_{\mathrm{ph}\sigma})m. \tag{9.3.21}$$

Note the above procedure is in accord with that of (8.1.38); here $\Sigma^{\rm ph}(k, \omega)$ does not depend on ξ_k. As can be seen from (9.3.17), $\lambda_{{\rm ph}\sigma} > 0$; near the Fermi surface, the effective mass of an electron is enhanced by the effect of EPI. As we will see shortly, generally $\lambda_{{\rm ph}\sigma} \cong O(1)$. Note that in the ferromagnetic state we have

$$\lambda_{\rm ph+} \neq \lambda_{\rm ph-}. \tag{9.3.22}$$

The electron mass enhancement due to EPI is spin dependent in the ferromagnetic state. In the paramagnetic state we have

$$\lambda_{\rm ph+} = \lambda_{\rm ph-} = \lambda_{\rm ph}. \tag{9.3.23}$$

The electron mass enhancement of (9.3.21) implies that the *density of energy levels* of electron at the Fermi surface is enhanced to

$$\overline{N}_\sigma(0) = \left(1 + \lambda_{{\rm ph}\sigma}\right) N_\sigma(0)\,. \tag{9.3.24}$$

However, if we consider the effect of the renormalization factor Z defined in (8.1.34)–(8.1.35), we note that $\overline{N}_\sigma(0)$ does not coincide with the *density of states* of electron [9.4].

9.3.3 $\lambda_{{\rm ph}\sigma}$ in the Ferromagnetic State

In this subsection we carry out some numerical calculation on (9.3.17) for the ferromagnetic state [9.6]. Like in Chapter 5, we adopt the jellium model ex-

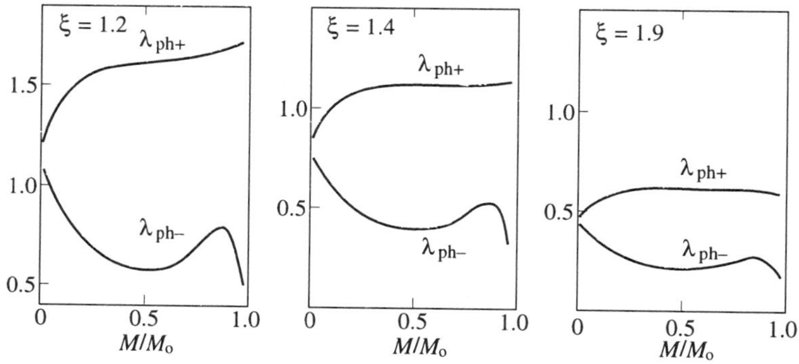

Fig.9.17 The spin dependent electron-phonon coupling constants at the absolute zero in the ferromagnetic state of an electron gas with $k_{\rm F} = 10^8$/cm as a function of magnetization, $M/M_{\rm o}$, for different values of ξ.

tended by introducing the parameter ξ of (5.3.9). To make the calculation simple we use the Debye approximation of (6.1.5) with the sound velocity $s(M)$ in the ferromagnetic state as given by (5.4.27) and (5.4.29). As for the $\alpha(\boldsymbol{q})$, we use (5.2.41) together with (5.2.38') or (5.3.5), namely,

$$2\Omega_q\,|\,\alpha(\boldsymbol{q})\,|^2\,/\,\mathrm{v}(\boldsymbol{q}) \;=\; \Omega_{\mathrm{pl}}^2. \tag{9.3.25}$$

For electrons, we assume the parabolic free electron dispersion with $k_F=10^8$/cm. The exchange interaction is assumed to take the form, $\tilde{V}(\boldsymbol{q}) = \tilde{V}\,/\,[1 + (q/k_F)^2\,]$.

Then, how $\lambda_{\mathrm{ph}\sigma}$ will depend on magnetization M is calculated for various values of ξ in Fig. 9.17. For simplicity, we considered at $T = 0$ and the change in magnetization is produced by changing $\tilde{V}$. The ferromagnetic state is assumed to be with $n_+ > n_-$ as in Fig.3.5(a). As anticipated, the size of $\lambda_{\mathrm{ph}\sigma}$ is $O(1)$, but it depends rather sensitively on the value of ξ. The difference between $\lambda_{\mathrm{ph}+}$ and $\lambda_{\mathrm{ph}-}$ is quite large. It will be interesting if we can actually observe such a spin dependence in $\lambda_{\mathrm{ph}\sigma}$. By what kind of experiment can we possibly observe such an effect? In this respect, note that the spin wave coupling constant which similarly enhances the effective mass of electron also is spin dependent: see 9.5.1. As for λ_{ph} for the paramagnetic state, we will give a numerical example in 9.5.

9.3.4 Migdal's Theorem

Note that the contribution from diagrams such as Fig.9.18(b) is not included in the electron self-energy of (9.3.3), whereas that of Fig.9.18(a) is included. It is known as Migdal's theorem [9.1] that the effect of the contribution of Fig.9.18(b) on the electron self-energy amounts to a relative correction of $O(\omega_D/\varepsilon_F)$ and, thus, can be neglected. Let us briefly describe how such conclusion comes out.

The problem is how the electron-phonon interaction constant $\alpha(\boldsymbol{q})$ will be effectively modified by the process given in Fig.9.19(b). we can express the effect as $\alpha(\boldsymbol{q}) \rightarrow \alpha(\boldsymbol{q})\,[1 + \gamma_{\mathrm{ph}}]$. γ_{ph} is called the *vertex correction due to EPI.* Let us compare γ_{ph} with the vertex correction due to the Coulomb interaction, $\gamma_C(\boldsymbol{q})$, given in Fig.9.19(a).

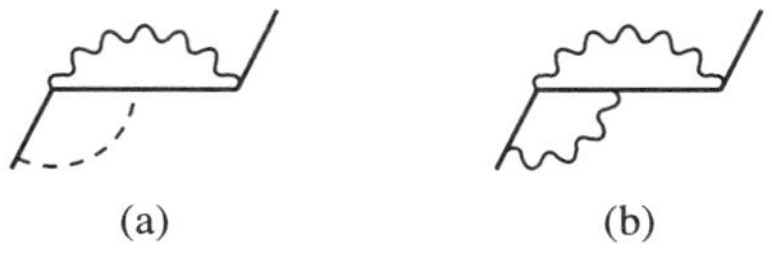

Fig.9.18 The lowest order vertex corrections by the Coulomb interaction, (a), and by the EPI, (b), in the electron self-energy due to the EPI.

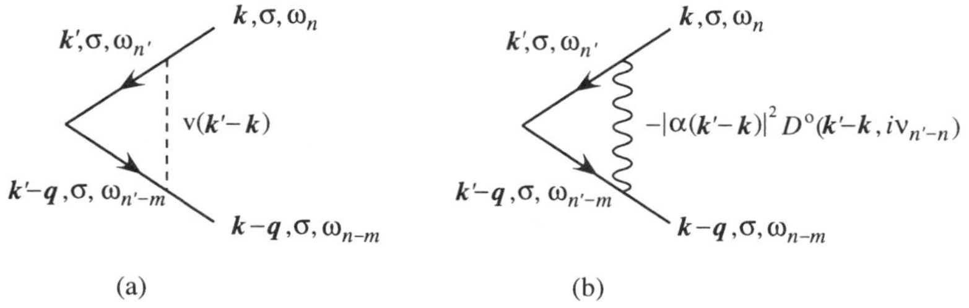

Fig.9.19 Fourier variables used in (9.3.26) and (9.3.27) for vertex corrections due to, respectively, the Coulomb interaction, (a), and the EPI, (b).

From the general rule given in 8.5.3, corresponding to Figs.9.19(a) and (b) we have, respectively,

$$\gamma_{\mathrm{C}}(\boldsymbol{k},\, \boldsymbol{q},\, i\nu_m) \;=\; -\frac{1}{\beta}\sum_{\boldsymbol{k}',n'} \mathrm{v}(\boldsymbol{k}'-\boldsymbol{k})\, G^{\mathrm{o}}_{\sigma}(\boldsymbol{k}',\, i\omega_{n'})\, G^{\mathrm{o}}_{\sigma}(\boldsymbol{k}'-\boldsymbol{q},\, i\omega_{n'}-i\nu_m), \tag{9.3.26}$$

$$\gamma_{\mathrm{ph}}(\boldsymbol{k},\, i\omega_n,\, \boldsymbol{q},\, i\nu_m) \;=\; \frac{1}{\beta}\sum_{\boldsymbol{k}',n'} \left[\, \left|\alpha(\boldsymbol{k}'-\boldsymbol{k})\right|^2 \frac{2\Omega_{\boldsymbol{k}'-\boldsymbol{k}}}{\Omega^2_{\boldsymbol{k}'-\boldsymbol{k}} - \left(i\omega_{n'} - i\omega_n\right)^2} \right]$$

$$\times\, G^{\mathrm{o}}_{\sigma}(\boldsymbol{k}',\, i\omega_{n'})\, G^{\mathrm{o}}_{\sigma}(\boldsymbol{k}'-\boldsymbol{q},\, i\omega_{n'}-i\nu_m). \tag{9.3.27}$$

First, in (9.3.26) the summation over $i\omega_{n'}$ can be carried out by the procedure of 8.4.5, as in (9.1.14)–(9.1.15):

$$\gamma_{\mathrm{C}}(\boldsymbol{k},\, \boldsymbol{q},\, i\nu_m) \;=\; \sum_{\boldsymbol{k}'} \mathrm{v}(\boldsymbol{k}'-\boldsymbol{k})\, \frac{f(\xi_{\boldsymbol{k}'-\boldsymbol{q},\sigma}) - f(\xi_{\boldsymbol{k}'\sigma})}{-i\nu_m - \left(\xi_{\boldsymbol{k}'-\boldsymbol{q}} - \xi_{\boldsymbol{k}'}\right)}$$

$$=\; \widetilde{V}(\boldsymbol{q})\, F_{\sigma}(\boldsymbol{q},\, i\nu_m). \tag{9.3.28}$$

Thus we find

$$\gamma_{\mathrm{C}} \cong O\!\left(\widetilde{V}N(0)\right) \;=\; O(1). \tag{9.3.29}$$

Next, as for γ_{ph}, if we can put for all the $\omega_{n'}$

$$\left|\alpha(\boldsymbol{k}'-\boldsymbol{k})\right|^2 \frac{2\Omega_{\boldsymbol{k}'-\boldsymbol{k}}}{\Omega_{\boldsymbol{k}'-\boldsymbol{k}}^2-\left(i\omega_{n'}-i\omega_n\right)^2} \cong \frac{|\alpha|^2}{\Omega_{\boldsymbol{k}'-\boldsymbol{k}}} \cong \frac{|\alpha|^2}{\omega_{\rm D}} \tag{9.3.30}$$

in (9.3.27), we would have $O(\gamma_{\rm ph}) \cong |\alpha|^2 F_\sigma(\boldsymbol{q}, i\nu_m)/\omega_{\rm D} \cong |\alpha|^2 N(0)/\omega_{\rm D} \cong O(1)$ (see (9.3.34)); $|\alpha|$ represents the order of magnitude of $|\alpha(\boldsymbol{k}'-\boldsymbol{k})|$. However, for the validity of (9.3.30) we require

$$\left|\omega_{n'}-\omega_n\right| \le \omega_{\rm D}. \tag{9.3.31}$$

Outside of this region, the quantity on the left most side of (9.3.30) rapidly decreases as $\left|\omega_{n'}-\omega_n\right|^{-2}$ with increasing n'. Thus, the summation over n' in (9.3.27) may be approximated by restricting it within the region, (9.3.31), with the simplification, (9.3.30).

In summing over n' in (9.3.27) note the following relation,

$$\frac{2\pi}{\beta}\sum_n g(i\omega_n) \cong \int d\varepsilon\, g(i\varepsilon). \tag{9.3.32}$$

Thus, by considering the restriction (9.3.31), $\gamma_{\rm ph}$ of (9.3.27) can be estimated as

$$\begin{aligned} O\left|\gamma_{\rm ph}(\boldsymbol{k}, i\omega_n, \boldsymbol{q}, i\nu_m)\right| &= O\left[\frac{|\alpha|^2}{\omega_{\rm D}}\frac{1}{\beta}\left(\frac{\omega_{\rm D}}{\varepsilon_{\rm F}}\right)\sum_{n'}\sum_{\boldsymbol{k}'} G_\sigma^0(\boldsymbol{k}', i\omega_{n'})\, G_\sigma^0(\boldsymbol{k}'-\boldsymbol{q}, i\omega_{n'}-i\nu_m)\right] \\ &= O\left[\frac{|\alpha|^2}{\omega_{\rm D}}\left(\frac{\omega_{\rm D}}{\varepsilon_{\rm F}}\right)N(0)\right] = O\left(\frac{\omega_{\rm D}}{\varepsilon_{\rm F}}\right), \end{aligned} \tag{9.3.33}$$

where we incorporated the restriction (9.3.31) by modifying $\Sigma_{n'}$ as $(\omega_{\rm D}/\varepsilon_{\rm F})\Sigma_{n'}$, and noted the relation (see (10.4.48))

$$O\left(|\alpha|^2 N(0)\right) = \omega_{\rm D}\,. \tag{9.3.34}$$

In this way we find the vertex correction due to EPI is of the order of $\omega_{\rm D}/\varepsilon_{\rm F} \cong 10^{-2}$, and, therefore, much smaller than the vertex correction due to the Coulomb interaction given in (9.3.29). This observation constitutes Migdal's theorem. For a more detailed discussion, see, for instance, Refs.[9.1–9.4, 1.3, 7.2].

Migdal's theorem is often invoked to justify the neglect the EPI vertex correction. However, now we all know that it causes superconductivity of a metal, as we will see in the next subsection. Also, as we saw in chapter 6, it can have vital effects on the magnetic properties of a metal; see 9.4.2. Thus, contrary to its original implication, it is misleading to invoke the Migdal's theorem in conjunction not only with superconductivity but also magnetism.

9.3.5 Superconducting Instability due to EPI

The EPI is the principal mechanism of the superconductivity of ordinary metals. Let us briefly review how the superconducting transition temperature is determined from the same electron-phonon coupling constant, λ_{ph}, which was discussed in 9.3.2.

According to the BCS theory [9.7,1.2], the occurrence of superconductivity is associated with the formation of the Cooper pairs of electrons, $(\boldsymbol{k},+; -\boldsymbol{k},-)$. Then, analogously to the case of ferromagnetic phase transition, the condition for superconductivity may be given by the divergence of the correlation of Cooper pairs.

The two-particle temperature Green's function to describe such correlation of Cooper's pairs may be introduced as

$$G_{\text{BCS}}(\boldsymbol{q}, \tau) = \sum_{\boldsymbol{k},\boldsymbol{k}'} \left\langle T_\tau\, a_{\boldsymbol{k}'+}(\tau)\, a_{-\boldsymbol{k}'+\boldsymbol{q},-}(\tau)\, a^+_{-\boldsymbol{k}+\boldsymbol{q},-}\, a^+_{\boldsymbol{k}+} \right\rangle, \quad (9.3.35)$$

where we assume $\boldsymbol{q} \neq 0$ for the moment. We pursue this Green's function by concentrating on the effect of the EPI ladders as shown in Fig.9.20.

The zero-th order term is given by

$$G^{\text{o}}_{\text{BCS}}(\boldsymbol{q}, \tau) = \sum_{\boldsymbol{k},\boldsymbol{k}'} \left\langle T_\tau\, a_{\boldsymbol{k}'+}(\tau)\, a^+_{\boldsymbol{k}+} \right\rangle_{\text{o}} \left\langle T_\tau\, a_{-\boldsymbol{k}'+\boldsymbol{q},-}(\tau)\, a^+_{-\boldsymbol{k}+\boldsymbol{q},-} \right\rangle_{\text{o}}$$

$$= \sum_{\boldsymbol{k}} G^{\text{o}}_+(\boldsymbol{k}, \tau)\, G^{\text{o}}_-(-\boldsymbol{k}+\boldsymbol{q}, \tau), \quad (9.3.36)$$

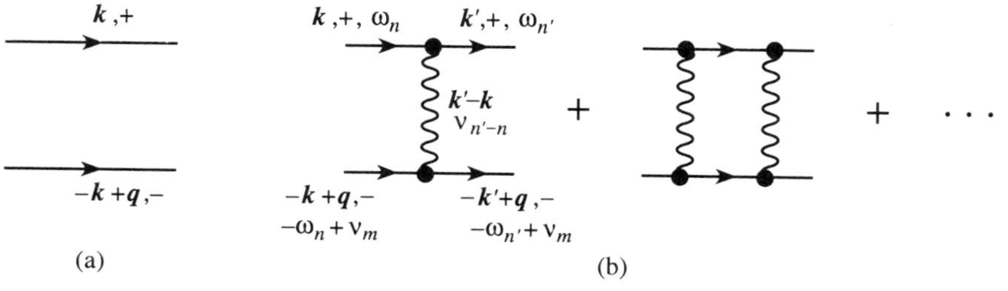

Fig.9.20 The two-electron (particle-particle) Green's function without any interaction effect, (a), and the contributions to it of the EPI ladders to infinite order, (b).

corresponding to Fig.9.20(a). Note the difference between Fig.9.1 or Fig.9.2 and Fig.9.20; while the former represents the propagation of an electron-hole pair, the latter represents that of an electron-electron pair. Also note the corresponding difference between (9.1.6) or (8.4.24) and (9.3.36) in the signs of times in Green's functions; while the former expression leads to the Lindhard function, the latter does not.

For higher order terms let us consider those processes given in Fig.9.20(b); these are known to lead to the BCS superconducting instability. Here both the phonon wavy lines and the electron-phonon interaction dots are the screened ones given, respectively, in Fig.9.10 and Fig.9.12. If we take various Fourier variables as in Fig.9.20(b), we can readily write down the contributions of these diagrams as

$$G_{\rm BCS}(\boldsymbol{q}, i\nu_m) = \frac{1}{\beta}\sum_{\boldsymbol{k},n} G_+^0(\boldsymbol{k}, i\omega_n)\, G_-^0(-\boldsymbol{k}+\boldsymbol{q}, -i\omega_n + i\nu_m)$$

$$\times \Lambda(\boldsymbol{k}, \boldsymbol{q}, i\omega_n, i\nu_m) \tag{9.3.37}$$

with

$$\Lambda(\boldsymbol{k}, \boldsymbol{q}, i\omega_n, i\nu_m) = 1 + \frac{1}{\beta}\sum_{\boldsymbol{k}',n'} |\overline{\alpha}(\boldsymbol{k}'-\boldsymbol{k})|^2\, D(\boldsymbol{k}'-\boldsymbol{k}, i\omega_{n'} - i\omega_n)$$

$$\times G_+^0(\boldsymbol{k}', i\omega_{n'})\, G_-^0(-\boldsymbol{k}'+\boldsymbol{q}, i\omega_{n'}+i\nu_m)\, \Lambda(\boldsymbol{k}', \boldsymbol{q}, i\omega_n, i\nu_m) \tag{9.3.38}$$

by using the rule given in 9.1; various notations are the same as before. A quantity like Λ is called the *vertex part*; it is related to the vertex correction such as introduced in 9.3.4 as $\Lambda = 1 + \gamma$. Note that the BCS instability is associated with divergence of the above $G_{\rm BCS}$ or Λ for $i\nu_m = 0$, and $\boldsymbol{q} = 0$.

In the summations in (9.3.37) and (9.3.38) involving EPI, we anticipate that the dominant contributions will come from the regions,

(i) $|\omega_n|, |\omega_{n'}| \leq \omega_{\rm D}$,

(ii) $|\xi_k|, |\xi_{k'}| \leq \omega_{\rm D}$, namely, $|\boldsymbol{k}|, |\boldsymbol{k}'| \cong k_{\rm F}$.

The region (i) represents where $D(\boldsymbol{k}'-\boldsymbol{k}, i\omega_{n'} - i\omega_n)$ can be put as $\sim 1/\omega_{\rm D}$; outside of this region $D(\boldsymbol{k}'-\boldsymbol{k}, i\omega_{n'} - i\omega_n)$ decreases rapidly as $1/(i\omega_{n'} - i\omega_n)^2$. The region (ii) is where the dominant contribution to the summation over $\boldsymbol{k}'$ of the factor

$$G^{\rm o}_{+}(\boldsymbol{k}', i\omega_{n'})\, G^{\rm o}_{-}(-\boldsymbol{k}', -i\omega_{n'}) \;=\; \frac{1}{\xi_{\boldsymbol{k}'}^{2} + \omega_{n'}^{2}} \tag{9.3.39}$$

with $\omega_{n'}$ within the above region (i), comes from. Within these regions (i) and (ii), for $\nu_m = 0$, and $\boldsymbol{q} = 0$, we may put $\Lambda(\boldsymbol{k}, 0, i\omega_n, 0) = \Lambda$, independent of $\boldsymbol{k}$ and $i\omega_n$. Then, (9.3.38) is rewritten as

$$\Lambda \cong \left[1 - \frac{1}{\beta}\sum_{\boldsymbol{k}', n'} |\overline{\alpha}(\boldsymbol{k}'-\boldsymbol{k}_{\rm F})|^2 D^{\rm o}_{+}(\boldsymbol{k}'-\boldsymbol{k}_{\rm F}, 0)\, G^{\rm o}_{+}(\boldsymbol{k}', i\omega_{n'})\, G^{\rm o}_{-}(-\boldsymbol{k}', -i\omega_{n'})\right]^{-1}$$

$$\tag{9.3.40}$$

where $\boldsymbol{k}_{\rm F}$ is a Fermi wave vector. The condition for the divergence of Λ and, therefore, $G_{\rm BCS}$ is

$$1 \;=\; \frac{1}{\beta}\sum_{\boldsymbol{k}', n'} \left|\,\overline{\alpha}(\boldsymbol{k}' - \boldsymbol{k}_{\rm F})\right|^2 \frac{2\Omega_{\boldsymbol{k}'-\boldsymbol{k}_{\rm F}}}{\omega^2_{\boldsymbol{k}'-\boldsymbol{k}_{\rm F}}}\, \frac{1}{\xi_{\boldsymbol{k}'}^{2} + \omega_{n'}^{2}}\,. \tag{9.3.41}$$

Now, by noting again that with regard to the summation over $\boldsymbol{k}'$ in (9.3.41) the dominant contribution comes from narrow region around the Fermi surface we approximate it as

$$\begin{aligned} 1 &\cong \frac{1}{\beta}\,\langle\langle\, |\overline{\alpha}(\boldsymbol{q})|^2 \frac{2\Omega_{\boldsymbol{q}}}{\omega_{\boldsymbol{q}}^2}\rangle\rangle \sum_{\boldsymbol{k}', n'} \frac{1}{\xi_{\boldsymbol{k}'}^{2} + \omega_{n'}^{2}} \\ &= \frac{1}{\beta}\,\frac{\lambda_{\rm ph}}{N(0)} \sum_{\boldsymbol{k}', n'} \frac{1}{\xi_{\boldsymbol{k}'}^{2} + \omega_{n'}^{2}}\,, \end{aligned} \tag{9.3.42}$$

where

$$\lambda_{\rm ph} \;=\; 2N(0)\,\langle\langle\, |\overline{\alpha}(\boldsymbol{q})|^2 \Omega_{\boldsymbol{q}} / \omega_{\boldsymbol{q}}^2\,\rangle\rangle \tag{9.3.43}$$

is the electron-phonon coupling constant for the paramagnetic state, corresponding to (9.3.17) of the ferromagnetic state; for the paramagnetic state we have $\langle\langle \cdots \rangle\rangle_+ = \langle\langle \cdots \rangle\rangle_- \equiv \langle\langle \cdots \rangle\rangle$. In (9.3.43), $\overline{\alpha}(\boldsymbol{q}) = \alpha(\boldsymbol{q})/\varepsilon(\boldsymbol{q}, 0)$ is the electron-phonon interaction constant screened by the screening constant, (4.3.29) and $\omega_{\boldsymbol{q}}$ is the screened phonon frequency for the paramagnetic state, (5.3.1) or (9.2.6); it is important to note that $\overline{\alpha}(\boldsymbol{q})$ and $\omega_{\boldsymbol{q}}{}^2$ are screened by different screening constants.

If we neglect the difference between Ω_q and ω_q in (9.3.43), it reduces to the familiar expression in the literature; with such a simplified expression, however, we fail to evaluate correctly the important constructive effect on $\lambda_{\rm ph}$ and, therefore, superconductivity of the exchange interaction between electrons which will be shown in 9.5.3.

For the last factor of (9.3.42), we first have

$$\sum_{k'} \frac{1}{\xi_{k'}^2 + \omega_{n'}^2} \cong N(0) \int_{-\omega_{\rm D}}^{\omega_{\rm D}} \frac{d\xi}{(\xi^2 + \omega_{n'}^2)} \cong N(0) \int_{-\infty}^{\infty} \frac{d\xi}{(\xi^2 + \omega_{n'}^2)}$$

$$= N(0)\,\pi\, \frac{1}{|\omega_{n'}|}, \tag{9.3.44}$$

where we noted $|\omega_{n'}| \leq \omega_{\rm D}$. Next, we have

$$\sum_{n,\, |\omega_n| \leq \omega_{\rm D}} \frac{1}{|\omega_n|} = \frac{\beta}{\pi} \sum_{n=0}^{\beta\omega_{\rm D}/2\pi} \frac{1}{n + \frac{1}{2}} = \frac{\beta}{\pi} \left(\gamma + \ln \frac{2\beta\omega_{\rm D}}{\pi} \right), \tag{9.3.45}$$

where γ is the Euler's constant [9.8].

Thus, (9.3.42) is now rewritten as

$$1 = \lambda_{\rm ph} \left[\gamma + \ln (2\beta \frac{\omega_{\rm D}}{\pi}) \right], \tag{9.3.46}$$

and from this we obtain the BCS superconducting transition temperature $T_{\rm C}$ as

$$T_{\rm C} = \frac{2\, e^{\gamma}}{\pi}\, \theta_{\rm D} e^{-1/\lambda_{\rm ph}} = 1.13\, \theta_{\rm D} e^{-1/\lambda_{\rm ph}}. \tag{9.3.47}$$

This result is quite familiar to us. However, we redrived it to convince us that what determines the superconducting transition temperature of a metal is $\lambda_{\rm ph}$ of (9.3.43) in which the magnetic properties of the metal is so deeply involved. We discuss this problem in more detail later in 9.5.3.

Now, going back to Fig.9.20, what would be the effect of the direct Coulomb interaction between electrons? The consideration of this effect, which is destructive to superconductivity, plays an important role in a quantitative discussion, as first shown by Bogoliubov et al. [9.9, 9.10].

9.4 THE EFFECT OF EPI ON THE SPIN SUSCEPTIBILITY

As for the effect of EPI on the spin susceptibility of a metal, we discussed it extensively in Chapter 6. Most of earlier discussions on this problems, however, started from questions different from that of Chapter 6. The first question raised was how the spin susceptibility will be affected by the effect of the electron mass enhancement due to EPI which is discussed in 9.3.2. The second question was addressed to the effect on the spin susceptibility of the EPI ladders which was discussed in 9.3.4 in connection with the Migdal's theorem. In the following two subsections, 9.4.1 and 9.4.2, we discuss these two questions. We, then, summarize our conclusion in 9.4.3.

9.4.1 The Effect of the Electron Mass Enhancement on the Spin Susceptibility

An important quantity which is directly proportional to the electronic density of states at the Fermi surface is the Pauli spin susceptibility of (3.1.14). The exchange enhanced Stoner susceptibility of (3.3.8) also is determined by the electronic density of states at the Fermi surface. Then, a question naturally arises is whether such magnetic susceptibility would be enhanced by the EPI effect of (9.3.24) [9.11]. Historically this question was the beginning of our inquiry on the possible effect of EPI on the magnetic properties of a metal and the negative answer to it is now well known [9.12, 3.10]; the Pauli or the Stoner magnetic susceptibility of a metal is not directly enhanced by the effect of (9.3.24).

The most often given explanation for such a conclusion may be through the illustration of Fig.9.21 [3.10]. The enhancement of the electronic density of states is confined to the region near the Fermi surface, as in Fig.9.21(a). When the electron bands are spin split by an external magnetic field H, the enhancement of the density of states now occurs around the new Fermi energy as shown in Fig.9.21(b). Thus, the difference between the shaded occupied areas of $\pm$ spin electronic density of states is not affected by the presence of the bumps around the Fermi energy in the electronic density of states.

I think, however, the above explanation is misleading. As we emphasized in conjunction with (9.3.24), it is the density of energy levels but not the density of states that is enhanced. If we consider the effect of the renormalization factor z, the electronic density of states near the Fermi surface is not at all enhanced. Therefore, the Pauli susceptibility is not enhanced by EPI.

A little more convincing analysis of the effect may be done in a manner similar to that in 6.1. As can be seen from Fig.9.16, the electrons to be affected

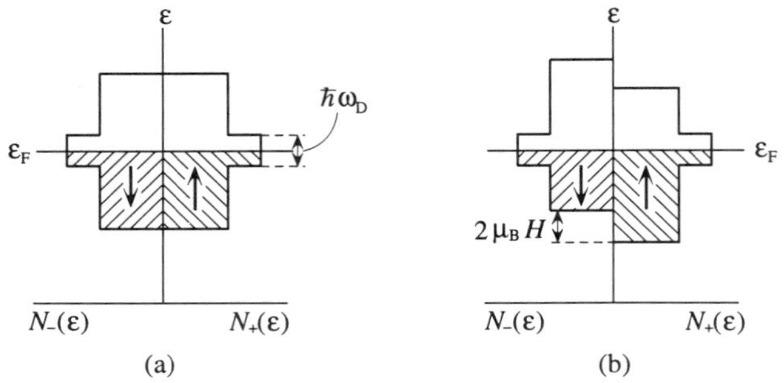

Fig.9.21 Suppose that the effect of the EPI, bumps are produced in the electronic density of states near the Fermi level of a metal as shown in (a). When the electronic density of states are spin split by an external magnetic field, the locations of the bumps also move to around the new Fermi levels. Thus, the produced magnetization, which is given by the difference between the shaded areas, is independent of the presence of the bumps in the electronic density of states. (As discussed in the text, however, the density of states may not be enhanced.)

by EPI are those within the energy range $\sim\omega_D$ from the Fermi surface, whose number being $\sim N(0)\omega_D$, and the size of the change in energy of each of such electrons is $\sim\omega_D$ (See Problem 9.5). Thus, the change in energy of the electron system due to EPI, let it denote as $E_{el\text{-}ph}$, is estimated as

$$|E_{el\text{-}ph}| \cong O[\omega_D\{N(0)\,\omega_D\}] \cong n\,\omega_D\,(\omega_D/\varepsilon_F). \tag{9.4.1}$$

The contribution to the spin susceptibility of this effect is determined from how $E_{el\text{-}ph}$ changes with the magnetization of the system. Since, as we saw in Chapter 6, $|\Delta\omega_D(M)/\omega_D| = O((M//M_o)^2)$, for the change in E_{el-ph} due to magnetization M we have

$$\left|\Delta E_{el\text{-}ph}(M)\right| \cong n\left(\omega_D{}^2/\varepsilon_F\right)(M/M_o)^2, \tag{9.4.2}$$

$M_o \cong n\mu_B$ being the maximum possible magnetization. The corresponding change in the electron kinetic energy, on the other hand, is given as

$$\Delta E_{el}^{o}(M) \cong n\,\varepsilon_F\,(M/M_o)^2. \tag{9.4.3}$$

Then by using (3.4.4), we have

$$\chi = \left[\frac{\partial^2 \Delta E_{\rm el}^{\rm o}(M)}{\partial M^2} + \frac{\partial^2 \Delta E_{\rm el\text{-}ph}(M)}{\partial M^2} \right]_{M=0}^{-1} = \chi_{\rm P} \left[1 \pm O\left(\frac{\omega_{\rm D}}{\varepsilon_{\rm F}}\right)^2 \right]. \tag{9.4.4}$$

The effect of the electron self-energy due to EPI may change the Pauli susceptibility $\chi_{\rm P} = 2\mu_{\rm B}{}^2 N(0)$, but only by the relative amount $O[(\omega_{\rm D}/\varepsilon_{\rm F})^2]$.

9.4.2 The Effect of the Phonon Ladders on the Spin Susceptibility

From the similarity of the roles of the Coulomb interaction and phonons, as we have seen in connection with Migdal's theorem, it is quite natural to ask the contribution of the diagram of Fig.9.22(a) to the spin susceptibility [9.3, 9.13]. Note that the effect we discussed in 9.4.1 corresponds to Fig.9.22(b).

We obtain Fig.9.22(a) simply by replacing the broken line in Fig.9.3(b) by a wavy line. Thus, corresponding to (9.1.14), the contribution to the spin susceptibility of a phonon ladder is written down as

$$\Delta_{\rm ph} G_{\sigma\sigma}(\boldsymbol{q}, i\nu_m) = -\frac{1}{\beta^2} \sum_{n,n',\boldsymbol{k},\boldsymbol{k}'} |\alpha(\boldsymbol{k}' - \boldsymbol{k})|^2 D^{\rm o}(\boldsymbol{k}' - \boldsymbol{k}, i\nu_{n'-n})$$

$$\times G_\sigma^{\rm o}(\boldsymbol{k}, i\omega_n)\, G_\sigma^{\rm o}(\boldsymbol{k}', i\omega_{n'})\, G_\sigma^{\rm o}(\boldsymbol{k}' - \boldsymbol{q}, i\omega_{n'-m})\, G_\sigma^{\rm o}(\boldsymbol{k} - \boldsymbol{q}, i\omega_{n-m})$$

$$= -\frac{1}{\beta} \sum_{n,\boldsymbol{k}} \gamma_{\rm ph}(\boldsymbol{k}, i\omega_n, \boldsymbol{q}, i\nu_m)\, G_\sigma^{\rm o}(\boldsymbol{k}, i\omega_n)\, G_\sigma^{\rm o}(\boldsymbol{k} - \boldsymbol{q}, i\omega_n - i\nu_m), \tag{9.4.5}$$

where $\gamma_{\rm ph}$ is the same vertex correction as given in (9.3.27). Then, from our discussion given concerning Migdal's theorem it is straightforward to conclude that

$$\left| \Delta_{\rm ph} G_{\sigma\sigma}(\boldsymbol{q}, i\nu_m) \right| = O[(\omega_{\rm D}/\varepsilon_{\rm F}) N(0)]. \tag{9.4.6}$$

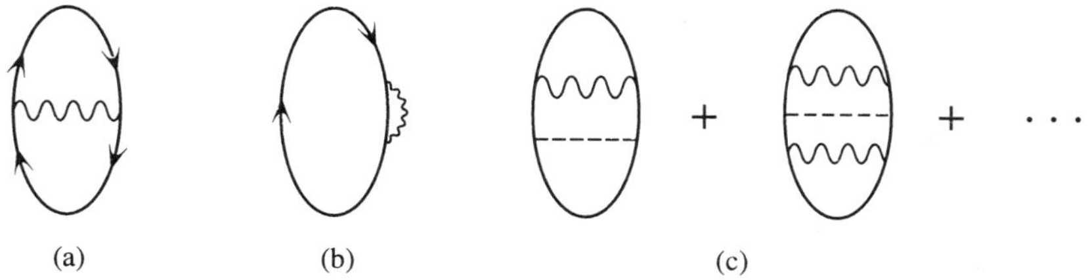

Fig.9.22 The possible effects of the EPI to the spin susceptibility of a metal, the phonon ladder, (a) and the self-energy effect, (b). (c) represents the higher order Coulomb and phonon ladder contributions.

The factor ω_D/ε_F originates from the condition (9.3.31).

Thus, if we sum all the contributions up to infinite order of the Coulomb and phonon ladders such as given in Fig.9.22(c), together with those of Figs. 9.3(b) and 9.22(a), the Stoner magnetic susceptibility is modified to

$$\chi = \frac{2\mu_B^2 N(0)}{1 - \left[\widetilde{V}N(0) + c\left(\frac{\omega_D}{\varepsilon_F}\right)\right]} . \tag{9.4.7}$$

Here, c is presumed to be a negative constant of $O(1)$ [9.3, 9.13].

This result agrees with the result of Chapter 6 in that the relative size of the EPI effect on the magnetic susceptibility is of the order of $\hbar\omega_D/\varepsilon_F$. However, at present it is difficult to go beyond the simple result of (9.4.7). Even it is not clear at all whether a more appropriate evaluation of the contributions of Figs.9.22(a) etc. would eventually reproduce the result of Chapter 6, say, (6.2.19). It is possible that we have to consider different kinds of diagrams.

Note that such phonon effect should also enter the shaded rings appearing in the Dyson equation for a phonon given in Fig.9.10. Then, the effective exchange interaction $\widetilde{V}$ is required to be replaced by $\widetilde{V} + J_{ph}$ in $\widetilde{P}$ appearing in (9.2.4)–(9.2.6). Not only that, in a more general treatment including the spin fluctuation effect, $\widetilde{V}$ will be replaced by $\widetilde{V} + J_{ph} + J_{sf}$. Note that such procedure results in replacing the Stoner susceptibility by the Curie-Weiss one in the sound velocity expression, (5.4.32), as we mentioned at the end of 5.4.3.

9.4.3 Overview of Previous Theories on the Effects of EPI on the Spin Susceptibilities

As discussed in the preceding two subsections, if we intend to calculate the effect of EPI on the magnetic susceptibility of a metal starting from the Kubo formula, there come up two possible physical effects, (i) the electron mass enhancement effect and, (ii) the phonon ladder effect. We have shown, then, that the relative size of the effects (i) and (ii) are, respectively, $O(\omega_D{}^2/\varepsilon_F{}^2)$ and $O(\omega_D/\varepsilon_F)$.

Note, however, that there are views different from the above conclusion. Enz and Matthias [9.14], for instance, raised the possibility of the phonon ladder effect having the size of $O(\lambda_{ph})$ in the presence of soft phonon modes.

On the other hand, Joshi and Rajagopal [9.3] claim that the effects (i) and (ii) have the same magnitudes of $O(\omega_D/\varepsilon_F)$ with, respectively, positive and negative signs, and therefore, the two effects cancel each other. A similar expectation was also presented by Fay and Appel [9.13]. In view of our discussion given in 9.4.1 such conclusions seem to require more scrutiny.

If we adopt the conclusion of 9.4.1, we can neglect the effect (i) compared to the effect (ii) and, then, we have the spin susceptibility of a metal in the form of (9.4.7). This result has the same structure as that of (6.2.8) with the same order of magnitude of the EPI effect. However, at present, it is not obvious whether we could fully reproduce the result of Chapter 6 on the effect of the EPI by evaluating more properly than, for instance, in Refs.[9.3, 9.13] the contributions of those diagrams given in Figs.22(a), (c) and 9.3(b).

Recently, in connection with the large temperature dependence of the magnetic susceptibility of A15 compounds, Pickett et al. [9.15, 9.16] claimed that when the electronic density of states near the Fermi energy is such as would change appreciably within the energy interval ω_D, the most important effect of EPI is to modify the shape of the electronic density of states. A similar view is also presented by Tripathi et al. [9.17]. Note, however, that even in such a situation, the temperature dependence coming from our J_{ph} is larger than that which Pickett et al. consider; while the former is $\propto (\varepsilon_F/\Delta)k_BT/\Delta$, the latter is $(k_BT/\Delta)^2$. Here Δ is the energy scale over which $N(\varepsilon)$ changes appreciably and, therefore, $(N'(0)/N(0))^2 \cong N''(0)/N(0) \cong 1/\Delta^2$ in (6.2.14)–(6.2.15) and (1.6.14).

It was Herring [3.9] and Hopfield [6.2] who first correctly noted that the relative size of the possible EPI effect on the Pauli spin susceptibility or the Stoner factor $\tilde{V}N(0)$ should be of the order of $O(\omega_D/\varepsilon_F)$. However, without any explicit expression for the magnetization dependence of phonon free energy, no one could go beyond such an order of magnitude estimation; even the sign of the effect could not be determined. As for the temperature dependence of J_{ph}, by which the phonon effect becomes important, no one before us noticed it at all.

Concerning our new result presented in Chapter 6, it seems to be Edwards [9.18] who first reexamined and reconfirmed it. More recently, Zverev and Silin [6.33, 6.34, and references therein] reproduced some of our basic results starting from a phonon free energy slightly different from ours, as we already mentioned in 6.12.

Now, we have a question: How can we reproduce the result of Chapter 6, the explicit expression for J_{ph}, in particular, by the method of Feynman diagrams? It is a big challenge. In this respect, recall our earlier experience with the calculation of elastic constants. At present, the even simpler result of Chapter 5, can not be reproduced from the total energy approach for a general situation. The relation between the dynamical and the total energy approaches is not simple and not fully clarified yet in various problems.

9.5 ELECTRON SELF-ENERGY DUE TO SPIN WAVES AND SPIN FLUCTUATIONS

9.5.1 Spin Wave Contribution to Electron Self-Energy

It is evident what kind of processes Fig.9.23 represents for the electron self-energy. It is straightforward to write down the self-energy as

$$\Sigma^{s}_{\sigma}(\boldsymbol{k}, i\omega_n) = \frac{1}{\beta}\sum_{\boldsymbol{q},m,\sigma'} \widetilde{V}(\boldsymbol{q})^2\, \widetilde{F}_{\sigma'\sigma}(\boldsymbol{q}, i\nu_m)\, G_{\sigma'}(\boldsymbol{k}-\boldsymbol{q}, i\omega_{n-m}), \qquad (9.5.1)$$

where $\widetilde{F}_{\sigma'\sigma}(\boldsymbol{q}, i\nu_m)$ is defined as in (9.1.27) and (9.1.28), with

$$F_{\sigma\sigma}(\boldsymbol{q}, i\nu_m) = F_{\sigma}(\boldsymbol{q}, i\nu_m); \quad \widetilde{F}_{\sigma\sigma}(\boldsymbol{q}, i\nu_m) = \widetilde{F}_{\sigma}(\boldsymbol{q}, i\nu_m)\,. \qquad (9.5.2)$$

In deriving the expression (9.5.1), first we note that the contribution of the first diagram of Fig.9.23 is given by replacing $\widetilde{F}_{\sigma'\sigma}$ with $F_{\sigma'\sigma}$ in (9.5.1). Then the procedure to arrive at the result of (9.5.1) by summing higher order contributions is the same as that which we used for Fig.9.4 and (9.1.18). The reason why $\widetilde{V}(\boldsymbol{q})$ appears in place of $\mathrm{v}(\boldsymbol{q})$ in (9.5.1) is the same as in (9.1.14) and (9.1.15).

The part of contribution for $\sigma' \neq \sigma$ in (9.5.1) can be rewritten in terms of the transverse spin susceptibility, (9.1.27) or (4.4.12), as

$$\Sigma^{st}_{\sigma}(\boldsymbol{k}, i\omega_n) \equiv \frac{1}{\beta}\sum_{\boldsymbol{q},m} \widetilde{V}(\boldsymbol{q})^2\, \overline{\chi}_{-\sigma,\,\sigma}(\boldsymbol{q}, i\nu_m)\, G_{-\sigma}(\boldsymbol{k}-\boldsymbol{q}, i\omega_{n-m})\,, \qquad (9.5.3)$$

where $\overline{\chi}_{-\sigma,\,\sigma}(\boldsymbol{q}, i\nu_m)$ is what one would obtain by putting $\mu_B{}^2 = 1$ in (9.1.27) (c.f. (5.3.2)). The superscript t on Σ^{st}_{σ} in (9.5.3) stands for the transverse component. It is the $\sigma' = -\sigma$ part of (9.5.1).

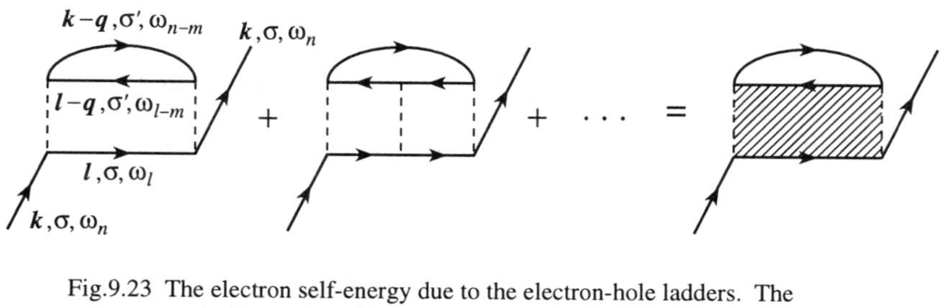

Fig.9.23 The electron self-energy due to the electron-hole ladders. The transverse spin fluctuation effect is represented by the case of $\sigma' = -\sigma$.

Let us compare (9.5.3) with (9.3.3). These two expressions are of exactly the same structure; we obtain (9.5.3), apart from a constant factor, simply by replacing the phonon Green's function by the transverse spin susceptibility in (9.3.3). As can be seen from (4.1.33) or (9.1.26), the spin susceptibility represents the correlation or propagation of spin density fluctuations. The susceptibility function $\chi_{-\sigma,\sigma}(\boldsymbol{q}, i\nu_m)$ is a Green's function as is $D(\boldsymbol{q}, i\nu_m)$, and in the ferromagnetic state the pole of $\chi_{-\sigma,\sigma}(\boldsymbol{q}, \omega + i0^+)$ gives us the spin wave spectrum(see 4.4.2), as the pole of $D(\boldsymbol{q}, \omega + i0^+)$ gives us the phonon spectrum. Thus, (9.5.3) is a direct analogue of (9.3.3).

We note that $i\omega_{n-m} = i\omega_n - i\nu_m$, in summing over ν_m on the right hand side of (9.5.3). Then, using the method of (8.4.38), we have

$$\Sigma_\sigma^{st}(\boldsymbol{k}, i\omega_n) = \frac{1}{2\pi i}\sum_{\boldsymbol{q}} \tilde{V}(\boldsymbol{q})^2 \int_C dz\, n(z)\, \overline{\chi}_{-\sigma,\sigma}(\boldsymbol{q}, z)\, \frac{1}{i\omega_n - z - \xi_{\boldsymbol{k}-\boldsymbol{q},-\sigma}}, \tag{9.5.4}$$

where for $G_\sigma(\boldsymbol{k}, i\omega_n)$ we used the mean field approximation result of (8.5.24). The integrand has two kinds of poles, that of $\overline{G}_{-\sigma}$, namely, $z = i\omega_n - \xi_{\boldsymbol{k}-\boldsymbol{q},-\sigma}$, and that of $\overline{\chi}_{-\sigma,\sigma}(\boldsymbol{q}, z)$. $\overline{\chi}_{-\sigma,\sigma}(\boldsymbol{q}, z)$ has, besides the spin wave poles, continuous poles on the real axis corresponding to the Stoner excitations (see 4.4.3). Considering such distribution of poles, we change the contour of the integration from C of Fig.8.6 to L_1 and L_2 of Fig.9.24. We, then, merge the two

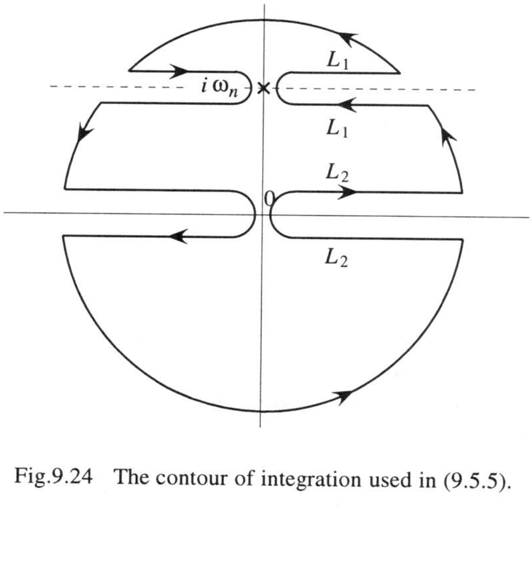

Fig.9.24 The contour of integration used in (9.5.5).

branches of the contours L_1 and L_2 on the imaginary axis, making both of them pairs of lines parallel to the real axis. Note that, in carrying out the contour integral over such L_1 and L_2 we should exclude the contributions from poles on the imaginary axis since they were originally not enclosed by L_1 or L_2.

Thus, we have for the contributions from the paths L_1 and L_2 of Fig.9.24, respectively, as

$$\Sigma_\sigma^{st}(\boldsymbol{k}, i\omega_n)_{L_1} = \sum_{\boldsymbol{q}} \widetilde{V}(\boldsymbol{q})^2 \overline{\chi}_{-\sigma,\sigma}(\boldsymbol{q}, i\omega_n - \xi_{\boldsymbol{k}-\boldsymbol{q},-\sigma}) \left(f(\xi_{\boldsymbol{k}-\boldsymbol{q},-\sigma}) - 1 \right), \tag{9.5.5}$$

$$\Sigma_\sigma^{st}(\boldsymbol{k}, i\omega_n)_{L_2} = -\sum_{\boldsymbol{q}} \widetilde{V}(\boldsymbol{q})^2 \frac{1}{\pi} \int_{-\infty}^{\infty} dz \frac{\mathrm{Im}\,\overline{\chi}_{-\sigma,\sigma}(\boldsymbol{q}, z + i0^+)}{z - i\omega_n + \xi_{\boldsymbol{k}-\boldsymbol{q},-\sigma}} n(z)$$

$$+ \frac{1}{\beta} \sum_{\boldsymbol{q}} \widetilde{V}(\boldsymbol{q})^2 \overline{\chi}_{-\sigma,\sigma}(\boldsymbol{q}, 0) \frac{1}{i\omega_n - \xi_{\boldsymbol{k}-\boldsymbol{q},-\sigma}}. \tag{9.5.6}$$

In (9.5.5) we used the relation

$$n(i\omega_n - \xi_{\boldsymbol{k}-\boldsymbol{q},-\sigma}) = f(\xi_{\boldsymbol{k}-\boldsymbol{q},-\sigma}) - 1, \tag{9.5.7}$$

and noted that the contour L_1 runs clockwise around the pole. In obtaining (9.5.6) we used the procedure to derive (8.4.38′) from (8.4.38).

We finally have to replace $i\omega_n$, by $\omega + i0^+$ in the self-energy. Thus, for $\Sigma_\sigma^{st}(\boldsymbol{k}, i\omega_n)_{L_2}$ we have

$$\Sigma_\sigma^{st}(\boldsymbol{k}, \omega + i0^+)_{L_2} = -\sum_{\boldsymbol{q}} \widetilde{V}(\boldsymbol{q})^2 \frac{1}{\pi} \int_{-\infty}^{\infty} dz \frac{\mathrm{Im}\,\overline{\chi}_{-\sigma,\sigma}(\boldsymbol{q}, z + i0^+)}{z - \omega + \xi_{\boldsymbol{k}-\boldsymbol{q},-\sigma} - i0^+} n(z), \tag{9.5.6'}$$

where the second term on the right hand side of (9.5.6), which is originated from the contribution of $m = 0$ in (9.5.3), turns out to be of $(\omega/\varepsilon_F)k_B T$ and, therefore, we neglected it.(See Problem 9.6) The result of (9.5.5) can also be rewritten in a form similar to (9.5.6′) by using the relation of (8.1.28). Thus, we have

$$\Sigma_\sigma^{\rm st}(\boldsymbol{k}, \omega + i0^+) = \Sigma_\sigma^{\rm st}(\boldsymbol{k}, \omega + i0^+)_{\rm L1} + \Sigma_\sigma^{\rm st}(\boldsymbol{k}, \omega + i0^+)_{\rm L2}$$

$$= -\sum_{\boldsymbol{q}} \int_{-\infty}^{\infty} dz\, \widetilde{V}(\boldsymbol{q})^2 \frac{1}{\pi} {\rm Im}\, \overline{\chi}_{-\sigma,\sigma}(\boldsymbol{q}, z + i0^+) \left[\frac{1 - f(\xi_{\boldsymbol{k}-\boldsymbol{q},-\sigma}) + n(z)}{z - \omega + \xi_{\boldsymbol{k}-\boldsymbol{q},-\sigma} - i0^+} \right]. \quad (9.5.8)$$

Note the similarity of this expression to that of the electron self-energy due to EPI, (9.3.8).

We assume $n_- > n_+$, for the ferromagnetic state as in 6.6.3; note that we assumed $n_+ > n_-$ in 4.4.2. Then, from (6.5.4) and (6.5.7), we have

$$1 - \widetilde{V}F_{-+}(\boldsymbol{q}, z + i0^+) = -\frac{1}{\overline{M}\,\widetilde{V}}(z - \omega_{\rm sw}(\boldsymbol{q}) + i0^+), \quad (9.5.9)$$

$$1 - \widetilde{V}F_{+-}(\boldsymbol{q}, z + i0^+) = \frac{1}{\overline{M}\,\widetilde{V}}(z + \omega_{\rm sw}(\boldsymbol{q}) + i0^+), \quad (9.5.10)$$

where $\omega_{\rm sw}(\boldsymbol{q})$ is the spin wave frequency which can be obtained from the pole of $\chi_{-+}(\boldsymbol{q}, \omega)$ as shown in 4.4.2. With these, from (9.5.8) we obtain

$$\Sigma_+^{\rm sw}(\boldsymbol{k}, \omega + i0^+) = \overline{M} \sum_{\boldsymbol{q}} \widetilde{V}(\boldsymbol{q})^2 \frac{1 - f(\xi_{\boldsymbol{k}-\boldsymbol{q},-}) + n(\omega_{\rm sw}(\boldsymbol{q}))}{\omega - \xi_{\boldsymbol{k}-\boldsymbol{q},-} - \omega_{\rm sw}(\boldsymbol{q}) + i0^+}, \quad (9.5.11)$$

$$\Sigma_-^{\rm sw}(\boldsymbol{k}, \omega + i0^+) = \overline{M} \sum_{\boldsymbol{q}} \widetilde{V}(\boldsymbol{q})^2 \frac{f(\xi_{\boldsymbol{k}-\boldsymbol{q},+}) + n(\omega_{\rm s}(\boldsymbol{q}))}{\omega - \xi_{\boldsymbol{k}-\boldsymbol{q},+} - \omega_{\rm sw}(\boldsymbol{q}) + i0^+}, \quad (9.5.12)$$

where the superscript sw stands for spin wave; we neglected the imaginary part of $F_{\sigma,-\sigma}(\boldsymbol{q}, \omega + i0^+)$ in the numerator of $\chi_{\sigma,-\sigma}(\boldsymbol{q}, \omega + i0^+)$ to have

$$F_{-+}(\boldsymbol{q}, \omega_{\rm sw}(\boldsymbol{q})) = F_{+-}(\boldsymbol{q}, -\omega_{\rm sw}(\boldsymbol{q})) = 1/\widetilde{V}(\boldsymbol{q}), \quad (9.5.13)$$

and, then,

$$\frac{1}{\pi} {\rm Im}\, \overline{\chi}_{-+}(\boldsymbol{q}, z + i0^+) = \overline{M}\, \delta(z - \omega_{\rm sw}(\boldsymbol{q})), \quad (9.5.14)$$

$$\frac{1}{\pi} {\rm Im}\, \overline{\chi}_{+-}(\boldsymbol{q}, z + i0^+) = -\overline{M}\delta(z + \omega_{\rm sw}(\boldsymbol{q})), \quad (9.5.15)$$

and used the relation, $n(-\omega) = -(n(\omega) + 1)$.

The structure of (9.5.11)–(9.5.12) is very similar to that of (9.3.11), the electron self-energy due to the EPI. There is, however, one difference: In the case of spin wave, it is only a down spin electron that can emit a spin wave, and a spin wave can be absorbed only by an up spin electron; note that we are assuming $n_- > n_+$. A down spin electron flips its spin upward by emitting a spin wave, and, an up spin electron flips its spin downward by absorbing a spin wave. Thus, the self-energy expressions are qualitatively different for an up spin electron and for a down spin electron. With this difference in mind, the implications of these results are understood parallelly to the case of phonon.

Similarly to the case of the EPI, let us discuss how an electron near the Fermi surface is affected by this spin wave effect. If we put $T = 0$ for simplicity, we have $n(\omega_s(\boldsymbol{q})) = 0$ in (9.5.11)–(9.5.12), and they reduce to

$$\mathrm{Re}\,\Sigma_+^{\mathrm{sw}}(\boldsymbol{k}, \omega + i0^+) = \overline{M}\sum_{\boldsymbol{q}} \widetilde{V}(\boldsymbol{q})^2 \frac{\theta(\xi_{\boldsymbol{k}-\boldsymbol{q},-})}{\omega - \xi_{\boldsymbol{k}-\boldsymbol{q},-} + \omega_{\mathrm{sw}}(\boldsymbol{q})}. \tag{9.5.16}$$

$$\mathrm{Re}\,\Sigma_-^{\mathrm{sw}}(\boldsymbol{k}, \omega + i0^+) = \overline{M}\sum_{\boldsymbol{q}} \widetilde{V}(\boldsymbol{q})^2 \frac{\theta(-\xi_{\boldsymbol{k}-\boldsymbol{q},+})}{\omega - \xi_{\boldsymbol{k}-\boldsymbol{q},+} - \omega_{\mathrm{sw}}(\boldsymbol{q})}, \tag{9.5.17}$$

The sum over $\boldsymbol{q}$ can be carried out by using the technique of (9.3.14), similarly to case of the EPI:

$$\mathrm{Re}\,\Sigma_+^{\mathrm{sw}}(\boldsymbol{k}, \omega) = \frac{V}{4\pi^2}\frac{m}{k}\overline{M}\int_{q_1}^{q_u} dq\, q\, \widetilde{V}(\boldsymbol{q})^2 \int_0^W \frac{d\xi_{\boldsymbol{k}'-}}{\omega - \xi_{\boldsymbol{k}'-} - \omega_{\mathrm{sw}}(\boldsymbol{q})}$$

$$= -\frac{V}{4\pi^2}\frac{m}{k}\overline{M}\int_{q_1}^{q_u} dq\, q\, \widetilde{V}(\boldsymbol{q})^2 \ln\left|\frac{\omega - W - \omega_{\mathrm{sw}}(\boldsymbol{q})}{\omega - \omega_{\mathrm{sw}}(\boldsymbol{q})}\right|, \tag{9.5.18}$$

$$\mathrm{Re}\,\Sigma_-^{\mathrm{sw}}(\boldsymbol{k}, \omega) = \frac{V}{4\pi^2}\frac{m}{k}\overline{M}\int_{q_1}^{q_u} dq\, q\, \widetilde{V}(\boldsymbol{q})^2 \int_{-\varepsilon_{\mathrm{F}+}}^{0} \frac{d\xi_{\boldsymbol{k}'+}}{\omega - \xi_{\boldsymbol{k}'+} + \omega_{\mathrm{sw}}(\boldsymbol{q})}$$

$$= \frac{V}{4\pi^2}\frac{m}{k}\overline{M}\int_{q_1}^{q_u} dq\, q\, \widetilde{V}(\boldsymbol{q})^2 \ln\left|\frac{\omega + \varepsilon_{\mathrm{F}+} + \omega_{\mathrm{sw}}(\boldsymbol{q})}{\omega + \omega_{\mathrm{sw}}(\boldsymbol{q})}\right|. \tag{9.5.19}$$

where W is the band width, and, $q_u = k_{\mathrm{F}+} + k_{\mathrm{F}-}$ and $q_1 = k_{\mathrm{F}+} - k_{\mathrm{F}-}$. For $\omega \cong 0$, that is, near the Fermi surface, this result is reduced to

$$\mathrm{Re}\,\Sigma_\sigma^{\mathrm{sw}}(\boldsymbol{k}_{\mathrm{F}\sigma}, \omega) = -\lambda_{\mathrm{sw}\sigma}\,\omega + \mathrm{Re}\,\Sigma_\sigma^{\mathrm{sw}}(\boldsymbol{k}_{\mathrm{F}\sigma}, 0), \tag{9.5.20}$$

with

$$\lambda_{\mathrm{sw}\sigma} = -\frac{\partial}{\partial\omega}\mathrm{Re}\,\Sigma_\sigma^{\mathrm{sw}}(\boldsymbol{k}_{\mathrm{F}\sigma}, \omega)\bigg|_{\omega=0}. \tag{9.5.21}$$

Thus we obtain

$$\begin{aligned}
\lambda_{\mathrm{sw}\sigma} &= \frac{V}{4\pi^2}\frac{m}{k_{\mathrm{F}\sigma}}\overline{M}\int_{q_1}^{q_u} dq\, q\, \widetilde{V}(\boldsymbol{q})^2 \frac{1}{\omega_{\mathrm{sw}}(\boldsymbol{q})} \\
&= \frac{1}{2}\widetilde{V}N_\sigma(0)\frac{\widetilde{V}(n_- - n_+)}{\omega_{\mathrm{sw}}(k_{\mathrm{F}\sigma})}\ln\left|\frac{k_{\mathrm{F}+} + k_{\mathrm{F}-}}{k_{\mathrm{F}-} - k_{\mathrm{F}+}}\right| \\
&= \frac{1}{2}\widetilde{V}N(0)\frac{\widetilde{V}(n_- - n_+)}{\hbar\omega_{\mathrm{sw}}(k_{\mathrm{F}})}\frac{N(0)}{N_\sigma(0)}\ln\left|\frac{k_{\mathrm{F}+} + k_{\mathrm{F}-}}{k_{\mathrm{F}-} - k_{\mathrm{F}+}}\right|,
\end{aligned} \tag{9.5.22}$$

where we put $\widetilde{V}(q) = \widetilde{V}$, independent of q. Clearly, $\lambda_{\mathrm{sw},\sigma}$, which we may call the *spin wave coupling constant*, is positive, with a magnitude of $O(1)$.

As we discussed in 9.3.2 for the case of the EPI, this spin wave coupling constant enhances the effective mass of an electron, similarly to (9.3.21), and, therefore, the electronic density of levels near the Fermi surface. Since the EPI and the spin wave effects are additive, for electrons near the Fermi surface we now have

$$m_\sigma^* = (1 + \lambda_{\mathrm{ph}\sigma} + \lambda_{\mathrm{sw}\sigma})\, m. \tag{9.5.23}$$

The effect of spin waves on an electron in a metal was previously discussed for the *s-d* exchange model (see 4.2.4); in that model the spin waves are those of the localized (d) spins, whereas the electrons which interact with them are itinerant (s) ones [9.19–9.22]. In our above discussion, spin waves arise from the same itinerant electrons with which the spin waves interact.

Our result of (9.5.7)–(9.5.12) and (9.5.22) is the outcome of the simplifications, (9.5.9)–(9.5.10); without the use of such simplifications, the discussion becomes quite complicated [9.23].

9.5.2 Spin Fluctuation Contribution to Electron Self-Energy

It is evident that the contribution of the processes shown in Fig.9.23 will be present also in the paramagnetic state of a metal. There is, however, a decisive difference from the case of the ferromagnetic state in the quantity, $1/\pi$ Im $\chi_{\sigma,\sigma'}(\boldsymbol{q}, \omega + i0^+)$, the spectral density of magnetic excitations. In the case of the ferromagnetic state, the main part of the spectral density consists of the delta function form of spin wave excitations, (9.5.14)–(9.5.15). In the case of the paramagnetic state, we have

$$\chi_{+-}(\boldsymbol{q}, \omega) = \chi_{-+}(\boldsymbol{q}, \omega) = 1/2\chi_{zz}(\boldsymbol{q}, \omega) \tag{9.5.24}$$

and the spectral density of magnetic excitations, which are the spin fluctuations, becomes of the following form,

$$\frac{1}{\pi}\,\mathrm{Im}\,\overline{\chi}_{+-}(\boldsymbol{q}, \omega + i0^+) = \frac{1}{\pi}\frac{I(q,\omega)}{\left[1 - \widetilde{V}R(q, \omega)\right]^2 + \left[\widetilde{V}I(q, \omega)\right]^2}$$

$$\cong \frac{1}{\pi}\frac{N(0)\dfrac{\omega}{\mathrm{v_F}q}}{\left[1 - \widetilde{V}F(0)\right]^2 + \overline{V}^2\left(\dfrac{\omega}{\mathrm{v_F}q}\right)^2}\,, \tag{9.5.25}$$

where $R(\boldsymbol{q}, \omega)$ and $I(\boldsymbol{q}, \omega)$ are, respectively, the real and imaginary parts of Lindhard function, and the final expression is for small q and $|\omega|$ such as $0 < \omega < \mathrm{v_F}q$ (see Appendix A). This spectral density is of the form of peaks centered around $\omega \cong \mathrm{v_F}q[1 - \widetilde{V}F(0)]$, with a width of, also, $\sim \mathrm{v_F}q[1 - \widetilde{V}F(0)]$. This spectral density of spin fluctuations can be measured from a neutron diffraction experiment; see (4.1.59).

With such spectral density of spin fluctuations the self-energy of an electron in the paramagnetic state is given as

$$\Sigma^{\mathrm{sf}}(\boldsymbol{k}, \omega + i0^+) = -\frac{3}{2}\sum_{\boldsymbol{q}}\int_{-\infty}^{\infty} dz\,\widetilde{V}(\boldsymbol{q})^2\,\frac{1}{\pi}\,\mathrm{Im}\,\overline{\chi}_{-+}(\boldsymbol{q}, z + i0^+)$$

$$\times\frac{1 - f(\xi_{\boldsymbol{k}-\boldsymbol{q}}) + n(z)}{z - \omega + \xi_{\boldsymbol{k}-\boldsymbol{q}} - i0^+}\,. \tag{9.5.26}$$

We dropped the subscript σ since we are considering the paramagnetic state. This expression is identical to (9.5.8) except the factor 3/2 which takes into account of (9.5.24) in summing all the components of spin fluctuations, χ_{+-}, χ_{-+} and χ_{zz}.

Let us calculate the enhancement of electron mass near the Fermi surface as we did for phonons and spin waves. As before we put $n(z) = 0$ in (9.5.26) by considering low temperatures. Then, (9.5.26) can be rewritten in the form of (9.5.5) with the factor 3/2. Then by carrying out the summation over $\boldsymbol{q}$ with the procedure of (9.3.14) we have

$$\Sigma^{\mathrm{sf}}(\boldsymbol{k}, i\omega_n) \cong \Sigma^{\mathrm{sf}}(\boldsymbol{k}, i\omega_n)_{\mathrm{L}_1}$$

$$= \frac{3}{2}\frac{V}{4\pi^2}\frac{m}{k}\int q\, dq\, \widetilde{V}(\boldsymbol{q})^2 \int d\xi\, (f(\xi) - 1)\, \overline{\chi}_{-+}(\boldsymbol{q}, i\omega_n - \xi). \quad (9.5.27)$$

The procedure to derive from here the electron mass enhancement factor is quite parallel to the case of the EPI and spin wave effects. Near the Fermi level, that is, for $\omega \cong 0$, (9.5.27) becomes as

$$\Sigma^{\mathrm{sf}}(k_{\mathrm{F}}, \omega) \cong -\lambda_{\mathrm{sf}}\omega + \Sigma^{\mathrm{sf}}(k_{\mathrm{F}}, 0), \quad (9.5.28)$$

$$\lambda_{\mathrm{sf}} = -\frac{\partial}{\partial\omega}\Sigma^{\mathrm{sf}}(k_{\mathrm{F}}, \omega)\bigg|_{\omega = 0}. \quad (9.5.29)$$

We call λ_{sf} the *spin fluctuation coupling constant.* The second term of (9.5.28) may be incorporated to the chemical potential. Then we find the electron mass is enhanced as

$$m^* = \left(1 + \lambda_{\mathrm{sf}}\right) m, \quad (9.5.30)$$

by the spin fluctuation effect.

Let us derive an explicit expression for λ_{sf}. In carrying out the differentiation of (9.5.29) on (9.5.27) with $i\omega_n = \omega + i0^+$, note

$$-\frac{\partial}{\partial\omega}\int d\xi\, (f(\xi) - 1)\, \overline{\chi}_{-+}(\boldsymbol{q}, \omega - \xi) = \int d\xi \left(f(\xi) - 1\right)\left(\frac{\partial}{\partial\xi}\overline{\chi}_{-+}(\boldsymbol{q}, \omega - \xi)\right)$$

$$= \int d\xi \left(-\frac{\partial f(\xi)}{\partial\xi}\right)\overline{\chi}_{-+}(\boldsymbol{q}, \omega - \xi) = \overline{\chi}_{-+}(\boldsymbol{q}, \omega).$$

Using this relation we immediately obtain

$$\lambda_{\mathrm{sf}} = \frac{3}{8\pi^2} V \frac{m}{k_{\mathrm{F}}} \int_0^{2k_{\mathrm{F}}} q\, dq\, \widetilde{V}(\boldsymbol{q})^2\, \overline{\chi}_{-+}(\boldsymbol{q}, 0)$$

$$= \frac{3}{2} N(0) \langle\langle\, \widetilde{V}(\boldsymbol{q})^2 \frac{F(\boldsymbol{q})}{1 - \widetilde{V}(\boldsymbol{q})F(\boldsymbol{q})} \rangle\rangle. \tag{9.5.31}$$

The upper bound of the q integration is put to be $2k_{\mathrm{F}}$ since the dominant contributions are expected to come from within that bound. Note, however, that the value of λ_{sf} depends on the value of the upper bound rather sensitively. The meaning of $\langle\langle \cdots \rangle\rangle$ is the same as in (9.3.18). According to (9.5.31), λ_{sf} is expected to become large when the exchange interaction is strong and the Stoner condition of (3.3.10) or (5.3.21) is nearly satisfied.

If we sum the effects of phonons and spin fluctuations, the electron mass at the Fermi surface is enhanced as

$$m^* = \left(1 + \lambda_{\mathrm{ph}} + \lambda_{\mathrm{sf}}\right) m, \tag{9.5.32}$$

where λ_{ph} is the electron-phonon coupling constant in the paramagnetic state given in (9.3.43). Note that λ_{ph} is enhanced by the exchange effect through the following two mechanisms. The first is the exchange softening of phonon frequency; the larger $\widetilde{V}(\boldsymbol{q})\, F(\boldsymbol{q})$ is, the smaller $\omega_{\boldsymbol{q}}{}^2$ becomes. Secondly, as can be seen from (4.3.29), the exchange effect makes $\varepsilon(\boldsymbol{q},0)$ smaller and, accordingly, $\overline{\alpha}(\boldsymbol{q})$ larger. Such a dependence of λ_{ph} on the exchange effect may be more clearly seen by rewriting (9.3.43) as

$$\lambda_{\mathrm{ph}} = N(0) \langle\langle \frac{\Omega_{\mathrm{pl}}^2\, \mathrm{v}(\boldsymbol{q})}{\varepsilon(\boldsymbol{q}, 0)^2} \Big/ \left(\xi\, s_0^2\, q^2 + \frac{\Omega_{\mathrm{pl}}^2}{1 + 2\mathrm{v}(\boldsymbol{q})\, \widetilde{F}(\boldsymbol{q})} \right) \rangle\rangle$$

$$= N(0) \langle\langle \frac{\mathrm{v_s}(\boldsymbol{q})}{1 - \widetilde{V}(\boldsymbol{q})\, F(\boldsymbol{q}) + \frac{1}{2} \frac{\xi}{\mathrm{v_s}(\boldsymbol{q})\, N(0)}} \rangle\rangle, \tag{9.5.33}$$

where we put

$$\mathrm{v_s}(\boldsymbol{q}) = \frac{\mathrm{v}(\boldsymbol{q})}{\varepsilon(\boldsymbol{q}, 0)} = \frac{\mathrm{v}(\boldsymbol{q})}{1 + \left(2\mathrm{v}(\boldsymbol{q}) - \widetilde{V}(\boldsymbol{q})\right) F(\boldsymbol{q})}, \tag{9.5.34}$$

with the screening constant of (4.3.29). (We have introduced screened Coulomb interactions in various ways with different screening constants. See (8.5.32) and (9.1.21).) Note that if we put $\xi = 0$ in (9.5.33), it becomes very similar to (9.5.31).

Now we have understood that both λ_{sf} and λ_{ph} are enhanced by the effect of the exchange interaction between electrons. Which of these two effects will be more strongly enhanced by the exchange effect ? To answer to this question we carry out numerical calculation on (9.5.31) and (9.5.33), respectively, for λ_{sf} and λ_{ph}. We use the same model and parameters as those we used in obtaining Fig.9.17, except the following two points. First, for the exchange interaction, here we assumed $\tilde{V}(\boldsymbol{q}) = \tilde{V}/[1 + (q/2k_F)^2]$. Secondly, here we did not use the Debye approximation unlike in Fig.9.17. Those differences, however, will be of minor consequence.

The result of our numerical calculation is given in Fig.9.25; the broken line gives λ_{sf} as the function of $\overline{V} = \tilde{V}N(0)$; values of λ_{ph} for different values of ξ are given by the solid lines. We find that λ_{ph} can be enhanced quite dramatically by the exchange effect. Compare the values of λ_{ph} for $\overline{V} = 0$ and $\overline{V} \to 1$. If $\xi < 0$, λ_{ph} can be more strongly exchange enhanced than λ_{sf}.

Here recall that, as discussed in 4.3.9, if the correlation effect between electrons with antiparallel spins is considered, the $\tilde{V}$'s appearing in λ_{ph} and λ_{sf} become different. We have to use a larger value of $\tilde{V}$ for λ_{ph} than that for λ_{sf}. If we use the same common $\tilde{V}$ in calculating both λ_{ph} and λ_{sf}, such situation could be effectively represented by introducing a negative constant in the denominator of the last expression of (9.5.33); this negative constant can be incorporated to ξ; the correlation effect effectively renders the value of ξ smaller.

9.5.3 Constructive Effect on Superconductivity of the Exchange Interaction between Electrons

As shown in 9.3.5, in the BCS theory of superconductivity the superconducting transition temperature T_C is given in terms of the electron-phonon coupling constant λ_{ph} as

$$T_C \cong \theta_D \exp(-1/\lambda_{ph}), \qquad (9.5.35)$$

where θ_D is the phonon Debye temperature.

Now, as shown in Fig.9.25, λ_{ph} becomes larger as $\overline{V} = \tilde{V}N(0)$ increases. Although θ_D decreases with increasing $\overline{V}$, the effect of the increase in λ_{ph} would dominate in (9.5.35) and raise T_C. Thus the superconducting transition temperature can be raised by the effect of the exchange interaction between electrons [9.6].

There is, however, another opposite effect of the exchange interaction. It was pointed out [6.3] that if the effect of the spin fluctuation coupling is considered we have to modify (9.5.35) as

$$T_C \cong \theta_D \exp\left(-\frac{1}{\lambda_{ph} - \lambda_{sf}}\right). \tag{9.5.36}$$

The interaction between electrons mediated by spin fluctuation is of ferromagnetic nature and, therefore, destructive to the formation of Cooper pairs. Note that for a fully quantitative discussion, (9.5.36) is not adequate and has to be modified. The reader is referred to Ref. [9.24] for details of such modification.

Within the result of (9.5.36), apart from details, whether a magnetic tendency is constructive or destructive to superconductivity is determined by which of λ_{ph} or λ_{sf} is more exchange enhanced. The model calculation of Fig.9.25 shows us that there certainly is the possibility of λ_{ph} being more enhanced than λ_{sf}. Superconductivity can be enhanced by the exchange effect if ξ is small or negative. In this respect it is interesting to note that the recently discovered high-temperature oxide superconductors [9.25, 9.26] are strongly magnetic materials; they show strong antiferromagnetic or SDW tendencies. Let us see how λ_{ph} can be exchange enhanced particularly strongly in such a situation [9.27].

If a paramagnetic metal has a tendency toward antiferromagnetism or SDW, there we have $\chi_{zz}(\boldsymbol{q}) \rightarrow \infty$, or

$$\widetilde{V}(\boldsymbol{q})\,F(\boldsymbol{q}) \cong 1 \tag{9.5.37}$$

for $|\boldsymbol{q}| = O(k_F)$, as we discussed in 5.3.2. In such a system the phonons to be softened are those for $|\boldsymbol{q}| \cong O(k_F)$. Note that such a region of $\boldsymbol{q}$ has generally much more weight than that of $\boldsymbol{q} \cong 0$ in the Fermi surface average of (9.3.17). Thus the exchange enhancement of λ_{ph} can be more effective in an antiferromagnetic system than in a ferromagnetic one.

Furthermore, in a system with such an antiferromagnetic tendency, the exchange effect in $|\widetilde{\alpha}(\boldsymbol{q})|^2 = |\bar{\alpha}(\boldsymbol{q})/\varepsilon(\boldsymbol{q},0)|^2$ also becomes important. Note that for $|\boldsymbol{q}| \cong k_F$, $\widetilde{V}(\boldsymbol{q}) \cong \mathrm{v}(\boldsymbol{q})$ and, therefore,

$$\varepsilon(\boldsymbol{q},0)\big|_{\widetilde{V}(q)F(q)\cong 1} = \left[1 + \left(2\mathrm{v}(\boldsymbol{q}) - \widetilde{V}(\boldsymbol{q})\right)F(\boldsymbol{q})\right]\big|_{\widetilde{V}(q)F(q)\cong 1} \cong 2, \tag{9.5.38}$$

when a system satisfies the condition of (9.5.37). On the other hand, if $\mathrm{v}(\boldsymbol{q})F(\boldsymbol{q}) \approx 1$, but $\widetilde{V}F(\boldsymbol{q}) \ll 1$ we have

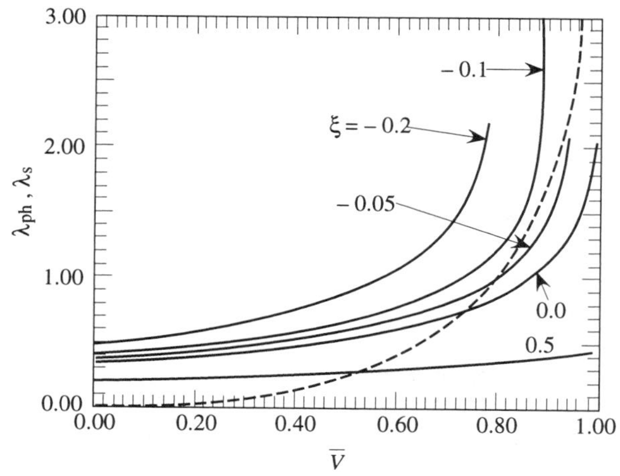

Fig.9.25 The exchange enhancement of the electron-phonon coupling constant λ_{ph} for different values of ξ. The broken line is for the spin fluctuation coupling constant λ_{sf}. Free electron mass and free electron energy dispersion with $k_F = 10^8$/cm are assumed.

$$\varepsilon(\boldsymbol{q}, 0)\,|_{\widetilde{V}(\boldsymbol{q})F(\boldsymbol{q}) \cong 0} \cong 3. \tag{9.5.39}$$

Thus in (9.3.43) we note

$$\frac{|\overline{\alpha}(\boldsymbol{q})|^2_{\widetilde{V}(\boldsymbol{q})F(\boldsymbol{q}) \cong 1}}{|\overline{\alpha}(\boldsymbol{q})|^2_{\widetilde{V}(\boldsymbol{q})F(\boldsymbol{q}) \cong 0}} = \left(\frac{3}{2}\right)^2. \tag{9.5.40}$$

This effect alone may enhance λ_{ph} by a factor of ~ 2. Such an important enhancement effect has been overlooked.

In the oxide superconductors, the electron–phonon interaction may not be the sole origin of the superconductivity. However, it may be worth noting that the electron–phonon coupling constant also can be particularly large in those systems. And, if magnetic excitations are directly responsible for the superconductivity, the EPI is involved in them as we have been seeing in this book.

PROBLEMS FOR CHAPTER 9

9.1 Obtain the Kubo formula for $\chi_{\mathrm{em}}(\boldsymbol{q}, \omega)$ which is defined in (4.3.20), and, then, reproduce the result of (4.3.21) by using the method of Green's functions and Feynman diagrams.

9.2 Make an order estimation of $\lambda_{\mathrm{ph}\sigma}$ which is given in (9.3.17), and confirm that $\lambda_{\mathrm{ph}\sigma} = O(1)$.

9.3 Confirm the result of Eq.(9.3.45).

9.4 In an improved approximation, $G^{0}_{\pm}$'s appearing in (9.3.37) and, according-ly, (9.3.40) should be replaced by $G_{\pm}$'s which include the effect of the electron self-energy due to EPI. If we treat the effect of the self-energy near the Fermi surface as in (9.3.19) we have

$$G_{+}(\boldsymbol{k}', i\omega_{n'})\, G_{-}(-\boldsymbol{k}', -i\omega_{n'}) = \left(\frac{1}{1+\lambda_{\mathrm{ph}}}\right)^{2} \frac{1}{\xi_{\boldsymbol{k}'}^{2}/(1+\lambda_{\mathrm{ph}})^{2} + \omega_{n'}^{2}}$$

in place of (9.3.39)(See 8.1.4.). Show that with such self-energy effect the superconducting transition temperature of (9.3.47) is modified to

$$T_{\mathrm{C}} = 1.13\, \theta_{\mathrm{D}} \exp\left[-\left(1+\lambda_{\mathrm{ph}}\right)/\lambda_{\mathrm{ph}}\right].$$

9.5 Show that near the Fermi surface, where $|\omega| < \omega_{\mathrm{D}}$,

$$\left|\mathrm{Re}\, \Sigma^{\mathrm{ph}}(k_{\mathrm{F}}, \omega)\right| = O(\omega_{\mathrm{D}}).$$

9.6 In (9.5.6), confirm that for $|\omega| \ll \varepsilon_{\mathrm{F}}$,

$$\left| \frac{1}{\beta} \sum_{\boldsymbol{q}} \widetilde{V}(\boldsymbol{q})^{2} \overline{\chi}_{-\sigma,\sigma}(\boldsymbol{q}, 0) \frac{1}{i\omega_{n} - \xi_{\boldsymbol{k}-\boldsymbol{q},-\sigma}} \right| = O(\omega\, k_{\mathrm{B}} T/\varepsilon_{\mathrm{F}}),$$

and, therefore, it can be neglected compared with the other term at low temperatures.

Chapter 10

CHARGE AND SPIN FLUCTUATIONS, PHONONS, AND ELECTRON CORRELATION

The fundamental collective excitations of a metal are known to be charge and spin fluctuations, in addition to phonons. In chapters 6 and 7, we saw aspects of their roles in magnetism, volume, and elasticity properties of a metal. In this chapter we show how the thermodynamic potential and the free energy of a metal can be obtained in terms of these three kinds of collective excitations. This result reconfirms the basis of our discussions given in Chapters 6 and 7. Also we point out some hitherto unknown problems concerning the treatment of spin waves and spin fluctuations.

10.1 CHARGE FLUCTUATIONS

It is widely believed that the charge and spin fluctuations, including the spin waves, are the most fundamental collective excitations in a metallic electron system. As for the charge fluctuation, it has been known to be the principal source of the correlation energy of an electron gas; we briefly mentioned this in deriving the RPA thermodynamic potential of an electron gas in 8.4. As for the spin fluctuations, in 9.5.2 and 9.5.3 we discussed respectively how its effect enhances the electronic mass and how it is destructive to superconductivity of BCS type. Further aspects of possible spin fluctuation effects were proposed more recently. It is claimed that the spin fluctuation effect is the principal mechanism to determine the temperature dependence of the magnetic susceptibility of a metal [6.6, 3.18, 3.19]. The origin of the Curie-Weiss type temperature dependence is attributed entirely to the effect of spin fluctuation [3.18, 3.19, 10.1], for instance. Most recently, the spin fluctuation effect is proposed as the origin of superconductivity in the heavy fermion systems and the high T_C oxide superconductors. In the case of heavy fermion systems, the spin fluctuation effect is also viewed as the principal origin of the heavy electron mass.

In view of such possible importance of the spin fluctuation effect, in this section we present a systematic discussion on this effect. The thermodynamic potential of an electron gas derived in 8.4 is not a satisfactory one since it apparently does not include the contribution of spin fluctuations. Thus, our goal is clear: Obtain the thermodynamic potential of a metal in such a way as to include the contributions of spin fluctuations, in addition to those of charge fluctuations. We achieve this goal by starting from the long range Coulomb interaction model we have been using in this book. We obtain a result [10.2] which is valid both for $T > T_C$ and $T < T_C$.

After having derived this pleasing result, however, we note that the inclusion of such spin fluctuation contribution to the thermodynamic potential causes drastically to violate the screening charge sum rule. This is a rather important question.

Note that our such conclusion is based on the long range Coulomb interaction model. Conclusions can be different for different models. In discussing spin fluctuation effects it is the Hubbard model that is predominantly used. Note, however, that the Hubbard model is rather inadequate for discussing charge fluctuation effects. The long range Coulomb interaction model seems to be more appropriate than the Hubbard model in simultaneously discussing charge and spin fluctuation effects. This might be, however, a matter of controversy.

10.1.1 Charge Fluctuations in Metals

In 8.4 we showed how to obtain the thermodynamic potential of an electron gas with the method of the linked cluster expansion. We will use this method more extensively in discussing the thermodynamic potentials of an electron gas and an interacting electron-phonon system. There is, however, another convenient way to obtain the thermodynamic potential of an interacting system by using the fluctuation-dissipation theorem [10.3, 10.4]. We introduce this method in this subsection and then use it to derive the thermodynamic potential of an electron gas in the next subsection.

Let us consider an electron gas whose Hamiltonian is given by

$$H_{\mathrm{el}} = H_{\mathrm{o}} + H_{\mathrm{C}} , \tag{10.1.1}$$

$$H_{\mathrm{o}} = \sum_{l} \varepsilon_l \, a_l^{+} a_l , \tag{10.1.2}$$

$$H_C = \frac{1}{2} \sum_{q,l,l'}{}' \mathrm{v}(\boldsymbol{q})\, a_l^+ a_{l'}^+ a_{l'+q} a_{l-q} \,. \tag{10.1.3}$$

The meanings of notations are the same as before. As in (8.5.34), here we introduce

$$H_{\mathrm{el}}(g) = H_o + g H_C = H_o + H_C(g). \tag{10.1.4}$$

Similarly, we define the thermodynamic potential corresponding to $H_{\mathrm{el}}(g)$ as

$$\Omega_{\mathrm{el}}(g) = -1/\beta \ln \mathrm{tr}\left(\mathrm{e}^{-\beta [H_{\mathrm{el}}(g) - \mu n]}\right). \tag{10.1.5}$$

Then we have

$$\frac{\partial}{\partial g} \Omega_{\mathrm{el}}(g) = \langle H_C \rangle_g, \tag{10.1.6}$$

where we defined

$$\langle A \rangle_g \equiv \mathrm{tr}\left(e^{-\beta [H_{\mathrm{el}}(g) - \mu n]} A\right) / \,\mathrm{tr}\left(e^{-\beta [H_{\mathrm{el}}(g) - \mu n]}\right). \tag{10.1.7}$$

If we integrate both sides of (10.1.6) over g, we obtain

$$\Omega_{\mathrm{el}} = \Omega_{\mathrm{el}}(1) = \Omega_{\mathrm{el}}^{\mathrm{o}} + \int_0^1 dg \langle H_C \rangle_g \equiv \Omega_{\mathrm{el}}^{\mathrm{o}} + \Delta\Omega_{\mathrm{el}}. \tag{10.1.8}$$

$\Omega_{\mathrm{el}}^{\mathrm{o}} = \Omega_{\mathrm{el}}(0)$ is the thermodynamic potential of the electron gas in the absence of the Coulomb interaction between electrons; it coincides with Ω_o of Chapter 8.

In applying (10.1.8) to an electron gas, note that (10.1.3) can be rewritten as

$$H_C = \frac{1}{2} \sum_{q}{}' \mathrm{v}(\boldsymbol{q})\, n(\boldsymbol{q})\, n(-\boldsymbol{q}) - \frac{1}{2} n \sum_{q}{}' \mathrm{v}(\boldsymbol{q}), \tag{10.1.9}$$

where $n(\boldsymbol{q})$ is the Fourier component of the electron density defined in (1.5.30) and n is the total number of electrons. For the moment we neglect the second term on the right hand side of (10.1.9) since it is a constant.

In obtaining $\langle H_C \rangle$ or $\langle H_C \rangle_g$, we recall the Kubo formula for the electron charge response function, (9.1.25), and the fluctuation-dissipation relation, (4.1.53), for $t = 0$. We obtain, thus,

$$\langle H_C \rangle = \frac{1}{2\pi} \sum_q{}' \mathrm{v}(\boldsymbol{q}) \int_{-\infty}^{\infty} d\omega \frac{1}{1 - \mathrm{e}^{-\beta\omega}} \mathrm{Im}\, \overline{\chi}_{\mathrm{ee}}(\boldsymbol{q}, \omega), \qquad (10.1.10)$$

where $\overline{\chi}_{\mathrm{ee}} = \chi_{\mathrm{ee}} / e^2$.

By putting (10.1.10) into (10.1.8), we arrive at the final result

$$\Omega_{\mathrm{el}} = \Omega^{\mathrm{o}}_{\mathrm{el}} + \Delta\Omega_{\mathrm{el}}$$

$$= \Omega^{\mathrm{o}}_{\mathrm{el}} + \frac{1}{2\pi} \sum_q{}' \mathrm{v}(\boldsymbol{q}) \int_{-\infty}^{\infty} d\omega \int_0^1 dg \frac{1}{1 - e^{-\beta\omega}} \mathrm{Im}\, \overline{\chi}_{\mathrm{ee}}(\boldsymbol{q}, \omega)_g \qquad (10.1.11)$$

$$= \Omega^{\mathrm{o}}_{\mathrm{el}} + \frac{1}{4\pi} \sum_q{}' \mathrm{v}(\boldsymbol{q}) \int_{-\infty}^{\infty} d\omega \int_0^1 dg \coth\frac{\beta\omega}{2} \mathrm{Im}\, \overline{\chi}_{\mathrm{ee}}(\boldsymbol{q}, \omega)_g \qquad (10.1.11')$$

$$= \Omega^{\mathrm{o}}_{\mathrm{el}} + \frac{1}{2\pi} \sum_q{}' \mathrm{v}(\boldsymbol{q}) \int_{-\infty}^{\infty} d\omega \int_0^1 dg \frac{1}{e^{\beta\omega} - 1} \mathrm{Im}\, \overline{\chi}_{\mathrm{ee}}(\boldsymbol{q}, \omega)_g \qquad (10.1.11'')$$

$$= \Omega^{\mathrm{o}}_{\mathrm{el}} + \frac{1}{2\pi} \sum_q{}' \mathrm{v}(\boldsymbol{q}) \int_0^1 dg \int_{-\infty}^{\infty} d\omega \left[\frac{1}{2} + n(\omega)\right] \mathrm{Im}\, \overline{\chi}_{\mathrm{ee}}(\boldsymbol{q}, \omega)_g, \qquad (10.1.11''')$$

where $\overline{\chi}_{\mathrm{ee}}(\boldsymbol{q}, \omega)_g$ is what we obtain by replacing $\mathrm{v}(\boldsymbol{q})$ by $g\mathrm{v}(\boldsymbol{q})$ in $\overline{\chi}_{\mathrm{ee}}(\boldsymbol{q}, \omega)$, $n(\omega)$ is the Bose distribution, and we noted the relation of (4.1.37) in rewriting the final expression. An important point to note is that the relation of (10.1.11) is an exact one. In practice, however, approximations enter through $\chi_{\mathrm{ee}}(\boldsymbol{q}, \omega)$; we do not have any exact expression for the electron charge response function.

Different approximations for χ_{ee} produce different results for $\Delta\Omega_{\mathrm{el}}$. First, if we neglect the effect of the exchange interaction ($\widetilde{V}(\boldsymbol{q}) = 0$) and consider the paramagnetic state ($F_+(\boldsymbol{q}, \omega) = F_-(\boldsymbol{q}, \omega) = F(\boldsymbol{q}, \omega)$), the charge response of (4.3.25) or (4.3.26) reduces to

$$\overline{\chi}^{\mathrm{RPA}}_{\mathrm{ee}}(\boldsymbol{q}, \omega) = \frac{2F(\boldsymbol{q}, \omega)}{1 + 2\mathrm{v}(\boldsymbol{q})F(\boldsymbol{q}, \omega)}, \qquad (10.1.12)$$

where the superscript RPA stands for the random phase approximation. If we put this into (10.1.11''), we obtain [10.3, 10.4]

$$\Delta\Omega_{\mathrm{el,RPA}} = \frac{1}{2\pi}\sum_{q}{}' \mathrm{v}(\boldsymbol{q}) \int_{-\infty}^{\infty} d\omega\, n(\omega) \int_{0}^{1} dg\, \mathrm{Im}\, \frac{2F(\boldsymbol{q},\omega)}{1 + 2g\,\mathrm{v}(\boldsymbol{q})\, F(\boldsymbol{q},\omega)}$$

$$= \frac{1}{2\pi}\sum_{q}{}' \int_{-\infty}^{\infty} d\omega\, n(\omega)\, \mathrm{Im}\ln\left[1 + 2\mathrm{v}(\boldsymbol{q})\, F(\boldsymbol{q},\omega)\right]. \qquad (10.1.13)$$

This result is the same as that of (8.4.43). It is straightforward to transform the final result of (8.4.43) into the form of (10.1.13) by changing the sum over $i\nu_m$ into the integral over real ω by the procedure (8.4.37'). Thus we have

$$\Delta\Omega_{\mathrm{el,ring}} = \Delta\Omega_{\mathrm{el,RPA}}. \qquad (10.1.14)$$

With a different χ_{ee} in (10.1.11) we obtain a different result for $\Delta\Omega_{\mathrm{el}}$. In the following we will discuss $\Delta\Omega_{\mathrm{el}}$ with χ_{ee} which includes the effects of the exchange interaction between electrons and the spin splitting of the electron energy bands in the ferromagnetic state. Before that, in the next subsection we explore what physics is contained in a result such as (10.1.13).

10.1.2 Charge Fluctuation and Electron Correlation Effect

Let us demonstrate that the result of (10.1.13) contains the effect of the existence of plasma oscillation and that of the screening of the Coulomb interaction between electrons.

If we follow the procedure given in 6.5, (10.1.13) is rewritten as

$$\Delta\Omega_{\mathrm{el,RPA}} = -\frac{1}{2\pi\beta}\sum_{q}{}' \left[\int_{0}^{\infty} d\omega \ln\left(1 - e^{-\beta\omega}\right) - \int_{-\infty}^{0} d\omega \ln\left(e^{-\beta\omega} - 1\right)\right]$$

$$\times\, \mathrm{Im}\left[\frac{\mathrm{v}(\boldsymbol{q})}{1 + \mathrm{v}(\boldsymbol{q})\sum_{\sigma} F_{\sigma}(\boldsymbol{q},\omega)} \sum_{\sigma} \frac{\partial}{\partial\omega} F_{\sigma}(\boldsymbol{q},\omega)\right]. \qquad (10.1.15)$$

Here, as we saw in Problem 4.9 and will again see in 10.2, we have

$$1 + \mathrm{v}(\boldsymbol{q})\sum_{\sigma} F_{\sigma}(\boldsymbol{q},\omega) = 1 + \frac{\omega_{\mathrm{pl}}^2}{(\omega_{\mathrm{pl}}^2(\boldsymbol{q}) - \omega_{\mathrm{pl}}^2) - \omega^2}, \qquad (10.1.16)$$

$$\text{for } \omega \gg k_F q / m$$

where $\omega_{\rm pl}(\boldsymbol{q})$ is the plasma frequency which does not include the exchange effects. Although in the present paramagnetic state $F_\sigma(\boldsymbol{q}, \omega) = F(\boldsymbol{q}, \omega)$, independent of spin, for later convenience we retained the spin subscript in the Lindhard function. Except at $\omega = \pm\omega_{\rm pl}(\boldsymbol{q})$, the integrand of (10.1.15) is slowly varying. Then we may separate the integral into two parts, that for the immediate vicinities of $\omega = \pm\omega_{\rm pl}(\boldsymbol{q})$, $\Delta\Omega_{\rm el,RPA1}$, and that for the remaining region, $\Delta\Omega_{\rm el,RPA2}$:

$$\Delta\Omega_{\rm el,RPA} = \Delta\Omega_{\rm el,RPA1} + \Delta\Omega_{\rm el,RPA2}. \qquad (10.1.17)$$

First, in integrating near $\omega = \pm\omega_{\rm pl}(\boldsymbol{q})$, if we recall that ω is to be replaced by $\omega + i0^+$, the integrand is rewritten as

$$\mathrm{Im}\left[\frac{\omega^2 - \omega_{\rm pl}^2(\boldsymbol{q}) + \omega_{\rm pl}^2}{\omega^2 - \omega_{\rm pl}^2(\boldsymbol{q})} \sum_\sigma \mathrm{v}(\boldsymbol{q}) \frac{\partial}{\partial\omega} F_\sigma(\boldsymbol{q}, \omega)\right]$$

$$\cong \mathrm{Im}\left[\frac{\omega_{\rm pl}^2}{(\omega - \omega_{\rm pl}(\boldsymbol{q}) + i0^+)(\omega + \omega_{\rm pl}(\boldsymbol{q}) + i0^+)} \frac{\partial}{\partial\omega}\left(-\frac{\omega_{\rm pl}^2}{\omega^2 - (\omega_{\rm pl}^2(\boldsymbol{q}) - \omega_{\rm pl}^2)}\right)\right]$$

$$\cong \mathrm{Im}\left[\frac{1}{\omega - \omega_{\rm pl}(\boldsymbol{q}) + i0^+} - \frac{1}{\omega + \omega_{\rm pl}(\boldsymbol{q}) + i0^+}\right]$$

$$= -\pi\left[\delta(\omega - \omega_{\rm pl}(\boldsymbol{q})) - \delta(\omega + \omega_{\rm pl}(\boldsymbol{q}))\right], \qquad (10.1.18)$$

where we noted (5.4.12). Then, putting (10.1.18) into (10.1.15), we obtain

$$\Delta\Omega_{\rm el,RPA1} = \frac{1}{2}\sum_{\boldsymbol{q}} \omega_{\rm pl}(\boldsymbol{q}) + \beta^{-1}\sum_{\boldsymbol{q}} \ln\left(1 - e^{-\beta\omega_{\rm pl}(\boldsymbol{q})}\right) \qquad (10.1.19)$$

This is of the standard form of the thermodynamic potential for bosons.

Next, in carrying out the integration outside the above plasma poles, we approximate the integrand as

$$\mathrm{Im}\left[\frac{\mathrm{v}(\boldsymbol{q})}{1 + \mathrm{v}(\boldsymbol{q})\sum_\sigma F_\sigma(\boldsymbol{q}, \omega)} \sum_\sigma \frac{\partial}{\partial\omega} F_\sigma(\boldsymbol{q}, \omega)\right]$$

$$\cong \frac{\mathrm{v}(\boldsymbol{q})}{1 + \mathrm{v}(\boldsymbol{q})\sum_\sigma F_\sigma(\boldsymbol{q}, 0)} \sum_\sigma \mathrm{Im}\frac{\partial}{\partial\omega} F_\sigma(\boldsymbol{q}, \omega)$$

$$= v_{sc}(\boldsymbol{q}) \sum_{\sigma} \text{Im} \frac{\partial}{\partial \omega} F_{\sigma}(\boldsymbol{q}, \omega), \tag{10.1.20}$$

where $v_{sc}(\boldsymbol{q})$ is the screened Coulomb potential which was defined in (8.5.32). Then by putting this in (10.1.15) and integrating by parts, we have

$$\Delta\Omega_{\text{el,RPA2}} = \frac{1}{2\pi} \sum_{\boldsymbol{q},\sigma} v_{sc}(\boldsymbol{q}) \int_{-\infty}^{\infty} d\omega\, n(\omega)\, \text{Im}\, F_{\sigma}(\boldsymbol{q}, \omega) \tag{10.1.21}$$

$$= \frac{1}{2\beta} \sum_{\boldsymbol{q},m,\sigma} v_{sc}(\boldsymbol{q})\, F_{\sigma}(\boldsymbol{q}, i\nu_m) \tag{10.1.21'}$$

$$= -\frac{1}{2} \sum_{\boldsymbol{k},\boldsymbol{q},\sigma} v_{sc}(\boldsymbol{q})\, f(\varepsilon_{\boldsymbol{k}\sigma})\, f(\varepsilon_{\boldsymbol{k}+\boldsymbol{q},\sigma}). \tag{10.1.21''}$$

The last two expressions were obtained by noting (8.4.42), and (8.4.42'). We immediately notice that (10.1.21') is exactly of the same form as $\Delta\Omega_1$ given in (8.4.42'), the first order exchange effect; simply the bare Coulomb potential in $\Delta\Omega_1$, is replaced by the screened one in $\Delta\Omega_{\text{el,RPA2}}$.

Thus we have confirmed that the physical consequences of the correlation effect are, firstly, to give rise to plasma oscillations and, secondly, to screen the Coulomb interaction between electrons. An important notice may be that these two consequences of the correlation effect are distinctly *different* things.

Throughout this book we have been writing $\widetilde{V}$ for the effective exchange interaction between electrons. Let us reconsider what $\widetilde{V}$ really is. A way to introduce $\widetilde{V}$ is to rewrite (10.1.21') as

$$\Delta\Omega_{\text{el,RPA2}} = -\frac{1}{2} \sum_{\boldsymbol{k},\boldsymbol{q},\sigma} v_{sc}(\boldsymbol{q})\, f(\varepsilon_{\boldsymbol{k}\sigma})\, f(\varepsilon_{\boldsymbol{k}+\boldsymbol{q},\sigma}). \tag{10.1.21''}$$

$$\cong -\frac{1}{2} \widetilde{V} \sum_{\sigma} n_{\sigma}^2. \tag{10.1.21a}$$

This is identical to (3.4.9), which we have been calling the mean field approximation result.

Alternatively, (3.4.9) or (10.1.21a) may be viewed to have been derived from the first order exchange interaction contribution, (8.4.42), or (9.1.4)–(9.1.15). However, in this book, especially when we were faced to choose the value of $\widetilde{V}$ or $\widetilde{V}N(0)$ in carrying out numerical calculations, we understood $\widetilde{V}$ as that which includes the screening effect. Thus, $\widetilde{V}$ is a quite complicated quantity. $\widetilde{V}$ emerged as the consequence of the two steps of rather drastic simplifications. The first is to approximate the dynamic screening by the static one; thus (10.1.21) came out from (10.1.15). The second is to simplify that (10.1.21") as (10.1.21a).

We should be, then, aware that it is rather drastic simplification to introduce such a $\widetilde{V}$ that is a constant independent of temperature etc. It is evident that such $\widetilde{V}$ can have only limited meaning.

10.1.3 Exchange Enhanced Charge Fluctuation: $\Delta\Omega_{\mathrm{el,GRPA}}$

Let us proceed to the next stage of approximation for $\Delta\Omega_{\mathrm{el}}$ by considering the effect of the exchange interaction in $\chi_{\mathrm{ee}}(\boldsymbol{q}, \omega)$. By putting (4.3.25) into (10.1.11") we obtain [10.5]

$$\Delta\Omega_{\mathrm{el,GRPA}}$$

$$= \frac{1}{2\pi}\sum_{\boldsymbol{q}}{}' \mathrm{v}(\boldsymbol{q})\int_{-\infty}^{\infty} d\omega\, n(\omega)\int_0^1 dg\, \mathrm{Im}\, \frac{\widetilde{F}_+(\boldsymbol{q}, \omega)_g + \widetilde{F}_-(\boldsymbol{q}, \omega)_g}{1 + g\,\mathrm{v}(\boldsymbol{q})\left[\widetilde{F}_+(\boldsymbol{q}, \omega)_g + \widetilde{F}_-(\boldsymbol{q}, \omega)_g\right]}, \tag{10.1.22}$$

with

$$\widetilde{F}_\sigma(\boldsymbol{q}, \omega)_g = \frac{F_\sigma(\boldsymbol{q}, \omega)}{1 - g\widetilde{V}(\boldsymbol{q})\, F_\sigma(\boldsymbol{q}, \omega)}. \tag{10.1.23}$$

The subscript GRPA stands for the *generalized random phase approximation*; unlike in the RPA, in the GRPA the exchange effect is considered. Note that this result is valid both for $T > T_{\mathrm{C}}$ and $T < T_{\mathrm{C}}$; for $T > T_{\mathrm{C}}$, it reduces to the result of Hubbard [4.30].

Compared to $\Delta\Omega_{\mathrm{el,RPA}}$ of (10.1.13), this result is understood as representing the contribution of the *exchange enhanced charge fluctuations*. Then from the analysis given in the preceding subsection, the exchange effect would show up both in the plasma frequency and the screening of the electron interaction. We will see later how important such a consideration of the exchange effect is. Before that, in the remaining parts of this section let us explore how the result (10.1.22) can be reproduced by using the Feynman diagrammatic method.

10.1.4 Shaded Ring Diagrams: $\Delta\widetilde{\Omega}_{\mathrm{el,ring}}$

Corresponding to the progress from $\Delta\Omega_{\mathrm{el,RPA}}$ to $\Delta\Omega_{\mathrm{el,GRPA}}$ in exploring the correlation (or, the charge fluctuation) effect based on the fluctuation-dissipation theorem, (10.1.11), now we pursue how $\Delta\Omega_{\mathrm{el,ring}}$ of (8.4.43) can be improved by considering the effect of the exchange interaction between electrons in the Feynman diagrammatic approach. From what we already discussed in 9.1, it is almost evident what to be done. We have only to shade all the ring or polarization diagrams appearing in Figs.8.2(b) and 8.5. We give in Fig.10.1 such *linked shaded ring diagrams* which contribute to the thermodynamic potential, $\Delta\widetilde{\Omega}_{\mathrm{el,ring}}$.

It is straightforward to write down the contribution of those shaded ring diagrams. We have only to replace $P(\boldsymbol{q}, i\nu_m)$ in (8.4.43) by $\widetilde{P}(\boldsymbol{q}, i\nu_m)$ defined in (9.1.20). Thus we obtain

$$\Delta\widetilde{\Omega}_{\mathrm{el,ring}} = \frac{1}{2\beta}\sum_{\boldsymbol{q},m}{}' \ln\left\{1 + \mathrm{v}(\boldsymbol{q})\,\widetilde{P}(\boldsymbol{q}, i\nu_m)\right\}$$

$$= \frac{1}{2\pi}\sum_{\boldsymbol{q}}{}' \int_{-\infty}^{\infty} d\omega\, n(\omega)\,\mathrm{Im}\ln\left[1 + \mathrm{v}(\boldsymbol{q})\left\{\widetilde{F}_+(\boldsymbol{q}, \omega) + \widetilde{F}_-(\boldsymbol{q}, \omega)\right\}\right]$$

$$= \frac{1}{2\pi}\sum_{\boldsymbol{q}}{}' \mathrm{v}(\boldsymbol{q})\int_{-\infty}^{\infty} d\omega\, n(\omega)\int_0^1 dg\,\mathrm{Im}\left[\frac{\widetilde{P}(\boldsymbol{q}, \omega)}{1 + g\,\mathrm{v}(\boldsymbol{q})\,\widetilde{P}(\boldsymbol{q}, \omega)}\right]. \qquad (10.1.24)$$

Comparing this result with (10.1.22), we immediately note

$$\Delta\widetilde{\Omega}_{\mathrm{el,ring}} \neq \Delta\Omega_{\mathrm{el,GRPA}}. \qquad (10.1.25)$$

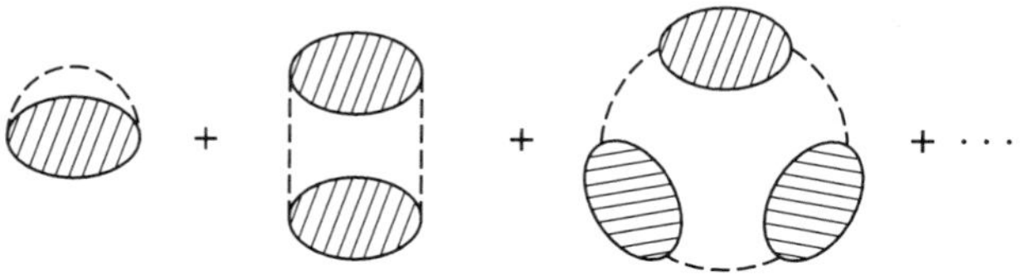

Fig.10.1 The contribution of the linked shaded ring diagrams to the thermodynamic potential, $\Delta\widetilde{\Omega}_{\mathrm{ring}}$.

Thus, unlike in the case of (10.1.14), the physical nature of $\Delta\widetilde{\Omega}_{\text{el,ring}}$ is not transparent; it does not directly represent the contribution of the exchange enhanced charge fluctuations.

As for the effect of the spin fluctuations, neither $\Delta\Omega_{\text{el,GRPA}}$ nor $\Delta\Omega_{\text{el,ring}}$ seems to contain their contribution properly. The transverse spin response function $\chi_{\sigma,-\sigma}(\boldsymbol{q}, \omega)$ (see (4.4.12) or (9.1.27)) which represents the effect of transverse spin fluctuations, for instance, does not appear in either of $\Delta\widetilde{\Omega}_{\text{el,ring}}$ and $\Delta\Omega_{\text{el,GRPA}}$. The situation is the same also with respect to the longitudinal spin fluctuations; χ_{zz} of (4.3.16) or (9.1.24) does not show up in either of $\Delta\widetilde{\Omega}_{\text{el,ring}}$ and $\Delta\Omega_{\text{el,GRPA}}$. The spin fluctuation is now widely believed to be one of the most important excitations in a metallic electron system. How can we proceed to improve the result of $\Delta\Omega_{\text{el,GRPA}}$ or $\Delta\widetilde{\Omega}_{\text{el,ring}}$ so as to include such spin fluctuation contributions?

As will be shown in the next subsection, fortunately we succeed to obtain by the Feynman diagrammatic approach the thermodynamic potential of an electron gas in a form which includes the spin fluctuation effect as well as the charge fluctuation effect. We arrive at such a pleasing result by including the ladder diagram contributions in addition to the already considered ring diagram contributions.

10.1.5 Ladder Diagrams: Spin Fluctuation Contributions to $\Delta\Omega_{\text{el}}$

The first kind of the ladder diagrams is shown in Fig.10.2(a). They correspond to the electron self-energy due to transverse spin fluctuations of Fig.9.23. The contributions of these diagrams, which we call the ladder 1, are summed similarly to (8.4.36) as

$$\Delta\Omega_{\text{el,lad1}} = \frac{1}{2\beta}\sum_{\boldsymbol{q},m,\sigma}\left[\ln\left\{1-\widetilde{V}(\boldsymbol{q})\,F_{\sigma,-\sigma}(\boldsymbol{q},i\nu_m)\right\}+\widetilde{V}(\boldsymbol{q})\,F_{\sigma,-\sigma}(\boldsymbol{q},i\nu_m)\right] \tag{10.1.26}$$

$$= \frac{1}{2\beta}\sum_{\boldsymbol{q},m}\left[-\int_0^1 dg\,\widetilde{V}(\boldsymbol{q})\left\{\overline{\chi}_{+-}(\boldsymbol{q},i\nu_m)_{\text{g}}+\overline{\chi}_{-+}(\boldsymbol{q},i\nu_m)_{\text{g}}\right\}\right.$$
$$\left.+\;\widetilde{V}(\boldsymbol{q})\left\{F_{+-}(\boldsymbol{q},i\nu_m)+F_{-+}(\boldsymbol{q},i\nu_m)\right\}\right] \tag{10.1.26'}$$

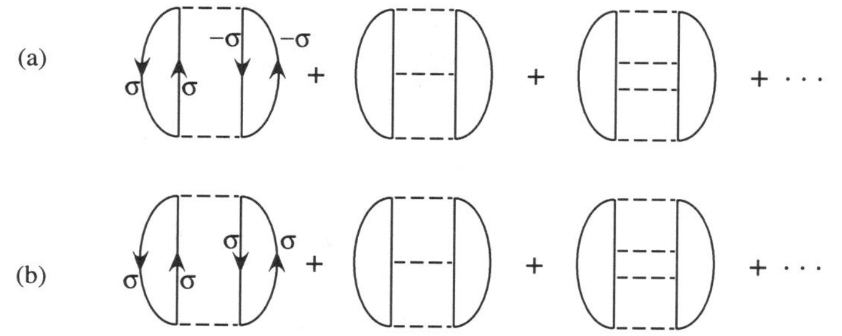

Fig.10.2 The two kinds of ladder diagrams, the ladder 1, (a), and the ladder 2, (b).

$$= \frac{1}{2\pi}\sum_{q,\sigma}\int_{-\infty}^{\infty} d\omega\ n(\omega)\ \mathrm{Im}\Big[\ \ln\big\{1 - \widetilde{V}(\boldsymbol{q})F_{\sigma,-\sigma}(\boldsymbol{q},\omega\,)\big\} + \widetilde{V}(\boldsymbol{q})F_{\sigma,-\sigma}(\boldsymbol{q},\omega)\Big] \qquad (10.1.26'')$$

$$= \frac{1}{\pi}\sum_{q,\sigma}\int_{0}^{\infty} d\omega\Big\{n(\omega) + \frac{1}{2}\Big\}\ \mathrm{Im}\Big[\ln\big\{1 - \widetilde{V}(\boldsymbol{q})\,F_{\sigma,-\sigma}(\boldsymbol{q},\omega\,)\big\} + \widetilde{V}(\boldsymbol{q})\,F_{\sigma,-\sigma}(\boldsymbol{q},\omega)\Big], \qquad (10.1.26''')$$

where notations are the same as in 9.5.1 and $\overline{\chi}_{\sigma,-\sigma}(\boldsymbol{q}, i\nu_m)_{\mathrm{g}}$ is defined similarly to (10.1.23); in deriving (10.1.26") from (10.1.26) we used (8.4.38') with the contour $C\,''$ of Fig.8.6(c) together with (4.1.46) and (9.3.9).

In our electron gas model with the long range Coulomb interaction, there is no reason to exclude those diagrams given in Fig.10.2(b) which we call the ladder 2; They do not appear in the Hubbard model since there direct interaction takes place only between electrons of opposite spins.

The contribution to the thermodynamic potential of Fig.10.2(b) is readily obtained, similarly to (10.1.26), as

$$\Delta\Omega_{\mathrm{el,lad2}} = \frac{1}{2\beta}\sum_{q,m,\sigma}\Big[\ln\big\{1 - \widetilde{V}(\boldsymbol{q})\,F_{\sigma}(\boldsymbol{q}, i\nu_m)\big\} + \widetilde{V}(\boldsymbol{q})\,F_{\sigma}(\boldsymbol{q}, i\nu_m)\Big]\,. \qquad (10.1.27)$$

Note that the second order diagrams in Figs.10.2(a) and 10.2(b) are of the same structure as those of Fig.8.5 which are already contained in $\Delta\widetilde{\Omega}_{\mathrm{el,ring}}$. Thus the second order diagrams are overcounted. The treatment of the second order terms in the ring and the ladder diagrams, however, are different, the former being proportional to $\mathrm{v}(\boldsymbol{q})^2$ whereas the latter to $\widetilde{V}(\boldsymbol{q})$.

Unlike in the case of $\Delta\Omega_{\text{el,lad1}}$, the physical implication of $\Delta\Omega_{\text{el,lad2}}$ is not transparent apart from the fact that in the paramagnetic state, where $F_{\sigma,-\sigma} = F_{\sigma}$ and $F_+ = F_-$, practically $\Delta\Omega_{\text{el,lad1}}$ and $\Delta\Omega_{\text{el,lad2}}$ contribute similarly. Remember that $\Delta\widetilde{\Omega}_{\text{el,ring}}$ of (10.1.24) also did not correspond to any single physical process directly; it does not represent the effect of the exchange enhanced electron density fluctuations as given by (10.1.22). Further note that so far the longitudinal spin fluctuation involving $\chi_{zz}(\boldsymbol{q}, \omega)$ of (4.3.16) or (9.1.24) did not appear at all.

Here we have a surprise. The sum of these two physically unclear contributions, $\Delta\Omega_{\text{el,lad2}}$ and $\Delta\widetilde{\Omega}_{\text{el,ring}}$, results in the desired two contributions, the exchange enhanced electron charge fluctuation and the longitudinal spin fluctuation. We will see this in the next subsection.

10.1.6 Sum of the Shaded Ring and the Ladder Contributions to $\Delta\Omega_{\text{el}}$: Charge and Spin Fluctuations

We find the sum of the results of (10.1.24) and (10.1.27) can be rewritten as

$$\Delta\widetilde{\Omega}_{\text{el,ring}} + \Delta\Omega_{\text{el,lad2}} = \frac{1}{2\beta}\sum_{\boldsymbol{q},m}\left[\ln\left\{1 + \mathrm{v}\left(\frac{F_+}{1-\widetilde{V}F_+} + \frac{F_-}{1-\widetilde{V}F_-}\right)\right\}\right.$$
$$\left. + \ln\left\{\left(1-\widetilde{V}F_+\right)\left(1-\widetilde{V}F_-\right)\right\} + \widetilde{V}(F_+ + F_-)\right]$$

$$= \frac{1}{2\beta}\sum_{\boldsymbol{q},m}\left[\ln\left\{1 + \left(\mathrm{v}-\widetilde{V}\right)(F_+ + F_-) - \widetilde{V}\left(2\mathrm{v}-\widetilde{V}\right)F_+F_-\right\} + \widetilde{V}(F_+ + F_-)\right]$$

$$= \frac{1}{2\beta}\sum_{\boldsymbol{q},m}\left[\int_0^1 dg\,\frac{\left(\mathrm{v}-\widetilde{V}\right)(F_+ + F_-) - 2g\widetilde{V}\left(2\mathrm{v}-\widetilde{V}\right)F_+F_-}{1 + g\left(\mathrm{v}-\widetilde{V}\right)(F_+ + F_-) - g^2\widetilde{V}\left(2\mathrm{v}-\widetilde{V}\right)F_+F_-} + \widetilde{V}(F_+ + F_-)\right]$$

$$= \frac{1}{2\beta}\sum_{\boldsymbol{q},m}\left[\int_0^1 dg\left(\mathrm{v}-\frac{\widetilde{V}}{2}\right)\frac{\widetilde{F}_{+g} + \widetilde{F}_{-g}}{1 + g\mathrm{v}\left(\widetilde{F}_{+g} + \widetilde{F}_{-g}\right)}\right.$$
$$\left. - \frac{\widetilde{V}}{2}\int_0^1 dg\,\frac{\widetilde{F}_{+g} + \widetilde{F}_{-g} + 4g\,\mathrm{v}\,\widetilde{F}_{+g}\widetilde{F}_{-g}}{1 + g\mathrm{v}\left(\widetilde{F}_{+g} + \widetilde{F}_{-g}\right)} + \widetilde{V}(F_+ + F_-)\right]. \qquad (10.1.28)$$

We abbreviated $F_\sigma(\boldsymbol{q}, i\nu_m)$ by F_σ, $\mathrm{v}(\boldsymbol{q})$ and $\widetilde{V}(\boldsymbol{q})$ by v and $\widetilde{V}$, and $\widetilde{F}_\sigma(\boldsymbol{q}, i\nu_m)_g$, defined similarly to (10.1.23), by $\widetilde{F}_{\sigma g}$.

In the first term of the last expression, we can immediately identify a quantity $\overline{\chi}_{\rm ee}(\boldsymbol{q}, i\nu_m)_g$ that appears in (10.1.22),

$$\overline{\chi}_{\rm ee}^{\rm GRPA}(\boldsymbol{q}, i\nu_m)_g = \frac{\widetilde{F}_+(\boldsymbol{q}, i\nu_m)_g + \widetilde{F}_-(\boldsymbol{q}, i\nu_m)_g}{1 + g\mathrm{v}(\boldsymbol{q})\left\{\widetilde{F}_+(\boldsymbol{q}, i\nu_m)_g + \widetilde{F}_-(\boldsymbol{q}, i\nu_m)_g\right\}}. \qquad (10.1.29)$$

Also, in the second term we can identify the longitudinal spin susceptibility $\overline{\chi}_{zz}(\boldsymbol{q}, i\nu_m)_g$ to be defined similarly to (10.1.29). Thus, we can rewrite (10.1.28).

$$\Delta\widetilde{\Omega}_{\rm el,\, ring} + \Delta\Omega_{\rm el,\, lad2} = \frac{1}{2\beta}\sum_{\boldsymbol{q},m}\left[\int_0^1 dg\left\{\left(\mathrm{v}(\boldsymbol{q}) - \frac{\widetilde{V}(\boldsymbol{q})}{2}\right)\overline{\chi}_{\rm ee}^{\rm GRPA}(\boldsymbol{q}, i\nu_m)_g\right.\right.$$

$$\left.\left. - \frac{\widetilde{V}(\boldsymbol{q})}{2}\overline{\chi}_{zz}(\boldsymbol{q}, i\nu_m)_g\right\} + \widetilde{V}(\boldsymbol{q})\left\{F_+(\boldsymbol{q}, i\nu_m) + F_-(\boldsymbol{q}, i\nu_m)\right\}\right]. \qquad (10.1.30)$$

The sum of two physically unclear contributions results in the desired form of contributions, the exchange enhanced charge fluctuations and the longitudinal spin fluctuations.

Adding to (10.1.30) the contribution of the ladder 1 diagrams, now we have the final result for the effect of the Coulomb interaction between electrons on the thermodynamic potential of an electron gas,

$$\Delta\Omega_{\rm el} = \Delta\widetilde{\Omega}_{\rm el,\, ring} + \Delta\Omega_{\rm el,\, lad1} + \Delta\Omega_{\rm el,\, lad2}$$

$$= \frac{1}{2\pi}\sum_{\boldsymbol{q}}\int_{-\infty}^{\infty} d\omega\, n(\omega)\left[\int_0^1 dg\left(\left\{\mathrm{v}(\boldsymbol{q}) - \frac{\widetilde{V}(\boldsymbol{q})}{2}\right\}\mathrm{Im}\,\overline{\chi}_{\rm ee}^{\rm GRPA}(\boldsymbol{q}, \omega)_g\right.\right.$$

$$\left. - \widetilde{V}(\boldsymbol{q})\,\mathrm{Im}\{\overline{\chi}_{+-}(\boldsymbol{q}, \omega)_g + \overline{\chi}_{-+}(\boldsymbol{q}, \omega)_g + \frac{1}{2}\overline{\chi}_{zz}(\boldsymbol{q}, \omega)_g\}\right)$$

$$\left. + \widetilde{V}(\boldsymbol{q})\,\mathrm{Im}\left\{F_+(\boldsymbol{q}, \omega) + F_-(\boldsymbol{q}, \omega) + F_{+-}(\boldsymbol{q}, \omega) + F_{-+}(\boldsymbol{q}, \omega)\right\}\right]. \qquad (10.1.31)$$

Our final result of (10.1.31), which is valid both in the ferromagnetic and the paramagnetic states, has a pleasing structure. It consists of the contributions from the exchange enhanced charge fluctuations and the spin fluctuations. In the ferromagnetic state, the most important components of the transverse spin fluctuations are the spin wave excitations as we have seen in 6.6.2 and 9.5.1. In the paramagnetic state where $\chi_{xx} = \chi_{yy} = \chi_{zz}$, the spin fluctuation effect is rotationally invariant; note that $2[\chi_{+-} + \chi_{-+}] = \chi_{xx} + \chi_{yy}$ (see (9.5.24) and (1.1.23)). An interesting point to note may be that the interaction constant appearing in the electron charge fluctuation part is not $\mathrm{v}(\boldsymbol{q})$ but $[\mathrm{v}(\boldsymbol{q}) - \widetilde{V}(\boldsymbol{q})/2]$.

10.2 THE EFFECT OF THE CHARGE FLUCTUATIONS ON THE MAGNETIC PROPERTIES OF A METAL

What effect would the charge fluctuations have on the magnetic behavior of a metal? In this section we try to answer to this question. It seems that this subject is not yet satisfactorily clarified.

By the charge fluctuation effect we may mean the role of either the first term on the right hand side of (10.1.31), or, more simply, $\Delta\widetilde{\Omega}_{\mathrm{el,ring}}$ given in (10.1.24); often it is more convenient to handle $\Delta\widetilde{\Omega}_{\mathrm{el,ring}} + \Delta\Omega_{\mathrm{el,lad2}}$ in its original form as is given by the right hand side of the first equality of (10.1.28) rather than that of the final equality. For definitness, throughout this section we take $\Delta\widetilde{\Omega}_{\mathrm{el,ring}}$ itself for the charge fluctuation effect.

We recall that if we neglect the exchange effect and consider only the paramagnetic state, $\Delta\widetilde{\Omega}_{\mathrm{el,ring}}$ reduces to $\Delta\Omega_{\mathrm{el,RPA}}$ which was discussed in 10.1.2. We then notice that $\Delta\widetilde{\Omega}_{\mathrm{el,ring}}$ can be handled quite parallelly to $\Delta\Omega_{\mathrm{el,RPA}}$; we only have to replace every $F_\sigma(\boldsymbol{q}, \omega)$ appearing in 10.1.2 by $\widetilde{F}_\sigma(\boldsymbol{q}, \omega)$.

Thus, correspondingly to (10.1.17), (10.1.19), and (10.1.21') we have

$$\Delta\widetilde{\Omega}_{\mathrm{el,ring}} = \Delta\widetilde{\Omega}_{\mathrm{el,ring1}} + \Delta\widetilde{\Omega}_{\mathrm{el,ring2}}, \tag{10.2.1}$$

$$\Delta\widetilde{\Omega}_{\mathrm{el,ring1}} = \frac{1}{2}\sum_{\boldsymbol{q}} \omega_{\mathrm{pl}}(\boldsymbol{q}, M) + \beta^{-1}\sum_{\boldsymbol{q}} \ln[1 - \exp(-\beta\omega_{\mathrm{pl}}(\boldsymbol{q}, M))], \tag{10.2.2}$$

$$\Delta\widetilde{\Omega}_{\mathrm{el,ring2}} = \frac{1}{2\beta}\sum_{\boldsymbol{q},m,\sigma} \hat{\mathrm{v}}_{\mathrm{sc}\sigma}(\boldsymbol{q})\, F_\sigma(\boldsymbol{q}, i\nu_m), \tag{10.2.3}$$

where the new magnetization dependent plasma frequency $\omega_{\mathrm{pl}}(\boldsymbol{q}, M)$ is to be obtained from

$$1 + \mathrm{v}(\boldsymbol{q})\left(\frac{F_+(\boldsymbol{q}, \omega)}{1 - \widetilde{V}F_+(\boldsymbol{q}, \omega)} + \frac{F_-(\boldsymbol{q}, \omega)}{1 - \widetilde{V}F_-(\boldsymbol{q}, \omega)}\right) = 0, \tag{10.2.4}$$

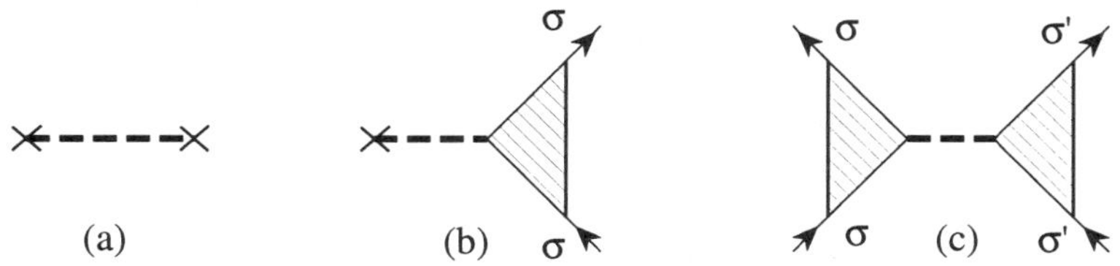

Fig.10.3 Three different screenings of the Coulomb interactions between external charges, (a), between an external charge and an electron, (b), and between electrons, (c). The corresponding screening constants are denoted, respectively, by $\varepsilon_e(\boldsymbol{q}, \omega)$, $\varepsilon_\sigma(\boldsymbol{q}, \omega)$, and $\hat{\varepsilon}_\sigma(\boldsymbol{q}, \omega)$ or $\hat{\varepsilon}_{\sigma\sigma'}(\boldsymbol{q}, \omega)$, and given, respectively by (4.3.53), (4.3.33), and (10.2.6) or (10.2.7).

and the new spin dependent screened Coulomb interaction $\hat{\mathrm{v}}_{\mathrm{sc}\sigma}(\boldsymbol{q})$ is given by

$$\hat{\mathrm{v}}_{\mathrm{sc}\sigma}(\boldsymbol{q}) = \frac{\mathrm{v}(\boldsymbol{q})}{\hat{\varepsilon}_\sigma(\boldsymbol{q}, 0)}, \tag{10.2.5}$$

with a new screening constant of the electron–electron interaction,

$$\hat{\varepsilon}_\sigma(\boldsymbol{q}, \omega) = \left(1 - \tilde{V}F_\sigma(\boldsymbol{q}, \omega)\right)^2 \left[1 + \mathrm{v}(\boldsymbol{q})\left(\frac{F_+(\boldsymbol{q}, \omega)}{1 - \tilde{V}F_+(\boldsymbol{q}, \omega)} + \frac{F_-(\boldsymbol{q}, \omega)}{1 - \tilde{V}F_-(\boldsymbol{q}, \omega)}\right)\right]. \tag{10.2.6}$$

This result together with Fig.10.3(c) immediately suggests a more general screening constant,

$$\hat{\varepsilon}_{\sigma\sigma'}(\boldsymbol{q}, \omega) = \left(1 - \tilde{V}F_\sigma(\boldsymbol{q}, \omega)\right)\left(1 - \tilde{V}F_{\sigma'}(\boldsymbol{q}, \omega)\right)$$
$$\times \left[1 + \mathrm{v}(\boldsymbol{q})\left(\frac{F_+(\boldsymbol{q}, \omega)}{1 - \tilde{V}F_+(\boldsymbol{q}, \omega)} + \frac{F_-(\boldsymbol{q}, \omega)}{1 - \tilde{V}F_-(\boldsymbol{q}, \omega)}\right)\right], \tag{10.2.7}$$

which screens the Coulomb interaction between electrons with σ and σ' spins.

In 4.3.7 we pointed out the screening constant of the Coulomb interaction *between an electron and an external charge* in the metal, $\varepsilon_\sigma(\boldsymbol{q}, \omega)$, and that *between two external charges*, $\varepsilon_e(\boldsymbol{q}, \omega)$, are different if the exchange effects are considered. Now we are finding the third screening constant, $\hat{\varepsilon}_\sigma(\boldsymbol{q}, \omega)$ or $\hat{\varepsilon}_{\sigma\sigma'}(\boldsymbol{q}, \omega)$, for the Coulomb interaction *between electrons*. We illustrate those three kinds of screening in Fig.10.3. The thick broken lines in all those diagrams are defined in Fig.9.6, and the shaded electrons lines in the figures (b) and (c) are given in Fig.9.12(c).

We will find both of above two contributions to $\Delta\widetilde{\Omega}_{\mathrm{el,ring}}$ to have interesting consequences. Let us discuss them separately in the following.

10.2.1 $\Delta\widetilde{\Omega}_{\mathrm{el,ring2}}$: Exchange Effect on the Screening of the Electron-Electron Interaction

First we discuss $\Delta\widetilde{\Omega}_{\mathrm{el,ring2}}$. If we replace $\widehat{\mathrm{v}}_{\mathrm{sc}\sigma}(\boldsymbol{q})$ by $\mathrm{v}(\boldsymbol{q})$ in (10.2.3), it reduces to the simple Hartree-Fock, or the 1st order exchange energy; see (8.4.42). Thus we reconfirm that one of two aspects of the charge fluctuation effect is to screen the Coulomb interaction between electrons as given by (10.2.5)–(10.2.7).

In 4.3.7 we showed that if we take into account the effect of the exchange interaction, the screening constants of the interaction between two external (non-electronic) charges, $\varepsilon_{\mathrm{e}}(\boldsymbol{q}, \omega)$, and that between an electron and an external charge, $\varepsilon_\sigma(\boldsymbol{q}, \omega)$, become different; see (4.3.54) and (4.3.33). Now we are presenting the third screening constant $\widehat{\varepsilon}_\sigma(\boldsymbol{q}, \omega)$ which is to be used for the screening of the electron-electron interaction.

This new screening constant brings about interesting consequences in both of the paramagnetic and ferromagnetic states. Let us discuss it for the paramagnetic and the ferromagnetic states separately.

A. Paramagnetic state

In the paramagnetic state, where $F_+(\boldsymbol{q}, \omega) = F_-(\boldsymbol{q}, \omega) = F(\boldsymbol{q}, \omega)$, (10.2.6) or (10.2.7) reduces to

$$\widehat{\varepsilon}_\sigma(\boldsymbol{q}, \omega) = \left(1 - \widetilde{V} F(\boldsymbol{q}, \omega)\right)^2 + 2\,\mathrm{v}(\boldsymbol{q}) \left(1 - \widetilde{V} F(\boldsymbol{q}, \omega)\right) F(\boldsymbol{q}, \omega)$$

$$\equiv \widehat{\varepsilon}(\boldsymbol{q}, \omega). \tag{10.2.8}$$

This result shows that for $\widetilde{V}F(0) \to 1$, either from above or from below, the screened electron-electron interaction tends to diverge. Let us discuss this interesting observation in more detail.

With the static screening constant, $\widehat{\varepsilon}(\boldsymbol{q}, 0)$, the Coulomb interaction between electrons in the paramagnetic state is screened as

$$\widehat{\mathrm{v}}_{\mathrm{sc}\sigma}(\boldsymbol{q}) = \widehat{\mathrm{v}}_{\mathrm{sc}}(\boldsymbol{q})$$

$$= \frac{4\pi e^2/(V\boldsymbol{q}^2)}{\left(1 - \widetilde{V} F(\boldsymbol{q}, 0)\right)^2 + (8\pi e^2/(V\boldsymbol{q}^2))\left(1 - \widetilde{V} F(\boldsymbol{q}, 0)\right) F(\boldsymbol{q}, 0)}. \tag{10.2.9}$$

Here, for $(q/k_F) << 1$ we can expand the static Lindhard function as (see (A.6)),

$$F(\boldsymbol{q}, 0) = F(0)\left[1 - a\,(q/k_F)^2 + \cdots\right],$$

where $a = 1/12$ for the present parabolic electron energy dispersion. From (10.2.9) we obtain

$$\hat{\mathrm{v}}_{sc}(\boldsymbol{q}) = \frac{4\pi\, e^{*2}}{V\left[\boldsymbol{q}^2 + \kappa_X^2\right]}, \tag{10.2.10}$$

with

$$e^{*2} = \frac{e^2}{\left(1 - \widetilde{V}F(0)\right)^2 + \frac{2\pi e^2}{3Vk_F^2}\, F(0)\left(2\widetilde{V}F(0) - 1\right)}, \tag{10.2.11}$$

$$\kappa_X^2 = \frac{\frac{8\pi e^2}{V}\, F(0)\left(1 - \widetilde{V}F(0)\right)}{\left(1 - \widetilde{V}F(0)\right)^2 + \frac{2\pi e^2}{3Vk_F^2}\, F(0)\left(2\widetilde{V}F(0) - 1\right)}. \tag{10.2.12}$$

If we put $\widetilde{V} = 0$, in (10.2.11) and (10.2.12), we have $e^* = e$, and κ_X reduces to the Thomas-Fermi result of (4.3.44).

Now consider the situation of $\widetilde{V}F(0) \to 1$. There we have

$$e^{*2} = \frac{3Vk_F^2}{2\pi F(0)}, \tag{10.2.13}$$

$$\kappa_X^2 = 12\left\{1 - \widetilde{V}F(0)\right\} k_F^2. \tag{10.2.14}$$

We assume $\widetilde{V}F(0) < 1$ for the paramagnetic state.

By Fourier inverse transforming (10.2.10) we have

$$\hat{\mathrm{v}}_{sc}(\boldsymbol{r}) = \frac{e^{*2}}{r}\, e^{-\kappa_X r}. \tag{10.2.15}$$

According to (10.2.14), then we have

$$\hat{\mathrm{v}}_{sc}(r) = \frac{e^{*2}}{r} \qquad \text{for} \quad \widetilde{V}F(0) \to 1\,. \tag{10.2.16}$$

As a paramagnetic itinerant electron system approaches the ferromagnetic instability by having $\widetilde{V}F(0) \to 1$, the screened Coulomb interaction returns to the long range, unscreened Coulomb interaction. There is a change, however; the electronic charge is modified to e^* as given in (10.2.13). Note that e^* becomes entirely independent of the original electronic charge e.

The size of e^* is estimated as following.

$$\frac{e^*}{e} = 6\left(\frac{4\pi e^2}{Vk_F^2}N(0)\right)^{-1} = 3k_F a_B$$

$$= 3\left(\frac{9\pi}{4}\right)^{1/3}\frac{1}{r_s} \qquad \text{for} \qquad \widetilde{V}F(0) \to 1, \tag{10.2.13'}$$

where a_B is the Bohr radius and r_s is the mean distance between electrons in units of a_B. For $2 < r_s < 6$, corresponding to the case of ordinary metals, we certainly have

$$e^* > e\,. \tag{10.2.17}$$

With such an e^* in (10.2.16), the Coulomb interaction between electrons is enhanced by the screening effect.

B. Ferromagnetic state

In the ferromagnetic state where $F_+(\boldsymbol{q}) \neq F_-(\boldsymbol{q})$, we first notice that $\widehat{\varepsilon}_\sigma(\boldsymbol{q})$ and, therefore, $\widehat{\mathrm{v}}_{\mathrm{sc}\sigma}(\boldsymbol{q})$ becomes spin dependent. Secondly, we notice that $\widehat{\mathrm{v}}_{\mathrm{sc}\sigma}(\boldsymbol{q})$ for both spins can become negative; as we experienced in obtaining, for instance, the result of Fig.4.6, the second factor on the right hand side of (10.2.6) can become negative when the electron energy bands are spin split.

Here we recall $\widehat{\mathrm{v}}_{\mathrm{sc}\sigma}(\boldsymbol{q})$ is what is to replace the $\mathrm{v}(\boldsymbol{q})$ in (8.5.22). Then, from a procedure similar to (10.1.21)–(10.1.21a) we obtain a screened effective exchange interaction, $\widehat{V}_\sigma$, which is spin dependent in the ferromagnetic state.

With such spin and magnetization dependence in the effective exchange interaction, the spin splitting of the electron energy bands is given as

$$\Delta(M) = \widehat{V}_+(M)\,n_+ - \widehat{V}_-(M)\,n_- + \sum_\sigma \frac{\partial \widehat{V}_\sigma(M)}{\partial M} n_\sigma^2 \tag{10.2.18}$$

corresponding to (3.4.17), where we assumed $n_- > n_+$ for the ferromagnetic state. The spin splitting is no longer simply proportional to the magnetization of a system.

What would happens when $\widehat{v}_{sc\sigma}(\boldsymbol{q})$ becomes negative? If $\widehat{v}_{sc\sigma}(\boldsymbol{q})$ of a system is negative, the system can not be in the ferromagnetic state; with a negative $\widehat{V}_\sigma$, ferromagnetic state can not be stable.

As can be seen from (10.2.6), it crucially depends on the electronic structure near the Fermi surface whether $\widehat{\varepsilon}_\sigma(\boldsymbol{q})$ would become negative or not under a small magnetization. In this respect T_C's of transition metals and alloys are plotted against electron concentrations (often denoted as e/a), that is, the number of electrons outside the closed shell per atom, they sit on a universal curve, and we observe sudden disappearances of ferromagnetism together with structural instabilities for $e/a \cong 8.5$. If we take the rigid band model, these observations can be understood from our above scenario: If the Fermi energy is located at a position in the electric density of states curve as would correspond to such a value of e/a, $\widehat{\varepsilon}_\sigma(\boldsymbol{q})$ becomes negative when the bands are infinitesimally spin split. Note also that the second factor on the right hand side of (10.2.6), which determines the sign of $\widehat{\varepsilon}_\sigma(\boldsymbol{q})$, is the very screening constant that determines phonon frequency. Thus an anomaly in the screened electron-electron interaction is closely related with that in phonon behavior; the breakdown of a ferromagnetic state by having a negative $\widehat{V}_\sigma$ or $\widehat{V}$ is concurrent with the occurrence of phonon softening; see (5.3.28).

Within the Stoner model, such disappearance of ferromagnetism may be attributed simply to having $\widetilde{V}F(0) < 1$ for $e/a \cong 8.5$. With such Stoner theory, however, the concurrent structural instability can not be understood.

If it is possible to have the paramagnetic state even with $\widetilde{V}F(0) > 1$ as we saw in the above, then an interesting possibility arises. With $\widetilde{V}F(0) > 1$, in (10.2.10)–(10.2.12), the screened Coulomb interaction between electrons becomes attractive. Can such a situation be possible? Here let us reconsider what $\widetilde{V}$ is, and what $\widehat{v}_{sc}(\boldsymbol{q})$, and, accordingly, $\widehat{V}_\sigma$ are. As can be seen from Fig.10.1 and Fig.9.4, the origin of $\widetilde{V}$ is the ladders, v($\boldsymbol{q}$)'s, which shade a ring (or, bubble) diagram; a dotted line in the ring corresponds to v($\boldsymbol{q}$), the bare Coulomb interaction. $\widehat{v}_{sc\sigma}(\boldsymbol{q})$ (or $\widehat{v}_{sc}(\boldsymbol{q})$) emerged by screening v($\boldsymbol{q}$), as we showed in this section. Then a question naturally arises. If we require self-consistency, shouldn't the electron interactions which shade a ring diagram be $\widehat{v}_{sc\sigma}(\boldsymbol{q})$ rather than v($\boldsymbol{q}$)? If that is the case, every $\widetilde{V}$ in (10.2.6) should be replaced by $\widehat{V}$, and that in (10.2.8) by $\widehat{V}_\sigma$. We put $\widehat{V}_\sigma = \widehat{V}$ for the paramagnetic state.

If we take such a view, it becomes impossible to have a negative $\widehat{v}_{sc}(\boldsymbol{q})$. Negative $\widehat{v}_{sc}(\boldsymbol{q})$ results in a negative $\widehat{V}$ and, then, $1 - \widehat{V}F(\boldsymbol{q})$, which replaces $1 - \widetilde{V}F(\boldsymbol{q})$ in (10.2.8), can not be negative. Thus, it seems to be impossible to have self-consistently an attractive electron interaction by invoking the exchange effects on screening.

10.2.2 Magnetization Dependence of Plasma Frequency

Next, in the remaining parts of this section we discuss the role of $\Delta\widetilde{\Omega}_{\mathrm{el,ring1}}$ given in (10.2.2). The plasma frequency appearing in $\Delta\widetilde{\Omega}_{\mathrm{el,ring}}$ is magnetization dependent since it is to be obtained from (10.2.4). Then $\Delta\widetilde{\Omega}_{\mathrm{el,ring1}}$ becomes magnetization dependent and would be involved in determining the magnetic properties of a metal. This is the subject of our discussion in this section.

First, in this subsection we study how the plasma frequency changes with the spin splitting of the electron energy bands. In dealing with (10.2.4) to obtain the magnetization dependent plasma frequency $\omega_{\mathrm{pl}}(\boldsymbol{q}, M)$, we expand $F_\sigma(\boldsymbol{q}, \omega)$ as follows,

$$\begin{aligned} F_\sigma(\boldsymbol{q}, \omega) &= \sum_{\boldsymbol{k}} \frac{f(\varepsilon_{\boldsymbol{k}\sigma}) - f(\varepsilon_{\boldsymbol{k}+\boldsymbol{q},\sigma})}{\varepsilon_{\boldsymbol{k}+\boldsymbol{q},\sigma} - \varepsilon_{\boldsymbol{k}\sigma} + \omega} \\ &= \frac{1}{\omega}\sum_{\boldsymbol{k}} \left[f(\varepsilon_{\boldsymbol{k}\sigma}) - f(\varepsilon_{\boldsymbol{k}+\boldsymbol{q},\sigma})\right]\left[\,1 + (\varepsilon_{\boldsymbol{k}+\boldsymbol{q},\sigma} - \varepsilon_{\boldsymbol{k}\sigma})/\omega\,\right]^{-1} \\ &= \frac{1}{\omega}\sum_{\boldsymbol{k}} \left[f(\varepsilon_{\boldsymbol{k}\sigma}) - f(\varepsilon_{\boldsymbol{k}+\boldsymbol{q},\sigma})\right]\Big[\,1 - (\varepsilon_{\boldsymbol{k}+\boldsymbol{q},\sigma} - \varepsilon_{\boldsymbol{k}\sigma})/\omega \\ &\qquad + (\varepsilon_{\boldsymbol{k}+\boldsymbol{q},\sigma} - \varepsilon_{\boldsymbol{k}\sigma})^2/\,\omega^2 + \cdots\Big] \end{aligned} \tag{10.2.19}$$

for $\left|(\varepsilon_{\boldsymbol{k}+\boldsymbol{q},\sigma} - \varepsilon_{\boldsymbol{k}\sigma})/\omega\right| < 1$. Obviously terms of odd orders in $1/\omega$ vanish, since $\mathrm{Re}\,F_\sigma(\boldsymbol{q}, -\omega) = \mathrm{Re}\,F_\sigma(\boldsymbol{q}, \omega)$ (see Appendix A). If we assume the parabolic energy dispersion for electrons, $\varepsilon_{\boldsymbol{k}} = \boldsymbol{k}^2/2m^*$, it is straightforward to obtain

$$F_\sigma(\boldsymbol{q}, \omega) = -\frac{2n_\sigma}{\omega^2}\frac{\boldsymbol{q}^2}{2m^*} - \frac{1}{\omega^4}\left[\frac{24}{5}\,n_\sigma\,\varepsilon_{\mathrm{F}\sigma}\left(\frac{\boldsymbol{q}^2}{2m^*}\right)^2 + 2n_\sigma\left(\frac{\boldsymbol{q}^2}{2m^*}\right)^3\right], \tag{10.2.19'}$$

where n_σ and $\varepsilon_{\mathrm{F}\sigma}$ are, respectively, the total number and Fermi energy of σ spin electrons. We used the zero temperature approximation for the Fermi distribution, and retained terms up to the 4-th order.

We rewrite (10.2.4) as

$$\begin{aligned} 1 + \widetilde{V}\left(\mathrm{v}(\boldsymbol{q}) - \widetilde{V}\right)&\left\{F_+(\boldsymbol{q}, \omega) + F_-(\boldsymbol{q}, \omega)\right\} \\ &- \widetilde{V}\left(2\,\mathrm{v}(\boldsymbol{q}) - \widetilde{V}\right)F_+(\boldsymbol{q}, \omega)\,F_-(\boldsymbol{q}, \omega) = 0. \end{aligned} \tag{10.2.4'}$$

Then, by putting (10.2.19') into (10.2.4'), and multiplying with ω^4, we obtain

$$\omega^4 - \left(\omega^*(\boldsymbol{q})\right)^2 \omega^2 - \left(\omega^*(\boldsymbol{q})\right)^2 c(\boldsymbol{q}, M) = 0, \qquad (10.2.20)$$

where we put

$$\left(\omega^*(\boldsymbol{q})\right)^2 = \left[1 - \widetilde{V} / \mathrm{v}(\boldsymbol{q})\right] \omega_{\mathrm{pl}}^2,$$

$$\omega_{\mathrm{pl}}^2 = \frac{4\pi n e^2}{m^* V}, \qquad (10.2.21)$$

$$c(\boldsymbol{q}, M) = \left[\frac{12}{5} \frac{n_+ \varepsilon_{\mathrm{F}+} + n_- \varepsilon_{\mathrm{F}-}}{n} + 4\, \widetilde{V} \frac{n_+ n_-}{n}\right] \frac{\boldsymbol{q}^2}{2m^*}. \qquad (10.2.22)$$

Note that ω_{pl} is the plasma frequency to be obtained from (10.1.16) by retaining only the first term on the right hand side of (10.2.19'). By solving (10.2.20) we obtain

$$\omega_{\mathrm{pl}}^2(\boldsymbol{q}, M) = \frac{\left(\omega^*(\boldsymbol{q})\right)^2}{2} + \frac{1}{2}\sqrt{\left(\omega^*(\boldsymbol{q})\right)^4 + 4\, c(\boldsymbol{q}, M)\left(\omega^*(\boldsymbol{q})\right)^2}$$

$$\cong \left(\omega^*(\boldsymbol{q})\right)^2 + c(\boldsymbol{q}, M). \qquad (10.2.23)$$

The magnetization dependence of plasma frequency is contained in $c(\boldsymbol{q}, M)$.

For the present parabolic free electron energy dispersion, $c(\boldsymbol{q}, M)$ is rewritten in terms of $\overline{M}/n = (n_+ - n_-)/n$ as

$$c(\boldsymbol{q}, M) = \varepsilon_{\mathrm{F}} \frac{\boldsymbol{q}^2}{2m^*}\left[\frac{6}{5}\left\{\left(1 + \frac{\overline{M}}{n}\right)^{5/3} + \left(1 - \frac{\overline{M}}{n}\right)^{5/3}\right\} + \frac{4}{3}\, \widetilde{V} N(0)\left(1 - \left(\frac{\overline{M}}{n}\right)^2\right)\right].$$

(10.2.22')

For $M = 0$ we have

$$\omega_{\mathrm{pl}}(\boldsymbol{q}, 0)^2 = \omega_{\mathrm{pl}}(\boldsymbol{q})^2$$

$$= \omega_{\mathrm{pl}}^2\left[1 + \left(\frac{12}{5} - \frac{4}{3}\, \widetilde{V} N(0)\right) \frac{\varepsilon_{\mathrm{F}}}{\omega_{\mathrm{pl}}^2} \frac{1}{2m^*} \boldsymbol{q}^2\right]. \qquad (10.2.24)$$

Note that in the present model we have

$$\frac{\varepsilon_F}{\omega_{pl}} = \frac{\hbar k_F^2}{4}\left(\frac{V}{\pi e^2 n m^*}\right)^{1/2} = \frac{1.063}{r_s^{1/2}}\left(\frac{m^*}{m}\right)^{-1/2} \tag{10.2.25}$$

where r_s is the inter-electronic distance in units of Bohr radius (see (8.4.44)). Since for metals generally $2 \leq r_s \leq 5$, we may assume $\omega_{pl} \cong 2\varepsilon_F$. Then, from (10.2.23) and (10.2.22') we have

$$\omega_{pl}(\boldsymbol{q}, M) = \omega_{pl}(\boldsymbol{q}, 0)\left[1 \pm O\left\{\frac{1}{\omega_{pl}}\frac{\boldsymbol{q}^2}{2m^*}\left(\frac{M}{n}\right)^2\right\}\right]. \tag{10.2.26}$$

The magnetization dependent part of the zero-point oscillation contribution, the first term on the right hand side of (10.2.2), is estimated as

$$\pm\ O\left\{n\ \varepsilon_F\left(\frac{k_c}{k_F}\right)^2\right\}\left(\frac{M}{n}\right)^2, \tag{10.2.27}$$

where k_c is the cut off wave number in summing over plasma modes. k_c is determined as the minimum wave number of plasma oscillations that can decay into an electron-hole pair. This condition may be put as

$$\varepsilon_{k_F+k_c} - \varepsilon_{k_F} = \omega_{pl}(k_c),$$

or

$$\frac{k_F k_c}{m^*} + \frac{k_c^2}{2m^*} = \omega_{pl}(k_c). \tag{10.2.28}$$

By putting $\omega_{pl}(k_c) \cong \omega_{pl}$ into the above, we obtain

$$k_c/k_F = (1 + \omega_{pl}/\varepsilon_F)^{1/2} - 1. \tag{10.2.28'}$$

If we note that $\omega_{pl}/\varepsilon_F < 3$, as can be seen from (10.2.25), we find $k_c/k_F \leq 1$. It is known (Pines [10.6]) that the following is a good approximation,

$$k_c = \omega_{pl}/v_F. \tag{10.2.29}$$

If we can put $k_c/k_F = 1$ in (10.2.27), the size of this energy is of the same order as that of the magnetization dependent part of electron energy (see(6.1.2')). Then, plasma oscillations can have significant effect on the magnetic properties of a metal. Let us discuss this in the next subsection, first, concerning the magnetic susceptibility.

10.2.3 Plasma Oscillation Effect on Magnetic Susceptibility

We use (3.4.4) to calculate the spin susceptibility of a metal. For simplicity we neglect the EPI and the spin fluctuation effects, and assume the free energy of the system is given by

$$F(M) = F_{\mathrm{el,m}} + \Delta\widetilde{\Omega}_{\mathrm{el,ring1}}, \tag{10.2.30}$$

where the mean field part of the electron free energy is given by (3.4.10), and, we assume that the effect of $\Delta\widetilde{\Omega}_{\mathrm{el,ring2}}$ is appropriately taken in by renormalizing (screening) $\widetilde{V}$ as we saw in 10.1. Then, for the spin susceptibility in the paramagnetic state we have (see (3.4.24)),

$$\frac{1}{\chi} = \frac{1}{2\mu_B^2}\left[\frac{1}{F(0)} - \widetilde{V} - J_{\mathrm{pl}}\right] \tag{10.2.31}$$

or

$$\chi = \frac{2\mu_{\mathrm{B}}^2}{1 - \left(\widetilde{V} + J_{\mathrm{pl}}\right) F(0)}, \tag{10.2.31'}$$

with the effective exchange interaction due to plasma oscillation,

$$J_{\mathrm{pl}} = -2\frac{\partial^2}{\partial \overline{M}^2} \Delta\widetilde{\Omega}_{\mathrm{el,ring1}}\Big|_{M=0} \tag{10.2.32}$$

$$\cong -\sum_q \frac{\partial^2}{\partial \overline{M}^2} \omega_{\mathrm{pl}}(q, M)\,\big|_{M=0}. \tag{10.2.33}$$

In the last expression we noted that at ordinary temperatures it suffices to consider only the zero-point oscillation contributions in (10.2.2).

From (10.2.22)–(10.2.23), we have

$$\frac{\partial \omega_{\mathrm{pl}}(\boldsymbol{q}, M)}{\partial \overline{M}} = \frac{\varepsilon_{\mathrm{F}}}{n\omega_{\mathrm{pl}}(\boldsymbol{q}, M)}\left[\left\{\left(1 + \frac{\overline{M}}{n}\right)^{2/3} - \left(1 - \frac{\overline{M}}{n}\right)^{2/3}\right\} - \frac{4}{3}\widetilde{V}N(0)\right]\frac{\boldsymbol{q}^2}{2m^*}, \tag{10.2.34}$$

$$\frac{\partial^2 \omega_{\rm pl}(\boldsymbol{q}, M)}{\partial \overline{M}^2} = \frac{\varepsilon_{\rm F}}{n^2 \omega_{\rm pl}(\boldsymbol{q}, M)} \left[\frac{2}{3} \left\{ \left(1 + \frac{\overline{M}}{n}\right)^{-1/3} + \left(1 - \frac{\overline{M}}{n}\right)^{-1/3} \right\} - \frac{4}{3} \widetilde{V} N(0) \right] \frac{\boldsymbol{q}^2}{2m^*}$$

$$- \frac{\varepsilon_{\rm F}}{n \omega_{\rm pl}^2(\boldsymbol{q}, M)} \left[\left\{ \left(1 + \frac{\overline{M}}{n}\right)^{2/3} - \left(1 - \frac{\overline{M}}{n}\right)^{2/3} \right\} - \frac{4}{3} \widetilde{V} N(0) \frac{\overline{M}}{n} \right] \frac{\boldsymbol{q}^2}{2m^*} \frac{\partial \omega_{\rm pl}(\boldsymbol{q}, M)}{\partial \overline{M}} . \tag{10.2.35}$$

Thus we obtain

$$\begin{aligned} J_{\rm pl} &= -\frac{4}{3}\left(1 - \widetilde{V}N(0)\right) \frac{\varepsilon_{\rm F}}{n^2} \sum_{\boldsymbol{q}} \frac{1}{\omega_{\rm pl}(\boldsymbol{q}, 0)} \frac{\boldsymbol{q}^2}{2m^*} \\ &\cong -\frac{4}{3}\left(1 - \widetilde{V}N(0)\right) \frac{\varepsilon_{\rm F}}{n^2 \omega_{\rm pl}} \sum_{\boldsymbol{q}} \frac{\boldsymbol{q}^2}{2m^*} \\ &= \frac{4}{3}\left(\widetilde{V}N(0) - 1\right) \frac{\varepsilon_{\rm F}}{n^2 \omega_{\rm pl}} \frac{3}{10} n\, \varepsilon_{\rm F} \left(\frac{k_{\rm c}}{k_{\rm F}}\right)^5 \\ &\cong \frac{2}{5}\left(\widetilde{V}N(0) - 1\right) \frac{\varepsilon_{\rm F}}{\omega_{\rm pl}} \left(\frac{\varepsilon_{\rm F}}{n}\right), \end{aligned} \tag{10.2.35}$$

where in the final result we assumed $k_{\rm c}/k_{\rm F} \cong 1$. With this result, in (10.2.31') we have

$$J_{\rm pl} F(0) \cong J_{\rm pl} N(0) = \frac{3}{10}\left(\widetilde{V}N(0) - 1\right) \frac{\varepsilon_{\rm F}}{\omega_{\rm pl}}, \tag{10.2.37}$$

where we noted that $N(0) = 3n/4\varepsilon_{\rm F}$. This size of $J_{\rm pl}N(0)$ agrees with our preliminary estimation made in the preceding subsection.

We note that $J_{\rm pl}$ can take either signs depending upon the magnitude of $\widetilde{V}N(0)$; the plasma oscillation effect can enhance the magnetic tendency of a metal. If we do not consider the role of the exchange interaction in the plasma oscillation, however, we would always have $J_{\rm pl} < 0$; this seems to be the currently prevailing view which originates from the works of Pines [10.6] and Brueckner and Sawada [10.7].

An interesting observation is that with the form of (10.2.37) the effect of plasma oscillation does not modify the Stoner condition for a ferromagnetic instability.

We should not of course neglect there is another totally different, important aspect of the charge fluctuation effects on the spin susceptibility coming from

$\Delta\widetilde{\Omega}_{\rm el,ring2}$ that is responsible for the screening of $\widetilde{V}$, which we discussed in the preceding subsection.

10.2.4 Plasmon Effect on the Spin Splitting of Electron Energy Bands and Magnetization

For simplicity we again restrict our consideration within (10.2.30) similarly to the preceding subsection. Following the discussion given in 6.8, the effect of the charge fluctuation on the spin splitting of the electron energy bands is obtained from the condition to determine the equilibrium magnetization, $M(T)$,

$$\frac{\partial F_{\rm el,m}(M)}{\partial M} + \frac{\partial}{\partial M}\Delta\widetilde{\Omega}_{\rm el,ring1} = 0, \qquad (10.2.38)$$

as

$$\mu_-(M(T)) - \mu_+(M(T)) = \widetilde{V}M/\mu_{\rm B} + 2\mu_{\rm B}H_{\rm pl}(M(T)), \qquad (10.2.39)$$

with

$$H_{\rm pl}(M(T)) = -\frac{\partial}{\partial M}\Delta\widetilde{\Omega}_{\rm el,ring1}(M)\bigg|_{M=M(T)}$$

$$= -\frac{1}{2\mu_{\rm B}}\sum_{\boldsymbol{q}}\frac{\partial\omega_{\rm pl}(\boldsymbol{q},M)}{\partial M}\bigg|_{M=M(T)}, \qquad (10.2.39')$$

where in the final result we retained only the zero-point oscillation contribution of (10.2.2). $H_{\rm pl}$ is understood as the effective magnetic field due to charge fluctuations.

From (10.2.22)–(10.2.23) we have

$$2\mu_{\rm B}H_{\rm pl}(M) \cong \frac{\varepsilon_{\rm F}}{n\omega_{\rm pl}}g(M)\sum_{\boldsymbol{q}}\frac{q^2}{2m}$$

$$= \frac{3}{10}\varepsilon_{\rm F}\frac{\varepsilon_{\rm F}}{\omega_{\rm pl}}\left(\frac{k_{\rm c}}{k_{\rm F}}\right)^5 h(M) \qquad (10.2.40)$$

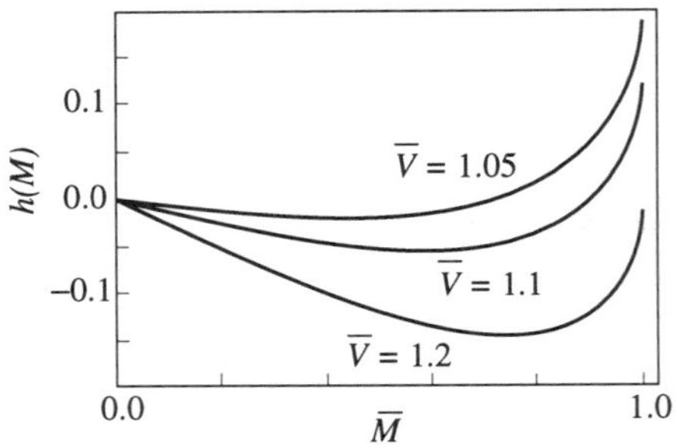

Fig.10.4 The magnetization dependence of $h(M)$, (10.2.41), which is the effect of the charge fluctuations on the spin splitting of the electron bands. We put $\overline{V} = N(0)\,\widetilde{V}$ and $M = -\mu_B \overline{M}$.

with

$$h(M) = \left\{ \left(1 + \frac{\overline{M}}{n}\right)^{2/3} - \left(1 - \frac{\overline{M}}{n}\right)^{2/3} \right\} - \frac{4}{3}\,\widetilde{V} N(0)\,\frac{\overline{M}}{n}\,. \qquad (10.2.41)$$

let us assume $k_c/k_F \cong 1$. Also from (10.2.25) we may put $\varepsilon_F/\omega_{pl} = 1/2$. Then $2\mu_B H_{pl}$, which depends on magnetization through $h(M)$ as shown in Fig.10.4, is order estimated as

$$2\mu_B H_{pl} = \frac{3}{20}\,\varepsilon_F\, h(M). \qquad (10.2.41')$$

This order of magnitude is as we anticipated. However as shown in Fig.10.4, the magnitude of $h(M)$ is of the order of 1/10. Thus we have $|\, 2\mu_B H_{pl} \,| \cong 10^{-2}\, \varepsilon_F$. This is much smaller than the effects of the spin wave and phonon which were discussed in 6.8. An interesting point, however, may be that $2\mu_B H_{pl}$ can take either sign; it can be positive.

In concluding this section, we note that the most important effect of charge fluctuations is to screen the effective exchange interaction. Another effect that derives from the magnetization dependence of plasma oscillation frequency is quantitatively not significant.

10.3 THE EFFECTS OF SPIN FLUCTUATIONS ON SPECIFIC HEATS AND SCREENING

In this section we discuss some aspects of spin fluctuation effects. Although spin fluctuation effects are believed to be well understood we will find it not to be quite so.

10.3.1 Contribution of Spin Fluctuations to Specific Heat

In 9.5.2, we discussed how the spin fluctuations would affect the energy spectrum of electrons in a metal. In this subsection let us discuss how such effects would show up in specific heats based on the result (10.1.31).

In the paramagnetic state, the spin fluctuation contribution to the thermodynamic potential can be written as,

$$\Delta\Omega_{\rm sf} = -3\frac{1}{2\pi}\sum_{q}\widetilde{V}(\boldsymbol{q})\int_{-\infty}^{\infty}d\omega\, n(\omega)\int_{0}^{1}dg\,{\rm Im}\,\overline{\chi}_{zz}^{\rm RPA}(\boldsymbol{q},\omega)_g$$

$$= -\frac{3}{\pi}\sum_{q}\int_{0}^{\infty}d\omega\left[\frac{1}{2}+n(\omega)\right]\tan^{-1}\left[\frac{\widetilde{V}(\boldsymbol{q})I(\boldsymbol{q},\omega)}{1-\widetilde{V}(\boldsymbol{q})R(\boldsymbol{q},\omega)}\right], \qquad (10.3.1)$$

where $\overline{\chi}_{zz}^{\rm RPA} = 2\overline{\chi}_{+-}^{\rm RPA} = 2\overline{\chi}_{-+}^{\rm RPA}$ is the susceptibility of the RPA given by (4.3.17) with $\mu_B^2 = 1$; the subscript g indicates to replace $\widetilde{V}(\boldsymbol{q})$ by $g\widetilde{V}(\boldsymbol{q})$ as in (10.1.23), and $R(\boldsymbol{q},\omega)$ and $I(\boldsymbol{q},\omega)$ are the real and imaginary parts of the Lindhard function (see Appendix A); as regards the change in the interval of the integration over ω, note (A.3) and (A.10). Noting that at low temperatures, the dominant contribution to the temperature dependent part of $\Delta\Omega_{\rm sf}$ comes from the region $\widetilde{V}(\boldsymbol{q})I(\boldsymbol{q},\omega) << 1$, we expand the arctangential function of the integrand up to the 3rd order terms. Then, as the most important temperature dependent contribution at low temperatures we have

$$\Delta\Omega_{\rm sf} \cong -\frac{\pi^2}{3}\lambda_{\rm sf}N(0)k_{\rm B}^2T^2 - \frac{\pi^6}{640}\left(\frac{\overline{V}}{1-\overline{V}}\right)^3 n\varepsilon_{\rm F}\left(\frac{k_{\rm B}T}{\varepsilon_{\rm F}}\right)^4\ln\left(\frac{k_{\rm B}T}{4\varepsilon_{\rm F}}\right), \qquad (10.3.2)$$

with $\lambda_{\rm sf}$ given in (9.5.31). For the procedure to derive this result, see the similar procedure to obtain the effect of the EPI on specific heat which will be given in 10.4.4.

From this result the contribution of spin fluctuations to the specific heat of a metal is obtained as

$$\Delta C_{\rm sf} \cong \frac{2\pi^2}{3}\lambda_{\rm sf}N(0)k_{\rm B}^2 T + \frac{3\pi^6}{160}\left(\frac{\overline{V}}{1-\overline{V}}\right)^3 nk_{\rm B}\left(\frac{k_{\rm B}T}{\varepsilon_{\rm F}}\right)^3\ln\left(\frac{k_{\rm B}T}{4\varepsilon_{\rm F}}\right), \qquad (10.3.3)$$

where we used the following thermodynamic relations

$$S = -\left(\frac{\partial \Omega}{\partial T}\right)_{V,\mu}; \quad C_V = T\left(\frac{\partial S}{\partial T}\right)_{V,\mu} = -T\left(\frac{\partial^2 \Omega}{\partial T^2}\right)_{V,\mu}, \tag{10.3.4}$$

where S is the entropy. The first term on the right hand side of (10.3.3) can be understood as representing the effect of the enhancement of the electron mass at the Fermi surface, as we already saw in 9.5.2. The following expression

$$\lambda_{\rm sf} = -9/2\,\overline{V}^2 \ln\left(1-\overline{V}\right), \tag{10.3.5}$$

which can be derived by putting $F(\boldsymbol{q}) = R(\boldsymbol{q}, 0) = N(0)\,[1 - 1/12\,(q/k_{\rm F})^2]$ and $\widetilde{V}(\boldsymbol{q}) = \widetilde{V}$, from (9.5.31), is often used. According to (10.3.5), $\lambda_{\rm sf}$ becomes large when $\overline{V} \to 1$. Such a large enhancement of the T-linear specific heat, however, was not observed in a number of systems where $\overline{V}$ are believed to be very close to unity, as pointed out by de Chatel et al. [10.8]. Note, however, that this mechanism is considered to be a possible origin of the large effective electron mass observed in the heavy fermion systems.

The observation of the $T^3 \ln T$ term in (10.3.3), also seems to be still controversial. For a recent experiment and references, see Dhar et al. [10.9]. Also note that the EPI also contributes a $T^3 \ln T$ term to the specific heat of a metal in a similar size but with opposite sign as we will see in 10.4.4.

10.3.2 Screening and Spin Fluctuations

If we recall the exact relation of (10.1.11), the result of (10.1.31) is equivalent to having the electron charge response function in the following form,

$$\begin{aligned}\overline{\chi}^{*}_{\rm ee}(\boldsymbol{q}, \omega) &= \left[1 - \frac{\widetilde{V}(\boldsymbol{q})}{2{\rm v}(\boldsymbol{q})}\right] \overline{\chi}^{\rm GRPA}_{\rm ee}(\boldsymbol{q}, \omega) \\ &\quad - \left\{\widetilde{V}(\boldsymbol{q})/{\rm v}(\boldsymbol{q})\right\}\left[\overline{\chi}_{+-}(\boldsymbol{q}, \omega) + \overline{\chi}_{-+}(\boldsymbol{q}, \omega) + 1/2\,\overline{\chi}_{zz}(\boldsymbol{q}, \omega)\right] \\ &\quad + \left\{\widetilde{V}(\boldsymbol{q})/{\rm v}(\boldsymbol{q})\right\}\left[F_{+}(\boldsymbol{q}, \omega) + F_{-}(\boldsymbol{q}, \omega) + F_{+-}(\boldsymbol{q}, \omega) + F_{-+}(\boldsymbol{q}, \omega)\right], \end{aligned} \tag{10.3.6}$$

where the spin susceptibilities $\overline{\chi}_{\sigma,-\sigma}$ and $\overline{\chi}_{zz}$ are those of the mean field approximation given by (9.1.27) and (9.1.24). This $\chi^{*}_{\rm ee}$ may be viewed as the electron density response function which contains the effect of spin fluctuations.

For the paramagnetic state (10.3.6) is rewritten as

$$\overline{\chi}^{*}_{ee}(\boldsymbol{q},\omega) = \left[1 - \frac{\widetilde{V}(\boldsymbol{q})}{2\mathrm{v}(\boldsymbol{q})}\right] \frac{2F(\boldsymbol{q},\omega)}{1+\left\{2\mathrm{v}(\boldsymbol{q}) - \widetilde{V}(\boldsymbol{q})\right\}F(\boldsymbol{q},\omega)}$$

$$- 3\frac{\widetilde{V}(\boldsymbol{q})}{2\mathrm{v}(\boldsymbol{q})} \frac{2F(\boldsymbol{q},\omega)}{1-\widetilde{V}(\boldsymbol{q})F(\boldsymbol{q},\omega)} + 4\frac{\widetilde{V}(\boldsymbol{q})}{\mathrm{v}(\boldsymbol{q})}F(\boldsymbol{q},\omega)$$

$$\equiv \frac{2F(\boldsymbol{q},\omega)}{\varepsilon^{*}_{e}(\boldsymbol{q},\omega)}, \tag{10.3.7}$$

where $\varepsilon^{*}_{e}(\boldsymbol{q},\omega)$ is understood as the screening constant including the effect of spin fluctuations which screens the interaction *between an external charge and an electron* (see 4.3.7). From (10.3.6) we have

$$\frac{1}{\varepsilon^{*}_{e}(\boldsymbol{q},\omega)} = \left\{1 - \frac{\widetilde{V}(\boldsymbol{q})}{2\mathrm{v}(\boldsymbol{q})}\right\} \frac{1}{1+\left\{2\mathrm{v}(\boldsymbol{q}) - \widetilde{V}(\boldsymbol{q})\right\}F(\boldsymbol{q},\omega)}$$

$$- \frac{3}{2}\frac{\widetilde{V}(\boldsymbol{q})}{\mathrm{v}(\boldsymbol{q})} \frac{1}{1-\widetilde{V}(\boldsymbol{q})F(\boldsymbol{q},\omega)} + 2\frac{\widetilde{V}(\boldsymbol{q})}{\mathrm{v}(\boldsymbol{q})}. \tag{10.3.8}$$

Apart from the factor $\{1 - \widetilde{V}(\boldsymbol{q})/2\mathrm{v}(\boldsymbol{q})\}$, the first term on the right hand side of (10.3.8) can be identified with the screening constant of (4.3.29). The second and third terms are the consequences of the spin fluctuation effect. Note that the sign of the contribution of the spin fluctuation effect can be negative. Then, if $\widetilde{V}(\boldsymbol{q}) \neq 0$, there is the possibility of having

$$\varepsilon^{*}_{e}(\boldsymbol{q},\omega) < 0 \tag{10.3.9}$$

in the paramagnetic state; the sign of the Coulomb potential of an external charge can be changed by the spin fluctuation effect. In the next subsection, however, we show that such a result of the spin fluctuation effects seems to require a fundamental reexamination.

10.3.3 Screening Charge Sum Rule and the Spin Fluctuation Effect

In the preceding subsection we saw that having the contribution of spin fluctuations to the electron thermodynamic potential in the form of (10.1.31) is equivalent to having the electron density response function in the form of (10.3.7). Note that we have a very simple test for the electron density response function, the screening charge sum rule discussed in 4.2.3.

If we take the position of the point charge Ze at the origin of the coordinates, from (4.2.38) the screening charge sum rule is written down as

$$\lim_{q \to 0} \bar{\chi}_{ee}(\boldsymbol{q}, 0) \frac{4\pi Z e^2}{V q^2} = Z . \qquad (10.3.10)$$

With χ^*_{ee} of (10.3.7), however, we have

$$\lim_{q \to 0} \bar{\chi}^*_{ee}(\boldsymbol{q}, 0) \frac{4\pi Z e^2}{V q^2} = Z - 3Z \frac{\widetilde{V} F(0)}{1 - \widetilde{V} F(0)} + 4Z \widetilde{V} F(0). \qquad (10.3.11)$$

Owing to the last two terms on the right hand side of (10.3.11), especially, the 2nd one, which reflects the spin fluctuation effects, the sum rule is broken in a drastic way.

How should we understand this situation? There seem to be fundamental problems concerning the current treatment of the spin fluctuation effects in a metal. The contributions of spin fluctuations to the thermodynamic potential and, accordingly, to the various physical properties of a metal, such as specific heats and magnetic susceptibility, might be significantly different from what we will anticipate from a result such as (10.1.31). The difficulty in observing the expected effect on specific heats of spin fluctuations as mentioned in 10.3.1 might be related to this problem.

Note that although predominantly the Hubbard model is used in discussing the spin fluctuation effects, the result is essentially the same as in the present electron gas model. The Hubbard model result of the spin fluctuation contribution to the thermodynamic potential is reproduced simply by replacing $\widetilde{V}(\boldsymbol{q})$ by I/N in the spin fluctuation parts of (10.1.31), I and N being respectively the Hubbard intra-atomic electron repulsion and the total number of atoms (see 3.5). Concerning the charge fluctuation effect, however, the Hubbard model gives quite different and inadequate result, as is well known. Recently, with the Hubbard model, there have been made various attempts to improve the estimation of the spin fluctuation effects beyond the result of (10.1.31) [3.19, 3.21, 10.1, 10.10]. Incorporation of those improvements, which amounts effectively to modify the spin susceptibilities to be used in (10.1.31), however, does not help recover the screening sum rule. For a recent review and references, see [10.11].

10.4 THE THERMODYNAMIC POTENTIAL OF A METALLIC ELECTRON–PHONON SYSTEM

The central theme of the present book is to explore the role of the EPI in the magnetic behavior of metals. The basis of our study in Chapters 6 and 7 was to note that the free energy of a metal is given as the sum of electron and phonon parts, as in (6.1.1), and that the phonon part also depends on the spin splitting of the electron energy bands in a manner as important as the electron part. In this final section of this chapter we explicitly derive such a free energy of an interacting electron-phonon system starting from the microscopic Hamiltonian [10.12]. As for earlier efforts to derive microscopically the free energy of a metallic electron-phonon system, see, for instance, Refs.[10.13, 10.14, 7.2].

The Hamiltonian of our system consists of the electron part of (10.1.2) and (10.1.3) and the phonon and the EPI parts which are given, respectively, in (5.2.25) and (5.2.40). If there is neither the Coulomb interaction between electrons, H_C, nor the EPI, $H_{el\text{-}ph}$, the thermodynamic potential of a system consisting of electrons and phonons is given as

$$\Omega^o = \Omega_{el}^o + \Omega_{ph}^o , \qquad (10.4.1)$$

$$\Omega_{el}^o = -\frac{1}{\beta} \ln \mathrm{tr} \left(\exp\left[-\beta\left(H_o - \mu_+ n_+ - \mu_- n_-\right)\right]\right)$$

$$= -\frac{1}{\beta} \sum_{l,\sigma} \ln \left[1 + \exp\left(-\beta \xi_{l\sigma}\right)\right], \qquad (10.4.2)$$

$$\Omega_{ph}^o = -\frac{1}{\beta} \ln \mathrm{tr}\left(e^{-\beta H_{ph}}\right)$$

$$= \sum_q \left[\frac{1}{2}\Omega_q + k_B T \ln\{1 - \exp(-\beta\Omega_q)\}\right]. \qquad (10.4.3)$$

In Chapter 8 we denoted Ω_{el}^o as Ω_o. In (10.4.2) we made it explicit that we are to calculate the thermodynamic potential, or the free energy as the function of magnetization, as is required in the Landau procedure such as (3.4.4) or (6.2.1); we vary the magnetization at our will and, then, determine μ_σ so as to realize such magnetization or n_σ. The effect of the Coulomb interaction between electrons and the EPI modifies (10.4.1) into

$$\Omega = \Omega^o + \Delta\Omega . \qquad (10.4.4)$$

Our goal is to obtain $\Delta\Omega$ by the method of the linked cluster expansion by extending (8.4.16) by replacing H_C in it with

$$H_{\text{int}} = H_C + H_{\text{el-ph}}. \tag{10.4.5}$$

10.4.1 EPI in the Ring and the Ladder Diagrams

In calculating $\Delta\Omega$ by the method of the linked cluster expansion it is essential to note that the role of EPI is to mediate an effective interaction between electrons as shown in Fig.10.5(b). According to the rule of the linked cluster expansion we attach a negative sign to each interaction, either the Coulomb interaction of Fig.10.5(a), or the EPI; in Fig.10.5(b) we have positive sign since it is second order in EPI. Let us set

$$-V_{\text{ph}}(\boldsymbol{q}, i\nu_m) = |\alpha(\boldsymbol{q})|^2 D^{\circ}(\boldsymbol{q}, i\nu_m). \tag{10.4.6}$$

Note, then, that

$$V_{\text{ph}}(\boldsymbol{q}, \omega) = |\alpha(\boldsymbol{q})|^2 \frac{2\Omega_{\boldsymbol{q}}}{\omega^2 - \Omega_{\boldsymbol{q}}^2}, \tag{10.4.6'}$$

which becomes attractive for $|\omega| < \Omega_{\boldsymbol{q}}$, is the familiar Fröhlich interaction that is responsible for superconductivity [10.15].

Let us sum the electron interactions of Fig.10.5(a) and (b) in the form

$$W(\boldsymbol{q}, i\nu_m) = v(\boldsymbol{q}) + V_{\text{ph}}(\boldsymbol{q}, i\nu_m) = v(\boldsymbol{q}) - |\alpha(\boldsymbol{q})|^2 D^{\circ}(\boldsymbol{q}, i\nu_m), \tag{10.4.7}$$

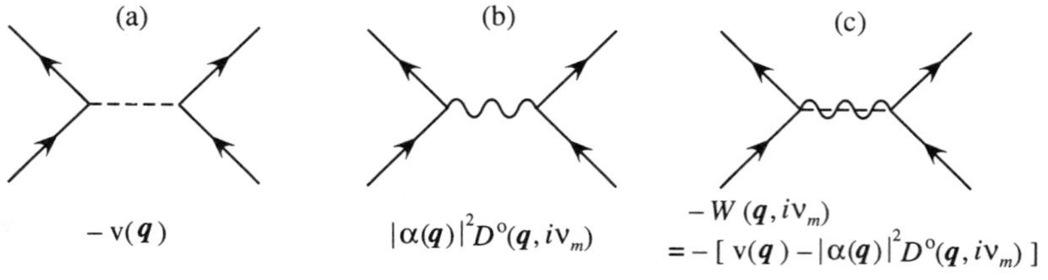

Fig.10.5 The Coulomb interaction, (a), and the interaction mediated by the EPI, (b), between electrons. The sum of the above two interactions can be represented as in (c).

which we represent by a double line given in Fig.10.5(c). Then, the application of the linked cluster expansion to the present electron-phonon system becomes straightforward. We have only to replace every $v(\boldsymbol{q})$ by $W(\boldsymbol{q}, i\nu_m)$ in the diagrams appeared in the linked cluster expansion for the electron gas [7.2].

In the case of the electron gas, $v(\boldsymbol{q})$ appeared in two qualitatively different manners; one is as the link of two rings, and the other is as the ladders within a shaded ring such as in Fig.9.4, or in the two kinds of ladder diagrams of Fig.10.2. As discussed in 9.4.2, however, at present we can not explicitly obtain the possible effect of phonon ladders, particularly, for non-vanishing $\boldsymbol{q}$ and ν_m. We know only that the effect of phonon ladders is smaller than that of Coulomb (exchange) ladders by a factor $O(\omega_D/\varepsilon_F)$. Thus, for the moment we neglect the contribution of V_{ph} to all these ladder diagrams. Thus, as to the effect of EPI on $\Delta\Omega$, we take it into account only in the ring diagrams of Fig.10.6.

Then, from our discussions given in 10.1, for the contribution of the EPI and the Coulomb repulsion between electrons to the thermodynamic potential we have

$$\Delta\Omega = \Delta\Omega_{\text{ring}} + \Delta\Omega_{\text{lad}}, \tag{10.4.8}$$

$$\Delta\Omega_{\text{ring}} = \frac{1}{2\beta}\sum_{\boldsymbol{q},m} \ln[1 + W(\boldsymbol{q}, i\nu_m)\widetilde{P}(\boldsymbol{q}, i\nu_m)] \tag{10.4.9}$$

$$= \frac{1}{2\beta}\sum_{\boldsymbol{q},m}\int_0^1 dg \frac{W(\boldsymbol{q}, i\nu_m)\widetilde{P}(\boldsymbol{q}, i\nu_m)}{1 + g\,W(\boldsymbol{q}, i\nu_m)\widetilde{P}(\boldsymbol{q}, i\nu_m)}, \tag{10.4.9'}$$

$$\Delta\Omega_{\text{lad}} = \Delta\Omega_{\text{el,lad}} = \Delta\Omega_{\text{el,lad1}} + \Delta\Omega_{\text{el,lad2}}. \tag{10.4.10}$$

The result of (10.4.9) is obvious if we note that it reduces to (10.1.24) if we replace $W(\boldsymbol{q}, i\nu_m)$ by $v(\boldsymbol{q})$ in it. Our task from now on is, then, to analyze the content of (10.4.9) or (10.4.9').

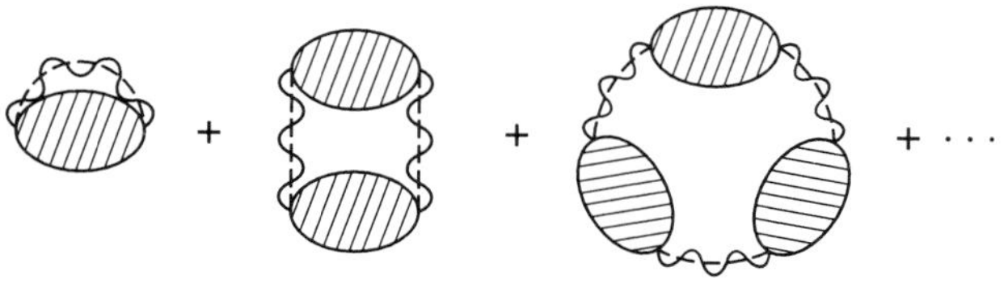

Fig.10.6 Ring diagrams with the Coulomb interaction and the EPI.

10.4.2 The Effect of EPI on the Phonon Thermodynamic Potential

If we explicitly put (10.4.6) into $W(\boldsymbol{q}, i\nu_m)$, (10.4.9) is rewritten as

$$\Delta\Omega_{\text{ring}} = \frac{1}{2\beta}\sum_{\boldsymbol{q}}\frac{\beta}{2\pi i}\int_{C'} dz\, n(z)\ln[1 + W(\boldsymbol{q}, z)\widetilde{P}(\boldsymbol{q}, z)] \tag{10.4.11}$$

$$= \frac{1}{2\pi}\sum_{\boldsymbol{q}}\int_{-\infty}^{\infty} d\omega\, n(\omega)\,\mathrm{Im}\ln\left[1 + W(\boldsymbol{q}, \omega + i0^{+})\widetilde{P}(\boldsymbol{q}, \omega + i0^{+})\right], \tag{10.4.11'}$$

where the contour C′ is given in Fig.8.6(b) and we used the relation of (8.4.38') with neglecting B(0) in it.

Integrating (10.4.11') by parts similarly to (10.1.15), we also have

$$\Delta\Omega_{\text{ring}} = -\frac{1}{2\pi\beta}\sum_{\boldsymbol{q}}\left[\int_{0}^{\infty} d\omega \ln\left(1 - e^{-\beta\omega}\right) - \int_{-\infty}^{0} d\omega \ln\left(e^{-\beta\omega} - 1\right)\right]$$

$$\times\,\mathrm{Im}\left[\frac{\frac{d}{d\omega}\left\{1 + W(\boldsymbol{q}, \omega)\widetilde{P}(\boldsymbol{q}, \omega)\right\}}{1 + W(\boldsymbol{q}, \omega)\widetilde{P}(\boldsymbol{q}, \omega)}\right]. \tag{10.4.11''}$$

Here in $W(\boldsymbol{q}, \omega)$ and $\widetilde{P}(\boldsymbol{q}, \omega)$, ω is understood to be $\omega + i0^{+}$. If we recall (9.2.5)–(9.2.6), in (10.4.11") we have

$$1 + W(\boldsymbol{q}, \omega)\widetilde{P}(\boldsymbol{q}, \omega) = \frac{\omega^2 - \omega_{\boldsymbol{q}}^2}{\omega^2 - \Omega_{\boldsymbol{q}}^2}\left(1 + \mathrm{v}(\boldsymbol{q})\widetilde{P}(\boldsymbol{q}, \omega)\right). \tag{10.4.12}$$

It is important to note that here we actually used the adiabatic approximation, (5.3.12), in obtaining the screened phonon frequency.

With (10.4.12) we have

$$\frac{\frac{d}{d\omega}\left\{1 + W(\boldsymbol{q}, \omega)\widetilde{P}(\boldsymbol{q}, \omega)\right\}}{1 + W(\boldsymbol{q}, \omega)\widetilde{P}(\boldsymbol{q}, \omega)} = \frac{\frac{d}{d\omega}\left[\frac{\omega^2 - \omega_{\boldsymbol{q}}^2}{\omega^2 - \Omega_{\boldsymbol{q}}^2}\left(1 + \mathrm{v}(\boldsymbol{q})\widetilde{P}(\boldsymbol{q}, \omega)\right)\right]}{\left[\frac{\omega^2 - \omega_{\boldsymbol{q}}^2}{\omega^2 - \Omega_{\boldsymbol{q}}^2}\left(1 + \mathrm{v}(\boldsymbol{q})\widetilde{P}(\boldsymbol{q}, \omega)\right)\right]}$$

$$= \left[\frac{2\omega}{\omega^2 - \omega_q^2} - \frac{2\omega}{\omega^2 - \Omega_q^2} \right] + \frac{\mathrm{v}(\boldsymbol{q}) \frac{d}{d\omega} \widetilde{P}(\boldsymbol{q}, \omega)}{1 + \mathrm{v}(\boldsymbol{q}) \widetilde{P}(\boldsymbol{q}, \omega)}$$

$$\equiv I(\boldsymbol{q}, \omega)_{\mathrm{a}} + I(\boldsymbol{q}, \omega)_{\mathrm{b}}, \qquad (10.4.13)$$

where ω_q is the screened phonon frequency that is given in (9.2.6).

Of the two different kinds of contributions in (10.4.13), first we consider only the first one. Then, if note the relation that

$$\mathrm{Im} \frac{1}{\omega^2 - \omega_q^2} = \frac{1}{2\omega_q} \mathrm{Im} \left[\frac{1}{\omega - \omega_q + i0^+} - \frac{1}{\omega + \omega_q + i0^+} \right]$$

$$= -\frac{\pi}{2\omega_q} \left[\delta(\omega - \omega_q) - \delta(\omega + \omega_q) \right], \qquad (10.4.14)$$

It is straightforward to obtain

$$\Delta\Omega_{\mathrm{ring,a}} = -\frac{1}{2\pi\beta} \sum_q \left[\int_0^{\infty} d\omega \ln\left(1 - e^{-\beta\omega}\right) - \int_{-\infty}^{0} d\omega \ln\left(e^{-\beta\omega} - 1\right) \right] \mathrm{Im}\, I(\boldsymbol{q}, \omega)_{\mathrm{a}}$$

$$= \sum_q \left[\frac{1}{2} \omega_q + \frac{1}{\beta} \ln\left(1 - e^{-\beta\omega_q}\right) \right] - \sum_q \left[\frac{1}{2} \Omega_q + \frac{1}{\beta} \ln\left(1 - e^{-\beta\Omega_q}\right) \right]. \qquad (10.4.15)$$

Now let us go back to (10.4.1)–(10.4.3). Then we find that

$$\Omega_{\mathrm{ph}} \equiv \overset{\mathrm{o}}{\Omega}_{\mathrm{ph}} + \Delta\Omega_{\mathrm{ring,a}}$$

$$= \sum_q \left[\frac{1}{2} \omega_q + \frac{1}{\beta} \ln\left(1 - e^{-\beta\omega_q}\right) \right]. \qquad (10.4.16)$$

This result justifies what we have been doing in this book. The effect of EPI on the phonon free energy is to replace Ω_q by ω_q in $\overset{\mathrm{o}}{\Omega}_{\mathrm{ph}}$.

Next we turn to the contribution of $I(\boldsymbol{q}, \omega)_{\mathrm{b}}$ given in (10.4.13). We immediately notice that its contribution in (10.4.11"), $\Delta\Omega_{\mathrm{ring,b}}$ gives us $\Delta\widetilde{\Omega}_{\mathrm{el,ring}}$ of (10.2.1):

$$\Delta\Omega_{\mathrm{ring,b}} = \Delta\widetilde{\Omega}_{\mathrm{el,ring}}$$

$$= \Delta\widetilde{\Omega}_{\text{el,ring1}} + \Delta\widetilde{\Omega}_{\text{el,ring2}} \tag{10.4.17}$$

Thus summing up the result of this subsection, for the total thermodynamic potential of an interacting electron-phonon system we have

$$\Omega = \Omega_{\text{el}} + \Omega_{\text{ph}}, \tag{10.4.18}$$

$$\Omega_{\text{el}} = \Omega^{\text{o}}_{\text{el}} + \Delta\Omega_{\text{el,lad}} + \Delta\widetilde{\Omega}_{\text{el,ring}}, \tag{10.4.19}$$

$$\Omega_{\text{ph}} = \Omega^{\text{o}}_{\text{ph}} + \Delta\Omega_{\text{ring,a}} \, . \tag{10.5.20}$$

Note that this final result is exactly what we have been using in this book.

Finally note that the free energy F of the system is related to the thermodynamic potential as

$$F = F_{\text{el}} + F_{\text{ph}}, \tag{10.4.21}$$

$$F_{\text{el}} = \Omega_{\text{el}} + \sum_{\sigma} n_\sigma \mu_\sigma, \tag{10.4.22}$$

$$F_{\text{ph}} = \Omega_{\text{ph}}. \tag{10.4.23}$$

An important point to note here is that, to be exact, the phonon frequency in the expression for the phonon free energy (10.4.16) should be the fully dynamical one which is to be determined from (9.2.6), or (5.1.1)–(5.1.3) without using the adiabatic approximation,

$$P(\boldsymbol{q}, \omega) \cong P(\boldsymbol{q}, 0). \tag{10.4.24}$$

In Chapters 6 and 7, however, we used phonon frequencies obtained with the adiabatic approximation in treating the phonon part of the free energy of a metal. Then, we often encountered with the divergence of phonon frequencies (sound velocity) and, therefore, the phonon free energy as we change the magnetization of the metal in carrying out the Landau procedure to obtain the equilibrium magnetization [6.1]. As we noted in 6.7, such a divergence of phonon free energy is an artifact of the adiabatic approximation; the phonon free energy will not diverge if we abandon the adiabatic approximation in computing the phonon frequency.

10.4.3 The Effects of EPI and Plasmons on Electrons

According to the final result of preceding subsection, the effect of EPI appears only on phonon, by changing $\Omega_{\rm ph}^{\rm o}$(or $F_{\rm ph}^{\rm o}$) to $\Omega_{\rm ph}$(or $F_{\rm ph}$). In $\Omega_{\rm el}$ or $F_{\rm el}$, given respectively in (10.4.19) and (10.4.22), we don't see any effect of EPI. Such result is contrary to our anticipation; we saw in 9.3 that electrons near the Fermi surface are significantly affected by EPI. Where have such EPI effect on electrons gone? Another important question is concerned with the contribution of the electron-plasmon interaction. Is it contained in (10.4.18)–(10.4.20)?

The answer to these questions is as follows: In the original $\Delta\Omega_{\rm ring}$, both of such EPI and plasmon effects on electrons are contained. However we lost EPI effect by using (10.4.12) which amounts to using the adiabatic approximation, (10.4.24), in obtaining $\omega_{\boldsymbol q}$. The plasmon effect is contained in $\Delta\Omega_{\rm ring}$ or $\Delta\Omega_{\rm el,ring}$. Let us show how we obtain such an answer.

In exploring the EPI and plasmon effects on electrons contained in $\Omega_{\rm ring}$, since we are going to calculate their contributions to specific heat of electrons in the next subsection, let us calculate the corresponding entropy (see (10.3.4)),

$$\Delta S_{\rm ring} \;=\; -\frac{\partial\Delta\Omega_{\rm ring}}{\partial T}\,. \tag{10.4.25}$$

Then we will find the entropy to consist of two parts, the fermion and the boson parts, with the fermion part in the following form

$$\Delta S_{\rm F} \;=\; -\sum_{\boldsymbol k,\sigma}\frac{\partial f_{\boldsymbol k,\sigma}}{\partial T}\;{\rm Re}\,\Sigma_\sigma^{\rm ph}(\boldsymbol k,\,\varepsilon_{\boldsymbol k,\sigma}), \tag{10.4.26}$$

with exactly the same electron self-energy due to EPI of (9.3.3). Let us see how such a result comes out.

By carrying out the differentiation of (10.4.25) on (10.4.11) we obtain

$$\Delta S_{\rm ring} = -\frac{1}{4\pi i}\sum_{\boldsymbol q}\int_{C'} dz\, n(z)\,\frac{\partial}{\partial T}\ln\left[1 + W(\boldsymbol q, z)\,\widetilde P(\boldsymbol q, z)\right]$$

$$-\frac{1}{4\pi i}\sum_{\boldsymbol q}\int_{C'} dz\,\frac{\partial n(z)}{\partial T}\ln\left[1 + W(\boldsymbol q, z)\,\widetilde P(\boldsymbol q, z)\right] \;\equiv\; \Delta S_1 + \Delta S_2. \tag{10.4.27}$$

As contrasted to (10.4.26) of fermion contribution, the second term, ΔS_2, is of the form of a boson contribution. Hereafter we concentrate on ΔS_1:

$$\Delta S_1 = -\frac{1}{4\pi i}\sum_{q}\int_{C'} dz\, n(z)\,\frac{W(\boldsymbol{q}, z)\,\partial\widetilde{P}(\boldsymbol{q}, z)/\partial T}{1 + W(\boldsymbol{q}, z)\,\widetilde{P}(\boldsymbol{q}, z)}, \qquad (10.4.28)$$

where

$$\frac{\partial\widetilde{P}(\boldsymbol{q}, z)}{\partial T} = \sum_{\sigma}\frac{1}{\left(1 - \widetilde{V}(\boldsymbol{q})\,F_\sigma(\boldsymbol{q}, z)\right)^2}\,\frac{\partial}{\partial T}F_\sigma(\boldsymbol{q}, z), \qquad (10.4.29)$$

$$\frac{\partial F_\sigma(\boldsymbol{q}, z)}{\partial T} = \sum_{\boldsymbol{k}}\frac{1}{z - \xi_{\boldsymbol{k},\sigma} + \xi_{\boldsymbol{k}-\boldsymbol{q},\sigma}}\left(\frac{\partial f_{\boldsymbol{k},\sigma}}{\partial T} - \frac{\partial f_{\boldsymbol{k}-\boldsymbol{q},\sigma}}{\partial T}\right). \qquad (10.4.30)$$

If we set

$$\widehat{W}_\sigma(\boldsymbol{q}, z) \equiv \frac{1}{\left(1 - \widetilde{V}(\boldsymbol{q})\,F_\sigma(\boldsymbol{q}, z)\right)^2}\,\frac{W(\boldsymbol{q}, z)}{1 + W(\boldsymbol{q}, z)\,\widetilde{P}(\boldsymbol{q}, z)}, \qquad (10.4.31)$$

(10.4.28) is reduced to

$$\begin{aligned}
\Delta S_1 &= -\frac{1}{2\pi}\sum_{\boldsymbol{q},\sigma}\int_{-\infty}^{\infty} d\omega\, n(\omega)\,\mathrm{Im}\left[\widehat{W}_\sigma(\boldsymbol{q}, \omega + i0^+)\,\frac{\partial}{\partial T}F_\sigma(\boldsymbol{q}, \omega + i0^+)\right] \\
&= -\frac{1}{2\pi}\sum_{\boldsymbol{k},\boldsymbol{q},\sigma}\int_{-\infty}^{\infty} d\omega\, n(\omega)\,\frac{\mathrm{Im}\,\widehat{W}_\sigma(\boldsymbol{q}, \omega + i0^+)}{\omega - \xi_{\boldsymbol{k},\sigma} + \xi_{\boldsymbol{k}-\boldsymbol{q},\sigma}}\left(\frac{\partial f_{\boldsymbol{k},\sigma}}{\partial T} - \frac{\partial f_{\boldsymbol{k}-\boldsymbol{q},\sigma}}{\partial T}\right) \\
&\quad + \frac{1}{2}\sum_{\boldsymbol{k},\boldsymbol{q},\sigma} n(\xi_{\boldsymbol{k},\sigma} - \xi_{\boldsymbol{k}-\boldsymbol{q},\sigma})\,\mathrm{Re}\,\widehat{W}_\sigma(\boldsymbol{q}, \xi_{\boldsymbol{k},\sigma} - \xi_{\boldsymbol{k}-\boldsymbol{q},\sigma} + i0^+)\left(\frac{\partial f_{\boldsymbol{k},\sigma}}{\partial T} - \frac{\partial f_{\boldsymbol{k}-\boldsymbol{q},\sigma}}{\partial T}\right).
\end{aligned}$$

(10.4.32)

Here, if we note the relation

$$-f_{\boldsymbol{k},\sigma}(1 - f_{\boldsymbol{k}-\boldsymbol{q},\sigma}) = (f_{\boldsymbol{k},\sigma} - f_{\boldsymbol{k}-\boldsymbol{q},\sigma})\,n(\xi_{\boldsymbol{k},\sigma} - \xi_{\boldsymbol{k}-\boldsymbol{q},\sigma}),$$

or

$$n(\xi_{\boldsymbol{k},\sigma} - \xi_{\boldsymbol{k}-\boldsymbol{q},\sigma})\left(\frac{\partial f_{\boldsymbol{k},\sigma}}{\partial T} - \frac{\partial f_{\boldsymbol{k}-\boldsymbol{q},\sigma}}{\partial T}\right)$$

$$= -\frac{\partial}{\partial T}\left[f_{k,\sigma}\left(1 - f_{k-q,\sigma}\right)\right] - \left(f_{k,\sigma} - f_{k-q,\sigma}\right) \frac{\partial}{\partial T} n\left(\xi_{k,\sigma} - \xi_{k-q,\sigma}\right), \quad (10.4.33)$$

and the Kramers-Kronig relation

$$\mathrm{Re}\, \widehat{W}_{\sigma}(\boldsymbol{q}, \xi_{k,\sigma} - \xi_{k-q,\sigma} + i0^{+}) = \frac{1}{\pi} \int_{-\infty}^{\infty} d\omega \, \frac{\mathrm{Im}\, \widehat{W}_{\sigma}(\boldsymbol{q}, \omega + i0^{+})}{\omega - \xi_{k,\sigma} + \xi_{k-q,\sigma}}, \quad (10.4.34)$$

we can rewrite (10.4.32) in the form

$$\Delta S_1 = -\sum_{k,\sigma} \frac{\partial f_{k,\sigma}}{\partial T} \Sigma_{\sigma}(\boldsymbol{k}, \xi_{k,\sigma}), \quad (10.4.35)$$

with

$$\Sigma_{\sigma}(\boldsymbol{k}, \xi_{k\sigma}) = \sum_{q} \frac{1}{2\pi} \int_{-\infty}^{\infty} d\omega \, \mathrm{Im}\, \widehat{W}_{\sigma}(\boldsymbol{q}, \omega + i0^{+})$$

$$\times \left[\frac{n(\omega) + 1 - f_{k-q,\sigma}}{\omega - \xi_{k,\sigma} + \xi_{k-q,\sigma}} - \frac{n(\omega) + f_{k-q,\sigma}}{\omega - \xi_{k-q,\sigma} + \xi_{k,\sigma}} \right]. \quad (10.4.36)$$

Comparing with (9.3.3) we anticipate this Σ_{σ} to represent an electron self-energy. By analyzing this result we demonstrate that it contains both of the plasmon and the phonon effects on electron.

We consider $\mathrm{Im}\, \widehat{W}_{\sigma}$ separately in (10.4.36) for two regions. **A**: $|\omega| < \omega_{\mathrm{D}}$, and **B**: $|\omega| \gg \omega_{\mathrm{D}}$. In the region **A** we obtain the phonon effect, and in the region **B** we obtain the plasmon effect.

A. $|\omega| < \omega_{\mathrm{D}}$: *The EPI effect*

We rewrite $\widehat{W}_{\sigma}$ of (10.4.31) as

$$\widehat{W}_{\sigma}(\boldsymbol{q}, z) = D(\boldsymbol{q}, z)\, \varphi_{\sigma}(\boldsymbol{q}, z), \quad (10.4.37)$$

with

$$\varphi_\sigma(\boldsymbol{q}, z) = \frac{1}{\left(1 - \widetilde{V}(\boldsymbol{q})\, F_\sigma(\boldsymbol{q}, z)\right)^2} \left[\frac{\mathrm{v}(\boldsymbol{q})}{1 + \mathrm{v}(\boldsymbol{q})\, \widetilde{P}(\boldsymbol{q}, z)} D^{\mathrm{o}}(\boldsymbol{q}, z)^{-1} - \frac{|\alpha(\boldsymbol{q})|^2}{1 + \mathrm{v}(\boldsymbol{q})\, \widetilde{P}(\boldsymbol{q}, z)} \right], \tag{10.4.38}$$

where $D^{\mathrm{o}}(\boldsymbol{q}, z)$ and $D(\boldsymbol{q}, z)$ are the Green's functions of the bare and the screened phonons given, respectively, by (9.2.1) and (9.2.5). In dealing with

$$\begin{aligned} \mathrm{Im}\, \widehat{W}_\sigma(\boldsymbol{q}, \omega + i0^+) &= \mathrm{Im}\, D(\boldsymbol{q}, \omega + i0^+)\, \mathrm{Re}\, \varphi_\sigma(\boldsymbol{q}, \omega + i0^+) \\ &\quad + \mathrm{Re}\, D(\boldsymbol{q}, \omega + i0^+)\, \mathrm{Im}\, \varphi_\sigma(\boldsymbol{q}, \omega + i0^+), \end{aligned} \tag{10.4.39}$$

as it appears in (10.4.36), in this region we use the approximation of (10.4.24). Then the second term on the right hand side vanishes ($\mathrm{Im}\varphi_\sigma = 0$). Note that if we identify Σ_σ of (10.4.36) with an electron self-energy, the most essential dynamics of electrons is contained in the last factor, and that the factor came from outside of the approximation of (10.4.24), as can be seen from (10.4.28)–(10.4.30). Thus, if we set

$$\frac{1}{\pi}\, \mathrm{Im}\, D(\boldsymbol{q}, \omega + i0^+) \cong \frac{\Omega_q}{\omega_q} [\delta(\omega - \omega_q) - \delta(\omega + \omega_q)] \tag{10.4.40}$$

in (10.4.39), from (10.4.36) we obtain

$$\begin{aligned} \Sigma_\sigma(\boldsymbol{k}, \xi_{k,\sigma})_{\mathrm{A}} &= -\sum_{q} \frac{\Omega_q}{\omega_q} |\overline{\alpha}_\sigma(\boldsymbol{q})|^2 \\ &\quad \times \left[\frac{n(\omega_q) + 1 - f_{k-q,\sigma}}{\omega_q - \xi_{k,\sigma} + \xi_{k-q,\sigma}} - \frac{n(\omega_q) + f_{k-q,\sigma}}{\omega_q - \xi_{k-q,\sigma} + \xi_{k,\sigma}} \right] \end{aligned} \tag{10.4.41}$$

with the screened electron-phonon interaction $\overline{\alpha}_\sigma(\boldsymbol{q})$ which is the same as that given in (9.3.6); we noted $n(-\omega_q) = -\, n(\omega_q) - 1$. This is identical to Re $\Sigma_\sigma^{\mathrm{ph}}$ of (9.3.8). $\Delta\Omega_{\mathrm{ring}}$ contains the entropy of electrons in the form of (10.4.26) with the electron self-energy in the precisely desired form of (9.3.7) or (10.4.41).

B. $|\,\omega| \gg \omega_{\mathrm{D}}$: *The plasmon effect*

In this region we have

$$W(\boldsymbol{q}, \omega) \cong \mathrm{v}(\boldsymbol{q}) \tag{10.4.42a}$$

and, therefore,

$$\widehat{W}(\boldsymbol{q},\,\omega) \cong \frac{1}{(\,1-\widetilde{V}F_\sigma(\boldsymbol{q},\,\omega)\,)^2}\,\frac{\mathrm{v}(\boldsymbol{q})}{1+\mathrm{v}(\boldsymbol{q})\,\widetilde{P}(\boldsymbol{q},\,\omega)} \qquad (10.4.42\mathrm{b})$$

Then, if we use (10.1.16) for the denominator of the last factor of $\widehat{W}$, from (10.4.36), we have

$$\Sigma_\sigma(\boldsymbol{k},\xi_{\boldsymbol{k},\sigma})_{\mathrm{B}} = \frac{1}{2}\sum_{\boldsymbol{q},\,\sigma}\frac{\mathrm{v}(\boldsymbol{q})}{\left(1-\widetilde{V}F_\sigma(\boldsymbol{q},\omega)\right)^2}\,\frac{\omega_{\mathrm{pl}}^2}{\omega_{\mathrm{pl}}(\boldsymbol{q})}$$

$$\times\left[\frac{n(\omega_{\mathrm{pl}}(\boldsymbol{q}))+1-f_{\boldsymbol{k}-\boldsymbol{q},\sigma}}{\xi_{\boldsymbol{k},\sigma}-\xi_{\boldsymbol{k}-\boldsymbol{q},\sigma}-\omega_{\mathrm{pl}}(\boldsymbol{q})}-\frac{n(\omega_{\mathrm{pl}}(\boldsymbol{q}))+f_{\boldsymbol{k}-\boldsymbol{q},\sigma}}{\xi_{\boldsymbol{k}-\boldsymbol{q},\sigma}-\xi_{\boldsymbol{k},\sigma}-\omega_{\mathrm{pl}}(\boldsymbol{q})}\right]. \qquad (10.4.43)$$

It is clear that this result represents the effect of the electron-plasmon interaction on an electron; compare with (9.3.8) for the case of the electron-phonon interaction.

10.4.4 The Effect of EPI on Electronic Specific Heats

In order to see of the effect of EPI on electrons contained in $\Delta\Omega_{\mathrm{ring}}$, particularly the result (10.4.26) for ΔS_{F} or ΔS_1, let us calculate its contribution to specific heats. Note that an explicit expression of $\mathrm{Re}\,\Sigma^{\mathrm{ph}}(\boldsymbol{k},\,\omega)$ for $k_{\mathrm{B}}T <<$ $\hbar\omega_{\mathrm{D}}$ is already given in (9.3.15). Before inserting it into (10.4.26) we expand the logarithmic function in it up to the 3rd order in $(\xi_{\boldsymbol{k}}\,/\omega_{\boldsymbol{q}})$ as,

$$\ln\left|\frac{\xi_{\boldsymbol{k}}+\omega_{\boldsymbol{q}}}{\xi_{\boldsymbol{k}}-\omega_{\boldsymbol{q}}}\right| = \begin{cases} 2\,\dfrac{\xi_{\boldsymbol{k}}}{\omega_{\boldsymbol{q}}}+\dfrac{2}{3}\left(\dfrac{\xi_{\boldsymbol{k}}}{\omega_{\boldsymbol{q}}}\right)^3 & \text{for } |\xi_{\boldsymbol{k}}| << \omega_{\boldsymbol{q}}, \\[2ex] 0 & \text{for } |\xi_{\boldsymbol{k}}| >> \omega_{\boldsymbol{q}}\,. \end{cases} \qquad (10.4.44)$$

Let us separately discuss the contributions of the 1st and 3rd order terms, which we set $\Delta S_1{}^{(1)}$ and $\Delta S_1{}^{(3)}$, respectively. For simplicity, we consider the paramagnetic state where $\Sigma_+^{\mathrm{ph}} = \Sigma_-^{\mathrm{ph}} = \Sigma^{\mathrm{ph}}$.

First, for the first order term we can readily derive the result,

$$\Delta S_1^{(1)} = 2\,\lambda_{\mathrm{ph}}\sum_{\boldsymbol{k}}\frac{\partial f(\xi_{\boldsymbol{k}})}{\partial T}\,\xi_{\boldsymbol{k}}$$

$$= \frac{2\pi^2}{3} N(0)\, \lambda_{\rm ph}\, k_{\rm B}^2 T\,, \tag{10.4.45}$$

with $\lambda_{\rm ph}$ given in (9.3.43). In deriving this result we noted that

$$\frac{\partial f(\xi_k)}{\partial T} = -\frac{\partial f(\xi_k)}{\partial \xi_k}\left(\frac{\xi_k}{T} + \frac{\partial \mu}{\partial T}\right), \tag{10.4.46}$$

and used (1.6.11) and (1.6.12). Together with the free electron contribution of (1.6.17) and the spin fluctuation contribution of (10.3.3), the part of the electronic specific heat linear in T is given as $\gamma(1 + \lambda_{\rm ph} + \lambda_{\rm sf})T$.

Next, the contribution corresponding to the third order term in (10.4.44) is given as

$$\Delta S_1^{(3)} = 2\sum_{k} \frac{\partial f(\xi_k)}{\partial T}\, \frac{2}{3}\, \frac{Vm}{4\pi^2}\, \frac{1}{k}\int_0^{2k_{\rm F}} dq\, q\,|\overline{\alpha}(\boldsymbol{q})|^2\, \frac{\Omega_q}{\omega_q}\left(\frac{\xi_k}{\omega_q}\right)^3$$

$$= \frac{2}{3} N(0)\, \frac{1}{k_{\rm F}^2}\int_0^{2k_{\rm F}} dq\, q\,|\,\overline{\alpha}(\boldsymbol{q})|^2\, \frac{\Omega_q}{\omega_q^4} \sum_{k} \frac{\partial f(\xi_k)}{\partial T}\, \xi_k^3\,. \tag{10.4.47}$$

Here, with the Bohm-Staver sound velocity of (5.3.10), for small q ($q/k_{\rm F} \ll 1$) we have

$$2\Omega_q |\overline{\alpha}(\boldsymbol{q})|^2 \cong (s_{\rm o} q)^2/(2N(0)). \tag{10.4.48}$$

We expect that in the above integral over $\boldsymbol{q}$, predominant contribution comes from small q region. If we then note that $\omega_q = sq$, s being the sound velocity given by (5.4.32), the integral over q of (10.4.47) can be transformed into that over ω_q as

$$\int_0^{2k_{\rm F}} dq\, q\,|\,\overline{\alpha}(\boldsymbol{q})|^2\, \frac{\Omega_q}{\omega_q^4} = \frac{1}{4N(0)}\, \frac{s_{\rm o}^2}{s^4} \int_{\xi_k}^{\omega_{\rm D}} d\omega_q\, \frac{1}{\omega_q}$$

$$= \frac{1}{4N(0)}\, \frac{s_{\rm o}^2}{s^4}\, \ln\left|\frac{\omega_{\rm D}}{\xi_k}\right|. \tag{10.4.49}$$

The lower bound of the integral reflects the condition on the real part of the self-energy given in (10.4.44). Thus we have

$$\Delta S_a^{(3)} = -\frac{1}{6}\frac{1}{k_F^2}\frac{s_o^2}{s^4}\sum_k \frac{\partial f(\xi_k)}{\partial T}\,\xi_k^3 \ln\left|\frac{\xi_k}{\omega_D}\right|. \qquad (10.4.50)$$

The sum over $\boldsymbol{k}$ can be carried out in the following way:

$$\sum_k \frac{\partial f(\xi_k)}{\partial T}\,\xi_k^3 \ln\left|\frac{\xi_k}{\omega_D}\right| = \frac{N(0)}{T}\int_{-\infty}^{\infty} d\xi\,\xi^4 \ln\left|\frac{\xi}{\omega_D}\right|\left(-\frac{\partial f(\xi)}{\partial \xi}\right)$$

$$= \frac{N(0)}{T}(k_B T)^4\left[\ln\left(\frac{k_B T}{\omega_D}\right)\int_{-\infty}^{\infty} dx\left(-\frac{\partial f}{\partial x}\right)x^4\right.$$

$$\left. + \int_{-\infty}^{\infty} dx\left(-\frac{\partial f}{\partial x}\right)x^4 \ln|x|\right], \qquad (10.4.51)$$

where we put $x = \beta\xi = \xi/k_B T$. The first and second terms in the last result contribute as $\{nk_B(k_B T)^3/(\omega_D{}^2\varepsilon_F)\}\ln(k_B T/\omega_D)$ and $nk_B\,(k_B T)^3/(\omega_D{}^2\varepsilon)$, respectively, to the specific heat. At low temperatures, $(k_B T/\omega_D) << 1$, the first contribution is more important than the second one. Thus, neglecting the second one, we have

$$\Delta S_1^{(3)} \cong -\frac{7\pi^2}{90}N(0)\left(\frac{q_D}{k_F}\right)^2\left(\frac{s_o}{s}\right)^2 k_B\frac{(k_B T)^3}{(\hbar\omega_D)^2}\ln\left(\frac{k_B T}{\hbar\omega_D}\right), \qquad (10.4.52)$$

where we recovered $\hbar$ in the final result and used

$$\int_{-\infty}^{\infty} dx\,x^4\left(-\frac{\partial f(x)}{\partial x}\right) = \frac{7\pi^4}{15}.$$

Finally, summing the contributions of (10.4.45) and (10.4.52), we obtain the effect of the EPI on the specific heat of a metal as

$$\Delta C_{\text{el-ph}} = \frac{2\pi^2}{3} N(0)\,\lambda_{\text{ph}} k_{\text{B}}^2 T - \frac{7}{40}\pi^2 \left(\frac{q_{\text{D}}}{k_{\text{F}}}\right)^2 \left(\frac{s_{\text{o}}}{s}\right)^2 n k_{\text{B}} \frac{(k_{\text{B}}T)^3}{\varepsilon_{\text{F}}(\hbar\omega_{\text{D}})^2} \ln\left(\frac{k_{\text{B}}T}{\hbar\omega_{\text{D}}}\right). \tag{10.4.53}$$

We immediately note that this EPI contribution has a structure very similar to that due to spin fluctuations which is given in (10.3.3). Note that the sign of the above $T^3 \ln T$ term is opposite to that of the spin fluctuation effect. Note also that the size of the phonon contribution to the $T^3 \ln T$ term can be larger than that of the spin fluctuation contribution. There seems to be an additional mechanism, however, as to the EPI contribution to the $T^3 \ln T$ term [10.16, 10.17] besides the above one.

PROBLEMS FOR CHAPTER 10

10.1 Obtain the expression (10.1.11‴).

10.2 Derive the result (10.3.3) on the spin fluctuation contribution to specific heat of a metal. (See the similar procedure to obtain the EPI effect on specific heat given in 10.4.4).

10.3 Derive the result (10.3.5).

10.4 Reproduce the result (10.4.11) for the thermodynamic potential of an interacting electron-phonon system by using the procedure of (8.4.35).

10.5 Carry out the calculation to derive the result (10.4.45).

10.6 Compare the contributions of spin fluctuations and EPI to the $T^3 \ln T$ term of specific heats. Assume appropriate values for the parameters in (10.3.3) and (10.4.53).

APPENDIX

Appendix A Lindhard Function

The Lindhard response function for the paramagnetic state of an electron system in the paramagnetic state is given as

$$F(\boldsymbol{q}, \omega) = \sum_{\boldsymbol{k}} \frac{f(\varepsilon_{\boldsymbol{k}}) - f(\varepsilon_{\boldsymbol{k}+\boldsymbol{q}})}{\varepsilon_{\boldsymbol{k}+\boldsymbol{q}} - \varepsilon_{\boldsymbol{k}} - \hbar\omega - i0^{+}}$$

$$= R(\boldsymbol{q}, \omega) + i\, I(\boldsymbol{q}, \omega), \tag{A.1}$$

see (4.2.14) and (5.4.10). In the following we present explicit expressions for the real and imaginary parts of the Lindhard function for an electron gas at the absolute zero of temperature $T = 0$.

(i) The real part, $R(\boldsymbol{q}, \omega)$

At $T = 0$, we have

$$R(\boldsymbol{q}, \omega) = \sum_{\boldsymbol{k}} \theta(\varepsilon_{\mathrm{F}} - \varepsilon_{\boldsymbol{k}}) \left[\frac{1}{\varepsilon_{\boldsymbol{k}+\boldsymbol{q}} - \varepsilon_{\boldsymbol{k}} - \hbar\omega} - \frac{1}{\varepsilon_{\boldsymbol{k}} - \varepsilon_{\boldsymbol{k}+\boldsymbol{q}} - \hbar\omega} \right], \tag{A.2}$$

where we used $\varepsilon_{-\boldsymbol{k}} = \varepsilon_{\boldsymbol{k}}$ and $\theta(x)$ is the step function defined in (8.1.4). From this expression it is obvious that

$$R(\boldsymbol{q}, -\omega) = R(\boldsymbol{q}, \omega). \tag{A.3}$$

Carrying out the integral in (A.2) (see, for instance, Ref.[1.2]) we have

$$R(\boldsymbol{q}, \omega) = \frac{Vmk_{\mathrm{F}}}{2\pi^2\hbar^2} \left[\frac{1}{2} - \frac{1}{8x} \left\{ 1 - \left(\frac{u}{4x} - x \right)^2 \right\} \ln \left| \frac{1 + (u/(4x) - x)}{1 - (u/(4x) - x)} \right|$$

$$+ \frac{1}{8x}\left\{ 1 - \left(\frac{u}{4x} + x\right)^2 \right\} \ln \left| \frac{1 + (u/(4x) + x)}{1 - (u/(4x) + x)} \right| \Bigg], \tag{A.4}$$

where we set

$$x = q/2k_F; \quad u = \hbar\omega/\varepsilon_F. \tag{A.5}$$

For $x = q/2k_F << 1$ and $u/x = 4\omega/(v_F q) << 1$, (A.4) can be approximated as

$$R(\boldsymbol{q}, \omega) \cong N(0)\left[1 - \frac{1}{3}\left(\frac{q}{2k_F}\right)^2 - \frac{1}{4}\left(\frac{k_F}{q}\right)^2\left(\frac{\hbar\omega}{\varepsilon_F}\right)^2 + \cdots\right], \tag{A.6}$$

where $N(0) = Vmk_F/2\pi^2\hbar^2$ is the electronic density of states per spin at the Fermi surface.

Setting $\omega = 0$ in (A.4), we have

$$R(\boldsymbol{q}, 0) = N(0)\left[\frac{1}{2} + \frac{1 - x^2}{4x} \ln \left|\frac{1 + x}{1 - x}\right|\right]. \tag{A.7}$$

Particularly, for $q = 0$, we have

$$R(0,0) = N(0). \tag{A.8}$$

Note that Lindhard function is often defined with a negative sign and a factor 2 multiplied on the right hand side of (A.1). The definition given in (A.1) was adopted so as to have the result of (A.8).

(ii) The imaginary part, $I(\boldsymbol{q}, \omega)$

The imaginary part of the Lindhard function, (A. 1), is given as (see (5.4.11)),

$$I(\boldsymbol{q}, \omega) = \pi \sum_{\boldsymbol{k}} \left[f(\varepsilon_{\boldsymbol{k}}) - f(\varepsilon_{\boldsymbol{k}+\boldsymbol{q}}) \right] \delta(\varepsilon_{\boldsymbol{k}+\boldsymbol{q}} - \varepsilon_{\boldsymbol{k}} - \hbar\omega). \tag{A.9}$$

From this expression it is obvious that

$$I(\boldsymbol{q}, -\omega) = -I(\boldsymbol{q}, \omega). \tag{A.10}$$

Thus, it suffices to know $I(\boldsymbol{q}, \omega)$ for $\omega > 0$.

By carrying out the $\boldsymbol{k}$ integral in (A.9) for various situations, we obtain the following result (see Refs.[1.2, 5.8]):

$$I(\boldsymbol{q}, \omega) = 0, \quad \text{for } \omega > v_F q + \hbar q^2/2m, \tag{A.11}$$

$$= N(0)\frac{\pi k_F}{4q}\left[1 - (\omega - \hbar q^2/2m)^2\right],$$

$$\text{for } v_F q - \hbar q^2/2m < \omega < v_F q + \hbar q^2/2m, \tag{A.12}$$

$$= N(0)\frac{\pi}{2}\frac{\omega}{v_F q}, \quad \text{for } 0 < \omega < v_F q - \hbar q^2/2m. \tag{A.13}$$

APPENDIX B A Derivation of the Wick's Theorem

Let A_i represent the interaction representation of a fermion annihilation or creation operator, $\hat{a}_{k_i}(\tau_i)$ or $\hat{a}^\dagger_{k_i}(\tau_i)$. First we assume, for simplicity, that

$$W_n = \langle T_\tau A_1 A_2 \cdots A_{2n} \rangle_0 = \langle A_1 A_2 \cdots A_{2n} \rangle_0, \tag{B.1}$$

which corresponds to assuming that

$$\tau_1 \geq \tau_2 \geq \cdots \geq \tau_{2n}. \tag{B.2}$$

For $W_n \neq 0$, of the $2n$ operators, $A_1, A_2, \cdots, A_{2n}$, a half are creation operators and another half are annihilation operators. As we can easily show by using (B.6) given below, these operators satisfy the following anticommutation relation,

$$A_i A_j + A_j A_i = (i, j), \tag{B.3}$$

with

$$(i, j) = e^{\pm \xi_i (\tau_i - \tau_j)}, \quad \text{or} \quad 0, \tag{B.4}$$

where in the double sign we take + when A_i is a creation operator and A_j is the corresponding annihilation operator, and – when A_i is an annihilation operator and A_j is the corresponding creation operator; it is clear when we have $(i, j) = 0$.

Using (B.3), we transform W_n of (B.1) in the following way:

$$\begin{aligned} W_n &= (1,2)\langle A_3 A_4 A_5 \cdots A_{2n}\rangle_0 - \langle A_2 A_1 A_3 A_4 \cdots A_{2n}\rangle_0 \\ &= (1,2)\langle A_3 A_4 A_5 \cdots A_{2n}\rangle_0 \\ &\quad - (1,3)\langle A_2 A_4 A_5 \cdots A_{2n}\rangle_0 + \langle A_2 A_3 A_1 A_4 A_5 \cdots A_{2n}\rangle_0 \\ &= (1,2)\langle A_3 A_4 A_5 \cdots A_{2n}\rangle_0 - (1,3)\langle A_2 A_4 A_5 \cdots A_{2n}\rangle_0 \end{aligned}$$

$$+ \ (1,4)\langle A_2 A_3 A_5 \cdots A_{2n}\rangle_o - \cdots$$

$$+ (1,2n)\langle A_2 A_3 A_4 \cdots A_{2n-1}\rangle_o - \langle A_2 A_3 A_4 \cdots A_{2n} A_1\rangle_o. \tag{B.5}$$

Here we note the following relations

$$\begin{aligned} \widehat{a}_k(\tau) &= e^{\tau H_o} a_k e^{-\tau H_o} = e^{-\xi_k \tau} a_k, \\ \widehat{a}_k^{+}(\tau) &= e^{\xi_k \tau} a_k^{+}(\tau), \end{aligned} \tag{B.6}$$

which can be proved similarly to the case of bosons as given in (8.5.11)–(8.5.13). Using (B.6), we can rewrite the last term in (B.5) into the following form,

$$\begin{aligned} \langle A_2 A_3 A_4 \cdots A_{2n} A_1\rangle_o &= \frac{\operatorname{tr}\left(A_2 A_3 A_4 \cdots A_{2n} A_1 e^{-\beta H_o}\right)}{\operatorname{tr} e^{-\beta H_o}} \\ &= \frac{\operatorname{tr}\left(A_2 A_3 A_4 \cdots A_{2n} e^{-\beta H_o} e^{\beta H_o} A_1 e^{-\beta H_o}\right)}{\operatorname{tr} e^{-\beta H_o}} \\ &= \frac{\operatorname{tr}\left(A_1 A_2 A_3 \cdots A_{2n} e^{-\beta H_o}\right) e^{\pm\beta\xi_1}}{\operatorname{tr} e^{-\beta H_o}}. \end{aligned} \tag{B.7}$$

The double sign in the last result is to be chosen either + or –, depending upon whether A_1 is a creation or annihilation operator.
In rearranging (B.5) by putting (B.7) in it, we note that

$$\frac{(1,j)}{1 + e^{\pm\beta\xi_1}} = \left\{ \begin{matrix} f(\xi_1) \\ 1 - f(\xi_1) \end{matrix} \right\} (1,j) = \langle A_1 A_j \rangle_o. \tag{B.8}$$

Then we have

$$\begin{aligned} W_n &= \langle A_1 A_2 A_3 \cdots A_{2n}\rangle_o \\ &= \langle A_1 A_2\rangle_o \langle A_3 A_4 A_5 \cdots A_{2n}\rangle_o - \langle A_1 A_3\rangle_o \langle A_2 A_4 A_5 \cdots A_{2n}\rangle_o \\ &\quad + \langle A_1 A_4\rangle_o \langle A_2 A_3 A_5 \cdots A_{2n}\rangle_o - \cdots \\ &\quad + \langle A_1 A_{2n}\rangle_o \langle A_2 A_3 A_4 \cdots A_{2n-1}\rangle_o. \end{aligned} \tag{B.9}$$

Note that in the above there are terms which vanish since

$$\left\langle \widehat{a}_1(\tau_1)\widehat{a}_j(\tau_j)\right\rangle_o = \left\langle \widehat{a}_1^{+}(\tau_1)\widehat{a}_j^{+}(\tau_j)\right\rangle_o = 0. \tag{B.10}$$

Next, we carry out the above procedure on the second factors, $\langle A_3 A_4 A_5 .. A_{2n} \rangle_o$ etc., each consisting of a product of $2(n-1)$ operators, of all the terms on the last result of (B.9).
By repeating the above procedure we finally obtain

$$W_n = \sum_{P_i} \delta(P_i) \langle A_{i_1} A_{i_2} \rangle_o \langle A_{i_3} A_{i_4} \rangle_o \cdots \langle A_{i_{2n-1}} A_{i_{2n}} \rangle_o , \tag{B.11}$$

where P_i represents a permutation, $(1, 2, \cdots, 2n) \to (i_1, i_2, \cdots i_{2n})$, with $\delta(P_i) = 1$ for an even permutation, and $\delta(P_i) = -1$ for an odd permutation. Here note that if the rearrangement of the operators is made in a manner shown in (B.9), in (B.11) we have

$$\tau_{i_1} > \tau_{i_2} , \ \tau_{i_3} > \tau_{i_4} , \cdots , \tag{B.12}$$

and

$$\tau_{i_1} > \tau_{i_3} \cdots > \tau_{i_{2n-1}} . \tag{B.13}$$

Thus, if we recall (B.1), (B.11) can be rewritten in the form of (8.4.3), namely,

$$\langle T_\tau A_1 A_2 \cdots A_{2n} \rangle_o$$

$$= \sum_{P_i} \delta(P_i) \langle T_\tau A_{i_1} A_{i_2} \rangle_o \langle T_\tau A_{i_3} A_{i_4} \rangle_o \cdots \langle T_\tau A_{i_{2n-1}} A_{i_{2n}} \rangle_o . \tag{B.14}$$

This is Wick's theorem.

The above derivation of Wick's theorem was done under the assumption of (B.2). However, in an integral such as (8.3.9), the order of "times" is not always like in (B.2). Thus we have to confirm that the result (B.11) holds for an arbitrary ordering of τ_i's.

For simplicity, let us consider the following situation in (B.1),

$$\tau_2 \geq \tau_1 \geq \tau_3 \geq \tau_4 \geq \cdots \geq \tau_{2n}, \tag{B.15}$$

where the order of only τ_1 and τ_2 is reversed in (B.2). The question is whether (B.11) will hold also in this situation.

Suppose we reduce (B.1) with (B.15) into the form of (B.14) through procedures like (B.5) etc. Then, let us examine the consequence in (B.14) of the change in the ordering of times from (B.2) to (B.15). First, the left hand side of (B.14) is modified as

$$\langle A_1 A_2 A_3 \cdots A_{2n} \rangle_o = \langle T_\tau A_1 A_2 A_3 \cdots A_{2n} \rangle_o$$

$$\Rightarrow - \langle A_2 A_1 A_3 \cdots A_{2n} \rangle_o = \langle T_\tau A_1 A_2 A_3 \cdots A_{2n} \rangle . \tag{B.16}$$

Next, on the right hand side of (B.14), those terms which contain the factor $\langle T_\tau A_1 A_2\rangle_o$ are modified as

$$\langle T_\tau A_1 A_2\rangle_o \cdots = \langle A_1 A_2\rangle_o \cdots$$

$$\Rightarrow - \langle A_2 A_1\rangle_o \cdots = \langle T_\tau A_1 A_2\rangle \cdots, \quad (B.17)$$

and those terms which do not contain the factor $\langle T_\tau A_1 A_2\rangle_o$ are modified, for instance, as

$$\langle T_\tau A_1 A_3\rangle_o \langle T_\tau A_2 A_4\rangle_o \cdots - \langle T_\tau A_1 A_4\rangle_o \langle T_\tau A_2 A_3\rangle_o \cdots$$

$$= \langle A_1 A_3\rangle_o \langle A_2 A_4\rangle_o \cdots - \langle A_1 A_4\rangle_o \langle A_2 A_3\rangle_o \cdots$$

$$\Rightarrow - [\langle A_2 A_3\rangle_o \langle A_1 A_4\rangle_o \cdots - \langle A_2 A_4\rangle_o \langle A_1 A_3\rangle_o \cdots]$$

$$= \langle T_\tau A_1 A_3\rangle_o \langle T_\tau A_2 A_4\rangle_o \cdots - \langle T_\tau A_2 A_3\rangle_o \langle T_\tau A_1 A_4\rangle_o \cdots. \quad (B.18)$$

Thus we find that (B.14) holds also under the condition (B.15).

We can extend the above confirmation to a more general time ordering than (B.15).

Also from the above demonstration it is obvious how we can obtain Wick's theorem for bosons.

APPENDIX C A Proof of the Relation between Thermodynamic Potential and the Green's Function

Under the one particle energy, H_o, of (1.5.14) and the Coulomb interaction between electrons, H_C, of (1.5.24), we obtain the following equation of motion for the electron Green's function of (8.2.1),

$$\frac{\partial}{\partial \tau} G(k,\tau,\tau') = -\delta(\tau-\tau') - \xi_k G(k,\tau,\tau')$$

$$-\sum_{\kappa,l} v(\kappa) \langle T_\tau a_k^+(\tau') a_l^+(\tau) a_{l+\kappa}(\tau) a_{k-\kappa}(\tau)\rangle, \quad (C.1)$$

where $a_k(\tau)$ etc. are the Heisenberg representations of a_k etc. which are defined as in (8.2.2). This equation of motion is derived as follows.

Differentiating the both side of (8.2.2) with respect to τ, we obtain

$$\frac{\partial}{\partial \tau} a_k(\tau) = [K, a_k(\tau)] = [K(\tau), a_k(\tau)],$$

$$= \; -\,\xi_k a_k(\tau) \;-\sum_{\kappa,\, l} \mathrm{v}(\kappa)\; a_l^{+}(\tau) a_{l+\kappa}(\tau)\, a_{k-\kappa}(\tau). \tag{C.2}$$

After multiplying $a_k^{+}(\tau')$ from the right and attaching the operator, $-T_\tau$, on both side of (C.2), we take their thermal average. Then, we arrived at the result (C.1) for $\tau \neq \tau'$, namely, (C.1) without the delta function on the right hand side.

Let us proceed to the case of $\tau = \tau'$. From the definition of (8.2.1), we have

$$G(k,\tau' + i0^{+}, \tau') - G(k,\tau' - i0^{+}, \tau')$$

$$= -\langle a_k(\tau')\, a_k^{+}(\tau')\rangle - \langle a_k^{+}(\tau')\, a_k(\tau')\rangle = -1, \tag{C.3}$$

and, therefore,

$$\int_{\tau' - 0^{+}}^{\tau' + 0^{+}} \frac{\partial}{\partial \tau} G(k,\tau,\, \tau')\, d\tau \;=\; -1. \tag{C.4}$$

In order to incorporate this discontinuity at $\tau = \tau'$, we need the delta function on the right hand side of (C.1).

Now, let us proceed to calculate by (10.1.8) the effect on the thermodynamic potential of the Coulomb interaction between electrons. From (C.1), the quantity required is given in terms of the one-particle Green's functions of electrons as

$$\langle H_C\rangle = \frac{1}{2} \lim_{\tau' \to \tau + 0^{+}} \sum_{k,\, l} \sum_{\kappa}{}' \mathrm{v}(\kappa) \langle T_\tau\, a_k^{+}(\tau')\, a_l^{+}(\tau)\, a_{l+\kappa}(\tau)\, a_{k-\kappa}(\tau)\rangle$$

$$= \frac{1}{2} \lim_{\tau' \to \tau + 0^{+}} \sum_{k} \left[-\left(\frac{\partial}{\partial \tau} + \xi_k\right) G(k,\tau,\, \tau') - \delta(\tau - \tau')\right]. \tag{C.5}$$

Next, we rewrite the electron Green's function and the delta function in the above by their Fourier series as defined in (8.2.12)–(8.2.13). If we note that

$$G(k,\, i\omega_n) = \frac{1}{i\omega_n - \xi_k - \Sigma(k,\, i\omega_n)} \tag{C.6}$$

with $\Sigma(k,\, i\omega_n)$ the electron self-energy (see (8.5.24) or (8.5.31)), and that

$$\delta(\tau - \tau') = \frac{1}{\beta} \sum_{n} e^{-i\omega_n(\tau - \tau')}, \tag{C.7}$$

the quantity within the square bracket in the last expression of (C.5) is rewritten as

$$\frac{1}{\beta}\sum_n \left[(i\omega_n - \xi_k)\frac{1}{i\omega_n - \xi_k - \Sigma(k, i\omega_n)} - 1\right] e^{-i\omega_n(\tau-\tau')}$$

$$= \frac{1}{\beta}\sum_n \frac{\Sigma(k, i\omega_n)}{i\omega_n - \xi_k - \Sigma(k, i\omega_n)} e^{-i\omega_n(\tau-\tau')}. \tag{C.8}$$

Thus we have

$$\langle H_C \rangle = \frac{1}{2\beta}\sum_{k,n} \Sigma(k, i\omega_n) G(k, i\omega_n) e^{i\omega_n 0^+}. \tag{C.9}$$

Using this result in (10.1.8), finally we obtain

$$\Omega = \Omega_0 + \frac{1}{2\beta}\sum_{k,n}\int_0^1 dg\, \frac{1}{g}\, \Sigma(k, i\omega_n)_g\, G(k, i\omega_n)_g, \tag{C.10}$$

where $\Sigma(k, i\omega_n)_g$ and $G(k, i\omega_n)_g$ are those which are obtained by replacing every Coulomb interaction $v(\kappa)$ by $gv(\kappa)$ in $\Sigma(k, i\omega_n)$ and $G(k, i\omega_n)$.

This result can be shown to be valid also for the effect of the EPI.

References

Chapter 1

1.1 C. Kittel, *Quantum Theory of Solids* (John Wiley & Sons, New York, 1963).
1.2 J. R. Schrieffer, *Theory of Superconductivity* (Addison-Wesley, Redwood City, 1964).
1.3 A. L Fetter and J. D. Walecka, *Quantum Theory of Many-Particle Systems* (McGraw-Hill, New York, 1971).
1.4 S. Doniach and E. H. Sondheimer, *Green's Functions for Solid State Physicists* (Addison-Wesley, Redwood City, 1974).
1.5 R. P. Feynman, *Statistical Mechanics* (Addison-Wesley, Mass. 1972).
1.6 C. P. Enz, *A Curse on Many-Body Theory Applied to Solid-State Physics* (World Scientific, Singapore, 1992).

Chapter 2

2.1 J. H. Van Vleck, *The Theory of Electric and Magnetic Susceptibilities* (Oxford Univ. Press, London, 1932).
2.2 P. W. Anderson, in *Magnetism*, vol.1, eds. G. Rado and H. Suhl (Academic Press, New York, 1963) p.25.
2.3 C. Kittel, *Introduction to Solid State Physics*, 5th ed. (John Wiley & Sons, New York, 1976).
2.4 A. Herpin, *Theorie du Magnetisme* (Presses Universitaires de France, 1968).
2.5 W. Jones and N. H. March, *Theoretical Solid State Physics,* vol.1 (Wiley-Interscience, London, 1973).
2.6 J. Callaway, *Quantum Theory of the Solid State*, Part A (Academic Press, New York, 1974).
2.7 S. V. Vonsovskii, *Magnetism* (John Wiley & Sons, New York, 1974).
2.8 D. C. Mattis, *The Theory of Magnetism*, I and II (Springer, Berlin, 1981, 1985).
2.9 R. M. White, *Quantum Theory of Magnetism* (McGraw-Hill, New York, 1970).
2.10 B. Barbara, D. Gignoux and C. Vettier, *Lectures on Modern Magnetism* (Science Press, Beijin, and Springer, Berlin, 1988).
2.11 G. Parisi, *Statistical Field Theory* (Addison-Wesley, 1988).
2.12 P. Fulde, *Electron Correlations in Molecules and Solids* (Springer, Berlin,1991).
2.13 K. Yoshida, *Theory of Magnetism* (Springer, Berlin, 1996).
2.14 W. Heisenberg, Z. Phys. 49 (1928) 619.
2.15 G. G. Low, Proc. Phys. Soc. 82 (1963) 992.

Chapter 3

3.1 A. H. Wilson, *The Theory of Metals* (Cambridge U. Press, Cambridge, 1936).
3.2 J. Friedel, in *The Physics of Metals,* 1, ed. J. M. Ziman (Cambridge U. Press, London, 1969) p.340.
3.3 A. Blandin, in *Magnetism, Selected Topics*, ed. S. Foner (Gordon and Breach, New York, 1976) p.1.
3.4 E. P. Wohlfarth, in *Magnetism, Selected Topics*, ed. S. Foner (Gordon and Breach, New York, 1976) p.59.
3.5 E. C. Stoner, Rep. Prog. Phys. 11 (1947) 43.
3.6 M. Shimizu, Rep. Prog. Phys. 44 (1981) 329.
3.7 Y. Nakagawa, J. Phys. Soc. Jpn. 11 (1956) 855.
3.8 J. Hubbard, Proc. Roy. Soc. A 276 (1963) 238.
3.9 C. Herring, Phys. Rev. 85 (1952) 1003; ibid 87 (1952) 60.
3.10 C. Herring, in *Magnetism*, eds. G. T. Rado and H. Suhl (Academic Press, New York, 1966 Vol.4).
3.11 T. Izuyama, Prog. Theor. Phys. 23 (1960) 969.
3.12 F. Englert and M. M. Antonoff, Physica 30 (1964) 429.
3.13 R. Kubo, T. Izuyama, D. J. Kim and Y. Nagaoka, J. Phys. Soc. Jpn. 17 Suppl. B.I (1962) (Proc. of International Conf. on Magnetism and Crystallography, Kyoto, 1961, vol.1) 67.
3.14 T. Izuyama, D. J. Kim and R. Kubo, J. Phys. Soc. Jpn. 18 (1963) 1025.
3.15 R. D. Lowde and C. G. Windsor, Adv. Phys. 19 (1970) 813.
3.16 M. K. Wilkinson and C. G. Shull, Phys. Rev. 103 (1956) 516.
3.17 W. Marshall, *Lectures on Neutron Diffraction* (Harvard Univ., 1959, unpublished).
3.18 K. K. Murata and S. Doniach, Phys. Rev. Lett. 29 (1972) 285.
3.19 T. Moriya and A. Kawabata, J. Phys. Soc. Jpn. 34 (1973) 639; *ibid.* 35 (1973) 669.
3.20 F. Gautier, in *Magnetism of Metals and Alloys*, ed. M. Cyrot (North-Holland, 1982), p.1.
3.21 T. Moriya, *Spin Fluctuations in Itinerant Electron Magnetism* (Springer, Berlin, 1985).
3.22 V. Korenman, in *Metallic Magnetism*, ed. H. Capellmann (Springer, Berlin, 1987) p.109.
3.23 D. J. Kim, Phys. Rept. 171 (1988) 129.

Chapter 4

4.1 R. Kubo, J. Phys. Soc. Jpn. 12 (1957) 570.
4.2 D. Pines and P. Nozieres, *Theory of Quantum Liquids* volume I, (Addison-Wesley, New York, 1989).
4.3 L. E. Reichl, *A Modern Course in Statistical Physics* (Univ. of Texas Press, Austin, 1980).
4.4 L. van Hove, Phys. Rev. 95 (1954) 1374.

4.5 M. A. Ruderman and C. Kittel, Phys. Rev. 96 (1954) 99.
4.6 T. Kasuya, Prog. Theor. Phys. 16 (1956) 45.
4.7 K. Yosida, Phys. Rev. 106 (1957) 893.
4.8 J. Friedel, Nuovo Cimento, Suppl. 7 (1958) 287.
4.9 K. Yosida, in *Progress in Low Temperature Physics,* vol.4, ed. G.J. Gorter (North-Holland, Amsterdam, 1964) p.263; T. Nagamiya, in *Solid State Physics*, vol.20, eds. F. Seitz, D. Turnbull and H. Ehrenreich (Academic, New Yowk,1967) p.305.
4.10 P. W. Anderson, Phys. Rev. 124 (1961) 41.
4.11 D. J. Kim and Y. Nagaoka, Prog. Theor. Phys. 30 (1963) 743.
4.12 J. R. Schrieffer and P. A. Wolf, Phys. Rev. 149 (1966) 491.
4.13 K. Yosida, Phys. Rev. 107 (1957) 396.
4.14 J. Kondo, Prog. Theor. Phys. 32 (1964) 37.
4.15 D. J. Kim, Bussei Kenkyu (Kyoto) 3 (1965) 435; Phys. Rev. 149 (1966) 434.
4.16 J. Bardeen and D. Pines, Phys. Rev. 99 (1955) 1140.
4.17 H. Fröhlich, Phys. Rev. 79 (1950) 845.
4.18 J. Kondo, Prog. Theoret. Phys. 33 (1965) 575.
4.19 M. Bailyn, Advan. Phys. 15 (1966) 179.
4.20 A. J. Heeger, in *Solid State Physics*, vol.23 (Academic Press, New York, 1969) p.284.
4.21 N. Andrei, K. Furuya, and J. H. Lowenstein, Rev. Mod. Phys. 55 (1983) 331.
4.22 R. E. Peierls, *Quantum Theory of Solids* (Clarendon, Oxford 1955).
4.23 D. J. Kim, H. C. Praddaude and B. B. Schwartz, Phys. Rev. Lett. 23 (1969) 419.
4.24 D. J. Kim, B. B. Schwartz and H. C. Praddaude, Phys. Rev. B. 7 (1973) 205.
4.25 B. B. Schwartz, in *Magnetism, Selected Topics*, ed. S. Foner (Gordon and Breach, New York, 1976) p.93.
4.26 A. K. Rajagopal, H. Brooks, and N. R. Ranganathan, Nuovo Cimento Suppl. 10 (1967) 807.
4.27 B. Giovannini, M. Peter and J. R. Schrieffer, Phys. Rev. Lett. 12 (1964) 736.
4.28 D. J. Kim and B. B. Schwartz, Phys. Rev. Lett. 20 (1968) 201; ibid 21 (1968) 1744.
4.29 G. G. Low and T. M. Holden, Proc. Phys. Soc. 89 (1966) 119.
4.30 J. Hubbard, Proc. Roy. Soc. A 243 (1957) 336.
4.31 D. J. Kim and B. B. Schwartz, Phys. Rev. Lett. 28 (1972) 310.
4.32 A. A. Gomes and I. A. Campbell, J. Phys. C 1 (1968) 253.
4.33 D. J. Kim and B. B. Schwartz, Phys. Rev. B 15 (1977) 377.
4.34 N. F. Mott and H. Jones, *The Theory of the Properties of Metals and Alloys* (Dover Publications, New York, 1958).
4.35 S. C. Lin, J. Appl. Phys. 40 (1969) 2173.
4.36 D. J. Gillespie, C. A. Mackliet, and A.I. Schindler, in *Amorphous Magnetism,* eds. H. O. Hooper and A. M. de Graaf (Plenum, New York, 1973), p.343.
4.37 R. Hasegawa and J. A. Dermon, Phys. Lett. A 42 (1973) 407.
4.38 S. Ogawa, S. Waki and T. Teranishi, Int. J. Magn. 5 (1974) 349.

4.39 T. Yoshiie, K. Yamakawa, and E. Fujita, J. Phys. Soc. Jpn. 37 (1974) 572.
4.40 V. Bänniger, G. Busch, M. Campagna, and H. C. Siegmann, Phys. Rev. Lett. 25 (1970) 585 ; G. Busch, M. Campagna, and H. C. Siegmann, Phys. Rev. B 4 (1971) 746.
4.41 P. M. Tedrow and R. Meservey, Phys. Rev. Lett. 26 (1971) 192 ; Phys. Rev. B 7 (1973) 318.
4.42 Y. Shapira, N. F. Oliviera, Jr., D. H. Ridgley, R. Kershaw, K. Dwight, and A. Wold, Phys. Rev. B 34 (1986) 4187; Y. Shapira, N. F. Oliveira, Jr., P. Becla and T. Q. Yu, Phys. Rev. B 41 (1990) 5931.
4.43 S. Ukon, in *Proc. Int. Conf. on Magnetism,* ed. D. Givord (Les Editions de Physique, Paris, 1988) p.79.
4.44 C. A. Kukkonen and A. W. Overhauser, Phys. Rev. B 20 (1979) 550.
4.45 A. W. Overhauser, Phys. Rev. 128 (1962) 1437.
4.46 A. W. Overhauser, Phys. Rev. 167 (1968) 691.
4.47 J. F. Cooke, J. A. Blackman and T. Morgan, J. Magn. Magn. Mat 54–57 (1986) 1115.
4.48 H. A. Mook, in *Spin Waves and Magnetic Excitations* vol.1, eds. A. S. Borovik-Romanov and S. K. Sinha (Elsevier Science Publishers, 1988), p.425.

Chapter 5

5.1 M. Hayase, M. Shiga and Y. Nakamura, J. Phys. Soc. Jpn. 34 (1973) 925.
5.2 Y. Nakamura, K, Sumiyama and M. Shiga, J. Magn. Magn. Mat. 12 (1979) 127.
5.3 G. Hausch and H. Warlimont, Phys. Lett. 41A (1972) 437.
5.4 G. Hausch, J. Phys. Soc. Jpn. 37 (1974) 819.
5.5 G. A. Alers, J. R. Neighbours and H. Sato, J. Phys. Chem. Solids 13 (1960) 40.
5.6 D. J. Kim, J. Phys. Soc. Jpn. 40 (1976) 1244, 1250.
5.7 M. Born and K. Huang, *Dynamical Theory of Crystal Lattices* (Clarendon, Oxford, 1954).
5.8 D. Pines, *Elementary Excitations in Solids* (Benjamin, New York, 1964).
5.9 A. A. Maradudin, E. W. Montroll, G. H. Weiss and I. P. Ipatova, *Theory of Lattice Dynamics in the Harmonic Approximation.* Second Edition. *Solid State Physics Suppl.*3 (Academic, New York, 1971).
5.10 W. Jones and N. H. March, *Theoretical Solid State Physics*, vol.1 (Wiley-Interscience, London, 1973).
5.11 P. B. Bruesch, *Phonons: Theory and Experiments* I (Springer, Berlin, 1982).
5.12 L. J. Sham and J. M. Ziman, in *Solid State Physics*, eds. F. Seitz and D. Turnbull (Academic, New York, 1963) vol.15.
5.13 S. K. Chan and V. Heine, J. Phys. F. 3 (1973) 795.

5.14 D. Bohm and I. Staver, Phys. Rev. 84 (1951) 836.
5.15 S. Kagoshima, H. Nagasawa and T. Sambongi, *One-Dimensional Conductors* (Springer, Heiderberg, 1988).
5.16 [4.2] p.101.
5.17 Kazunori Suzuki, M. S. thesis, Aoyama Gakuin University (1980) (unpublished).
5.18 D. J. Kim, Phys. Rev. Lett. 39 (1977) 98.
5.19 R. M. Bozorth, *Ferromagnetism* (Van, Nostrand, Princeton, 1951)
5.20 E. P. Wohlfarth, Private communication (1986).
5.21 M. Shiga, Y. Kusakabe, Y. Nakamura, K. Makita and M. Sagawa, Physica B 161 (1989) 206; J. A. J. Lourens and H. L. Alberts, *Proc. of International Conference on the Physics of Transition Metals.* eds. P. M. Oppeneer and J. Küber (World Scientific, Singapore, 1993) P. 630.
5.22 W. Rehwald, M. Rayl, R. W. Cohen and G. D. Cody, Phys. Rev. B. 6 (1972) 363.
5.23 W. Dieterich and P. Fulde, Z. Phys. 248 (1971) 154.
5.24 M. Weger, Rev. Mod. Phys. 36 (1964) 175.
5.25 J. Labbe and J. Friedel, J. Phys. 27 (1966) 153.
5.26 B. M. Klein, L. L. Boyer, D. A. Papaconstantopoulos and L. F. Mattheiss, Phys. Rev. B. 18 (1978) 6411.
5.27 D. J. Kim and I. Yoshida, Phys. Rev. B 52 (1995) 6588.
5.28 D. J. Kim, Solid State Commun. 36 (1980) 373.
5.29 D. J. Kim, Solid State Commun. 38 (1981) 441.
5.30 R. D. Lowde, R. T. Harley, G. A. Saunders, M. Sato, R. Scherm and C. Underhill, Proc. Roy. Soc. A 374 (1980) 87; M. Sato, R. D. Lowde, G. A. Saunders, and M. M. Hargreaves, Proc. Roy. Soc. A 374 (1980) 115.
5.31 E. Fawcett, Rev. Mod. Phys. 60 (1988) 209.
5.32 S. Ami, N. A. Cade and W. Young, J. Magn. Magn. Mat. 31–34 (1983) 59; W. Yeung, Physica B 149 (1988) 185.
5.33 H. Yamada, Physica B 119 (1983) 99.

Chapter 6

6.1 D. J. Kim, Phys. Rev. B 25 (1982) 6919.
6.2 J. J. Hopfield, Phys. Lett. 27A (1968) 397.
6.3 N. F. Berk and J. R. Schrieffer, Phys. Rev. Lett. 17 (1966) 433.
6.4 S. Doniach and S. Engelsberg, Phys. Rev. Lett. 17 (1966) 750.
6.5 W. F. Brinkman and S. Engelsberg, Phys. Rev. 169 (1968) 417.
6.6 M. T. Béal-Monod, S. K.-Ma and D. R. Fredkin, Phys. Rev. Lett. 20 (1968)
6.7 C. Tanaka and H. Shiina, Phys. Lett. A 141 (1989) 307.
6.8 S. G. Mishra and T. V. Ramakrishnan, Phys. Rev. B 18 (1978) 2308.
6.9 G. G. Lonzarich and L. Taillefer, J. Phys. C 18 (1985) 4339.

6.10 P. C. E. Stamp, J. Phys. F 15 (1985) 1829.
6.11 D. J. Kim and C. Tanaka, J. Magn. Magn. Mat. 58 (1986) 254.
6.12 H. Kojima, R. S. Tebble and D. E. G. Williams, Proc. Roy. Soc. A 260 (1961) 237.
6.13 V. L. Moruzzi, J. F. Janak and A. R. Williams, *Calculated Electronic Properties of Metals* (Pergamon, New York, 1978).
6.14 M. Shimizu, J. Magn. Magn. Mat. 31–34 (1983) 299; for related references, see S. Misawa et.al or T. Tanaka and K. Tsuru, Europhys. Lett., 14 (1991) 377.
6.15 S. Misawa and K. Kanematsu, J. Phys. F 6 (1976) 2119.
6.16 W. D. Weiss and R. Kohlhaas, Z. Naturf. a 19 (1964) 1631.
6.17 J. de Launay, in *Solid State Physics*, vol.2, eds. F. Seitz and D. Turnbull, (Academic, New York, 1956).
6.18 F. Y. Fradin, D. D. Koelling, A. J. Freeman and T. J. Watson-Yang, Phys. Rev. B 12 (1975) 5570.
6.19 Y. Baer, P. F. Heden, J. Hedman, M. Klasson, N. Nordling and K. Siegbahn, Physica Scripta 1 (1970) 55.
6.20 S. Arajs and R. V. Colvin, J. Phys. Chem. Solids 24 (1963) 1233.
6.21 S. Hirooka, Physica B 149 (1988) 156.
6.22 B. D. Budworth, F. E. Hoare and J. Preston, Proc. Roy. Soc. A257 (1960) 250.
6.23 Y. Ishikawa, S. Onodera and K. Tajima, J. Magn. Magn. Mat. 10 (1979) 183.
6.24 N. Rosov, J. W. Lynn, J. Kästner, E. F. Wassermann, T. Chattopadhyay and H. Bach, J. Magn. Magn. Mat. 140-144 (1995) 235.
6.25 P. J. Brown, B. Roessli, J. G. Smith, K.-U. Neumann and K. R. A. Ziebeck, J. Phys. Condens. Matter 8 (1996) 1527.
6.26 D. J. Kim, Phys. Rev. Lett. 47 (1981) 1213.
6.27 S. Fukumoto, S. Ukon and D. J. Kim, J. Magn. Magn. Mat. 90-91 (1990) 743.
6.28 D. J. Kim and I. Yoshida, J. Magn. Magn. Mat. 178 (1997) 266.
6.29 D. J. Kim, S. Fukumoto, M. M. Antonoff and K. H. Lee, J. Magn. Magn. Mat. 137 (1994) 249.
6.30 D. J. Kim, and I. Yoshida, Molecular Physics Reports 17 (1997) 105.
6.31 I. Yoshida, D. Isoda and D. J. Kim, Physica B 237-238 (1997) 509.
6.32 I. D. Moore and J. B. Pendry, J. Phys. C 11 (1978) 4615.
6.33 E. Kisker in *Metallic Magnetism*, ed. H. Capellmann (Springer, Berlin, 1987) p.57.
6.34 I. Yoshida and D, J, Kim, Physica B, 219-220 (1996) 115
6.35 G. S. Knapp, E. Corenzwit and C. W. Chu, Solid State Commun. 8 (1970) 639.
6.36 V. M. Zverev and V. P. Silin, Pis'ma Zh. Eksp. Teor. Fiz. 45 (1987) 178 [Sov. Phys. JETP Lett. 45 (1987) 220].
6.37 V. M. Zverev and V. P. Silin, Physica B 159 (1989) 43.
6.38 J. Appel and D. Fay, Phys. Rev. B 22 (1980) 1461.

6.39 P. W. Anderson, Private communication via L. Gruenberg.

Chapter 7

7.1 E. G. Brovman, Yu. Kagan and A. Kholas, Zh. Eksp. Teor. Fiz. 57 (1969) 1635 [Sov. Phys. JETP 30 (1970) 883].
7.2 G. D. Mahan, *Many Particle Physics* (Plenum, New York, 1990).
7.3 M. W. Long and W. Yeung, J. Phys. C. 19 (1986) 5077.
7.4 V. L. Moruzzi, A. R. Williams and J. F. Janak, Phys. Rev. B. 15 (1977) 2854.
7.5 D. J. Kim, M. W. Long and W. Yeung, Phys. Rev. B. 36 (1987) 429.
7.6 See for instance [7.2] p.411.
7.7 For a recent review and references, I. A. Campbell and G. Creuzet, in *Metallic Magnetism*, ed. H. Capellmann (Springer, Berlin, 1987). p.207; E. F. Wassermann, in *Ferromagnetic Materials*, Vol. 5 eds. E. P.Wohlfarth and K. H. J. Busschow (North-Holland, Amsterdam, 1990) p.237; P. E. Brommer and J. J. M. Franse, *ibid.* p. 323; *Proceeding of the International Symposium on Magnetoelasticity in Transion Metals and Alloys,* eds. M. Shimizu, Y. Nakamura and J. J. M. Franse, Physica 119B+C (1983) 1; *Proceeding of the International Symposium on Magnetoelasticity and Electronic Structure of Transition Metals, Alloys and Films,* eds. E. F. Wassermann, K. Usadel and D.Wagner, Physica B 167 (1989).1.
7.8 R. J. Weiss, Proc. Phys. Soc. London 82 (1963) 281.
7.9 V. L. Moruzzi, Physica B161 (1989) 99. ; Phys. Rev. B 41 (1990) 6939.
M. Podgórny, Physica B161 (1989) 110.
E. G. Moroni and T. Jarlborg, Phys. Rev. B 41 (1990) 9600.
P. Entel, E. Hoffmann, P. Mohn, K. Schwarz and V. L. Moruzzi, Phys. Rev. B 47 (1993) 8706, and references therein.
7.10 D. Wagner, J. Phys. Condens. Matter 1 (1989) 4635.
M. Schröter, P. Entel and S. G. Mishra, J. Magn. Magn. Mat. 87 (1990) 163.
7.11 M. Shiga, Solid State Commun. 10 (1972) 1233.
W. F. Schlosser, Phys. Status. Solidi. A 17 (1973) 199.
Y. Kakehashi, J. Phys. Soc. Jpn. 49 (1980) 2421; *ibid.* 50 (1981) 2236.
Y. Kakehashi and J. H. Samson, Phys. Rev. B 34 (1986) 1734.
7.12 T. Moriya and K. Usami, Solid State Commun. 34 (1980) 95.
H. Hasegawa, Physica 119 B (1983) 15.
A. J. Holden, V. Heine and J. H. Samson, J. Phys. F 14 (1984) 1005.
7.13 A. Z. Menshikov, Physica B 161 (1989) 1.
P. J. Brown, I. K. Jassim, K.-U. Neumann and K. R. A. Ziebeck, Physica B 161 (1989) 9.

7.14 V. M. Zverev and V. P. Silin, Zh. Eksp. Teor. Fiz. 93 (1987) 709 [Sov. Phys. JETP 66 (1987) 401].
7.15 D. J. Kim, Phys. Rev. B 39 (1989) 6844.
7.16 D. J. Kim and H. Koyama, Physica B 161 (1989) 165.
7.17 D. J. Kim, T. Kizaki, N. Miyai, J. Iwai, N. Hino and S. Fukumoto, J. Korean Phys. Soc. (Proc. Suppl.) 28 (1995) S203.
7.18 J. Friedel and C. M. Sayers, J. de Phys. (Paris) 38 (1977) L–263.
7.19 V. L. Moruzzi and P. M. Marcus, Phys. Rev. B 48, 7665 (1993)
7.20 D. J. Kim, S. Fukumoto, T. Kizaki, N. Miyai, H. Saita and H. Hino, in *The Invar Effect: A Centennial Symposium*, ed. J. Witternauer (TMS, Pennsylvania, 1997) p.75.
7.21 T. Kizaki, N. Miyai, H. Saita, N. Hino, D. Isoda, T. Maehashi, H. Shibuya and D. J. Kim, Physica B 237-238 (1997) 506.
7.22 J. Iwai, N. Miyai, T. Kizaki and D. J. Kim, J. Magn. Magn. Mat. 140-144 (1995), 243.
7.23 Ll. Manosa, G. A. Saunders, H. Radhi, U. Kawald, J. Pelzl and H. Bach, J. Phys. Condens. Matter 3 (1991), 2273; G. A. Saunders, H. B. Senin, H. A. A. Sidek and J. Pelzl, Phys. Rev. B 48 (1993), 15801.

Chapter 8

8.1 L. P. Kadanoff and G. Baym, *Quantum Statistical Mechanics* (Addison-Wesley, Redwood City, 1989).
8.2 A. A. Abrikosov, L. P. Gorkov, and I. E. Dzyaloshinski, *Methods of Quantum Field Theory in Statistical Physics* (Prentice Hall, Englewood Cliff., N. J., 1963).
8.3 R. Abe, *Statistical Mechanics*, transl. Y. Takahashi (Univ. of Tokyo Press, Tokyo, 1975).
8.4 G. Rickayzen, *Green's Functions and Condensed Matter* (Academc Press, London, 1980).
8.5 J.-P. Blaizot and G. Ripka, *Quantum Theory of Finite Systems* (MIT Press, Cambridge, 1986).
8.6 J. W. Negele and H. Orland, *Quantum Many-Particle Systems* (Addison-Wesley, Redwood City, 1988).
8.7 M. Gell-Mann and K. A. Brueckner, Phys. Rev. 106 (1957) 364.

Chapter 9

9.1 A. B. Migdal, Zh. Eksp. Teor. Fiz. 34 (1958) 1438 [Sov. Phys. JETP 34 (1958) 996].
9.2 D. J. Scalapino, in *Superconductivity*, ed. R. D. Parks (Marcel Dekker, New York, 1969) vol.1, p.449.
9.3 S. K. Joshi and A. K. Rajagopal, in *Solid State Physics*, eds. F. Seitz, D. Turnbull, and H. Ehrenreich (Academic, New York, 1968) vol.22, p.159.

9.4 G. Grimvall, *The Electron-Phonon Interaction in Metals* (North-Holland, Amsterdam, 1981).
9.5 K. Shinha, in *Dynamical Properties of Solids*, vol.3, eds. G.K. Horton and A. A. Maradudin (North-Holland, Amsterdam, 1980) p.1.
9.6 D. J. Kim, Phys. Rev. B. 17 (1978) 468; D. J. Kim and S. Ukon, Prog. Theor. Phys. Suppl. 69 (1980) 281.
9.7 J. Bardeen, L. N. Cooper, and J. R. Schrieffer, Phys. Rev. 108 (1957) 1175.
9.8 M. Abramowitz and I. A. Segun, *Handbook of Mathematical Functions* (Dover, New York, 1965).
9.9 N. N. Bogoliubov, V. V. Tolmachev and D. V. Shirkov, *A New Method in the Theory of Superconductivity*, (Academy of Science, Moscow, 1958 ; Consultant Bureau, New York, 1959).
9.10 P. Morel and P. W. Anderson, Phys. Rev. 125 (1966) 1263.
9.11 M. J. Buckingham and M. R. Schafroth, Proc. Phys. Soc. (London) A67 (1954) 828.
9.12 J. J . Quinn, in *The Fermi Surface*, eds. W. A. Harrison and M. B. Webb (Wiley, New York, 1960), p.58.
9.13 D. Fay and J. Appel, Phys. Rev. B 20 (1979) 3705 ; J. de Phys. 42 (1981) C6-490.
9.14 C. P. Enz and B. T. Matthias, Z. Physik B 33 (1979) 19.
9.15 W. E. Pickett, Phys. Rev. B 26 (1982) 1186.
9.16 B. Mitrović and W. E. Pickett, Phys. Rev. B35 (1987) 3415.
9.17 G. S. Tripathi, C. M. Misra, P. Tripathi and P. K. Misra, Phys. Rev. B 39 (1989) 94.
9.18 D. M. Edwards, in *Magnetic Phase Transitions*, edited by M. Ausloos and R. J. Elliot, (Springer, Berlin, 1983) p.25.
S. Nakajima, Progr. Theot. Phys. (Kyoto) 38 (1967) 23.
9.20 L. C. Davis and S. H. Liu. Phys. Rev. 163 (1967) 503
9.21 H. S. D. Cole and R. E. Turner, Phys. Rev. Lett. 19 (1967) 501
9.22 D. J. Kim, Phys. Rev. 167 (1968) 545.
9.23 W. F. Brinkman and S. Engelsberg, Phys. Rev. 169 (1968) 417.
9.24 S. V. Vonsovsky, Yu. A. Izyumov and E. Z. Kurmaev, *Superconductivity of Transition Metals* (Springer, Berlin, 1982).
9.25 J. G. Bednorz and K. A. Müller, Z. Phys. B 64 (1986) 189.
9.26 M. K. Wu. J. R. Ashburn, C. J. Torng, D. H. Hor, R. L. Meng, L. Gao, Z. J. Huang, Y. Q. Wang and C. W. Chu, Phys. Rev. Lett. 58 (1987) 908.
9.27 D. J. Kim, Physica 148 B (1987) 278.

Chapter 10

10.1 T. V. Ramakrishnan, Phys. Rev. B 10 (1974) 4114.
10.2 D. J. Kim, Phys. Rev. B 37 (1988) 7643.
10.3 P. Nozières and D. Pines, Nuovo Cimento 9 (1958) 470.
10.4 F. Englert and R. Brout, Phys. Rev. 120 (1960) 1085.
10.5 D. J. Kim, Phys. Rev. B 9 (1974) 3307.

10.6 D. Pines, Phys. Rev. 92 (1953) 626.
10.7 K. Sawada, K. A. Brueckner, N. Fukuda and R. Brout, Phys. Rev. 108 (1957) 507.
10.8 D. J. Kim, J. Phys. Soc. Jpn. 66 (1997) 1583.
10.9 P. F. de Chatel, F. R. de Boer, W de Wood, J. H. J. Fluitman and C. J. Schinkel, J. de Physique, 32 (1971) C 999; W. de Dood and P. F. de Chatel, J. Phys. F3 (1973) 1039.
10.10 S. K. Dhar, K. A. Gschneidner, Jr., L. L. Miller and D. C. Johnston, Phys. Rev. B 40 (1989) 11489, and references therein.
10.11 G. G. Lonzarich and L. Taillefer, J. Phys. C. 18 (1985) 4339.
10.12 P. C. E. Stamp, J. Phys. F 15 (1985) 1829.
10.13 D. J. Kim and C. Tanaka, Phys. Rev. B 3 7(1988) 3948.
10.14 G. M. Eliashberg, Zh. Eksp. Teor. Fiz. 43 (1962) 1005 [Sov. Phys. JETP 16 (1963) 780].
10.15 G. Grimval, Phys. Kondens. Materie 9 (1969) 283.
10.16 H. Fröhlich, Proc. Roy. Soc. (London) A 215 (1952) 291.
10.17 M. Danino and A. W. Overhauser, Phys. Rev. B 26 (1982) 1569.
10.18 D. Coffey and C. J. Pethick, Phys. Rev. B 37(1988) 442.

INDEX